SOIL AND WATER CONTAMINATION

Soil and Water Contamination

2nd Edition

Marcel van der Perk

Faculty of Geosciences, Utrecht University, The Netherlands

CRC Press
Taylor & Francis Group
Boca Raton London New York Leiden

CRC Press is an imprint of the
Taylor & Francis Group, an **informa** business

A BALKEMA BOOK

Front cover: M.C. Escher's "Puddle"
© 2012 The M.C. Escher Company B.V. – The Netherlands. All rights reserved.

CRC Press/Balkema is an imprint of the Taylor & Francis Group, an informa business
Copyright © 2014 Taylor & Francis Group plc, London, UK

Published by: CRC Press/Balkema
       P.O. Box 11320, 2301 EH, Leiden, The Netherlands
       e-mail: Pub.NL@taylorandfrancis.com
       www.crcpress.com – www.taylorandfrancis.com

ISBN: 978-0-415-89343-5 (Pbk)
ISBN: 978-0-203-76889-1 (eBook PDF)

Printed in Great-Britain

Contents

Preface to the first edition

This book is based on an undergraduate course-book written for students of the 'Soil and Water Pollution' course at the Faculty of Geosciences of Utrecht University, the Netherlands. These students, studying Earth Sciences or Environmental Sciences, are taught the concepts of transport and fate processes of environmental contaminants, so that they understand, and can predict, contaminant patterns in soil, groundwater, and surface water.

The book is in four parts: 1) An introduction to soil and water pollution; 2) Source, role, and behaviour of substances in soil and water; 3) Transport and fate processes of substances in soil and water; and 4) Patterns of substances in soil and water. Part 1 introduces the fundamentals of environmental pollution, environmental chemistry, and the basic physical and chemical properties of soil, groundwater, and surface water. Part 2 covers the natural and anthropogenic sources of nutrients, heavy metals, radionuclides, and organic pollutants, as well as their physico-chemical characteristics, behaviour, and toxicity in the natural environment. Part 3 introduces the processes of transport, exchange, and transformation, such as advection, dispersion, and kinetics of adsorption and biochemical decay. Particular attention is paid to the mathematical description and modelling of these processes. Part 4 develops the information presented in Parts 1 and 2, by presenting practical applications: spatial and temporal patterns of pollutants in soil, groundwater, and surface water are described by means of a number of recent case studies. Each chapter concludes with a set of exercises to consolidate the chapter content. It is assumed that readers already understand the fundamentals of chemistry, soil science, and hydrology, although a brief review of basic concepts is given and there is a list of recommended reading for those wishing to acquire more background knowledge.

I am grateful to colleagues for their support during the writing of this text. In particular, I would like to thank Dr. Pauline van Gaans and Professor Wladimir Bleuten of the Department of Physical Geography, Utrecht University for their ideas and suggestions and for reviewing some of the chapters. I thank Professor Peter Burrough, Dr. Hans Middelkoop, Dr. Job Spijker and Dr. Marc Vissers of the Department of Physical Geography, Utrecht University; Dr. Paolo DeZorzi and Sabrina Barbizzi MSc. of the Agenzia per la Protezione dell'Ambiente e per i servizi Technici, Rome; Dr. Cathy Ryan of the Department of Geology and Geophysics, University of Calgary; and Prof. Gene Likens of the Institute of Ecosystem Studies, Millbrook NY; who supplied data, figures or other materials used in this book. Dr. Joy Burrough-Boenisch (Unclogged English, Goring-on-Thames) advised on the English. Lastly, I would like to acknowledge Margriet Ganzeveld, Fred Trappenburg, Johan van der Wal, Margot Stoete, and Rien Rabbers of GeoMedia, Faculty of Geosciences, Utrecht University for assisting with the graphics and preparing the camera-ready copy.

Marcel van der Perk
Utrecht, April 2006

Preface to the second edition

It has been seven years since the first edition of this book, *Soil and Water Contamination – from molecular to catchment scale*, was published. Although the fundamentals of contaminant behaviour and transport have not changed much since then, considerable progress has been made in understanding the catchment-scale processes controlling the spatial and temporal patterns of contaminants. Therefore, the main changes in this second edition are in Chapters 16–18 (Part 4), which present case studies of spatial and temporal patterns of pollutants in soil, groundwater, and surface water, respectively. Several new example studies that reflect the new insights have been added to these chapters. In Part 2, new sections have been added to cover additional contaminants and contaminant groups such as asbestos, nanomaterials, selenium, and emerging substances of concern. Additionally, some figures have been updated throughout the entire text to include more recent data where appropriate.

I am grateful to a number of people for their help, inspiration, and support during the preparing of this second edition. First of all, I would like to thank my students who provided helpful and constructive feedback. They identified and commented on minor inconsistencies, errors, and typos in the previous edition and made suggestions for improvements to the text. I would also like to re-thank my colleagues who helped me with the first edition. In addition, I am grateful to Dr. Aaldrik Tiktak of the PBL Netherlands Environmental Assessment Agency for supplying material used in this book. As with the first edition, Dr. Joy Burrough-Boenisch (Unclogged English, Renkum, the Netherlands) advised on the English, which I very much appreciate. Lastly, I would like to acknowledge Drs. Ton Markus, Faculty of Geosciences, Utrecht University for drawing the new graphics in this edition.

Marcel van der Perk
Utrecht, July 2013

Part I
An introduction to soil and water contamination

The improvements in the socioeconomic status of many people have come about thanks to expansion of agricultural and industrial production. Some of the activities associated with the expansion, however, have adversely affected soil and water quality, impacting on people's health and quality of life and on the environment. This book sets out to explain the physical, chemical, and biological processes that govern the environmental fate of pollutants at landscape scale, i.e. at a scale ranging from several tens of metres to hundreds of kilometres. We need this knowledge to be able to develop appropriate solutions for a broad range of environmental problems.

1

General introduction

1.1 HISTORICAL PERSPECTIVE

Rapid growth of the world population and the pursuit of material prosperity have generated a massive expansion in industrial and agricultural production in recent decades. The associated increase in energy consumption and the generation of waste have enormously increased the pressure on the natural environment and have led to changes in the composition of the atmosphere, soil, fresh water resources, seas, and oceans. This, in turn, has led to destabilisation of natural ecosystems and a deterioration of environmental quality (i.e. the ability of the environment to support all appropriate beneficial uses by humans and wildlife). The increasing population density has made human society increasingly vulnerable to the natural variability of the environment and, especially, to environmental change.

Much pollution goes unnoticed and the resulting environmental deterioration is often difficult to detect. As a consequence, environmental issues have long been ignored. In the past 40 years, however, public awareness and concern about the state of the global and local environment has grown dramatically. The pollution of air, water, and soil has attracted particular interest, because of its direct adverse impact on landscapes and ecosystems (e.g. rivers, lakes, wetlands, heaths, woodland, pasture), cultural heritage (e.g. listed buildings and heritage sites), and human health. The first prominent publication that called attention to the abuse of persistent, hazardous pesticides such as DDT was *Silent Spring*, by the biologist Rachel Carson (1962). Ten years later, the publication of the book *Limits to Growth* by the Club of Rome (Meadows *et al.*, 1972) fuelled the debate about environmental issues. Based on one of the first computer-based simulation models, the authors predicted that the exponential increase of the world population and the accompanying growth of consumption and environmental pollution would cause a massive reduction in the Earth's ability to sustain future life within a time span of 100 years. The pessimistic forecasts presented in *Limits to Growth* have not come about, but the book's assumptions, methods, and results generated a vigorous debate among scientists and the general public. And whereas in the 1970s the environmental debate focused on population growth, heavy metals, and persistent pesticides, during the 1980s and 1990s it expanded to include smog, acid rain, radioactivity, the ozone hole, the greenhouse effect, and biodiversity.

In the scientific world, many subdisciplines of the natural sciences (e.g. ecology, physical geography, geology, geochemistry, hydrology, soil science) and engineering (civil engineering, agricultural engineering) started environmentally-oriented research, which developed into a new interdisciplinary branch of environmental science. This led to the founding of a number of specialised international journals in the field of environmental pollution, such as *Environmental Pollution* (1970; Elsevier), *Water, Air and Soil Pollution* (1971; Kluwer, now Springer), and the *Journal of Environmental Quality* (1972; The American Society of Agronomy, Crop Science Society of America, and Soil Science Society of America).

In the meantime, many countries had created government departments or agencies to provide decision-makers with the appropriate information needed for making effective policies to protect and improve the quality of the environment by preventing, controlling, and abating environmental pollution. In 1972, the United Nations established the United Nations Environmental Programme (UNEP) with its headquarters in Nairobi, Kenya. In 1991, the European Union set up the European Environmental Agency in Copenhagen, Denmark. Since the 1970s and 1980s, environmental protection and pollution control measures have increasingly been incorporated into national and supranational legislation, for example the Nitrate Directive (91/676/EC) and the Water Framework Directive (2000/60/EC) of the European Commission, implemented in 1991 and 2000, respectively. Although much has been achieved to abate and control environmental pollution in the western world, the United Nations Millennium Ecosystem Assessment (2005) has demonstrated that at the global scale, substantial changes are required to mitigate the negative consequences of growing pressures on ecosystems. However, pressures on land resources have continued to increase during recent years, despite international commitment and resolve to improve the management of the resources. Similarly, many water bodies are still affected by pollution, and many emerging contaminants have poorly understood effects (UNEP, 2012).

1.2 ENVIRONMENTAL POLLUTION

Before further discussing the issues related to environmental pollution, it would be useful to clarify some terminology. ***Pollution*** and ***contamination*** are used synonymously to mean the introduction into the environment by humans of substances that are harmful or poisonous to people and ecosystems. These substances (called pollutants or contaminants) are, therefore, anthropogenic, that is, they result from human activities. However, anthropogenic does not mean that all pollutants are man-made or synthetic chemicals, such as DDT or plutonium; chemical compounds that occur naturally in the environment can also be anthropogenic. In fact, the most widespread environmental pollution involves 'natural' compounds (for example, carbon dioxide) and fertilisers (such as nitrate). Furthermore, pollution is not restricted to substances, but can also refer to energy wastes, such as heat, light, and noise. In all cases, pollution alters the chemical, physical, biological, or radiological integrity of soil and water by killing species or changing their growth rate, interfering with food chains, or adversely affecting human health and well-being.

Note that some experts distinguish between pollution and contamination. They use the term contamination for situations where a pollutant is present in the environment, but not causing any harm, while they use the term pollution for situations where harmful effects are apparent (see Alloway and Ayres, 1997). However, the distinction between contamination and pollution may not be clear, because harmful effects may be present but unobserved. The above definition of pollution and contamination avoids this problem, so in this book the two terms are used interchangeably.

Human activity is unevenly distributed over the world; the Earth's surface, too, is variable by nature. As a result, the intensity and consequences of environmental pollution vary from place to place. Of the 136 million km² land surface on Earth, about 10 percent is used as arable land and 25 percent consists of productive pasture and of forests that might be converted into agricultural land. The pursuit of ever-higher agricultural yields, which has been made possible through technological innovation and the cultivation of marginal areas, has resulted in widespread degradation of agricultural land. In the past half-century, 40 percent of the world's agricultural land has been degraded by accelerated erosion by wind and water, salinisation, compaction, nutrient exhaustion, pollution, and urbanisation (Millennium Ecosystem Assessment, 2003). Moreover, the excess application of fertilisers and pesticides contaminates groundwater and surface water via leaching and contaminates the soil of natural areas next to agricultural land via atmospheric deposition. The increased loads of nutrients in fertilisers leads to the eutrophication of soil and surface waters: the excessive enrichment of nutrients, associated with rapid growth of vegetation. In the past four decades, excessive nutrient loading has emerged as one of the most important direct drivers of ecosystem change in terrestrial, freshwater, and marine ecosystems (Millennium Ecosystem Assessment, 2005). In many cases, groundwater polluted by excess fertiliser application is endangering supplies of drinking water. This is of particular concern in the countries in which groundwater is the main source of drinking water.

The pressure on the natural environment is clearly highest in urbanised regions. Cities and urban areas with their concentrated population, large energy consumption, transport and industrial activities, generally display the worst environmental pollution problems. Some industrial and commercial activities in urban areas may have brought about local soil contamination through spillage or dumping of chemicals. If these activities have been abandoned and the contaminated or potentially contaminated land is available for building development, the land is generally referred to as a brownfield site or, simply, a **brownfield**. Examples of brownfields are the sites of abandoned factories, petrol stations, and dry-cleaning establishments. The redevelopment of such sites may be complicated by the actual or potential contamination. Urbanised areas also typically consume more energy and other natural resources and produce more waste per capita than rural areas. Industrial emissions of polluting substances to the atmosphere or into surface water and the dumping of industrial and domestic wastes on waste disposal sites cause environmental pollution problems far beyond the city limits. Furthermore, the exploitation of natural resources required to maintain the standard of living in urban areas brings about spills and emissions of environmental pollutants elsewhere, for example during the mining and smelting of metal ores.

How far an emitted pollutant spreads in the environment depends on the physical and chemical characteristics of the pollutant itself, the soil's ability to retain it, and the nature and transport rate of the medium by which it is transported (air, water). The complex behaviour of pollutants in the environment is primarily caused by the variation of processes at or near the Earth's surface in both space and time. Nevertheless, they can be elucidated by the basic principles of physics and chemistry.

1.3 ENVIRONMENTAL POLLUTANTS

1.3.1 Classification of pollutants

In general, there are two basic types of pollutants. **Primary pollutants** cause harmful effects in the form in which they are released into the environment and **secondary pollutants** are formed as a result of a chemical process in the environment, often from less harmful precursors (Alloway and Ayres, 1997). As defined above, substances become pollutants if they are harmful or poisonous and have been introduced into the environment by humans, or as a result of human activities. There are many substances that are not commonly considered as pollutants but may cause pollution if they are released into the environment in excessive amounts or in an unfortunate place and time. Milk, fruit juice, and sugar, for example, are generally not considered as pollutants, but if directly released into surface water they are harmful to aquatic life, since the oxidation of the organic substances depletes dissolved oxygen in the water. On the other hand, many substances that are generally thought of as pollutants may occur naturally in soil and water: heavy metals, nitrate, and polycyclic aromatic hydrocarbons (PAHs), for example.

Pollutants can be classified in numerous ways based on their physical and chemical properties, abundance, persistence in the environment, effect on ecosystems, or toxicity.

This book follows a common subdivision based on multiple criteria, which distinguishes the following pollutant groups: solid phase constituents, major dissolved phase constituents, nutrients, heavy metals, radionuclides, and organic pollutants. The major dissolved constituents are comprised of inorganic substances that are abundant as dissolved ions in soil moisture, groundwater, and surface water, and which make up the major part of the total dissolved solids. Nutrients are those compounds that are essential for plant and animal life. Heavy metals include metals and metalloids with a high atomic mass, which are associated with contamination and potential toxicity. Radionuclides are elements having an unstable nucleus which spontaneously disintegrates, thereby emitting ionising radiation. Organic pollutants are substances made up of carbon (C), hydrogen (H), and oxygen (O), and a few other elements. Except for man-made radionuclides and synthetic organic compounds, all these substances occur naturally in the environment, but they are considered to be major pollutants because they are highly concentrated and widely dispersed in the environment as a result of human activities. The natural occurrence of substances in the environment will be discussed in the next sections. The properties, role, and effects of the above pollutant groups will be dealt with further in Part II of this book.

1.3.2 Background concentrations

Substances present in the environment are of either natural or anthropogenic (i.e. man-made) origin. The presence of detectable concentrations of chemical substances in the environment does not necessarily indicate the existence of pollution. Many substances, such as nutrients and heavy metals, occur naturally in soils, groundwater, and surface waters as an inevitable consequence of their natural occurrence in the Earth's crust. A substance's natural (i.e. without human interference) concentration in soil or water is called the ***background concentration***. Background concentrations may vary in both space and time. It is therefore important to distinguish between the natural occurrence of substances and the extent to which this has been augmented by human activities. Although often difficult to make, such distinctions are essential to ensure that informed decisions are made about the management of soil and water.

Geochemists have long studied the relative abundance of elements in the Earth's crust. The average composition of continental crustal rocks is given in Table 1.1. The atomic number, atomic weights and full names of the elements are given in Appendix I.

The composition of natural soils and water is, however, only indirectly related to the average concentrations listed in Table 1.1. Below a depth of 16 km the Earth's crust consists predominantly of igneous rocks, so therefore the average composition of the crust approaches that of igneous rocks. In the crust near the Earth's surface, however, sedimentary rocks predominate over igneous rocks. Sedimentary rocks can be subdivided geochemically into resistates, hydrolysates, precipitates, and evaporites. Resistates are rocks composed largely of residual minerals not chemically altered by the weathering of the parent rock (mainly quartz–SiO_2), such as sandstone, conglomerate, and graywacke. Hydrolysates are rocks composed mainly of relatively insoluble minerals produced during weathering of the parent rock (mainly clay minerals), such as shales. Precipitates are rocks produced by chemical precipitation of dissolved minerals from aqueous solution, such as limestone ($CaCO_3$) or dolomite ($Mg,CaCO_3$). Evaporites are formed when soluble minerals are deposited because the water in which they were dissolved has evaporated, for example halite (NaCl) or gypsum($CaSO_4 \cdot 2H_2O$). Because chemical precipitation is the basic process of the formation of both precipitates and evaporites, the distinction between precipitate rock and evaporite rock is rather arbitrary and evaporites often occur interbedded with precipitates.

The composition of both igneous rocks and sedimentary rocks may vary greatly from place to place. The variation in the composition of igneous rocks is the result of geological

Table 1.1 The average composition of the Earth's continental crust* (after Wedepohl, 1995).

Element	Average Concentration	Element	Average Concentration
O	47.2 %	Hf	4.9 ppm
Si	28.8 %	Gd	4.0 ppm
Al	7.96 %	Dy	3.8 ppm
Fe	4.32 %	Cs	3.4 ppm
Ca	3.85 %	Be	2.4 ppm
Na	2.36 %	Sn	2.3 ppm
Mg	2.2 %	Er	2.1 ppm
K	2.14 %	Yb	2.0 ppm
Ti	4010 ppm	As	1.7 ppm
C	1990 ppm	U	1.7 ppm
P	757 ppm	Ge	1.4 ppm
Mn	716 ppm	Eu	1.3 ppm
S	697 ppm	Ta	1.1 ppm
Ba	584 ppm	Mo	1.1 ppm
F	525 ppm	Br	1.0 ppm
Cl	472 ppm	W	1.0 ppm
Sr	333 ppm	I	800 ppb
Zr	203 ppm	Ho	800 ppb
Cr	136 ppm	Tb	650 ppb
V	98 ppm	Tl	520 ppb
Rb	78 ppm	Lu	350 ppb
Zn	65 ppm	Tm	300 ppb
N	60 ppm	Sb	300 ppb
Ce	60 ppm	Se	120 ppb
Ni	56 ppm	Cd	100 ppb
La	30 ppm	Bi	85 ppb
Nd	27 ppm	Ag	70 ppb
Cu	25 ppm	In	50 ppb
Co	24 ppm	Hg	40 ppb
Y	24 ppm	Te	5 ppb
Nb	19 ppm	Au	2.5 ppb
Li	18 ppm	Pd	0.4 ppb
Sc	16 ppm	Pt	0.4 ppb
Ga	15 ppm	Re	0.4 ppb
Pb	14.8 ppm	Ru	0.1 ppb
B	11 ppm	Rh	0.06 ppb
Th	8.5 ppm	Os	0.05 ppb
Pr	6.7 ppm	Ir	0.05 ppb
Sm	5.3 ppm		

* ppm = parts per million and ppb = parts per billion (see Section 2.2). This table does not include elements with an average concentration of less than 0.05 ppb (e.g. Ne, Kr, Xe), and short-lived radioactive elements.

processes during formation. By contrast, the variation in the composition of sedimentary rocks is principally the result of geomorphological processes of differential weathering and mechanical sorting during transport of the weathering products. At the same time, the natural variation in the composition of natural waters is largely attributable to differences in the chemical composition and weatherability of the bedrock. Whereas the soil composition reflects the more or less stable solid residue of the bedrock, the composition of natural waters reflects the easily weatherable fraction of the bedrock. Accordingly, the background concentrations of the different substances in soil and water can display a wide variation in geographical space. Nevertheless, the background concentrations are usually low compared to the concentrations due to anthropogenic inputs. Exceptions are locations near ore outcrops or volcanic springs, where the soil and water may be enriched in, for example, heavy metals or radioactivity.

1.3.3 Anthropogenic sources

Human activities may enhance the concentration of polluting substances in soil and water directly or indirectly. Direct releases of pollutants may occur from a single location or ***point source***, such as an effluent discharge pipe or a container leaking petrol into the soil, or a natural outcrop of a mineral vein rich in heavy metals. Pollution around a point source is mostly confined to a plume in the downstream direction from the source (e.g. Figure 1.1a). The extent to which the front of the plume spreads from the source depends on the age of the pollution source, the flow velocity, and the retention along the flow paths. The dispersal perpendicular to the flow direction is generally limited. In general, point sources are relatively easily monitored and controlled, because they can often be identified as being caused by a particular individual or organisation.

Chemicals may also enter the environment over a wide area. Examples are anthropogenically enhanced atmospheric deposition of acid compounds (acid rain), global atmospheric fallout of radionuclides following the nuclear bomb testing in the 1950s and 1960s and nuclear accidents, fertiliser leaching from agriculture fields (see Figure 1.1b), contaminated runoff from city streets, and deposition of sediment-associated pollutants on floodplains. A pollutant release that occurs over a wide area is said to have a non-point source or ***diffuse source***. The pollutant concentration gradients in soil and water originating from diffuse sources are usually gradual. In addition, relatively small point sources may occur at many different places that cannot be distinguished individually: for example, discharge of domestic wastewater into urban groundwater (e.g. via septic tanks) or surface water, or a large number of leaking containers in an extensive waste disposal site. The impact from any one location may be minimal, but the cumulative effect of all source locations can be substantial. The fact that the pollutants mix in groundwater or surface water makes it impossible to identify the contribution from each individual source, so the total of different point source locations acts as a diffuse source.

In addition to direct environmental pollution due to point source or diffuse source emissions, the release of certain substances into the environment may also lead to indirect pollution. For example, enhanced deposition of acid compounds may effectively increase the rate of natural weathering of bedrock. Consequently, this increases the rate at which minerals, including heavy metals, are released from the bedrock into soil and water.

The extent to which polluting substances in soil and water are enriched by human activities depends on the background concentration, the amount released, the source type, and the pollution's tendency to disperse in the environment. The enrichment ranges between less than one order of magnitude up to several orders of magnitude. Obviously, a substance with a background concentration close to zero is more sensitive to enrichment than a substance with a large background concentration. The enrichment is also greater if large

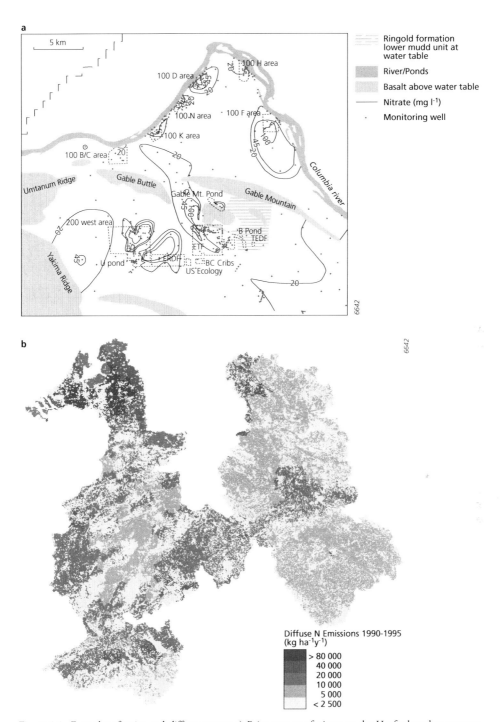

Figure 1.1 Examples of point and diffuse sources: a) Point sources of nitrate at the Hanford nuclear weapons complex, Washington state USA (Hartman *et al.*, 2002); b) Diffuse nitrogen emissions from agricultural land (kg ha⁻¹ y⁻¹) in the Rhine and Elbe catchments from 1990 through 1995 (De Wit, 1999).

quantities of a pollutant are released from a point source and if the pollutant is contained at
or near its location of release. If the pollutants are released over a larger area (diffuse source)
or if a pollutant disperses rapidly over a larger area or volume, the pollutant is diluted and the
enrichment is accordingly smaller.

1.4 ECOLOGICAL IMPACTS

The presence of pollutants in the environment may have adverse effects on the health of both
humans and ecosystems. The effects on individual organisms (including humans) and whole
ecosystems are studied by scientists from various related fields of natural sciences. ***Toxicology***
is the study of the nature and mechanisms of harmful effects of substances on living
organisms, usually *Homo sapiens*. ***Ecotoxicology*** is the study of the nature and mechanisms
of toxic effects of chemicals on living organisms, especially on populations and communities
within defined ecosystems (Butler, 1978). ***Radioecology*** is the study of the transfer of
radionuclides through natural and agricultural ecosystems and the effects of environmental
radioactivity on plants and animals, and humans (Alexakhin *et al.*, 2001). All three
disciplines consider the interactions between chemicals and their environment, their transfer
pathways, and their harmful effects. In addition, they assess the hazards (i.e. the potential to
cause harm) and risks (i.e. the probability that harm will be caused) related to exposure to
toxic chemicals, and they develop methods of diagnosis, prevention, and treatment.

The term ***toxicity*** is widely used to denote the capacity to cause harm to a living
organism. There are various ways in which toxicity can be measured, but they are nearly all
assessed relative to a particular outcome or end point. Initially, most toxicity tests measured
the number of organisms killed by a particular dose or concentration of the chemical
being tested. Dose is often used where the dietary dose of a test chemical can be accurately
determined; for example, in the case of terrestrial animals, the dose administered (taken
orally, applied to the skin, or injected) is usually recorded. For aquatic organisms or where
the test chemical is dosed onto the surrounding medium, the tests usually measure the
concentration of chemical in the surrounding water. The median lethal dose (LD50) and the
median lethal concentration (LC50) describe the level of exposure (dose or concentration)
that kills 50 percent of the population. In recent years, there has been a move away from
the use of lethal end points in toxicity testing towards the measurement of effects other
than death. Examples of such effects are changes in growth (e.g. biomass or body length),
reproduction (e.g. number of offspring), or biochemical or physiological processes (e.g.
enzyme synthesis or respiration). The median effect dose/concentration (ED50/EC50)
describes the level of exposure that causes a defined effect to 50 percent of the population.
The no observed effect level (NOEL) is the general term for the dose (NOED) or
concentration (NOEC) at which a test chemical does not cause an effect that is statistically
significantly different from the control.

In natural systems, organisms are often exposed to more than one pollutant at the same
time. It has often been assumed that the toxicity of combinations of chemicals is roughly
additive, and in many cases this is true. However, in some cases, the resulting toxicity of
more than one chemical is not additive. The presence of other chemicals may decrease
toxicity (antagonism), or increase the toxicity (synergism or potentiation) of a chemical.

When assessing the ecological impact of a chemical it is not enough to consider its
toxicity; it is also essential to consider its ***bioavailability***. Bioavailability means the rate
and extent to which a substance can be taken up or absorbed into the tissues of organisms
and so influence their physiology. It is possibly the most important factor determining the
extent to which a contaminant in soil or water will enter the food chain. Chemicals enter the
food chain via a variety of pathways. Primary producers (green plants and algae) and some

bacteria absorb pollutants directly from soil or water. Subsequently, when animals or humans eat these plants as part of their diet, the absorbed chemicals may be transferred to higher trophic levels. Foodstuffs may also become contaminated due to, for example, atmospheric deposition, application of pesticides, or food processing. In addition, contaminants may enter the bodies of humans and animals through drinking water, inhalation of contaminated airborne dust, or via the skin. Contaminants that are not stored in body tissues leave the body in urine and faeces. The extent to which pollutants accumulate over time in the tissues of organisms (e.g. leaves, roots, bones, body fat) through any route, including respiration, ingestion, or direct contact is referred to as *bioaccumulation*. Some pollutants have a short *biological half-life* (i.e. the time it takes to remove 50 percent of the quantity of a substance in a specified tissue, organ, or any other specified biota as a result of biological processes), which means that they are excreted soon after intake and do not have the opportunity to bioaccumulate. Other pollutants have a long biological half-life and they accumulate in the organism's tissues. Whereas bioaccumulation refers to the storage of pollutants in an individual organism, the term *biomagnification* is commonly used for the increasing concentrations as the pollutant passes through the food chain.

1.5 SPATIAL AND TEMPORAL VARIABILITY AND THE CONCEPT OF SCALE

This book is to do with the analysis and prediction of the formation and dynamics of spatial contaminant patterns in our environment and their impact on the environment. To study the response of the spatially complex and heterogeneous natural environment to the large-scale stresses of environmental pollutants the approach must be interdisciplinary, integrating various subdisciplines of earth and life sciences, such as geochemistry, hydrology, soil science, geomorphology, meteorology, ecotoxicology, classical and spatial statistics, and geographic information science. The water flow and solute transport models developed by hydrologists, soil scientists, and geochemists are essential simulation tools for assessing potential temporal and spatial changes in the fate and movement of pollutants in soil, groundwater, and surface water. Classical statistics is valuable for model validation and assessing data uncertainty, whereas spatial statistics is helpful for examining the spatial variability and spatial structure by analysing both spatial trends and spatial correlation. Geographical information systems (GIS) serve as a means of storing, manipulating, retrieving, and displaying spatial data associated with environmental pollution.

Spatial and temporal variability are key issues in studies of the environment. The characteristics of spatial and temporal variation influence which spatial and temporal *scale* is chosen for the systematic description of the sources and processes that govern the patterns of chemicals in the environment. The concept of scale arises from the notion that the properties of phenomena or processes vary when measured over different spatial or temporal extents or different levels of resolution (Bierkens *et al.*, 2000). The *extent* refers to the size or scope of the study area or duration of time; the *resolution* refers to the detail of measurement (Figure 1.2), which is determined by both sampling interval (distance or time lag between two samples) and so-called sample support, i.e. physical size or extent of the sample (area, volume, or mass). Extent and resolution are thus both characteristics of scale, but scale usually refers to the spatial or temporal extent. All environmental issues are defined within a spatial and temporal framework and can thus be investigated at different scales, depending on the objectives of the research: management, planning, or restoration. For example, river water quality can be monitored or simulated at a range of temporal scales from hours or days (e.g. studies of the response of phosphorus concentration to a heavy rainfall event in a small catchment), months and years (e.g. studies of seasonal or year-to-year variation of phosphorus load from a river basin). If historical records are used, the scale can be extended

a.   b.

Figure 1.2 Extent and resolution: floodplain area along the river Elbe near Bleckede, Germany: a) original resolution and extent; b) subarea with a smaller extent and a coarser resolution. Background: Digital Orthophoto 1:5000, reprinted with permission from the publisher: LGN – Landesvermessung und Geobasisinformation Niedersachsen – D10390.

to decades or, if proxies from sediment records are used, even to centuries or thousands of years or more (e.g. studies on climate change or global phosphorus cycles). Likewise, soil quality can be investigated at the scale of a few nanometres (reaction scale, e.g. studies of sorption phenomena at the edges of clay minerals), centimetres (e.g. studies of vertical distribution of soil nutrients over the soil profile), tens of metres (e.g. soil survey in a field or at a contaminated industrial site), or tens or hundreds of kilometres, (e.g. regional or supra-regional soil pollution mapping). Spatial scale is generally correlated with the persistence of attributes observed at that scale, and therefore with temporal scale. This means that system-level features such as the spatial variation of heavy metals in a drainage basin persist over longer temporal scales than microscale features such as heavy metal concentrations in soil pore water. Figure 1.3 depicts the general correlation between spatial and temporal scales.

It is relevant to note that different patterns and processes dominate at different scales. Patterns that seem very important at a particular scale may be insignificant at another scale. For example, on a clay pigeon shooting ground, the pattern of lead contamination is largely determined by the positions of the clay pigeon shooters relative to the locations of the traps (i.e. the devices that release clay pigeons into the air). However, this lead contamination pattern at the field scale may be of negligible importance at regional or catchment scale. Although the shooting grounds may be a 'hot spot' of soil contamination of lead, lead contamination derived from other sources, such as traffic, metal smelters, and lead mines, is likely to determine the spatial variation of soil contamination by lead at these larger scales. Likewise, short-term variations of in-stream nitrate concentrations in a small forested catchments are controlled by soil moisture and temperature status and rainfall intensity in the catchment, whilst long-term variations in nitrate transport from large river basins, such as the river Rhine basin, are determined by nitrogen emissions controlled by the number

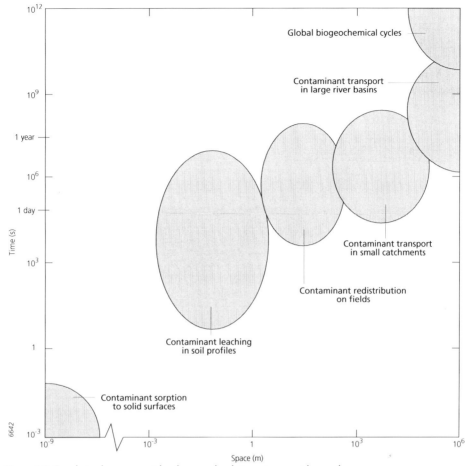

Figure 1.3 Correlation between spatial and temporal scale in environmental research.

of inhabitants, the state of wastewater treatment, the intensity of agriculture in the river basin, and nitrogen transfer controlled by lithology, soil texture, and net annual precipitation (De Wit, 1999). At the global scale, the mean annual nitrate concentration and the nitrate export to coastal areas is primarily controlled by the human population within the river basin (Peierls *et al.*, 1991).

The coupling of scales or the question how information at one scale can be implemented at another scale, often referred to as upscaling and downscaling, is an ongoing challenge in environmental sciences. This issue is, however, beyond the scope of this text; for further information on this topic, see the literature (e.g. Bierkens *et al.*, 2000).

1.6 OUTLINE AND RATIONALE OF THIS BOOK

This book aims to give a broad introduction to the basic concepts of the processes and patterns of contaminants in soil and fresh water (surface water and groundwater) at scales ranging from the microscale of molecules to the regional scale of ecosystems and catchments (typical synonyms: drainage basins, river basins, watersheds). It does not deal with air pollution. After a brief introduction to environmental chemistry and the features of the environmental compartments soil, groundwater, and surface water, Part II of this book addresses the sources and physical and chemical properties of the major contaminants in soil and water. This provides the basic knowledge needed to understand and assess the existence and behaviour of pollutants in the environment. Part III goes into more detail about the processes of pollutant dispersal and the mathematical modelling of pollution. Finally, Part IV discusses the resulting spatial and temporal pollution patterns in the landscape (i.e. at scales ranging from several tens of metres to hundreds of kilometres) using case studies from the literature. Key questions that are discussed in this part are:

- Which hydrological, geomorphological, geochemical, and biological processes are responsible for observed spatial contaminant patterns in the environment?
- How do these patterns evolve in time?
- How do landscape factors, such as soil properties, land use, slope gradient, river network, or vegetation structure control the fluxes of emitted contaminants through soil and water?
- What are the effects of environmental change on the processes controlling contaminant transport and fate?

These questions are important in both fundamental research and water and soil management. They also help us improve our understanding of spatial patterns of soil and water contamination, to predict their future changes, and to identify potential environmental hazards.

EXERCISES

1. Define the following terms:
   a. Pollution
   b. Contamination
   c. Secondary pollutants
   d. Background concentration
   e. Diffuse source
   f. LD50
   g. NOEC
   h. Bioavailability
   i. Bioaccumulation

2. Describe a method for the determination of background concentrations of heavy metals in soil.

3. Describe two different realistic scenarios for future spatial development in a radius of about 10 km around your home town and identify relevant issues with respect to soil and groundwater pollution.

4. Describe the role of physical geography and related disciplines of earth science in environmental research.

5. Why does spatial scale generally correlate with temporal scale in environmental research?

6. Explain why ploughing is an important process of transport of soil nutrients at the scale of a soil profile and not at the scale of an entire river basin.

Basic environmental chemistry

2.1 INTRODUCTION

Molecules, ions, and solid particles in the environment undergo physical and chemical transfer processes which distribute mass within or among phases; for example, the volatilisation of a chemical from water to air, or the dissolution of calcite ($CaCO_3$). Knowledge of the basic principles of chemistry is essential to be able to quantify these transfer processes and to assess the distribution of concentrations of chemicals in space and time and their chemical *speciation*. Speciation refers to the occurrence of different forms, whether chemical or physical, of an element. The speciation of an element determines its bioavailability and toxicity and controls its transport and fate in soil and water.

This section recapitulates the fundamentals of environmental chemistry which will be applied in subsequent chapters. Only the chemical interactions between relatively simple ions and molecules and between phases will be covered in this chapter. Chemical interactions on solid surfaces are dealt with in Chapters 4 and 13.

Chemical reactions result in the breakdown of one chemical and the formation of another. This process inevitably involves chemical bonds being broken or formed between atoms. The various types of chemical bonds include covalent bonds, in which atoms share one or more electrons, ionic bonds, in which opposite electric charges on neighbouring atoms result in the bonding force, and hydrogen bonds, in which a slightly positively charged hydrogen atom at the edge of a molecule loosely bonds with a slightly negatively charged atom of another molecule. In addition, weak, attractive Van der Waals forces between molecules may contribute to bonding as well.

Most inorganic chemicals are electrolytes which to varying extent can dissolve in water to form ions. Positively charged ions (e.g. Na^+ or Ca^{2+}) are called *cations* and negatively charged ions (e.g. Cl^- or HCO_3^-) are called *anions*. Organic substances can also dissolve in water to form organic cations or anions. However, most organic liquids are non-electrolytes and dissolve in water as non-ionic molecules.

2.2 UNITS OF ANALYSIS

We have already encountered the term *concentration* in the previous chapter, being used to specify how much of a substance is present in a given amount of soil or water. The concentration of chemicals in soil and water can be presented in various units. Though the official SI units for concentration are moles m^{-3} or moles kg^{-1}, these are not commonly used. The most common units are listed in Table 2.1.

For dilute fresh water samples, the unit $mg\ l^{-1}$ is numerically equal to $mg\ kg^{-1}$, since the density of the water is $1\ g\ cm^{-3}$. For samples of higher density, for example seawater (density $\approx 1.023\ g\ cm^{-3}$), recalculation is necessary to convert the results from concentration by volume to concentration by weight. To recalculate the results from $mg\ l^{-1}$ to $mmol\ l^{-1}$,

Table 2.1 Units of concentration of substances in soil and water

Unit	Description
mg l^{-1}	milligrams per litre water (aqueous solution)
mg kg^{-1}	milligrams per kilogram soil or H_2O
ppm	parts per million by weight
ppb	parts per billion by weight
mmol l^{-1}	millimoles per litre water
M	molality, moles per kilogram H_2O
mmol kg^{-1}	millimoles per kilogram soil or H_2O
N	normality, equivalents* per litre water
meq l^{-1}	milliequivalents* per litre water
meq/100 g	milliequivalents* per 100 g of soil

* The official SI unit for equivalents is moles of charge (mol$_c$). For example, meq l^{-1} is equal to mmol$_c$ l^{-1}.
In this text, the unit equivalents are used.

the figures should be divided by the molar mass or molecular weight, which is the mass in grams of 1 mol of molecules or atoms (1 mol = $6.022 \cdot 10^{23}$ molecules = Avogadro's number). The molar mass can be calculated by summing the atomic weights of the atoms that form the molecule (the atomic weights are given in Appendix 1). For example, the molar mass of nitrate (NO_3^-) is 1 × 14 (atomic weight of N) + 3 × 16 (atomic weight of O) = 62. To convert the results from mmol l^{-1} to meq l^{-1} the concentration in mmol l^{-1} is multiplied by the charge z of the ions. For example, 1 mmol l^{-1} of Ca^{2+} equals 2 meq l^{-1}.

The units mg l^{-1} (water) and mg kg^{-1} are mostly used to report environmental concentrations of chemicals. However, the molarity (mmol l^{-1}) or molality (mol kg^{-1}) is preferable for the chemical evaluation of the concentrations in solutions, since a balanced chemical equation gives stoichiometric information directly in terms of moles of reactants and products. The unit meq l^{-1} is often used to check whether the charges of the cations and anions balance each other (see Section 5.1).

Example 2.1 Conversion of concentration units

The PO_4^{3-} concentration is 23 μg l^{-1}. Calculate the concentration in meq l^{-1}.

Solution
First, calculate the molar mass of PO_4^{3-}: 1 mol PO_4^{3-} is made up of 30.97 g of P and 4 × 16.00 g of O (see Appendix I), making a total mass of 94.97 g mol^{-1}
Thus, 23 μg l^{-1} corresponds to 23/94.97 = 0.242 μmol l^{-1} = 2.42 10^{-4} mmol l^{-1}. To obtain the concentration in meq l^{-1}, multiply the concentration in mmol l^{-1} by the charge of the PO_4^{3-} ion. Thus, 2.42 10^{-4} mmol PO_4^{3-} l^{-1} × 3 = 7.26 meq PO_4^{3-} l^{-1}

In addition to the concentration units listed above, a number of additional parameters related to the concentrations of substances in water are relevant in environmental studies; they include pH, total dissolved solids, electrical conductivity (EC), and redox potential (Eh). These parameters are further specified in Table 2.2. Other common parameters related to soil and water, for example cation exchange capacity, alkalinity, biological oxygen demand, are covered in the chapters on solid phase constituents (Chapter 4) and major dissolved phase constituents (Chapter 5).

Table 2.2 Additional parameters related to concentrations in soil and water (after Appelo and Postma, 2005).

Parameter	Unit	Description
pH	pH units [-]	Logarithm of the H^+ activity; $-\log[H^+]$
TDS	mg l^{-1}	Total dissolved solids;
EC	μS cm^{-1}	Electrical conductivity of water; as the EC is temperature dependent, it is commonly reported as the EC at 25°C
Eh	Volt	Redox potential (see Section 2.10.3)
pe	-	Redox potential expressed as $-\log[e^-]$, where $[e^-]$ = the activity of electrons; $pe = Eh/0.059$ at 25 °C (see Section 2.10.4)

The pH is defined as the reciprocal of the logarithm of the 'effective concentration' (activity; see Section 2.3) of hydrogen ions (H^+). Water with a pH < 7 is considered acidic and with a pH > 7 is considered basic. The pH of pure water (H_2O) is 7 at 25 °C, but when exposed to the carbon dioxide in the atmosphere, the pH decreases to approximately 5.2. The normal range for pH in surface water is 6.5 to 8.5; for groundwater it is 6 to 8.5. Section 2.9 goes into more detail about the pH, acids, and bases.

The total dissolved solids (TDS) represents the total concentration of dissolved minerals in the water and is determined by evaporating a known volume of a filtered sample in an oven heated to 105 °C. The electrical conductivity (EC) is a measure of the water's ability to conduct electricity, and therefore a measure of the water's ionic concentration. The major dissolved phase constituents (see Chapter 3) contribute most to the EC; the larger their concentration, the higher the conductivity. Therefore, the EC is generally found to be a good measure of the concentration of total dissolved solids (TDS) and salinity. The EC expressed in μS cm^{-1} ranges from 1.0 to 2.0 times the TDS concentration in mg l^{-1} and can be estimated more accurately using:

$$EC \approx 100 \times meq \text{ (cations or anions) } l^{-1} \tag{2.1}$$

The EC increases substantially with temperature, which can have a confounding effect on attempts to compare the EC across different waters or different seasons. To eliminate this complication and allow comparisons to be made, the EC is normalised to a temperature of 25 °C.

The redox potential is defined as the ability of a system or species to consume or donate electrons; it is measured in volts or millivolts. The more negative the redox potential is relative to the standard potential, the greater a system's tendency to lose electrons and the greater that system's reducing potential. Conversely, a more positive redox potential indicates an oxidising environment, which implies a greater affinity for electrons and less tendency to donate electrons. The concept of redox potential is further explained in Section 2.10.

2.3 ACTIVITY

Because the dissolved ions in an aqueous solution influence each other through the electrostatic forces between them, the concentration of the dissolved ions is not the correct measure of their reactivity. Therefore, instead of the total concentrations, an 'apparent' concentration should be used for calculations of chemical reactions in aqueous solutions. This apparent concentration, called the ***activity***, is corrected for the non-ideal effects in aqueous solutions arising from the electrostatic forces between all dissolved ions in the water. It is related to the molar concentration:

$$|\mathbf{x}_i| = \gamma_i m_i \qquad\qquad (2.2)$$

where $[\mathbf{x}_i]$ = the activity of an ion i (mol l^{-1}), γ_i = the activity coefficient [-], and m_i is the molar concentration of the ion i (mol l^{-1}). The activity coefficient γ_i depends on the ionic strength, which is a measure of the total concentration of dissolved ions:

$$I = 0.5 \sum m_i z_i^2 \qquad\qquad (2.3)$$

where I = the ionic strength and z_i = the charge of the ion i. The ionic strength of fresh water is usually less than 0.02 and that of ocean water is about 0.7. The activity coefficient depends not only on the ionic strength but also on the specific charge of the ion, and on temperature and pressure. Activity coefficients are tabulated in chemical textbooks as function of ionic strength and specific charge, or can be estimated from several activity models, such as the extended Debye–Hückel equation:

$$\log \gamma_i = \frac{-A z_i^2 \sqrt{I}}{1 + B a_i \sqrt{I}} \qquad\qquad (2.4)$$

where A and B are temperature-dependent constants and a_i = the radius of the hydrated ion [L]. The values for A, B, and a_i are listed in Table 2.3. The extended Debye–Hückel equation can be used to estimate values of the activity coefficient to a maximum ionic strength of about 0.1 or a total dissolved solids concentration of approximately 5000 mg l^{-1} (Domineco and Schwarz, 1997). However, it is worth noting that in natural fresh waters the activity coefficient is usually close to one and the effect of ionic strength may be neglected for approximate calculations. In saline waters, such as ocean water, it is usually necessary to use activity coefficients in equilibrium calculations. Studies of more saline waters, such as brines, require other more sophisticated activity models, for example the Pitzer model (Pitzer, 1981). For further descriptions of chemical equilibrium modelling, see standard chemical textbooks, such as Stumm and Morgan (1996), Drever (2000), or Faure (1998).

Table 2.3 Values of A, B, and a_i used in the extended Debye–Hückel equation (after Domenico and Schwartz, 1997).

Temperature (°C)	A	B (× 10^8)	a_i (× 10^{-8} cm)	Ion
0	0.4883	0.3241	2.5	NH_4^+, Cs^+, Rb^+, Ag^+
5	0.4921	0.3249	3.0	K^+, NO_3^-, Cl^-, Br^-, I^-
10	0.4960	0.3258	3.0	OH^-, F^-, HS^-, BrO_3^-
15	0.5000	0.3262	4.0 – 4.5	Na^+, HCO_3^-, $H_2PO_4^-$, HPO_4^{2-}, PO_4^{3-}, SO_4^{2-}, HSO_3^-, Hg_2^{2+}
20	0.5042	0.3273	4.5	Pb^{2+}, CO_3^{2-}, MoO_4^{2-}
25	0.5085	0.3281	5.0	Sr^{2+}, Ba^{2+}, Ra^{2+}, Cd^{2+}, Hg^{2+}, S^{2-}, WO_4^{2-}
30	0.5130	0.3290	6.0	Li^+, Ca^{2+}, Cu^{2+}, Zn^{2+}, Sn^{2+}, Mn^{2+}, Fe^{2+}, Ni^{2+}, Co^{2+}
35	0.5175	0.3297	8.0	Mg^{2+}, Be^{2+}
40	0.5221	0.3305	9.0	H^+, Al^{3+}, Cr^{3+}, trivalent rare earth
50	0.5319	0.3321	11	Th^{4+}, Zr^{4+}, Ce^{4+}, Sn^{4+}
60	0.5425	0.3338		

Example 2.2 Calculation of ionic strength and activity

A groundwater sample contains 4.05 mmol HCO_3^- l^{-1}, 1.54 mmol Cl$^-$ l^{-1}, 1.17 mmol SO_4^{2-} l^{-1}, 2.1 mmol Na$^+$ l^{-1}, 0.2 mmol K$^+$ l^{-1}, 3.8 mmol Ca^{2+} l^{-1}, 0.43 mmol Mg^{2+} l^{-1}, 0.2 mmol Fe^{2+} l^{-1}, and 0.03 mmol Mn^{2+} l^{-1}. Calculate the activities of these ions at 20 °C.

Solution
The calculation of the activities is summarised in the table below. First, calculate the ionic strength using Equation (2.3). The ionic strength of the solution is $9.58 \cdot 10^{-3}$. Subsequently, calculate the activity coefficients using Equation (2.4). From Table 2.3, the values for A and B at 20 °C can be obtained, as well as the radii of the respective ions (ai). The activities can then be calculated by multiplying the activity coefficient by the molar concentration given above.

Ion	Charge	Molar concentration	Contribution to ionic strength	Radius of the hydrated ion	Log activity coefficient	Activity coefficient	Activity
	z	m_i	$0.5\, m_i \cdot z_i^2$	a_i	$\log \gamma_i$	γ_i	$[x_i]$
		(mol l^{-1})		(10^{-8} cm)			(mol l^{-1})
HCO_3^-	-1	$4.05 \cdot 10^{-3}$	$2.03 \cdot 10^{-3}$	4	-0.0437	0.90	$3.66 \cdot 10^{-3}$
Cl$^-$	-1	$1.54 \cdot 10^{-3}$	$0.77 \cdot 10^{-3}$	3	-0.0450	0.90	$1.39 \cdot 10^{-3}$
SO_4^{2-}	-2	$1.17 \cdot 10^{-3}$	$1.17 \cdot 10^{-3}$	4	-0.1749	0.67	$0.78 \cdot 10^{-3}$
Na$^+$	1	$2.1 \cdot 10^{-3}$	$1.05 \cdot 10^{-3}$	4	-0.0437	0.90	$1.90 \cdot 10^{-3}$
K$^+$	1	$0.2 \cdot 10^{-3}$	$0.10 \cdot 10^{-3}$	3	-0.0450	0.90	$0.18 \cdot 10^{-3}$
Ca^{2+}	2	$3.8 \cdot 10^{-3}$	$3.80 \cdot 10^{-3}$	6	-0.1655	0.68	$2.60 \cdot 10^{-3}$
Mg^{2+}	2	$0.43 \cdot 10^{-3}$	$0.43 \cdot 10^{-3}$	8	-0.1571	0.70	$0.30 \cdot 10^{-3}$
Fe^{2+}	2	$0.2 \cdot 10^{-3}$	$0.20 \cdot 10^{-3}$	6	-0.1655	0.68	$0.14 \cdot 10^{-3}$
Mn^{2+}	2	$0.03 \cdot 10^{-3}$	$0.03 \cdot 10^{-3}$	6	-0.1655	0.68	$0.02 \cdot 10^{-3}$
Total			$9.58 \cdot 10^{-3}$				

Note that the activity coefficients for the monovalent ions are much greater than those of the divalent ions.

2.4 BACKGROUND THERMODYNAMICS

In natural systems, energy may have various forms: for example, heat, radiation, motion, electricity, or chemical bonds. Thermodynamics is the science of the distribution of energy among substances in a system. The term system as used here refers to a body consisting of solids, gases, and liquids (the latter including dissolved material contained in them) that potentially interact with each other. Thermodynamic principles constitute a basis for a quantitative assessment of physical and chemical transfer processes. They allow us to predict the direction a system is tending to go, and the ultimate equilibrium concentrations and phases of the reactants and products in that system.

The chemical energy stored in a substance or molecule at constant temperature and pressure is called the ***enthalpy***. Enthalpy includes the internal energies associated with intramolecular forces due to chemical bonds and attractions within molecules, as well as the external energies associated with intermolecular forces due to chemical bonds and attractions between molecules. The enthalpy is expressed as:

$$H \equiv U + PV \tag{2.5}$$

where H = the enthalpy [M L^{2}T^{-2}], U = the internal energy [M L^{2}T^{-2}], P = the pressure [M L^{-1} T^{-2}], V = the volume [L^3]. The unit for enthalpy in Equation (2.5) is kJ kg^{-1}, but most often the enthalpy is expressed in kJ mol^{-1}.

The first law of thermodynamics, also known as the law of conservation of energy, states that the total energy of a system and its surroundings remains constant. This means that, in general, the formation of molecules takes energy and the reaction is **endothermic** (enthalpy increases); for example, the formation of glucose from carbon dioxide and water through photosynthesis in green plants requires energy in the form of sunlight. Conversely, the energy released by the breaking of chemical bonds is converted to other forms of energy (e.g. heat, electricity, or formation of other chemical bonds), and the reaction is **exothermic** (enthalpy decreases). There are a few exceptions to this general rule: for example, the formation of O_2 from two oxygen atoms and the formation of N_2 from two nitrogen atoms are exothermic reactions.

The breakdown of molecules also implies an increase in the level of disorder, which is referred to as **entropy**. Thus, a disorganised, more random system of molecules or atoms is reflected in an increased entropy. Gases have more entropy than liquids, which in turn have more entropy than solids. The second law of thermodynamics states that in a system at constant temperature and pressure, the entropy increases or remains the same. Eventually, the energy becomes evenly distributed over the system. Thus, the distribution of energy in a given system through reactions and phase changes seeks equilibrium between a minimum level of enthalpy and a maximum level of disorder (entropy). This equilibrium is represented by a minimum of the **Gibbs free energy**, which depends on the chemical composition, pressure, and temperature. At constant temperature and pressure, the Gibbs free energy (further referred to as 'free energy') is quantitatively expressed as the difference between enthalpy and the entropy, multiplied by the absolute temperature:

$$G \equiv H - TS \tag{2.6}$$

where G = the free energy expressed in Joules (J) [M L^{2}T^{-2}], H = the enthalpy [M L^{2}T^{-2}], T = the absolute temperature [θ] (in Kelvin (K) = 273.15 + °C), S = the entropy [ML^{2}T^{-2} θ^{-1}]. This is a general statement of the third law of thermodynamics, which stipulates that the entropy of a pure perfect crystal is zero at absolute zero (0 K). Entropy is more difficult to evaluate and observe quantitatively than the free energy that is released or consumed in a phase transition or chemical reaction. Therefore, it is more convenient to express Equation (2.6) in terms of the change in free energy (ΔG):

$$\Delta G = \Delta H - T\Delta S \tag{2.7}$$

At equilibrium, the change in free energy is zero. If the change in free energy is negative, the reaction or phase transition is spontaneous; if the change in free energy is positive, the reaction is non-spontaneous.

Example 2.3 Change in the Gibbs free energy

Solid calcium carbonate is formed from the reaction between solid calcium oxide and carbon dioxide gas:

CaO (s) + CO_2 (g) → $CaCO_3$ (g)

Calculate the free energy change for the formation of one mole of calcium carbonate at 25 °C according to this reaction, given ΔH = -178.2 kJ per mole and ΔS = -160.6 J per mole. Is this reaction spontaneous?

Solution
This is a straightforward application of Equation (2.7):

$$\Delta G \;=\; \Delta H - T\Delta S \;=\; -178.2 - 298.15 \times -0.1606 \;=\; -130 \text{ kJ mol}^{-1}$$

Note that the J–kJ conversion must be accounted for in the calculation.

The value of ΔG is negative, indicating that the reaction proceeds spontaneously. From Equation (2.7) it can also be seen that ΔG becomes zero at a temperature above 25 °C. Thus, if the temperature is sufficiently high, the above reaction will proceed in the reverse direction, i.e. calcium carbonate will decompose into calcium oxide and carbon dioxide gas.

2.5 PHASES AND PHASE TRANSITIONS

2.5.1 Phases

Substances can occur in three states: solid, liquid, and gas. A ***phase*** is a distinct and homogeneous state of a material with no visible boundary separating it into parts. As a rule, there is always only one gas phase, as gaseous substances mix fully. There may be more than one liquid phase (e.g. oil and water). Liquids of one particular substance always mix fully, except for liquid helium that may occur in two phases. Many substances may occur in different solid phases; for example, carbon may occur in both the diamond and graphite phase, and silicon dioxide may occur in a crystalline (quartz) or amorphous, glassy phase. Chemicals that are dissolved in water occur in the so-called ***dissolved phase***, which is also called the soluble phase or aqueous phase. Sometimes the term ***liquid phase*** is used to refer to the dissolved phase, but it should be noted that this term ignores the difference between pure liquids and aqueous solutions. The ***solid phase*** is a generic but ill-defined term for all chemicals that are in soil and sediment solids, or in suspended solids in water. There are also various other terms specifying particular components of the solid phase. The term ***adsorbed phase*** refers to chemicals adsorbed to solid surfaces or suspended matter. Substances associated with suspended solids in water, i.e. adsorbed onto or incorporated in suspended particles, are also referred to as the ***particulate phase*** or suspended phase.

2.5.2 Thermodynamic considerations on phase transitions

Chemicals may occur in multiple phases at the same time. The conversion between phases is called a phase transition. Table 2.4 shows the most important phase transitions of chemicals in soil, water, and air. Chemicals tend to establish equilibrium among the different phases, which ultimately results in an equilibrium distribution. As seen in the previous section, this equilibrium state is characterised by a zero change in free energy (ΔG = 0). From Equation (2.7) it follows that:

$$\Delta H \;=\; T\Delta S \tag{2.8}$$

The enthalpy changes primarily due to a change in volume (see Equation 2.5). Because gases have more entropy than liquids, which, in turn, have more entropy than solids, it takes energy (i.e. a change in enthalpy) to convert a solid into a liquid and, subsequently, into a gas with no temperature change. Conversely, during phase transitions towards a phase with less entropy, this energy is released again. This energy is called latent heat; it becomes manifest as an increase in temperature when the entropy decreases, for instance during condensation. The specific latent heat of fusion of a substance is the amount of heat required to convert a unit mass of a solid into a liquid without a change in temperature; the specific latent heat of vaporisation is the amount of heat required to convert a unit mass of a liquid into a vapour without a change in temperature. The latent heats expressed per mole of a substance are called enthalpies of fusion and vaporisation, respectively. Specific latent heats or enthalpies of fusion and vaporisation are tabulated in many physical and chemical textbooks. Phase transitions from or to the aqueous phase are commonly considered as chemical reactions. The thermodynamics of these transitions will be dealt with in Section 2.6.

2.5.3 Partition coefficient

It has often been observed experimentally that at small concentrations the ratio of concentrations in two phases in equilibrium is constant. This constant is also called the partition coefficient:

$$\frac{C_{phase1}}{C_{phase2}} = K \tag{2.9}$$

Table 2.4 Phase transitions and the compound property that determines equilibrium partitioning between two phases.

Phase 1	Phase transition	Phase 2	Compound property
Pure solid	melting $\longrightarrow$ / $\longleftarrow$ freezing	Pure liquid	Melting point
Pure solid	sublimation $\longrightarrow$ / $\longleftarrow$ condensation	Gas	Vapour pressure
Pure liquid	vaporisation $\longrightarrow$ / $\longleftarrow$ condensation	Gas	Vapour pressure
Pure solid	dissolution $\longrightarrow$ / $\longleftarrow$ precipitation	Aqueous solution	Aqueous solubility
Pure liquid	dissolution $\longrightarrow$ / $\longleftarrow$ exsolution	Aqueous solution	Aqueous solubility
Gas	dissolution $\longrightarrow$ / $\longleftarrow$ volatilisation	Aqueous solution	Air–water partition constant (Henry's law constant)

where C_{phase1} and C_{phase2} refer to the concentration in two different phases [M L^{-3}] and K = the partitioning coefficient [-]. The partition coefficient is also known as the **distribution coefficient**. For partitioning among the aqueous and gas phases, the equilibrium Equation (2.9) is known as **Henry's law** and K as Henry's law constant (K_H).

A special case of the partition coefficient is the **octanol–water partition coefficient** K_{ow}, which is commonly used for organic compounds. The octanol–water partition coefficient K_{ow} is defined as the ratio between the concentration of a chemical in octanol ($C_8H_{17}OH$) and in water at equilibrium and at a specified temperature and is a measure of the tendency of a chemical to partition itself between the aqueous phase and organic phase. Octanol is an organic solvent used as a surrogate for natural organic matter. Non-polar, hydrophobic organic compounds prefer octanol (large value of K_{ow}) but polar, hydrophilic organic compounds prefer water (small value of K_{ow}). Values of K_{ow} for many chemicals are available from the literature or chemical fact sheets (e.g. EPA, 2013; ATSDR, 2013) and are useful in the estimation of other parameters, such as the aqueous solubility and distribution coefficient for adsorption of organic compounds to organic matter. The distribution coefficient approach for sorption of substances to solids is discussed in more detail in the next section.

In addition, partition coefficients are also often used to describe the partitioning of chemicals between the various trophic levels of the food chain. The partition coefficient is then referred to as a concentration factor for water–biota transfer or a transfer factor for soil biota, or a food–biota transfer (Blust, 2001).

2.5.4 Partitioning between dissolved phase and adsorbed phase

When an aqueous solution of a chemical is mixed with a suspension of solids, the total mass of the chemical equilibrates between the dissolved phase and the adsorbed phase. If this experiment of mixing a solution with a solid medium is repeated for different initial concentrations C_i at the same temperature (a so-called batch experiment), the values of the concentration of adsorbed chemical on the solids C_s plotted against the equilibrium concentration C_w form an adsorption **isotherm**. Isotherms can be linear, convex, or concave, or a complex combination of these shapes. The most commonly adopted empirical relationships are the Freundlich isotherm:

$$C_s = K\,C_w^{\,n} \tag{2.10}$$

and the Langmuir isotherm:

$$C_s = \frac{Q^0 K C_w}{1 + K C_w} \tag{2.11}$$

where C_s = the concentration of the chemical adsorbed to the solid [M M^{-1}], C_w = the equilibrium concentration of the chemical in solution [M L^{-3}], K = a partition coefficient reflecting the extent of sorption, n = an exponent usually ranging between 0.7 and 1.2, and Q^0 = the maximum sorptive capacity of the solids. Figure 13.2 shows some examples of Freundlich and Langmuir isotherms.

A Freundlich isotherm with an exponent n = 1 is a special case, since the isotherm becomes linear. The resulting equation is analogous to Equation (2.9) and relates the concentration of the solids to the solute concentration, using a distribution coefficient:

$$K_d = \frac{C_s}{C_w} \tag{2.12}$$

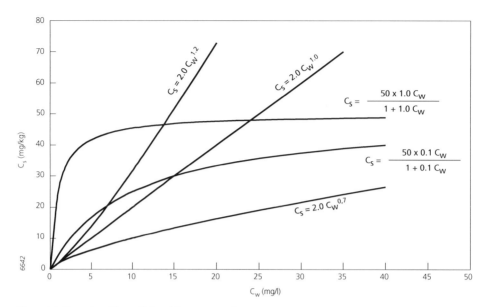

Figure 2.1 Examples of Freundlich and Langmuir isotherms.

where K_d = the distribution coefficient [L^3 M^{-1}]. Increasing values of the distribution coefficient K_d indicate a greater proportion adsorbed to solids. The value of K_d depends on the physico-chemical properties of the ***adsorbent*** (i.e. the substrate onto which a substance is sorbed) and is theoretically independent of the amount of adsorbent, but as noted in the previous section, it is only constant at small concentrations of the ***adsorbate*** (i.e. substance that becomes adsorbed).

Example 2.4 Distribution coefficient

Ten grams of sediment containing no cadmium is added to 10 litres of an aqueous solution containing 60 µg l^{-1} cadmium. The water is stirred, so that the sediment remains in suspension. After equilibrium has been reached the cadmium concentration in solution (C_w) is 20 µg l^{-1}. Calculate the distribution coefficient K_d.

Solution
The initial concentration is redistributed between the dissolved phase (C_w) and the adsorbed phase (C_s). The total concentration C_{tot} remains equal to the initial concentration of 60 µg l^{-1} (= 0.06 mg l^{-1}) and equals:

$$C_{tot} = C_w + C_s \, SS$$

where SS = the suspended sediment concentration [M L^{-3}]. Combining this equation with Equation (2.12) yields:

$$C_{tot} = C_w + K_d \, C_w \, SS$$

Hence,

$$K_d = \frac{C_{tot} - C_w}{C_w \, SS}$$

The suspended sediment concentration is 10 g per 10 l, which is 0.001 kg l^{-1}. So the distribution coefficient K_d is:

$$K_d = \frac{0.06 - 0.02}{0.02 \times 0.001} = 2000 \text{ l kg}^{-1}$$

2.5.5 Fugacity

Another way to predict the distribution of chemicals among the different phases is the concept of *fugacity*. Fugacity means the tendency of a substance to flee from the phase it is in. The concept of fugacity is similar to that of activity and has been based on the chemical potential, which is defined as the increase in free energy with each increment of a substance:

$$\mu_i = \frac{\partial G}{\partial n_i} \tag{2.13}$$

where μ_i = the chemical potential expressed in kJ mol^{-1} [M $L^2 T^2$ mol^{-1}], and n_i = the amount of substance i [mol]. At constant temperature, the incremental change in chemical potential of a gaseous compound is related to a corresponding change in pressure:

$$d\mu_i = \frac{V}{n_i} dP_i \tag{2.14}$$

where V = volume of the gas [L^3], and P_i = the *partial pressure* of i [M L^{-1} T^{-2}]. The partial pressure of a gas is that pressure it would exert if it occupied the entire volume by itself. The *ideal gas law* can be used to convert the partial pressure into corresponding moles per unit volume:

$$\frac{n_i}{V} = \frac{P_i}{RT} \tag{2.15}$$

where R = the gas constant (= 8.3144 J mol^{-1} K^{-1} = 0.0821 l atm mol^{-1} K^{-1}), and T = the temperature (K). Equation (2.14) becomes:

$$d\mu_i = \frac{RT}{P_i} dP_i \tag{2.16}$$

Integration of Equation (2.16) yields:

$$\mu_i = \mu_i^0 + RT \ln\left[\frac{P_i}{P_i^0}\right] \tag{2.17}$$

where μ_i^0 = standard chemical potential [M $L^2 T^2$ mol^{-1}], and P_i^0 = standard *vapour pressure* (i.e. the partial pressure that the chemical would have in a gas volume in equilibrium with the pure liquid or solid phase at 1 atmosphere pressure) [M L^{-1} T^{-2}]. Since chemists are mostly interested in conditions that prevail under normal environmental conditions near the Earth's surface, the standard conditions are commonly chosen at 25 °C (= 298.15 K) and 1 atmosphere pressure. It is difficult to quantify the absolute value for the standard chemical potential, and so for the chemical potential. However, this issue is less relevant, since the change and the differences in chemical potential are of most interest. Equation (2.17) applies for ideal gases. Since gases do not fully behave ideally under all circumstances, the fugacity, which is closely related to pressure, has been defined, so that Equation (2.17) becomes:

$$\mu_i = \mu_i^0 + RT \ln\left[\frac{f_i}{f_i^0}\right] \tag{2.18}$$

where f_i = the fugacity of compound i [M L^{-1} T^{-2}], and f_i^0 = the standard fugacity [M L^{-1} T^{-2}]. For gases, the fugacity is:

$$f_i \quad = \quad \theta_i \; x_i \; P \tag{2.19}$$

where θ_i = fugacity coefficient [-], x_i = the molar fraction of i in the mixture or solution [-], and P = the total pressure [M L^{-1} T^{-2}]. The fugacity coefficient corrects for the non-ideality of the gas, but under normal environmental conditions it is very close to 1, so:

$$f_i \quad \cong \quad P_i \tag{2.20}$$

The fugacity of a chemical in the liquid or solid phase is related to the vapour pressure by:

$$f_i \quad = \quad \gamma_i \; x_i \; P_i^0 \tag{2.21}$$

where γ_i = the activity coefficient [-]. For a pure liquid or solid, both the molar fraction x_i and the activity coefficient γ_i are equal to 1. The activity coefficient in a liquid solution (e.g. water) may well be different from 1. For solutions of non-polar organic compounds in polar solvents like water, the activity coefficients are much larger than 1 (Schwarzenbach *et al.* 1993). The following equation can be used to convert the mole fraction x_i to the molar concentration m_i:

$$m_i \quad = \quad \frac{x_i}{V_{mix}} \tag{2.22}$$

where V_{mix} = the molar volume of the mixture or solution (l mol^{-1}). Hence:

$$f_i \quad = \quad \gamma_i \; m_i \; V_{mix} P_i^0 \tag{2.23}$$

For aqueous solutions of organic compounds that are hardly or moderately soluble, the contribution of the organic solute to the molar volume may be neglected (Schwarzenbach *et al.*, 1993). This means we may assume the molar volume to be equal to the molar volume of water, which amounts to 0.018 l mol^{-1}. Organic liquids usually have molar volumes of the order of 0.2 l mol^{-1}.

From Equation (2.23) it can be seen that the fugacity is linearly related to the molar concentration. In general, the relation between the concentration and fugacity for a given phase can be written as:

$$m_{i,j} \quad = \quad Z_{i,j} \; f_i \tag{2.24}$$

where m_{p_j} = the molar concentration of compound i in phase j [M L^{-3}], $Z_{i,j}$ = the fugacity capacity of compound i in phase j [T^2 L^{-2}], usually expressed in units of mol atm^{-1} m^{-3}. Each substance tends to accumulate in compartments where the fugacity capacity Z is large. At equilibrium, the change in free energy is zero, which implies that the chemical potential is equal in each phase. This, in turn, implies that the fugacities in each phase are the same. The fugacity for the entire system, which consists of different phases, is (Hemond and Fechner-Levy, 2000):

$$f_i \quad = \quad \frac{M_i}{\displaystyle\sum_i (Z_{i,j} \; V_j)} \tag{2.25}$$

where M_i= the total number of moles of compound i in the system and V_j = the volume of phase j [L^3]. By combining the ideal gas law (Equation 2.15) and Equation (2.24) with $m_{i,j}$ = n_i/V, it can be shown that:

$$Z_{air} = \frac{1}{RT} \qquad (2.26)$$

where Z_{air} = the fugacity capacity constant for air. The fugacity capacity constant for other phases depends on the compound's partition coefficient between that phase and water and the compound's Henry's law constant. The fugacity capacity constant for water is:

$$Z_{water} = \frac{1}{K_H} \qquad (2.27)$$

where Z_{water} = the fugacity capacity constant for water, K_H = the Henry's law constant [$L^2 T^2$]. For sediment, the fugacity capacity constant is:

$$Z_{sed} = \frac{\rho_s K_d}{K_H} \qquad (2.28)$$

where Z_{sed} = the fugacity capacity constant for sediment, ρ_s = the bulk density of sediment [$M L^{-3}$] and K_d = the distribution coefficient [$L^3 M^{-1}$] (see Equation 2.12). The number of moles of the compound in each phase can then be calculated by:

$$M_{i,j} = Z_{i,j} f_i V_j \qquad (2.29)$$

For further details about the fugacity concept, see Mackay (1991) or Schwarzenbach *et al.* (1993).

Example 2.5 Fugacity

Consider a pond ecosystem comprising 10 000 000 m³ of air, 50 000 m³ of water, and 400 m³ of bottom sediment. Ten litres of the organic liquid benzene is spilled into it. Calculate the equilibrium partitioning among air, water, and sediment using the fugacity approach. Relevant data is given in the table below.

Benzene properties

Vapour pressure	0.125	atm
Henry's law constant	$5.55 \cdot 10^{-3}$	atm · m³ mol⁻¹
Molecular weight	78.11	g mol⁻¹
Specific density	873.81	kg m⁻³

Sediment properties

Bulk density	1100	kg m⁻³
Distribution coefficient for benzene	0.6	l kg⁻¹

Solution
First, calculate the total amount of benzene released expressed in moles:
$10\, l \times 873.81\ g\ l^{-1}/78.11\ g\ mol^{-1}$ = 112 mol
Second, calculate the fugacity capacity for the three phases considered. For the air phase, use Equation (2.26):

$$Z_{air} = \frac{1}{RT} = \frac{1}{8.21 \cdot 10^{-4} \cdot 298.15} = 40.9\ mol\ atm^{-1}\ m^{-3}$$

For the water phase (dissolved phase)) use Equation (2.27):

$$Z_{water} \quad = \quad \frac{1}{H} \quad = \quad \frac{1}{5.55 \cdot 10^{-3}} \quad = \quad 180 \text{ mol atm}^{-1} \text{ m}^{-3}$$

For the sediment phase (adsorbed phase), use Equation (2.28):

$$Z_{sed} \quad = \quad \frac{\rho_s \cdot K_d}{H} \quad = \quad \frac{1100 \times 0.6 \cdot 10^{-3}}{5.55 \cdot 10^{-3}} \quad = \quad 119 \text{ mol atm}^{-1} \text{ m}^{-3}$$

Third, calculate the fugacity for the whole system using equation (2.25):

$$f \quad = \quad \frac{112}{40.9 \times 10^7 + 180 \times 5 \cdot 10^4 + 119 \times 400} \quad = \quad 2.68 \cdot 10^{-7} \text{ atm}$$

Finally, from Equation (2.29) the amount of benzene in the three different phases can be calculated:

$$M_{air} \quad = \quad 2.68 \cdot 10^{-7} \times 10^7 \times 40.9 \quad = \quad 110 \text{ mol}$$

$$M_{water} \quad = \quad 2.68 \cdot 10^{-7} \times 5 \cdot 10^4 \times 180 \quad = \quad 2.4 \text{ mol}$$

$$M_{sed} \quad = \quad 2.68 \cdot 10^{-7} \times 400 \times 119 \quad = \quad 0.013 \text{ mol}$$

Thus, most of the benzene is present in the air. Nevertheless, the benzene concentrations in the different phases are proportional to the fugacity capacity and, therefore, the highest benzene concentration is found in the water (0.048 μmol l^{-1}).

2.6 CHEMICAL EQUILIBRIUM AND KINETICS

2.6.1 Equilibrium

Chemical equilibrium in a closed system describes the state of maximum thermodynamic stability, which means that there is no chemical energy available to alter the relative distribution of mass between the reactants and products in a reaction, and the final expected chemical composition has been reached. Consider a reversible reaction where components A and B (reactants) react to produce components C and D (products):

$$a\text{A} \quad + \quad b\text{B} \quad \leftrightarrow \quad c\text{C} \quad + \quad d\text{D} \tag{2.30}$$

where the uppercase letters represent the chemical compounds and the lowercase letters the stoichiometric coefficients. Similarly to Equation (2.17), the change in free energy per additional mole reacting as a system proceeds towards equilibrium can be expressed as:

$$\Delta G \quad = \quad \Delta G^0 \quad + \quad RT \ln Q \tag{2.31}$$

where ΔG^0 = the standard free energy change at 25 °C and 1 atm pressure [M L^2 T^{-2}], which is constant for a given reaction, and Q = the reaction quotient [-]The change of the standard free energy for a given reaction can be calculated from the free energies of formation (i.e. the energy needed to produce one mole of a substance from the elements in their most stable form) of the substances participating in the reaction:

$$\Delta G^0 \;=\; \sum \Delta G^0_{f \; products} \;-\; \sum \Delta G^0_{f \; reactants} \tag{2.32}$$

where ΔG^0_f = the free energy of formation, tabulated for standard conditions (25 °C and 1 atmosphere pressure) in many standard chemical, geochemical, or hydrochemical textbooks (e.g. Stumm and Morgan, 1996; Drever, 2000; Morel, 1983).
The reaction quotient Q is:

$$Q \;=\; \frac{[C]^c \, [D]^d}{[A]^a \, [B]^b} \tag{2.33}$$

where [A], [B], [C], and [D] refer to the activities of the chemicals A, B, C, and D. Note that the square brackets refer to activities. As remarked before, the effect of ionic strength may be neglected for dilute fresh waters, so that for approximate calculations, activities may be approximated by concentrations. Activities for dissolved species are expressed in mol l^{-1}; pure solids and the solvent (H_2O) have activities equal to 1; gases (whether in the gaseous phase or dissolved) are expressed in units of partial pressure. The reaction quotient is also referred to as the ion activity product. If ΔG is positive, the reaction (Equation 2.30) proceeds to the left until ΔG becomes zero. Conversely, if ΔG is negative, the reaction proceeds to the right until ΔG becomes zero. If ΔG is zero, the system is at equilibrium and the forward and reverse reactions in Equation (2.30) occur at the same rate, so the chemical composition of the system does not change. Thus at equilibrium, the following is valid:

$$\Delta G^0 \;+\; RT \ln Q \;=\; 0 \tag{2.34}$$

Hence,

$$\Delta G^0 \;=\; -RT \ln Q \tag{2.35}$$

At equilibrium, the reaction quotient Q equals the ***equilibrium constant*** K:

$$K \;=\; \frac{[C]^c \, [D]^d}{[A]^a \, [B]^b} \;=\; e^{-\Delta G^0 / RT} \tag{2.36}$$

This Equation (2.36) is also known as the ***mass action law***, and therefore some authors refer to the equilibrium constant K as the mass action constant. The equilibrium constant represents the final expected distribution of mass between the reactants and products at a given temperature and pressure. Depending on the reaction, we refer to the equilibrium constant K as a) an acidity or dissociation constant in acid–base reactions (see Section 2.9), b) a complexation constant in complexation reactions (see Section 2.8), c) a solubility constant in solid dissolution reactions (see Section 2.7), d) an adsorption constant in sorption reactions, or e) a Henry's law constant in gas dissolution (see Section 14.1). The values for these constants are typically derived from laboratory experiments and thermodynamic calculations.

Example 2.6 Calculation of the equilibrium constant from Gibbs free energy data

Consider the reaction of calcite with carbon dioxide:

$$CaCO_3 + CO_2 + H_2O \leftrightarrow Ca^{2+} + 2\,HCO_3^-$$

The Gibbs free energies of formation for the reactants and reaction products at 25 °C are

$\Delta G^0_{f\,CaCO_3} = -1128.8$ kJ mol^{-1}

$\Delta G^0_{f\,CO_2} = -394.4$ kJ mol^{-1}

$\Delta G^0_{f\,H_2O} = -237.1$ kJ mol^{-1}

$\Delta G^0_{f\,Ca^{2+}} = -553.6$ kJ mol^{-1}

$\Delta G^0_{f\,HCO_3^-} = -586.8$ kJ mol^{-1}

Calculate the equilibrium constant from the data given above.

Solution
The Gibbs free energy for the reaction is

$\Delta G^0_r = \Delta G^0_{f\,Ca^{2+}} + 2\,\Delta G^0_{f\,HCO_3^-} - \Delta G^0_{f\,CaCO_3} - \Delta G^0_{f\,CO_2} - \Delta G^0_{f\,H_2O}$

$\Delta G^0_r = -553.6 - 2 \times 586.8 + 1128.8 + 394.4 + 237.1 = 33.1$ kJ mol^{-1}

$\Delta G^0_r = -RT \ln K$

$33.1 = -(8.3144 \cdot 10^{-3})(298.15) \ln K$

$\ln K = -13.35$

$K = e^{-13.35} = 1.59 \cdot 10^{-6}$

Because natural environmental conditions usually deviate from the standard conditions at 25 °C and 1 atmosphere pressure, the equilibrium constant K needs to be corrected. Variations in pressure have only a small effect on the values of the equilibrium constant and therefore these variations are generally neglected. However, variations in temperature have a significant effect on the equilibrium constant. To correct the equilibrium constant K for temperature, the Van 't Hoff equation (which can be derived by combining Equations 2.7 and 2.36) is used:

$$\frac{d \ln K}{dT} = \frac{\Delta H^0_r}{RT^2} \tag{2.37}$$

where K = equilibrium constant [-], T = absolute temperature (K), ΔH^0_r = the standard reaction enthalpy (kJ mol^{-1}), and R is the gas constant (J mol^{-1} K^{-1}). Integration of Equation (2.37) and conversion of the natural logarithm to a base-10 logarithm yields:

$$\log K_T = \log K_{25\,°C} + \frac{\Delta H^0_r}{2.303\,R} \cdot \left(\frac{1}{298.15} - \frac{1}{T} \right) \tag{2.38}$$

where K_T = equilibrium constant at temperature T (in K) [-], T = absolute temperature [θ], ΔH^0_r = the standard reaction enthalpy (kJ mol^{-1}). The reaction enthalpy ΔH^0_r is negative for exothermic reactions and positive for endothermic reactions (see Table 2.4). For the derivation of Equation (2.38), it was assumed that temperature variations have a negligible effect on the reaction enthalpy, which is generally true for the temperature ranges normally found in nature. The reaction enthalpy ΔH_r^0 is calculated from the formation enthalpies, which are listed in thermodynamic tables.

Example 2.7 Temperature dependency of the equilibrium constant

Calculate the equilibrium constant for the reaction of calcite with carbon dioxide (see Example 2.6) at 15 °C

The enthalpies of formation for the reactants and reaction products are
$\Delta H_{f\,CaCO_3}^0 = -1206.9$ kJ mol^{-1}

$\Delta H_{f\,CO_2}^0 = -393.5$ kJ mol^{-1}

$\Delta H_{f\,H_2O}^0 = -285.8$ kJ mol^{-1}

$\Delta H_{f\,Ca^{2+}}^0 = -542.8$ kJ mol^{-1}

$\Delta H_{f\,HCO_3^-}^0 = -692.0$ kJ mol^{-1}

Solution
The reaction enthalpy is

$$\Delta H_r^0 = \Delta H_{f\,Ca^{2+}}^0 + 2\,\Delta H_{f\,HCO_3^-}^0 - \Delta H_{f\,CaCO_3}^0 - \Delta H_{f\,CO_2}^0 - \Delta H_{f\,H_2O}^0$$

$$\Delta H_r^0 = -542.8 - 2 \times -692.0 + 1206.9 + 393.5 + 285.8 = -40.6 \text{ kJ mol}^{-1}$$

Note that ΔH_r^0 is negative, so the reaction is exothermic, which implies that heat is lost when calcite dissolves. Use Equation (2.33) to calculate log K at 15 °C:

$$log\,K_{15\,°C} = log\,K_{25\,°C} + \frac{\Delta H_r^0}{2.303\,R} \cdot \left(\frac{1}{298.15} - \frac{1}{288.15}\right)$$

$$log\,K_{15\,°C} = log\left(1.59 \cdot 10^{-6}\right) + \frac{-40.6}{2.303 \cdot 8.3144 \cdot 10^{-3}} \cdot \left(\frac{1}{298.15} - \frac{1}{288.15}\right)$$

$$= -5.42$$

$K_{15°C} = 10^{-5.42} = 3.80 \cdot 10{\text -}6$

The larger value of K at 15 °C implies that the equilibrium shifts to the right with decreasing temperature. This means that the solubility of calcite increases with decreasing temperature.

2.6.2 Kinetics

Reaction kinetics refers to the rate at which chemical reactions occur. In the general chemical reaction in Equation (2.30), the reaction rate is usually assumed to be proportional to the product of the activity of each substance participating in the reaction raised to the power of an exponent. Consequently, the rates of the forward, reverse, and entire reactions can be written as:

Rate of forward reaction: $\quad r_1 = k_1[A]^{\alpha}[B]^{\beta}$ (2.39a)

Rate of reverse reaction: $r_2 = k_2 [C]^\gamma [D]^\delta$ (2.39b)

Net rate of the reaction: $r_n = k_1 [A]^\alpha [B]^\beta - k_2 [C]^\gamma [D]^\delta$ (2.39c)

where r = reaction rate [M L^{-3} T^{-1}], k_1= the rate constant of the forward reaction, k_2 = rate constant of reverse reaction (dimension depends on the exponents). The overall ***reaction order*** is defined as the sum of the exponents in the rate expression. For the rate in the forward reaction (Equation 2.39a), the reaction order is thus equal to $\alpha + \beta$. The exponents and the reaction order need to be derived experimentally.

In the absence of experimental evidence or other knowledge, the reactions in most environmental models are often assumed to follow ***first-order kinetics***, i.e. the reaction rate is proportional to the activity or concentration of the reactant to the first power:

$$\frac{d[A]}{dt} = -k[A]^{1.0} = -k[A] \qquad (2.40)$$

where k = the first-order reaction rate constant [T^{-1}]. Note that this kinetic equation represents an irreversible process and can be applied in cases in which the reverse reaction does not have a major effect, i.e. for a reaction of the form:

$$A \rightarrow products \qquad (2.41)$$

The analytical solution of this first-order differential Equation (2.40) reads:

$$A(t) = A_0 e^{-kt} \qquad (2.42)$$

where A_0 = the initial concentration (activity) of A at t = 0. The negative sign in front of the rate constant indicates that the chemical is being removed, i.e. the concentration decreases with time and goes to zero in the limit.

Example 2.8 Reaction order

Consider the decomposition of N_2O_5:
$2N_2O_5 \rightarrow 4NO_2 + O_2$

The rate law for this reaction is
rate = $-k$ [N_2O_5]

Thus, the reaction is first-order with respect to the N_2O_5 concentration. If the first-order reaction rate constant k = 0.062 s^{-1}, calculate the time needed to lower the initial concentration by 90 percent.

Solution
From Equation (2.42), we see that the N_2O_5 concentration decreases exponentially over time:
$[N_2O_5](t) = [N_2O_5](t = 0) \, e^{-kt}$

After a certain time, the concentration has decreased to 10 percent of its initial value, thus:
$[N_2O_5](t)/[N_2O_5](t=0) = 0.1 = e^{-kt} = e^{-0.062 \, t}$

–0.062 t = ln(0.1) = -2.303
t = 37.1 s

So, it takes 37.1 s to decrease the N_2O_5 concentration to 10 percent of its initial value.

2.7 DISSOLUTION–PRECIPITATION REACTIONS

Dissolution and precipitation reactions are an important control on the transition of chemicals between the solid and dissolved phases. As noted earlier, some compounds dissolve in water as molecules, whereas electrolyte minerals dissociate and dissolve as ions. Consider the following dissociation reaction:

$$M_xA_y \quad \leftrightarrow \quad xM^{m+} \quad + \quad yA^{a-} \tag{2.43}$$

where M = the cation with charge $m+$ and A = the anion with charge $a-$, and x and y refer to the stoichiometric constants. In an aqueous solution in contact with an excess of an undissolved, solid, mineral, the mineral will dissolve until the solution is saturated and no more solid can dissolve. The concentration of the saturated solution at a given temperature and pressure is termed the **solubility** of the substance. In such a saturated solution, equilibrium is established, which means that the forward reaction in Equation (2.43) has the same rate as the backward reaction. Since the activity of the solid M_xA_y is 1 by definition (see previous section), the equilibrium constant for this reaction is:

$$K_s \quad = \quad [M^{m+}]^x[A^{a-}]^y \tag{2.44}$$

where K_s = the equilibrium constant for dissolution–precipitation reactions, also referred to as the **solubility product**. The solubility product increases with the solubility of a mineral. It is important to stress that solubility and the solubility product refer to a reversible equilibrium condition. However, dissolution reactions are often characterised by slow reaction rates, so that 'non-equilibrium' conditions prevail. The extent and direction of departure from saturation can be evaluated using the saturation index (*SI*), which is defined as:

$$SI \quad = \quad \log\left(\frac{Q}{K_s}\right) \tag{2.45}$$

where Q = the reaction quotient (see Equation 2.33) [-], and K_s is the solubility product [-]. If *SI* is less than zero ($Q/K_s < 1$), the system is undersaturated and may move towards saturation by dissolution. Conversely, if *SI* is greater than zero ($Q/K_s > 1$), the system is supersaturated and should be moving towards saturation by precipitation.

Example 2.9 Saturation index

A leachate from coal ash contains 0.03 mg l^{-1} barium (Ba^{2+}) and 90 mg l^{-1} sulphate (SO_4^{2-}). The ionic strength is 0.06. Calculate the saturation index *SI* of barite ($BaSO_4$) given the log solubility product $\log K_s$ for barite = –9.97 at 25 °C.

Solution
First, calculate the concentrations of Ba^{2+} and SO_4^{2-} in mol l^{-1}:
The atomic weight of Ba^{2+} = 137.33 (see Appendix I), so the Ba^{2+} concentration is 0.03 mg l^{-1}/1000 mg g^{-1}/137.33 g mol^{-1} = 4.37 · 10^{-7} mol l^{-1}

The molar mass of SO_4^{2-} = 32.07 + 4 × 16.00 = 96.07, so the SO_4^{2-} concentration is 90 mg l^{-1}/1000 mg g^{-1}/96.07 g mol^{-1} = 1.56 · 10^{-3} mol l^{-1}

Second, calculate the activities of Ba^{2+} and SO_4^{2-} in mol l^{-1}:

Use Equation (2.4) to calculate the log activity coefficients. The values for A (= 0.5085) and B (= 0.3281) at 25 °C are read from table 2.3, as well as the ion-specific values for $a_{Ba^{2+}}$ (= 5.0), and $a_{SO_4^{2-}}$ (= 4.0). Filling in Equation (2.4) we obtain log$\gamma_{Ba^{2+}}$ = -0.355 and log$\gamma_{SO_4^{2-}}$ = -0.377. So, the activity coefficients are $\gamma_{Ba^{2+}}$ = 0.44 and $\gamma_{SO_4^{2-}}$ = 0.42. The Ba^{2+} activity is 0.44 × 4.37 · 10^{-7} mol l^{-1} = 1.92 · 10^{-7} mol l^{-1} and the SO_4^{2-} activity is 0.42 × 1.56 · 10^{-3} mol l^{-1} = 6.55 10^{-4} mol l^{-1}.

Third, calculate the reaction coefficient Q, also referred to as the ion activity product (*IAP*):

$IAP = Q = [Ba^{2+}][SO_4^{2-}] = 1.26\ 10^{-10}$

Finally, calculate the saturation index *SI* (Equation 2.41)

$$SI = \log\left(\frac{Q}{K_s}\right) = \log Q - \log K_s = \log(1.26 \cdot 10^{-10}) - (-9.97) =$$

$$-9.90 + 9.97 = 0.07.$$

This implies that the solution is lightly oversaturated with barite. Barite is likely to precipitate until equilibrium has been reached (*SI* = 0).

2.8 COMPLEXATION

Not only the solubility product but also the total concentration of an ion in an aqueous solution is controlled by the formation of inorganic and organic **complexes**. Such complexes may considerably reduce the activity of ions, thereby enhancing the total concentration of substances in solution. This section focuses on inorganic complexes; the formation of organic complexes is further elaborated upon in Section 4.3.2.

Examples of aqueous complexes are $CaSO_4^0$, $CaOH^+$, $CdCl^+$, $PbHCO_3^+$. The formation of aqueous complexes can be described by equilibria of the type:

$$xM^{m+} + yA^{a-} \leftrightarrow M_xA_y^{m-a} \tag{2.46}$$

The accompanying mass action law yields:

$$K = \frac{[M_xA_y^{m-a}]}{[M^{m+}]^x[A^{a-}]^y} \tag{2.47}$$

The equilibrium constant K for complexation reactions is also termed the stability constant. Stability constants can be obtained from extensive databases, such as the WATEQ/PHREEQE database (Nordstrom *et al.*, 1990) or hydrochemical textbooks (e.g. Appelo and Postma, 1996). The total concentration of the element M is equal to the sum of the molal concentrations of all species of M:

$$\sum M^{m+} = m_{M^{m+}} + m_{M_xA_y^{m-a}} + m_{M_pB_q^{m-b}} + \dots \tag{2.48}$$

where m refers to the molal concentrations [mol M^{-1}]. Even in normal fresh water, a significant part (up to tens of percents) of some dissolved ions, such as SO_4^{2-} and HCO_3^-, may be present in the form of aqueous complexes.

Example 2.10 Complexation

In a groundwater sample the total dissolved copper (Cu) concentration is 10^{-6} mol l^{-1} and the concentration of sulphate (SO_4^{2-}) is 1.5 mmol l^{-1}. The ion strength is 7.2 10^{-3}. Calculate the activity of the $CuSO_4^-$ ion at 25 °C given the $\log K = -2.31$ for the reaction $CuSO_4^- \leftrightarrow Cu^{2+} + SO_4^{2-}$

Disregard the formation of other complexes of copper and other cation complexes with sulphate and assume the activity coefficient for $CuSO_4^-$ to be unity.

Solution
First, use Equation (2.4) to calculate the activity coefficients for Cu^{2+} ($a_{Cu^{2+}} = 6.0$) and SO_4^{2-} ($a_{SO_4^{2-}} = 4.0$):

$$\log \gamma_{Cu^{2+}} = -0.148 \quad \Rightarrow \quad \gamma_{Cu^{2+}} = 0.711$$

$$\log \gamma_{SO_4^{2-}} = -0.155 \quad \Rightarrow \quad \gamma_{SO_4^{2-}} = 0.699$$

(It is not necessary to calculate the activity coefficient $CuSO_4^-$, since we assume that the activity coefficient for $CuSO_4^-$ equals 1).
The mass action expression for the dissociation of $CuSO_4^-$ is

$$K = \frac{[Cu^{2+}][SO_4^{2-}]}{[CuSO_4^-]} = \frac{\gamma_{Cu^{2+}}\{Cu^{2+}\}\gamma_{SO_4^{2-}}\{SO_4^{2-}\}}{\{CuSO_4^-\}} = 10^{-2.31} = 4.90 \cdot 10^{-3}$$

We may assume that the total amount of Cu is distributed between Cu^{2+} and $CuSO_4^-$. This means that the sum of the Cu^{2+} and $CuSO_4^-$ concentrations in mol l^{-1} equals 10^{-6} mol l^{-1}. Thus,

$$\{Cu^{2+}\} = 10^{-6} - \{CuSO_4^-\}$$

and

$$\frac{\gamma_{Cu^{2+}}\left(10^{-6} - \{CuSO_4^-\}\right)\gamma_{SO_4^{2-}}\{SO_4^{2-}\}}{\{CuSO_4^-\}} =$$

$$\frac{0.711 \times (10^{-6} - \{CuSO_4^-\}) \times 0.699 \times 0.0015}{\{CuSO_4^-\}} = 4.90 \cdot 10^{-3}$$

$$7.45 \cdot 10^{-4} \times (10^{-6} - \{CuSO_4^-\}) = 4.90 \cdot 10^{-3}\{CuSO_4^-\}$$

$$7.45 \cdot 10^{-10} - 7.45 \cdot 10^{-4}\{CuSO_4^-\} = 4.90 \cdot 10^{-3}\{CuSO_4^-\}$$

$$4.15 \cdot 10^{-3}\{CuSO_4^-\} = 7.45 \cdot 10^{-10}$$

$$\{CuSO_4^-\} = 1.79 \cdot 10^{-7} \text{ mol } l^{-1}$$

This means that almost 18 percent of the total dissolved Cu occurs in the $CuSO_4^-$ form.

2.9 ACID−BASE REACTIONS

2.9.1 Introduction

Acid–base reactions are of great significance for the formation, alteration, and dissolution of minerals and, therefore, the composition of natural waters. Usually, these reactions proceed very fast, so equilibrium is reached quickly. Therefore, equilibrium constants are very effective for predicting the composition of solutions susceptible to acid–base reactions. The strength of an acidic solution is usually measured in terms of its pH (= $-^{10}\log[H^+]$; see Section 2.2). The pH of the solution is a major variable that largely determines the direction and intensity of the alteration of minerals. The pH in an aqueous solution is the resultant equilibrium of all interrelated reactions involving conjugate acid–base pairs, including the dissociation, i.e. the self-ionisation, of water:

$$H_2O \quad \leftrightarrow \quad H^+ \quad + \quad OH^- \tag{2.49a}$$

$$K_w \quad = \quad [H^+][OH^-] \quad = \quad 10^{-14}; \qquad pK_w \quad = \quad 14 \tag{2.49b}$$

where K_w = equilibrium dissociation constant for water at 25 °C. An aqueous solution is said to be neutral if [H⁺] equals [OH⁻] = 10^{-7}, so the pH of a neutral solution is equal to 7. Since K_w changes with temperature, the pH of neutrality also changes slightly with temperature. As mentioned earlier, acid solutions have a pH less than 7 and basic solutions have a pH more than 7.

2.9.2 Acids

An *acid* is a chemical which produces hydrogen ions, i.e. protons (H⁺), when dissolved in water. When an acid loses a proton it forms its conjugate base:

$$HA \quad \leftrightarrow \quad H^+ \quad + \quad A^- \tag{2.50}$$

where HA = an acid and A- = the conjugate base. The equilibrium constant for the reaction (2.50) is:

$$K_a \quad = \quad \frac{[H^+][A^-]}{[HA]} \tag{2.51}$$

where K_a = the equilibrium dissociation constant for the acid, also referred to as the *acidity constant*, which, like all equilibrium constants, is usually reported for a constant temperature of 25 °C. Because acidity constants for different acids vary enormously (by over 10^{60}), a log scale is often used, analogous to the pH:

$$pK_a \quad = \quad -\log(K_a) \tag{2.52}$$

A strong acid dissociates completely in an aqueous solution (e.g. hydrochloric acid: HCl) and has a large K_a ($K_a \gg 1$; $pK_a \ll 0$), whereas a weak acid only dissociates partially (e.g. acetic acid; CH_3COOH) and has a small K_a ($K_a < 0.001$; $pK_a > 3$). The weaker an acid, the stronger is its conjugate base; the stronger an acid, the weaker is its conjugate base. At a pH above the pK_a of an acid, the conjugate base will predominate and at a pH below the pK_a the conjugate acid will predominate.

Example 2.11 Weak acids

In a solution of acetic acid (CH_3COOH), the equilibrium concentrations are found to be $[CH_3COOH] = 10$ mmol l^{-1} and $[CH_3COO^-] = 0.422$ mmol l^{-1}. Calculate the pH of this solution and the pK_a of acetic acid.

Solution
For the dissociation reaction of acetic acid
$$CH_3COOH \leftrightarrow H^+ + CH_3COO^-$$

H^+ and CH_3COO^- are produced in equal amounts, thus

$$[H^+] \ = \ [CH_3COO^-] \ = \ 0.422 \cdot 10^{-3} \text{ mol } l^{-1}$$

$$pH = -\log[H^+] = -\log[0.422 \cdot 10^{-3}] = 3.375$$

The acidity constant K_a is

$$K_a \ = \ \frac{[H^+][CH_3COO^-]}{[CH_3COOH]} \ = \ \frac{(0.422 \cdot 10^{-3})^2}{10 \cdot 10^{-3}} \ = \ 1.78 \cdot 10^{-5}$$

$$pK_a \ = \ -\log(K_a) \ = \ -\log(1.78 \cdot 10^{-5}) \ = \ 4.75$$

2.9.3 Bases

Conversely, the definition of a ***base*** is that it generates hydroxide ions (OH^-) or takes up H^+ ions when dissolved in water forming its conjugate acid:

$$B \ + \ H_2O \ \leftrightarrow \ OH^- \ + \ BH^+ \tag{2.53a}$$

$$K_b \ = \ \frac{[BH^+][OH^-]}{[B]} \tag{2.53b}$$

where B = a base, BH^+ = the conjugate acid, and K_b = the basicity constant. Analogous to the definitions of weak and strong acids, a strong base is a base that dissociates completely (e.g. sodium hydroxide: NaOH), and a weak base, a base that dissolves partially (e.g. magnesium hydroxide: $Mg(OH)_2$).

Bases neutralise acids in a neutralisation reaction, resulting in the production of water and a dissolved salt:

$$HA \ + \ B \ \leftrightarrow \ H_2O \ + \ A^- \ + \ BH^+ \tag{2.54}$$

In general, acid–base reactions proceed in the direction which yields the weaker acid and the weaker base.

Example 2.12 Weak bases

0.1 mmol NH_3 is dissolved in distilled water. Calculate the resulting pH given the $pK_b = 4.75$ for the reaction
$$NH_3 + H_2O \leftrightarrow NH_4^+ + OH^-$$

Solution

The remaining concentration NH_3 (in mmol l^{-1}) after the above dissociation reaction can be expressed as

$[NH_3] = 0.1 \cdot 10^{-3} - [OH^-]$

Because the dissociation of NH_3 produces equal amounts of NH^+ and OH^-, we may write for basicity constant K_b

$$K_b = \frac{[NH_4^+][OH^-]}{[NH_3]} = \frac{[OH^-]^2}{1.0 \cdot 10^{-4} - [OH^-]} = 10^{-4.75} = 1.78 \cdot 10^{-5}$$

$$[OH^-]^2 = 1.78 \cdot 10^{-5}(1.0 \cdot 10^{-4} - [OH^-])$$

$$[OH^-]^2 + 1.78 \cdot 10^{-5} \cdot [OH^-] - 1.78 \cdot 10^{-9} = 0$$

Subsequently, use the quadratic formula with a = 1; b = $1.78 \cdot 10^{-5}$; and c = $-1.78 \cdot 10^{-9}$ to solve this equation and calculate the OH^- concentration:

$$[OH^-] = \frac{-b + \sqrt{b^2 - 4ac}}{2a} =$$

$$\frac{-1.78 \cdot 10^{-5} + \sqrt{(1.78 \cdot 10^{-5})^2 + 4 \times 1.78 \cdot 10^{-5}}}{2} = 5.20 \cdot 10^{-5} \text{ mol } l^{-1}$$

Finally, calculate the H^+ concentration and the pH using Equation (2.45b):

$$K_w = [H^+][OH^-] = 10^{-14}$$

$$[H^+] = \frac{10^{-14}}{[OH^-]} = \frac{10^{-14}}{5.20 \cdot 10^{-5}} = 1.92 \cdot 10^{-10}$$

So, the pH is

$pH = -\log[H^+] = -\log[1.92 \cdot 10^{-10}] = 9.72$

2.9.4 Buffering

Buffered solutions are solutions that are barely sensitive to changes in their pH as a result of the addition of moderate quantities of strong acids or strong bases. The pH of such solutions is controlled by reversible equilibria involving weak acids (HA) and their conjugate salts (MA, where M is a cation). Because HA is a weak acid, it only partly dissociates at equilibrium. The solution also contains A^- from the dissolved salt, which suppresses the already slight dissociation of HA by shifting the equilibrium in Equation (2.50) to the left. The H^+ concentration of the buffer solution follows from Equation (2.51):

$$[H^+] = K_a \frac{[HA]}{[A^-]} \tag{2.55}$$

So,

$$pH = pK_a - \log\frac{[HA]}{[A^-]} \tag{2.56}$$

If a strong acid is added to the buffer solution, the reaction (2.50) is driven to the left; if a strong base is added, the reaction is driven to the right. Thus, since the solution can shift in response to either addition of acid or base, the change in pH will be small. The range of pH over which the buffering effects are effective depends on the concentration and nature of the weak acid and its conjugate salt. In general, a weak acid is an effective buffer at pH = $pK_a \pm 1$ pH unit. Most natural waters are buffered to some extent by carbonate equilibria, which will be dealt with further in Section 5.10.

Example 2.13 Buffer solution

A buffer solution contains 100 mmol l^{-1} acetic acid (CH_3COOH) and 100 mmol l^{-1} sodium acetate ($NaCH_3COO$). Two (2.0) ml of 10 mol l^{-1} HCl is added to this buffer solution. What will be the pH change? The pKa for acetic acid is 4.75.

Solution
First, calculate the pH of the buffer solution before addition of the strong acid. Use Equation (2.51):

$$pH = pK_a - \log\frac{[CH_3COOH]}{[CH_3COO^-]}$$

The acetate ion is derived from both the sodium acetate and the acetic acid. The salt (sodium acetate) completely dissociates:
$NaCH_3COO \rightarrow Na^+ + CH_3COO^-$

In contrast, the weak acid (acetic acid) is only slightly ionised:
$CH_3COOH \leftrightarrow H^+ + CH_3COO^-$

The high concentration of the acetate ion provided by the sodium acetate shifts the weak acid equilibrium to the left, decreasing the concentration of H^+. Thus, solutions that contain a weak acid plus a salt of the weak acid are always less acidic than solutions that contain the same concentration of the weak acid alone. The dissociation of the acetic acid is so small that we may approximate the CH_3COOH concentration by the initial concentration of the acetic acid (= 0.100 mol l^{-1}) and the CH_3COO^- concentration by the initial concentration of the $NaCH_3COO$ (= 0.100 mol l^{-1}). Because both concentrations are equal, the log term in Equation (2.50) becomes zero, so the pH equals the pK_a value. Thus, the pH is 4.75.

Second, calculate the pH after addition of the HCl solution. Since the added volume (2 ml) is negligible compared to the total volume (1 ml), we neglect the change in volume. The amount of HCl added is
0.0020 l × 10 mol l^{-1} = 0.020 mol HCl
The H^+ is nearly completely consumed by the acetate ion to form acetic acid. By approximation,

$[CH_3COOH] = 0.100 + 0.020 = 0.120$ mol l^{-1}
$[CH_3COO^-] = 0.100 - 0.020 = 0.080$ mol l^{-1}

Use Equation (2.50) to calculate the new pH

$$pH = pK_a - \log\frac{[CH_3COOH]}{[CH_3COO^-]} = 4.75 - \log\frac{0.120}{0.080} = 4.57$$

In comparison, addition of 0.020 mol HCl to 1.0 l distilled water would result in a pH of 1.70.

2.10 REDOX REACTIONS

2.10.1 Introduction

Oxidation and reduction (redox) reactions constitute a major class of geochemical and biochemical reactions that control the form of species such as oxygen, iron, sulphur, nitrogen, and organic materials, and their distribution in water and sediment. Redox reactions involve the transfer of electrons between atoms. **Reductants** and **oxidants** are defined as electron donors and electron acceptors, respectively, in an analogous way that acids and bases can be considered as donors and acceptors of protons. For example, the oxidation reaction of ferrous iron (Fe^{2+}) to ferric iron (Fe^{3+}) can be written as:

$$Fe^{2+} \rightarrow Fe^{3+} + e^- \tag{2.57}$$

where e^- represents the electron. There are no free electrons, so oxidation reactions are always coupled to a simultaneous reduction reaction: for example, the reduction of oxygen (O_2):

$$O_2 + 4H^+ + 4e^- \rightarrow 2H_2O \tag{2.58}$$

Adding the so-called half reactions (2.57) and (2.58) and balancing them according to the number of electrons on each side of the equation results in the following full redox equation of the oxidation of iron (reductant) by oxygen (oxidant):

$$4Fe^{2+} + O_2 + 4H^+ \rightarrow 4Fe^{3+} + 2H_2O \tag{2.59}$$

Thus, an oxidant is a substance that causes oxidation by accepting electrons from a reductant, while being reduced itself. Many redox reactions proceed extremely slowly, except for those mediated by microorganisms – though these too may be rather slow. This implies that the concentrations of oxidisable and reducible substances may very well deviate from equilibrium concentrations predicted thermodynamically and that, besides thermodynamic equilibrium, the reaction kinetics also have to be considered. This section, however, concentrates on redox equilibria.

2.10.2 Oxidation state

The electron transfer between atoms causes a change in the **oxidation state** of the reactants and products. The oxidation state represents the hypothetical charge that an atom would have if the molecule or ion were dissociated. This hypothetical assignment of electrons to an atom is carried out according to the following rules (Stumm and Morgan, 1996):

1. The oxidation state of a monoatomic substance equals its electronic charge
2. In a covalent compound, the oxidation state of each atom is the charge remaining on the atom if each shared electron pair is fully assigned to the more electronegative atom (i.e. the atom that has the largest tendency to accept electrons) of the two atoms sharing them. An electron pair shared by two atoms of the same electronegativity is split between them.
3. For molecules, the sum of the oxidation states equals zero and for ions it equals their formal charge.

The oxidation state is indicated by roman numbers: for example, Fe(II) for ferrous iron and Fe(III) for ferric iron. Since O is highly electronegative and H is electropositive, as a rule of thumb it may be assumed that O is always in the 2- oxidation state (except in O_2) and that H is always in the 1+ oxidation state (except in H_2), which simplifies the calculation of the oxidation state of other elements.

> **Example 2.14 Oxidation state**
>
> Derive the oxidation state of chromium (Cr) in chromate (CrO_4^{2-})
>
> *Solution*
> Because O is in the 2- oxidation state, the total negative charge due to the 4 oxygens is (8-). Subtracting the total negative charge of the chromate ion (2-) gives a net charge of (6-) for the four oxygen atoms. Hence, the oxidation state of the chromium atom must be 6+.

2.10.3 Redox potential

Natural systems are characterised by multiple equilibria and the tendency of the system to donate or accept electrons is represented by the ***redox potential***, analogous to the pH, which represents the tendency of a system to donate or accept protons. The voltage or electromotive force developed by a redox reaction is related to the free energy of a system:

$$E \;=\; \frac{-\Delta G}{nF} \tag{2.60}$$

where E = the potential [M $L^2 T^{-2}$ Q^{-1}], ΔG = the change in free energy [M $L^2 T^{-2}$], n = the amount of electrons transferred in the reaction [mol], F = the Faraday constant [Q mol^{-1}], i.e. the charge of 1 mol of electrons ($F = 96.490$ C mol^{-1}). Every redox reaction like Equation (2.59) can be written in its general form:

$$aA_{red} \;+\; bB_{ox} \;\leftrightarrow\; cC_{ox} \;+\; dD_{red} \tag{2.61}$$

where the subscripts *red* and *ox* denote respectively the reductants and oxidants participating in the reaction. The change of free energy for this reaction is (see Equations 2.35 and 2.36):

$$\Delta G \;=\; \Delta G^0 \;+\; RT \ln \frac{[C_{ox}]^c \, [D_{red}]^d}{[A_{red}]^a \, [B_{ox}]^b} \tag{2.62}$$

Combining Equations (2.60) and (2.62) yields:

$$E \;=\; E^0 \;-\; \frac{RT}{nF} \ln \frac{[C_{ox}]^c \, [D_{red}]^d}{[A_{red}]^a \, [B_{ox}]^b} \tag{2.63}$$

where E^0 = the standard potential [M $L^2 T^{-2}$ Q^{-1}], which is the potential with all substances present at unit activity at 25 °C and 1 atmosphere pressure. The redox potential for a given redox system is now defined as the potential relative to the potential of a hydrogen electrode, which is an electrode at which the equilibrium:

$$2H^+ \;+\; 2e^- \;\leftrightarrow\; H_2 \tag{2.64}$$

is established on a platinum surface pH = 0.0 and P_{H2} = 1 atm. The redox potential of a system relative to the hydrogen electrode is measured in a redox cell, as illustrated in Figure 2.2. In a solution of pH = 0, H_2 gas is bubbled over the platinum electrode on the left-hand side. In the compartment on the right-hand side, a platinum electrode is present in the solution whose redox potential is being measured. The redox potential is measured by a voltmeter connected to both electrodes, and the electrical circuit is closed by a salt bridge. The conditions at the reference electrode are:

$$[B_{ox}] \ = \ [H^+] \ = \ 1 \, mol \, l^{-1} \tag{2.65a}$$

and

$$[D_{red}] \ = \ P_{H_2} \ = \ 1 \, atm \tag{2.65b}$$

Since both $[B_{ox}]$ and $[D_{red}]$ are equal to unity, Equation (2.63) becomes:

$$Eh \ = \ E^0 \ - \ \frac{RT}{nF} \ln \frac{[C_{ox}]^c}{[A_{red}]^a} \tag{2.66}$$

Equation (2.66) is known as the Nernst equation. At the standard temperature of 25 °C and with base-10 logarithms instead of natural logarithms, Equation (2.66) becomes

$$Eh \ = \ E^0 \ - \ \frac{0.0592}{n} \log \frac{[C_{ox}]^c}{[A_{red}]^a} \tag{2.67}$$

where Eh = the redox potential. The standard redox potential E^0 has been tabulated for most half reactions. By convention, the sign of the standard redox potential corresponds to the log K of the reaction written as a reduction reaction (like Equation 2.58). Note that the sign

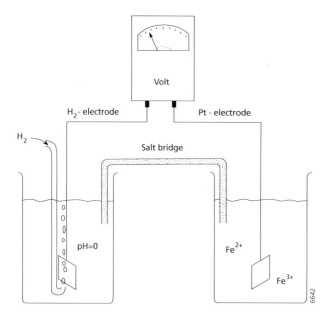

Figure 2.2 Schematic representation of a redox cell to measure the redox potential.

changes if the reaction is written as an oxidation reaction. Consequently, the sign of the redox potential is negative if the system is reducing, but positive if the system is oxidising. For example, the E^0 for half reaction 2.57 is -0.77 Volt and for half reaction 2.58 is +1.23 Volt. The E^0 for the complete redox reaction (Equation 2.59) is obtained by adding both values, which results in a standard potential of +0.46 Volt. The positive sign implies that the reaction (2.59) proceeds spontaneously to the right if all activities are equal to one.

In theory, the redox potential of an aqueous solution could be measured using an electrode system as shown in Figure 2.2. Unfortunately, in practice it appears that results from measurement using a redox probe are not unambiguous and may contain considerable errors. The major causes of these erroneous results are the lack of equilibrium and analytical difficulties using a platinum electrode due to, for example, insensitivity to some redox couples and contamination of the electrode (Appelo and Postma, 1996).

Example 2.15 Calculation of redox speciation using the Nernst equation

A groundwater sample with a pH of 6.0 contains $[Fe^{2+}] = 10^{-3.22}$ and $[Fe^{3+}] = 10^{-5.69}$ at 25 °C. Calculate the Mn^{2+} activity if the sample were in equilibrium with MnO_2. The standard redox potentials for the reactions involved are

$$Fe^{3+} + e^- \rightarrow Fe^{2+} \qquad\qquad E^0 = +0.77 \text{ V}$$

$$MnO_2 + 4H^+ + 2e^- \rightarrow Mn^{2+} + 2H_2O \qquad E^0 = -1.23 \text{ V}$$

Solution
First, calculate the Eh using the Nernst equation (Equation 2.66) and the activities of the Fe species:

$$Eh = 0.77 - 0.0592 \log\frac{[Fe^{2+}]}{[Fe^{3+}]} = 0.77 - 0.0592 \log\frac{10^{-3.22}}{10^{-5.69}} =$$

$$0.77 - 0.0592 \times 2.47 = 0.624 \text{ V}$$

The Nernst equation for the second reaction involving the Mn species is

$$Eh = 1.23 - \frac{0.0592}{2} \log\frac{[Mn^{2+}]}{[H^+]^4}$$

Note that MnO_2 and H_2O are not considered in this equation, since they have unit activity by definition. This equation can be rewritten as (note that pH = $-\log[H^+]$)

$$Eh = 1.23 - 0.030 (\log[Mn^{2+}] + 4 pH) = 1.23 - 0.030 \log[Mn^{2+}] - 0.118 pH$$

Thus,

$$1.23 - 0.030 \log[Mn^{2+}] - 0.118 pH = 0.624 \text{ V}$$

$$\log[Mn^{2+}] = \frac{0.624 - 1.23 + 0.118 \times 6.0}{-0.030} = -3.41$$

$$[Mn^{2+}] = 10^{-3.41} \text{ mol l}^{-1}$$

2.10.4 Redox reactions and pe

The equilibrium constant for the half-reaction reaction 2.53 is:

$$K = \frac{[Fe^{3+}][e^-]}{[Fe^{2+}]} = 10^{-13.05} \tag{2.68}$$

In contrast to the Nernst equation (Equation 2.66), the activity of the electrons, $[e^-]$, appears explicitly in Equation (2.68). Analogous to the pH, the pe is defined as:

$$pe = -\log[e^-] \tag{2.69}$$

The advantage of using the pe approach is that the calculations of redox reactions are similar to other equilibrium calculations. A disadvantage of the pe is that it cannot be measured directly, but fortunately the pe is directly related to the redox potential:

$$Eh = \frac{2.303RT}{F} pe \tag{2.70}$$

At 25 °C this is equal to

$$Eh = 0.0592\ pe \tag{2.71}$$

> **Example 2.16 Calculation of redox speciation using the pe concept**
>
> A groundwater sample contains $[Fe^{2+}] = 10^{-3.22}$ and $[Fe^{3+}] = 10^{-5.69}$ at 25 °C (see Example 2.15). Calculate the pe and the Eh given the equilibrium constant for the half reaction
>
> $$Fe^{3+} + e^- \rightarrow Fe^{2+} \qquad\qquad pK = 13.05$$
>
> *Solution*
> The mass action expression for the half reaction is
>
> $$K = \frac{[Fe^{2+}]}{[Fe^{3+}][e^-]} = 10^{13.05}$$
>
> Note that this equilibrium constant is the reciprocal of Equation (2.68), since the above half reaction has been written as a reduction reaction and the half reaction corresponding to Equation (2.68) has been written as an oxidation reaction (Equation 2.57).
>
> The above mass action expression can be rewritten as
>
> $$\log K = \log[Fe^{2+}] - \log[Fe^{3+}] + pe$$
>
> $$pe = \log K + \log[Fe^{3+}] - \log[Fe^{2+}]$$
>
> Filling in the values gives
>
> $$pe = 13.05 - 5.69 + 3.22 = 10.58$$
>
> The Eh is calculated using Equation (2.67):
>
> $$Eh = 0.0592\ pe = 0.626\ V$$
>
> Which is approximately the same as calculated in Example 2.15.

2.10.5 pH–Eh diagrams

The environmental conditions that control the stability and solubility of many substances can conveniently be summarised by two key variables, namely the pH and redox potential (Eh), and plotted in a ***pH–Eh diagram***. This is a two-dimensional graph in which the pH is plotted on the horizontal axis and the Eh (or pe) on the vertical axis. Subsequently, the fields of stability or dominance of both dissolved species and minerals as function of the pH and Eh are indicated, whereby the dominant stable species can be recognised directly and easily. Figure 2.3 shows the stability of water and the ranges of pH and redox conditions in natural environments. Box 2.I explains the main principles of the construction of pH–Eh diagrams. Further details about pH–Eh diagrams can be found in Hem (1989), Stumm and Morgan (1996), Appelo and Postma (1996), and Drever (2000).

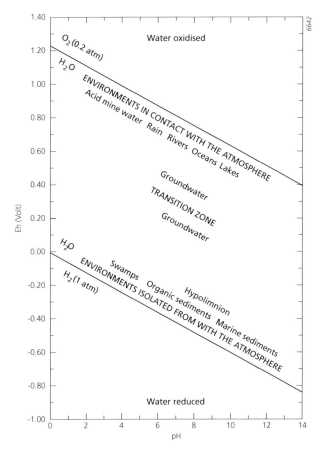

Figure 2.3 The stability of water and the ranges of pH and Eh for different natural environments (after Appelo and Postma, 1996).

Box 2.I Basic principles for the construction of pH–Eh diagrams

The field of the stability of water is confined between the conditions where H_2O is oxidised:

$$2H_2O\ (l)\ \leftrightarrow\ O_2\ (g)\ +\ 4H^+\ +\ 4e^- \tag{2.Ia}$$

and where H_2O is reduced:

$$2H^+\ +\ 2e^-\ \leftrightarrow\ H_2\ (g) \tag{2.Ib}$$

The Nernst equation applied to Reaction 2.Ia written as a reduction equation for standard conditions is (see Equation 2.63):

$$Eh\ =\ E^0\ -\ \frac{0.0592}{n}\ \log\left(\frac{1}{P_{O_2}\times[H^+]^4}\right) \tag{2.Ic}$$

The standard redox potential for the reaction 2.Ia is calculated from the standard change in free energy (see Equation 2.58), which in turn is calculated from thermodynamic data:

$$\Delta G_R^0\ =\ 2\Delta G^0{}_{H_2O(l)}\ -\ \Delta G^0{}_{O_2(g)}\ -\ 4\Delta G^0{}_{H^+}\ =$$
$$-474.36\ -\ 0\ -\ 0\ =\ -474.36\ \ kJ\ mol^{-1} \tag{2.Id}$$

Thus,

$$E^0\ =\ \frac{-\Delta G_R^0}{nF}\ =\ \frac{474.36}{4\cdot96.490}\ =\ 1.229\ V \tag{2.Ie}$$

The maximum partial pressure of oxygen permissible in the system as defined is 1 atmosphere, so Equation 2.Ic becomes:

$$Eh\ =\ 1.229\ -\ 0.0592\ pH \tag{2.If}$$

Thus, Equation 2.If defines the upper stability boundary for H_2O. The standard change in free energy for reaction 2.Ib equals 0 by definition, and the lower boundary calculated similarly as above is:

$$Eh\ =\ -0.0592\ pH \tag{2.Ig}$$

The boundaries between the stable species can be calculated in a similar way using the basic thermodynamic data and the Nernst equation. Sections 5.6 and 5.7 further elucidate the pH–Eh diagrams for iron and manganese species.

2.11 FURTHER READING ON BASIC CHEMISTRY

The following textbooks on basic chemistry cover similar but more detailed material at this level:
- American Chemical Society, 2005, *Chemistry, A General Chemistry Project of the American Chemical Society*, (New York: W.H. Freeman).
- Metiu, H., 2006, *Physical Chemistry, Thermodynamics*, (London: Taylor and Francis).
- Rosenberg, J. L., Epstein, L., and Krieger, P., 2013, *Schaum's Outline of College Chemistry*, 10th edition, (New York: McGraw-Hill).
- Timberlake, K.C. and Timberlake, W., 2014, Basic Chemistry, 4th edition, (Upper Saddle River, NJ: Prentice Hall).

EXERCISES

1. The concentration of Ca^{2+} in a freshwater sample is 100.2 ppm. Recalculate this concentration in
   a. $mg\ l^{-1}$
   a. $mmol\ l^{-1}$
   b. $meq\ l^{-1}$

2. Given the following composition of spring water:

Cations	Concentration mg l^{-1}	Anions	Concentration mg l^{-1}
Na^+	8.7	Cl^-	2.3
K^+	1.1	HCO_3^-	255
Ca^{2+}	43.7	SO_4^-	5.5
Mg^{2+}	21.6	NO_3^-	4.8
H^+	$10^{-4.5}$ (pH = 7.5)		

   a. Convert these concentrations into meq l^{-1}.
   b. Check whether the charges of the cations and anions balance each other. Which ions can be ignored for this check?
   c. Estimate the electrical conductivity of this solution.
   d. Calculate the ionic strength of this solution.
   e. Calculate the activity coefficients for the different ions at 25 °C. What can be concluded?

3. Describe the spontaneity of a chemical reaction in terms of enthalpy, entropy, and temperature.

4. a. Define fugacity.
   b. Explain why the fugacities in each phase are equal in each phase at equilibrium.
   c. Describe in your own words how the fugacity concept can be used to estimate partitioning of substances amongst phases.

5. Consider a simple lake ecosystem of 10 000 m^3. The lake contains 1000 kg sediment and 350 kg fish. An amount of 1 kg of cadmium is released into this lake. Calculate the concentrations in water (mg l^{-1}), sediment (mg kg^{-1}), and fish (mg kg^{-1}) given a

distribution coefficient of 70 l kg^{-1} for sediment–water partitioning and a concentration factor of 65 for fish–water partitioning.

6. The adsorption of caesium (Cs) to illite (a clay mineral) was studied by means of a batch experiment. During this experiment an amount of 8 mg of illite was added to 80 ml solution with a known initial Cs concentration. The solution was buffered at a pH of 8.0. After shaking for 12 days, the dissolved Cs concentration was measured. The results are shown in the table below

Experiment no.	Initial Cs concentration µg l^{-1}	Cs concentration after 12 days µg l^{-1}
1.	1	0.5
2.	2	1.1
3.	5	3.1
4.	10	7.0
5.	20	15.5

   a. Calculate the adsorbed concentration in µg g^{-1} for each of the five experiments after 12 days.
   b. Draw the isotherm of Cs.
   c. Is this a Freundlich isotherm or a Langmuir isotherm? Explain your answer.

7. The precipitation reaction of barite (barium sulphate; $BaSO_4$):
    $Ba^{2+} + SO_4^{2-} \leftrightarrow BaSO_4$ (s)
    is characterised by the following thermodynamic properties: ΔG^0 = -56.91 kJ mol^{-1}; ΔH^0 = -6.35 kJ mol^{-1}.
    a. Is the *dissolution* of barium sulphate endothermic or exothermic?
    b. Calculate the solubility product for barium sulphate at 25 °C.
    c. Calculate the Ba^{2+} concentration in a solution saturated with $BaSO_4$ at 25 °C.
    d. Calculate the Ba^{2+} concentration in a solution saturated with $BaSO_4$ at 10 °C.
    e. Explain what happens if a small amount of sulphuric acid (H_2SO_4) is added to the saturated solution.

8. A water sample has a calcium (Ca^{2+}) activity of 0.001 mol l^{-1} and a sulphate (SO_4^{2-}) activity of 0.02 mol l^{-1}. Evaluate the saturation index of gypsum ($CaSO_4 \cdot 2H_2O$) given the solubility product of gypsum = $10^{-4.61}$. What is your conclusion?

9. A solution in equilibrium with gypsum also contains a certain amount of the $CaSO_4^0$ complex. Discuss what the presence of aqueous $CaSO_4^0$ complexes implies for the total amount of calcium in solution.

10. Phosphoric acid (H_3PO_4) dissociates in three steps at different pHs and with different K_a values:
    First dissociation step: $H_3PO_4 \leftrightarrow H_2PO_4^- + H^+$ K_a = 6.92·10^{-3}
    Second dissociation step: $H_2PO_4^- \leftrightarrow HPO_4^{2-} + H^+$ K_a = 6.17·10^{-8}
    Third dissociation step: $HPO_4^{2-} \leftrightarrow PO_4^{3-} + H^+$ K_a = 2.09·10^{-12}
    What is the dominant phosphate species (i.e. H_3PO_4, $H_2PO_4^-$, HPO_4^{2-}, or PO_4^{3-}) at pH = 8, pH = 6, and pH= 4?

11. Consider the reaction: $NH_3 + H_2O \leftrightarrow OH^- + NH_4^+$ K_b = 1.8·10^{-5}
    Calculate the pH of a solution of 0.15 mol l^{-1} of ammonium chloride (NH_4Cl)
    (Hint: Note that $pK_a + pK_b = pK_w$ = 14).

12. A buffered solution is prepared by mixing 500 ml of 0.5 M acetic acid (CH_3COOH) and 500 ml of 0.5 M sodium acetate ($NaCH_3COO$).
   a. Calculate the pH of the buffered solution given the pK_a of acetic acid = 4.75.
   b. Calculate the pH after 0.05 mol of HCl has been added to the buffered solution.

13. The presence of dissolved iron (Fe^{2+}) is an appropriate indicator of the redox conditions in groundwater.
   a. Give the half reaction for the reduction of iron hydroxide ($Fe(OH)_3$) to dissolved Fe^{2+}.
   b. What is the effect of lowering the pH on the solubility of $Fe(OH)_3$?
   c. Give the relationship of Eh versus dissolved Fe^{2+} concentration and pH given the log K for the above half reaction = 16.43 (Hint: calculate the relationship between pe and Fe^{2+} and pH first).
   Draw the line between $Fe(OH)_3$ and Fe^{2+} in a pH–Eh diagram at a fixed dissolved Fe^{2+} concentration of $10^{-6.00}$ mol l^{-1}.
   d. In which natural environments is iron hydroxide reduced to dissolved Fe^{2+} (compare Figure 2.3)?

14. Given the following analysis of two groundwater samples:

Sample 1	Sample 1 Concentration mg l^{-1}	Sample 2 Concentration mg l^{-1}
pH	4.44	4.26
Na^+	8.27	6.38
K^+	6.15	1.14
Ca^{2+}	17.56	2.28
Mg^{2+}	3.09	0.93
Fe^{2+}	0.006	3.22
Mn^{2+}	0.170	0.077
Al^{3+}	5.08	4.04
Cl^-	28.0	11.0

Calculate the pe of both samples based on the Fe^{2+} concentrations (see also exercise 13).

3

Environmental compartments

3.1 INTRODUCTION

The basic chemical turnover processes treated in the previous chapter play a key role in the ultimate fate of pollutants in the environment. Since the nature and intensity of these processes are very variable in space, pollution studies are traditionally performed in units or environmental compartments that are more or less homogenous with respect to the prevailing physico-chemical conditions. A common subdivision into ***environmental compartments*** is based on the major phase present (gas, liquid, or solid phase) and distinguishes between soil, water, and air. Water is usually further classified in surface waters that are in contact with the free atmosphere and subsurface waters (groundwater). It is, however, important to realise that the gas, liquid, and solid phases are often all present in these environmental compartments. Although in principle the same wide range of chemical processes brings about the variation in the overall composition of soil and water, their direction and equilibrium state are different in the different compartments. The governing chemical processes include silicate weathering, carbonate dissolution and precipitation, redox processes, and sorption to solid surfaces. The redox processes in particular are often biologically mediated. These processes control amongst others the genesis of soil profiles, the total concentrations of substances in water, the decomposition of organic substances, and the retention of chemicals in sediment.

It should be noted that many physical and chemical processes occur not only within the compartments, but typically also at the interfaces between the compartments. Examples of such processes are the exchange of oxygen between the atmosphere and surface water, soil pollution due to deposition of atmospheric pollutants, the leaching of pollutants from polluted soils to groundwater, and the exfiltration of polluted groundwater to soil or surface water. In general, pollutants in the solid or adsorbed phase are rather immobile and are only translocated by bioturbation or by suspended or bedload transport in flowing surface waters, or when soil is displaced by people. On the other hand, pollutants in the gas or liquid phase are generally much more mobile, so pollutant transport within and between the different environmental compartments occurs mainly in these phases. The principal driving force behind the transfer between the environmental compartments is the ***hydrological cycle*** (Figure 3.1). Atmospheric water derived from evaporation of the oceans, and to a lesser extent, from evapotranspiration at the land surface, returns to the oceans and land surface through precipitation in the form of rain or snow. Part of the water that falls onto the land infiltrates into the soil and another part runs off directly into rivers and lakes. In soil, the water percolates towards the groundwater and, in the course of time, the groundwater discharges into rivers and lakes, which, in turn, discharge into the oceans. As water goes through this hydrological cycle it collects many solutes and takes them along its pathways.

The different environmental compartments are thus closely interlinked and act as a continuum. Nevertheless, the subdivision used in this book follows the general classification mentioned above, but where necessary, special attention is given to the interaction between the compartments. Because issues of atmospheric pollution are beyond the scope of this

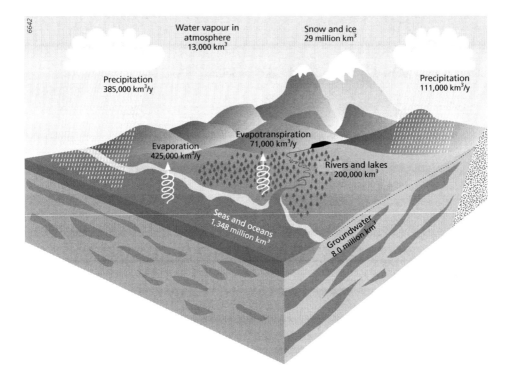

Figure 3.1 The hydrological cycle (pools and fluxes after Winter *et al.*, 1999)

book, the discussion on the environmental compartments in this chapter and in part IV of this book is restricted to soil, groundwater, and surface water. This chapter focuses on the physical and chemical characteristics of these environmental compartments and the main sources of pollution in them.

3.2 SOIL

3.2.1 Definition of soil

According to the US Department of Agriculture (USDA, 1999), soil is the collective term for a natural body comprised of solids (minerals and organic matter), liquid, and gases that occurs on the land surface, occupies space, and is characterised by one or both of the following: horizons, or layers, that are distinguishable from the initial material as a result of additions, losses, transfers, and transformations of energy and matter or the ability to support rooted plants in a natural environment. The upper limit of soil is the boundary between soil and air, shallow water, live plants, or plant materials that have not begun to decompose. Areas are not considered to have soil if the surface is permanently covered by water too deep (typically more than 2.5 meters) for the growth of rooted plants. The lower boundary that separates soil from the non-soil underneath is most difficult to define. Soil consists of horizons near the Earth's surface that, in contrast to the underlying parent material, have been altered by the interactions of climate, relief, and living organisms over time. Commonly, soil grades at its lower boundary to hard rock or to earthy materials virtually devoid of animals, roots, or other marks of biological activity. For purposes of classification, the lower boundary of soil is arbitrarily set at 200 cm.

Apart from this extensive definition of soil, there are many other definitions. From the viewpoint of pollution issues, it is opportune to also include all unconsolidated, granular mineral material down to the water table in the definition of soil, instead of restricting the definition to the top 2 metres. The unsaturated zone may extend from zero to several tens of metres thickness.

The soil is made up from porous, unconsolidated material consisting of autochthonous (i.e. native), weathered bedrock, called regolith, or of allochthonous (i.e. alien) sediment. It is characterised by intense activity of biota, including plant roots, fungi, bacteria, earthworms, nematodes, snails, beetles, spiders, and soil macrofauna (e.g. moles, voles), particularly in the top few decimetres. Biological activity is one of the soil-forming factors responsible for the physical and chemical nature of the soil by decomposing organic matter and freeing nutrients for reuse by plants. The other soil-forming factors are parent material, climate (mainly precipitation and temperature), topography (elevation, slope, aspect), and time. The interplay of these factors results in the development of a soil profile through gains, losses, transformations, and the translocation of soil material made up of inorganic minerals of various grain sizes and organic matter from dead plant material. Gains of soil material occur as a result of inputs of, for example, water from precipitation or irrigation, organic litter from plants and animals, nitrogen from microbial fixation, sediment from water or wind, salts from groundwater, and fertilisers from agriculture. Losses arise from evapotranspiration and percolation of water, leaching of substances soluble in percolate water, uptake of nutrients and other chemicals by harvested or grazed plants, oxidation of organic carbon to CO_2, and denitrification of nitrate. Transformations involve a wide range of biochemical reactions: for example, the decomposition of organic matter, alteration or dissolution of soil minerals, and the precipitation of newly formed minerals. Translocation takes place through the burrowing activity of animals and plant roots (bioturbation), the expansion and shrinkage of soil materials due to freezing–thawing or wetting–drying cycles, or the dissolution, mobilisation, and subsequent precipitation or deposition of clays, carbonates or iron oxides deeper in the profile. Soil formation is inherent in weathering, including the physical disintegration of particles to smaller sizes (physical weathering) and the dissolution or chemical alteration of minerals (chemical weathering).

Variation in the above soil-forming factors and process leads to the formation of distinct soil profiles that consist of different soil horizons as mentioned above. Based on, amongst others, the presence, thickness, and composition of the horizons, more than 15 000 different soils have been described throughout the world. However, there are many general features present in this broad range of different soil profiles. Figure 3.2 shows a generalised soil profile typical in a loamy soil in temperate humid regions. The uppermost topsoil layer is composed of fresh organic material, primarily from leaf litter and is called the O horizon. Below this is the thin A horizon that consists of mineral material rich in dark organic matter (humus). The zone below the A horizon, the E horizon, is often leached by water that has removed most of the organic matter and some clays. This leached material accumulates in the B horizon, below which the slightly altered parent material is present. Note that this is a generalisation of a soil profile in the temperate humid climate regions. There are many other possible soil profiles: for example, organic soils (peats and mucks) that are formed where stagnant water impedes the decomposition of organic matter, so that only the partially decomposed plant remains accumulate in the soil profile. In tropical regions, the abundant excess precipitation and high temperatures cause an intense weathering and leaching of the soil profile. Conversely, in arid and semi-arid regions, evaporation and the consequent predominantly upward movement of water results in the accumulation of soluble salts in the upper part of the soil profile. The study of soils is an entire field of science itself and for further details

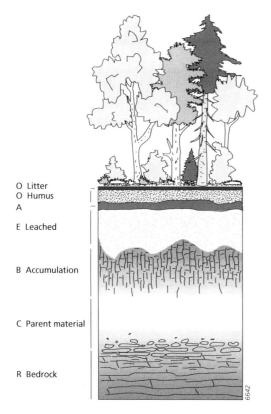

O Litter
O Humus
A
E Leached
B Accumulation
C Parent material
R Bedrock

Figure 3.2 A generalised soil profile in a loamy soil in temperate humid regions.

about soil formation and classification, the reader is referred to standard textbooks on soil science (e.g., Scheffer and Schachtschabel, 1989; Locher and De Bakker, 1990; Miller and Gardiner, 2008; and Gerrard, 2000).

3.2.2 Sources of soil pollution

Input of pollutants into soil occurs via many pathways. The deliberate dumping of solid or liquid wastes in heaps or tailing ponds is one of the most visible kinds of soil pollution. Municipal, industrial, or mine wastes, or dug-up contaminated soil or sediments may contain a wide range of pollutants, including metals, cyanide (CN^-), polycyclic aromatic hydrocarbons (PAHs), polychlorinated biphenyls (PCBs), asbestos, methane (CH_4), ammonia (NH_3), and hydrogen sulphide (H_2S), which may pose a threat to the surrounding, unpolluted soil. Nowadays, the dispersal of pollutants from waste disposal sites is restricted by means of a sustainable isolation of the site, but there are many former dump sites where this isolation is absent or malfunctioning. Releases of industrial pollutants into soil may also occur accidentally, for example as a result of road accidents or accidental spills from industrial installations. Liquid substances in particular can easily pollute large volumes of soil, because they easily penetrate and disperse in soil. The soils of military training grounds and battlefields are often contaminated by ammunition, bullets, explosives, fuels, and scrap metals containing substances such as PAHs, lead, and depleted uranium.

In agricultural areas, the application of pesticides and fertilisers constitutes an important source of pollutants in soil. Pesticides are often sprayed directly onto the crops, whereas fertilisers are either sprayed onto the soil surface or injected into the soil at a shallow depth. Pesticides comprise a variety of organic compounds (e.g. lindane, isoproturon, atrazine), which, in turn, may contain heavy metals (e.g. copper (Cu), mercury (Hg), arsenic (As)). Fertilisers promote plant growth by enhancing the supply of essential nutrients, such as nitrogen (N), phosphorus (P), and potassium (K). These nutrients are included in organic fertilisers, such as manure, compost, or sewage sludge, or in inorganic artificial fertilisers, such as for example diammonium phosphate $((NH_4)_2HPO_4)$ or potassium chloride (KCl). As well as containing the major plant nutrients, manure and fertilisers often contain heavy metals, such as cadmium (Cd) in phosphate fertilisers, and copper (Cu) in pig manure. As a consequence, the application of fertiliser has resulted in slight but widespread contamination of most agricultural soils in the western world. Sewage sludge and compost, which are used as fertilisers and to enhance the soil structure, also usually contain heavy metals as well as other pollutants; spreading sewage sludge or compost on agricultural land causes these pollutants to be added to the soil. In many western countries, under recent legislation, sludge may only be spread if it is relatively free of pollutants. Excess application of fertilisers causes a build-up of nutrients in soil, which in turn may cause enhanced volatilisation of ammonia (NH_3) to the atmosphere, and the leaching of nutrients to deeper horizons beyond the reach of crop roots. The chemical properties of both soil and nutrient species control the extent and rate at which leaching occurs. In the countryside where sewage treatment systems are lacking, septic tanks comprise another significant source of soil pollution by nutrients. In contrast to the application of fertilisers, septic tanks can be considered as point sources.

Soils are also affected by pollutant inputs from the atmosphere. The most notorious example is the atmospheric deposition of acidifying compounds originating from power plants, chemical plants, and traffic (e.g. sulphur dioxide (SO_2), nitrogen oxide (NO_x), which are transformed into sulphuric acid (H_2SO_4) and nitric acid (HNO_3), respectively, by atmospheric reactions), and agriculture (mainly ammonia (NH_3), which is transformed into nitric acid in soil). Before the introduction of unleaded fuel, traffic used to be an important source of lead (Pb) contamination. In the vicinity of roads and highways, traffic is still a source of atmospheric deposition of benzene and polycyclic aromatic hydrocarbons (PAHs), both of which are constituents of petrol. De-icers and rubber tyre particles (containing heavy metals) also contribute to soil pollution in the vicinity of roads. Atmospheric emission and subsequent deposition onto soil is also induced by incineration of wastes (primarily heavy metals, NO_x), PAHs, PCBs), and metal smelting (heavy metals). In the 1950s and 1960s nuclear bomb tests in the open atmosphere caused global atmospheric radioactive fallout of various radionuclides (e.g. radioiodine (^{131}I) radiocaesium (^{134}Cs and ^{137}Cs) and radiostrontium (^{90}Sr)). Also the Chernobyl accident in 1986 caused fallout of these radionuclides over vast areas of Europe. Other radiological accidents have caused radioactive fallout and soil pollution at a more regional scale. Furthermore, polluted soil particles resuspended in the air can be deposited elsewhere on the soil surface: for example wind-blown dust from mine tailings, smelters, or roads. A similar process occurs in rivers, where contaminated sediments can be deposited on the floodplain soils during flooding. Because sediment deposition also entails a substantial input of soil material, floodplain soils may have become contaminated over a considerable depth. Finally, upward seepage of contaminated groundwater may also be a source of contaminants in the soil profile.

3.2.3 Soil water

Almost the entire part of the soil as defined here belongs to the unsaturated zone, also called the ***vadose zone***. In contrast to the saturated zone, where almost all the interconnected pores between the grains or particles are filled with water, in the unsaturated zone the water occurs as a thin film on the surface of the particles. The sizes of soil pores can range from more than 1 cm in diameter to less than 0.001 mm in diameter. Pores larger than 0.1 mm in diameter are often referred to as macropores, those between 0.1 and 0.01 mm in size as mesopores, and those smaller than 0.01 mm as micropores (Beven and Germann, 1982). The zone below the water table, which is defined as the depth at which pore water pressure is equal to atmospheric pressure, belongs definitely to the saturated zone. In coarse-textured materials (sand or gravel) the transition between the saturated zone and the unsaturated zone is approximately the water table. In fine porous materials (clay, loam, and silt), capillary rise also causes complete saturation in a zone just above the water table. Although some organic liquid pollutants may move independently through soil, movement of water and its dissolved constituents is the most important vector of transport of pollutants in soil. In the vadose zone, the movement of water only occurs via the water-filled pores and is driven by gravitational force and gradients in pore water pressure (or suction because the pore water pressure in the unsaturated zone is negative due to the capillary forces between water and the soil grains). These gradients in pore water pressure result from both vertical and lateral variations in soil moisture content, which may be very large. The effect of gravity and evaporation on soil water transport is dominant, so the water moves mainly vertically through soil. The soil moisture is replenished by infiltrating rainwater and snowmelt water and by capillary rise from groundwater, and is depleted by evaporation from the soil surface and by plants transpiring water. If the soil moisture content in the unsaturated zone becomes sufficiently large, the excess water percolates downward to the groundwater.

Traditional studies assumed that pore water flow in the vadose zone was homogeneous in the horizontal dimension of space. However, the soil properties that govern soil water flow are often heterogeneous in nature, due to, for example, the presence of soil horizons, cracks, and root channels. Such heterogeneities may give rise to ***preferential flow***, which means that the percolating water does not tend to move as a horizontal wetting front. In general, the occurrence of preferential flow paths in soil is attributed two major phenomena, namely macropore flow (Beven and Germann, 1982) and finger flow (Hillel and Baker, 1988). Because of their large size, macropores generally do not participate in unsaturated flow of water in soil. Macropores start to transmit when they are completely filled with water. This only happens when the soil near the macropore becomes saturated or when ponded water on the soil surface can flow into open macropores. Finger flow occurs when infiltrating water breaks through the wetting front at isolated points. The rapid infiltration at these isolated points causes the formation of narrow fingers of wetted soil, while the bulk of soil remains dry. Because the fingers are narrow, the infiltrating water and the solutes it carries can potentially move rapidly to great depths. Preferential flow can thus be an important mechanism that causes contaminants to leach rapidly from the soil surface into the groundwater. Finger flow tends to be transient and the fingers usually disappear as soon as infiltration at the soil surface ceases.

3.2.4 Soil erosion

Precipitation that does not infiltrate into the soil is stored temporarily in puddles or as snow or ice cover, or runs off the land surface via overland flow. Overland flow may give rise to soil erosion if the shear stress aroused by the runoff water becomes sufficiently large to detach the topsoil particles. The direct impact of raindrops (splash) also detaches soil particles and contributes to soil erosion. Soil erosion is greatly enhanced by agricultural activities and is a major land degradation process in many regions in the world. The amount of soil erosion increases with increasing volumes of overland flow (which depends on the infiltration capacity of the soil and the upstream catchment area) and with increasing slope gradient.

Therefore, soil erosion typically occurs in hilly or mountainous regions. Nevertheless, in flat lowland areas with clay soils, the infiltration capacity may be so low that substantial overland flow and thereby soil erosion occur (see Bleuten, 1990).

Since many pollutants enter the soil at or near the soil surface (e.g. application of fertilisers or pesticides, atmospheric deposition), the topsoil layers that are susceptible to soil erosion also often contain the largest concentration of contaminants. In this manner, sediment-associated pollutants are transported laterally across the landscape and accumulate in valley bottoms or enter the river network. Sections 12.5 and 12.6 will further elucidate the processes of soil erosion.

3.2.5 Physico-chemical conditions in soil

The physical and chemical conditions occurring in soils are closely interlinked with the processes responsible for soil formation and vary as much as there are soil types over the world. One of the most important factors controlling physical and chemical processes in soil is the particle size distribution of soil material (***soil texture***). Together with the organic matter content, the soil texture determines the soil's capacity to hold water. Fine materials have a greater water-holding capacity than coarse materials. Soil texture also plays a

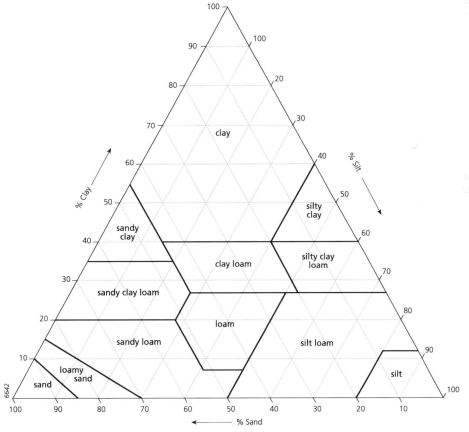

Figure 3.3 Texture classes adopted by the US Department of Agriculture (USDA, 1999).

significant role in the ability to exchange and retain substances that are transported in the soil solution. Both clays and organic matter have this important property and will be discussed further in Chapter 4.

The soil texture composition can be described and classified in many different ways. The soil texture classification adopted by the US Department of Agriculture (USDA, 1999), is one of the most common classifications based on the content of clay (particles < 2 μm), silt (2-50 μm) and sand (50 μm – 2 mm) (Figure 3.3).

Because most soils are not fully saturated with water (exceptions include peat soils, submerged paddy soils, and periodically saturated gley soils), the soil gas phase is in direct contact with the atmosphere. Respiration by microorganisms that are decomposing organic matter and plant roots reduces the oxygen concentration of the soil air and greatly increases its carbon dioxide concentration compared with the free atmosphere. The carbon dioxide dissolves in the soil water to form carbonic acid. Along with acid compounds derived from atmospheric deposition and organic acids formed as a result of organic matter decomposition, the principal acid-forming compound in soil is carbonic acid. These acids would lower the pore water pH, were it not that they promote the weathering and dissolution of minerals. These buffering processes also act against temporary pH changes. Whereas rainwater contains very few dissolved ions, the dissolution of soil minerals causes the soil water to become enriched in ions. Some ions (in particular, nutrients such as nitrogen compounds, phosphate, and potassium) are extracted selectively from the water by plants, via root uptake. On the other hand, evapotranspiration removes essentially pure water from the soil, so that the ion concentration always exceeds that of fresh precipitation and tends to build up in periods without rain. The increase in concentration may even result in chemical precipitation of solid salts, as occurs in many saline soils in arid and semi-arid regions.

Besides weathering and dissolution of soil minerals, the exchange of H^+ ions at the surfaces of minerals and organic matter also buffers the acidity of the soil solution. The total buffer capacity is determined by the existence of soluble and weatherable soil minerals. It is the calcium carbonate, salt, and clay contents in particular that control the buffer capacity of soils. Buffered soils are rich in such weatherable minerals and the pH of these soils ranges between 6.0 and 8.5. Poorly buffered soils, such as sandy podsolic soils, organic (peat) soils, and intensively leached soils in tropical regions, lack these weatherable minerals. The parent bedrock on which these soils have been formed often consists of inert materials, such as nearly pure quartz sand, quartzite, or granite. Poorly buffered soils have a low pH, with values ranging from 3.5 to 6.0. Note that under natural conditions the pH of the topsoil is often up to two pH units below that of the underlying soil, because the buffer capacity is first depleted in the topsoil. In contrast, the application of lime (calcium carbonate) on agricultural soils may have raised the pH of the topsoil above that of the subsoil.

The decomposition of organic matter further controls the redox potential of soils. The presence of oxygen gas in soil causes the redox potential to be high. As mentioned above, the oxygen consumption by microorganisms and plant roots depletes the oxygen in soil air. The oxygen is replenished by diffusive exchange between soil air and the free atmosphere at a rate that is greatly determined by the soil moisture content. Diffusive transport of gases occurs much faster in the gas phase than in the dissolved phase. So, in soils with high soil moisture content or in water-saturated soils the oxygen becomes rapidly depleted, which causes a considerable decrease of the redox potential of the soil. As a consequence, the redox potential is high in well-aerated sandy soils with low organic matter content. Conversely, the redox potential is low in saturated soils with a large organic matter content, such as peat soils.

3.3 GROUNDWATER

3.3.1 Definition of groundwater

Groundwater is defined as the water present in the saturated zone beneath the water table. As noted above, the pore spaces between the particles of the porous medium are completely or nearly completely filled with water, so that all water is in contact and the water pressure increases proportionally with depth. The groundwater slowly flows from places with high **hydraulic head**, i.e. a measure of potential energy consisting of an elevation term and a pressure term, to places with low hydraulic head. The hydraulic head is measured by determining the position of the water table in an unpumped well or **piezometer** (i.e. a pipe open at the top and bottom and placed vertically in the soil) relative to a reference surface.

A water-yielding rock formation that contains and is able to transmit sufficient groundwater to be a source of water supply is called an **aquifer**. Although aquifers are often comprised of a stratum or combination of strata of coarse unconsolidated rock, such as gravel or coarse sand, they can also consist of porous rock, such as sandstone, or non-porous but fractured rock, such as limestone or granite. Aquifers are bounded by a layer of low permeability (**aquitard**) or an impermeable body of rock (**aquiclude**). If there is no aquitard or aquiclude on the top of an aquifer, the aquifer is said to be unconfined or phreatic. A phreatic aquifer is bounded from above by a phreatic surface, which is defined as the surface at every point of which the pressure in the water equals the atmospheric pressure. A special case of a phreatic aquifer is the perched aquifer, which is a phreatic aquifer formed on a semi-permeable, or impermeable, layer below which unsaturated material is present. Confined aquifers are aquifers that are bounded by aquitard or aquiclude at both the top and bottom. The hydraulic pressure at the top of a confined aquifer may well be higher than atmospheric pressure. Artesian aquifers are a special case of confined aquifers and are characterised by a hydraulic head that is higher than the local ground surface level, so that water can flow out of wells without any need for pumps. Figure 3.4 depicts the different types of aquifers.

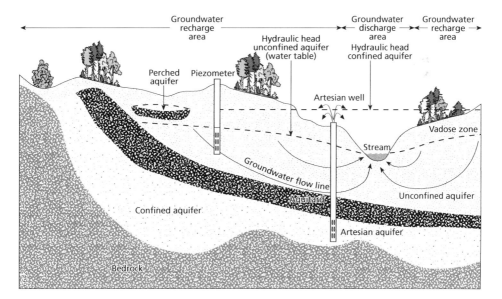

Figure 3.4 Different types of aquifers.

Groundwater is replenished by percolation water from the unsaturated zone. In some cases, surface water may also be a source of groundwater if the surface water level is higher than the water table, so the surface water infiltrates. The replenishment of groundwater is also referred to as **recharge**. As mentioned above, groundwater flows to areas of low hydraulic head, which are usually surface water bodies, such as rivers and lakes, and low-lying wetlands. In these areas, groundwater is discharged and becomes surface water. Groundwater can also be discharged into the atmosphere via evapotranspiration, leaving dissolved salts behind in the soil (see Section 3.2).

Groundwater flow velocities are slow and range between stagnant to tens of metres per day; they are proportional to the gradient in hydraulic head and the **hydraulic conductivity**. The hydraulic conductivity refers to the ability of the porous material to transmit water and is largely determined by the texture of the porous material. Sediments made up of coarse materials have a larger hydraulic conductivity than those made up of fine particles. At equal gradient in hydraulic head, the groundwater flow velocity is much slower in clay than in sand. The hydraulic conductivity also depends on the sorting of the particles. In poorly sorted materials, small particles block the pore space between larger particles, thereby reducing the hydraulic conductivity. Well-sorted soil material contains particles of similar size, giving the soil a relatively large hydraulic conductivity compared to material with a similar *average* size but poorly sorted. The apparent groundwater flow velocity is also inversely related to the **effective porosity** of the material, i.e. the volume of interconnected pores relative to the volume of the bulk sediment. Thus, groundwater flows faster in sediments with a small porosity than in sediments with a large porosity, because the same volume has to flow through a smaller fraction of the total cross-sectional area. Typical values of the porosity of sediments range between 0.2 and 0.4.

Groundwater flow in the different aquifer types as mentioned above obeys the same basic principles, but the different aquifers react differently to changes in hydraulic head. In the case of a lowering of the hydraulic head in an unconfined (phreatic) aquifer, pores are drained and partly filled by air in the zone that is vacated by the groundwater. This drainage cannot happen in confined aquifers because they are not in direct contact with the atmosphere. Instead, the pore space that is released due to a lowering of hydraulic head is compensated by compression of the aquifer or expansion of the water. Consequently, the amount of water released by a drop in hydraulic pressure is much larger in an unconfined aquifer than in a confined aquifer.

3.3.2 Sources of groundwater pollution

The major sources of groundwater pollution are principally the same as those of soil pollution and include landfills (waste dumps), accidental spills, agriculture, septic tanks, and atmospheric deposition. Dissolved pollutants move with the percolating soil water into groundwater, while organic liquid pollutants may reach the groundwater autonomously. In addition, in areas where surface water infiltrates to groundwater, surface water pollution is a potential source of groundwater contamination.

The input of pollutants into groundwater from the unsaturated zone and surface water is often closely related to fluctuations in groundwater level. A rise in groundwater level is often accompanied by an increased percolation by which pollutants are transferred from soil to groundwater. Furthermore, the raised water table may also capture pollutants present in soil for further transport. Conversely, a falling groundwater level results in changing redox conditions in soil, which may influence the leaching of pollutants from soil. Whether the leaching is enhanced or decreased depends on the chemical properties of the pollutant

Another source of pollutants in groundwater is the artificial input of contaminated water by deep well injection. One application of deep well injection is for storing liquid wastes in deep confined aquifers. Another is for artificially recharging groundwater near drinking water pumping stations, either to counteract the decline of ecosystems or crop yields following the fall in the water table and reduction of upward seepage, or to increase the aquifer's water yield. Deep well injection for the storage of liquid wastes is usually done into deep aquifers (up to several hundreds of metres deep) that are confined vertically by impermeable layers and contain water that is saline or otherwise unpotable. However, leakage or dispersal of these contaminants to neighbouring aquifers may result in unintentional contamination of other groundwater bodies. The target contaminant groups for deep well injection are volatile and semi-volatile organic compounds (VOCs and SVOCs), oil, explosives, and pesticides. In Russia and the USA, liquid radioactive wastes have also been disposed of in this manner. For artificial groundwater recharge near groundwater pumping stations, surface water is usually employed. The surface water is generally not pre-treated before injection and therefore any pollutants it contains may contaminate the aquifer. Furthermore, the aquifer may become contaminated due to chemical reactions between the recharge water and the native groundwater and/or the aquifer material.

3.3.3 Physico-chemical conditions in groundwater

Daily and seasonal variations in temperature at the soil surface are dampened in the soil profile. As a consequence, the groundwater temperatures at a given location 10 m below the water table are relatively stable. Groundwater temperature varies with latitude, being warmer near the equator and colder nearer the poles. In general, the average annual temperature of groundwater is about two degrees (°C) higher than the mean annual air temperature and increases by 1 to 5 °C (average about 2.5 °C) per 100 m depth. Larger increases may occur near local volcanic or geothermal activity.

As noted above, oxygen diffuses much more slowly in the saturated zone than in the unsaturated zone. As a consequence, the oxygen becomes depleted if it is consumed by the decomposition of organic matter faster than it is replenished by diffusion. Although the organic matter content in the saturated zone is usually much less than in soil (exceptions are peats and mucks, more than 80 percent of which consist of organic matter), all oxygen will be consumed after some time and so the redox potential decreases. Because groundwater flows, this process is reflected in a spatial separation of oxic (aerobic) groundwater and anoxic (anaerobic) groundwater. Therefore, oxic groundwater is usually found in the upper layers in infiltration areas, where the aquifer consists of coarse materials poor in organic matter. Nevertheless, reducing conditions tend to prevail in groundwater. As we saw in Section 2.10, the change in redox conditions in groundwater has an important effect on the speciation and solubility of substances, particularly of heavy metals.

The process of weathering and dissolution of minerals, which occurs in the unsaturated zone and is accelerated by the presence of acids in the soil solution, continues in the saturated zone. As a consequence, the concentration of dissolved substances in groundwater increases with time and thus in distance from the point of infiltration, until chemical equilibrium is reached. Some minerals, such as carbonates (e.g. calcite; $CaCO_3$) and evaporites (e.g. halite; NaCl) dissolve rapidly, significantly changing the mineral composition in the unsaturated zone. The weathering of other minerals, such as silicates (e.g. feldspars and clay minerals), which proceeds much more slowly, usually has only a minor effect on water composition. Thus, the amount of total dissolved substances in groundwater depends on rock type. Chapter 5 will further discuss the effect of rock type on water composition.

Furthermore, the sediment composition has a great influence on the retention of contaminants. Both fine-grained clay minerals and organic matter are able to sorb and hold large quantities of cations and organic pollutants. The mechanisms responsible for this sorption process will be further elucidated in Chapter 4.

3.4 SURFACE WATER

3.4.1 Definition of surface water

Surface water includes all water on the surface of the Earth found in rivers, streams, canals, ditches, ponds, lakes, marshes, wetlands, coastal and marine waters, and as ice and snow (EEA, 2006a). In this book, however, the definition of surface water is restricted to fresh waters, so therefore excludes the water in seas and oceans. The water–atmosphere interface constitutes the top boundary of the surface water compartment and, therefore, surface waters are generally well aerated. The bottom boundary of a permanent surface water body is formed by *bed sediments*, except where bare hard rock is present directly underneath the surface water.

A notable difference between surface water and groundwater is that sunlight can penetrate into surface water. This allows primary producers (green plants, including algae and macrophytes) to grow in surface water, thereby producing both oxygen and organic matter by means of photosynthesis. This process is crucial in the alteration of surface water chemistry.

Surface water is fed by inflow of rising groundwater, and occasional overland flow or throughflow (i.e. shallow, lateral flow of water through soil) from the surrounding uplands, and precipitation that falls directly on the water surface. In streams, groundwater comprises most of the base flow (i.e. dry weather flow). During and directly after rainfall events, a large proportion of stream water may be derived from overland flow or throughflow. Water is removed by discharge to downstream areas, infiltration (only at locations where the surface water level is higher than the water table), and evaporation. In addition, artificial surface water sources and sinks may be present: for example, discharge of wastewater (effluent), or surface water extraction for industrial processes or the production of drinking water. The flow velocity of surface water is positively related to water discharge and slope gradient of the water level, and is inversely related to the cross-sectional area of the water body perpendicular to the water flow and to roughness of the channel bed. The presence of ripples, dunes, or aquatic vegetation increases the roughness and slows down water flow. Lakes have a large cross-sectional area, so the flow velocity in lakes is generally low, varying between 0.001 and 0.01 m s^{-1} (near-surface values) (Meybeck and Helmer, 1996). Besides having a large cross-sectional area, lakes also have a relatively large surface area exposed to wind. Wind blowing across an open water surface generates waves and induces currents. This effect increases with the unobstructed distance over which the wind can blow (wind fetch), and thus increases with the size of an open water surface. Whereas the water flow in rivers is largely driven by gravity, usually the principal means of water transport in lakes is wind-induced currents. Transport rates in the surface water are usually much faster than in soil and groundwater: whereas soil water or groundwater may move only tens of metres per year, surface waters may cover this distance in a few seconds to hours.

3.4.2 Bed sediments

The submerged sediments present beneath or alongside rivers and lakes constitute a dominant link between the subsurface environment and the surface water. They consist of porous, hydraulically conductive, clastic (mineral) sediments and organic matter, which is either allochthonous (i.e. the sediment material originates from upstream areas with a possibly different geology) or autochthonous, (i.e. the material has been derived from local rocks). The zone of bed sediments is also referred to as the *hyporheic zone* (Edwards, 1998) (see Figure 3.5). Bed sediments are zones of intense biological activity and usually contain substantial quantities of organic matter. They encompass steep, vertical physical and chemical

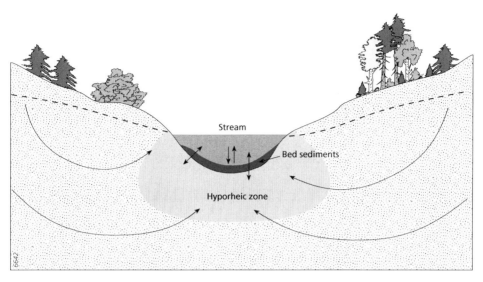

Figure 3.5 Interaction between surface water and groundwater in the hyporheic zone.

gradients and, as a consequence, they are able to retain, store, or release significant amounts of nutrients and heavy metals. Therefore, bed sediments may play a very important role in controlling surface water quality. Interaction between surface water and bed sediment also occurs through processes of sediment erosion and deposition. These processes will be further discussed in Section 4.1 and Chapter 12.

3.4.3 Sources of surface water pollution

Because much of the surface water has previously passed through the soil (surface or profile) or groundwater, many of the sources of soil pollution and groundwater pollution may be sources of surface pollution as well. Urban runoff in particular contains increased levels of contaminants including nitrogen, phosphorus, heavy metals, and organic compounds like PAHs. If surface water pollution sources are derived from groundwater or runoff water from upland soils, the pollutants will generally enter the surface water body over a large area. This implies that sources of pollutants that reach surface water via the soil surface (overland flow) or groundwater (upward seepage) are diffuse sources. In navigable surface water bodies, inland shipping may also be a significant diffuse source of pollutants as a result of wastewater releases, oil spills, and leaching from anti-fouling paints. These paints are applied to ship hulls to prevent organisms such as algae and molluscs from adhering to the hulls where they slow down the vessels and increase their fuel consumption.

Direct point source releases of pollutants include drainage from mine areas and effluent discharges of untreated or treated industrial and municipal wastewater. Sewage water is particularly rich in pathogenic bacteria and viruses, organic matter, and nutrients, but may also contain increased levels of heavy metals, and dissolved salts. As a result of oxidation of the organic matter, the discharge of untreated wastewater drastically reduces the dissolved oxygen concentration in the receiving water, causing massive fish kills and other health hazards. Wastewater treatment involves several steps, which are further explained in Box 3.I. Although it may improve the quality of the effluent water compared to the untreated water, effluent discharges may still place a considerable burden on surface water quality downstream from the outflow, especially if the wastewater has only undergone a limited treatment.

Box 3.1 Wastewater treatment plants

Wastewater purification in treatment plants occurs in multiple stages. The purpose of the primary mechanical treatment is to reduce the amount of organic solids in the water through settling and filtering. Primary treatment can reduce the solid particles by 30 to 60 percent and oxygen- demanding waste by 20 to 40 percent (Marsh and Grossa, 2002). Secondary biological treatment involves a further settling and filtering of the wastewater, plus aeration to accelerate the oxidation of organic matter by bacteria. Primary and secondary treatment together can remove up to 90 percent of the oxygen-demanding waste. However, most secondary treatment systems remove only 50 percent of nitrogen, 30 percent of phosphorus and even lower percentages of heavy metals and organic compounds. To remove these remaining pollutants, a tertiary treatment is applied, which includes the following actions (Marsh and Grossa, 2002):
• flocculation and settling to remove phosphorus and suspended solids;
• chemical adsorption of organic compounds;
• advanced filtering such as reverse osmosis to remove dissolved organic and inorganic substances;
• application of disinfectants such as chlorine, ozone, or UV light to kill pathogenic bacteria and some viruses.

Though tertiary treatment is effective and can usually remove up to 95–98 percent of the pollutants, it is very expensive.

In modern western countries, a high proportion of households and industrial premises are connected to sewage systems and wastewater treatment plants. In Northern and Western Europe the connection to wastewater treatment plants was 80 percent on average in 1987 (EEA, 1998, Farmer, 1999) (see Table 3.Ia) and this number is still increasing. The proportion of households connected to tertiary treatment was considerably lower, but EU legislation was passed in 1991 to promote the ongoing upgrade from secondary to tertiary treatment works.

Table 3.Ia Urban wastewater treatment in EU member states (in 1987) (Farmer, 1999).

Country	Total Connection Rate*: Sewerage and treatment in treatment works	Mechanical treatment only
Denmark	98 %	8 %
France	50 %	No data
Germany	90 %	2 %
Greece	No data	1 % (1985)
Italy	60 %	No data
Luxembourg	91 %	14 % (1985)
Netherlands	89 %	7 %
Portugal	11 %	4 %
Spain	29 % (1985)	13 % (1985)
United Kingdom	84 %	6 %

* Note that the households that are not connected to wastewater treatment plants do not necessarily discharge the untreated wastewater into the surface water. Most of the wastewater is treated locally in septic tanks.

Accidental spills can cause severe pollution of surface water. They may occur due to, for example, leaks of process chemicals and products from industrial installations, release of contaminated firefighting water during fire abatement, or failures of tailing dams of mine reservoirs. Notorious examples are the Sandoz accident in Basle (river Rhine), Switzerland in 1986, the cyanide spill at Baia Mare (Tisza river), Romania in 2000, and the benzene and nitrobenzene spill in Jilin (Songhua river), China in 2005 (see Section 10.2 for further details). Spills may also occur at a more local scale: for example, car accidents.

Pollutants discharged into surface water may partly be retained in the bed sediments. In this case, the bed sediments act as a sink for pollutants. A considerable stock of pollutants may accumulate over the years and as soon as the pollutant concentrations in the surface water fall, these pollutants may be re-released from the bed sediments. In such cases, bed sediments may act like a diffuse source of pollutants. This type of source is also refereed to as internal loading. Internal loading also encompasses the release of nutrients and other pollutants due to the decomposition of internally produced organic matter (e.g. litter from aquatic plants and algae).

3.4.4 Physico-chemical conditions in surface water

An important parameter in surface water quality studies is water temperature, since it affects many physical, chemical, and biological processes. It influences, amongst others, the dissolved oxygen concentration, the growth rate and rate of photosynthesis of algae and other aquatic plants, and the decomposition rate of organic matter. The temperature of surface water generally follows seasonal and local weather patterns, although the course of water temperature is subdued and lags behind the air temperature. Surface water bodies exposed to direct sunlight may be up to 10 °C warmer than shaded water bodies. In deep lakes, vertical *stratification* may occur due to small differences in density caused by differences in water temperature or solute content. Stratification occurs when the water in

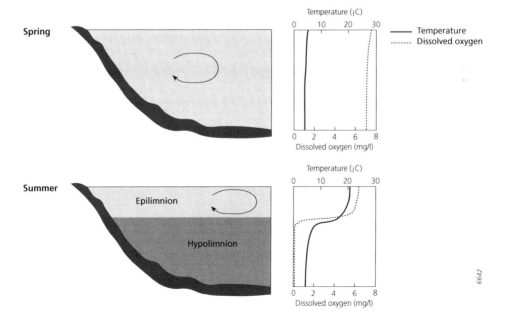

Figure 3.6 Thermal stratification of a lake.

the lower layer is denser and water currents in the upper layer usually generated by wind are unable to cause eddies strong enough to penetrate the boundary between the two layers. In regions with distinct cold and warm seasons, stratification is mostly the result of temperature differences, as the upper layer of the lake warms up in summer due to solar radiation. Since warm water is less dense, it tends to float, and the warm water concentrates at the surface. If this happens, the lake is said to be thermally stratified. The upper, well-mixed, warm layer is called the ***epilimnion*** and the lower, cold layer is called the ***hypolimnion*** (see Figure 3.6). Between the epilimnion and the hypolimnion is a zone of rapid temperature change, the ***thermocline*** or metalimnion, which may extend for several metres. The epilimnion is usually well mixed due to wind-driven circulation, while the hypolimnion is isolated from wind effects and is consequently rather quiescent. The temperature difference between epilimnion and hypolimnion may be fairly large. During the summer in the temperate zone, it is not exceptional for the epilimnion to warm up to between 15 °C and 25 °C, while the water temperature in the hypolimnion remains considerably below 10 °C. During stratification, little exchange of chemicals and heat occurs between the epilimnion and hypolimnion. The only exchange occurs through settling of suspended particles, which causes a material flux from the epilimnion to the hypolimnion. Thermal stratification is usually overturned in autumn, which causes the lake to be well mixed over the entire depth (Figure 3.6). During winter, lakes may stratify because ice covers the lake surface. In spring, as ice melts, this winter stratification is removed. Shallow lakes are usually well mixed and may warm throughout their depth and not exhibit a thermocline. In some shallow lakes, the bottom layer may be fed by colder spring water; in such cases the thermocline may be quite abrupt.

In surface water, just as in soil and groundwater, oxygen is consumed by the decomposition of organic materials, but because surface waters are in direct contact with the atmosphere, they are usually well aerated. Moreover, photosynthesis by aquatic vegetation and algae adds oxygen to the surface water. As a result, the redox potential in surface water is usually high (Eh > 700 mV). The maximum total dissolved oxygen concentration in water is determined by the atmospheric pressure and, as noted above, temperature. Cold water holds more dissolved oxygen than warm water, and surface water bodies at sea level contain more oxygen than water bodies at high altitudes. Water may become depleted in oxygen due to the presence of abundant amounts of easily degradable organic matter originating from inputs of domestic waste or extensive algal blooms (see Chapter 6). Because water in the hypolimnion

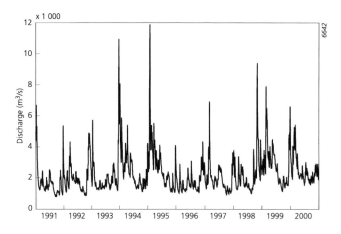

Figure 3.7 Discharge of the river Rhine near Lobith, the Netherlands between 1991 and 2000 (source: Waterbase, 2013).

of a deep lake is isolated from the atmosphere, it may also become depleted in oxygen if sufficient organic matter is available for decomposition. Under these conditions, the redox potential is considerably decreased.

Surface water is mostly less acidic than rainwater and groundwater. Primary producers extract dissolved carbon dioxide (carbonic acid) from solution for photosynthesis, which causes the pH to increase. When water contains high levels of carbonates (e.g. water in limestone catchments) and the growth of algae is not limited by nutrient shortage, the pH can rise to well above 8. Therefore, most surface waters are neutral to basic (compare Figure 2.3), except water bodies in catchments with poorly buffered soils, for example catchments draining upland moors or granite or sandstone bedrock.

River flow is highly dynamic. During periods of little precipitation, most rivers carry a minimum amount of water through the river channel. This condition is called baseflow and is in most cases controlled by groundwater discharge. Therefore, during baseflow, stream waters generally reflect the composition of near-surface groundwater, which is, in turn, determined by local geology. Considerable increases in discharge above the baseflow occur during rainfall or snowmelt events. The increase in discharge in response to rainfall events is a function of the rainfall intensity and duration, the surface area and permeability of the upstream catchment, antecedent soil moisture conditions, groundwater levels, topography, and surface resistance to flow. Surface water discharge may vary by one order of magnitude in large drainage basins, as illustrated in Figure 3.7. In small catchments, discharge can be rather flashy and may rapidly increase by even two or more orders of magnitude during periods of prolonged heavy rain or snowmelt. During such events, the stream water becomes diluted with rainwater and overland flow water, which usually contain less dissolved substances.

Increased stream flow during rainfall events may also cause erosion of bed sediments, which may bring sediment and associated chemicals into suspension. Sediment is therefore an important vector in contaminant transport in surface water. Suspended sediments may come from erosion of hillslopes, stream banks, and stream beds. Their source exerts an important influence on the features of the suspended sediment load (e.g. grain size distribution, geochemical composition) (Walling and Webb, 1996). Fine sediments transported in suspension can be transported across particularly long distances. Considerable amounts of relatively insoluble contaminants contained in fine mineral particles or living or dead organic matter can be transported in particulate form. However, particulate matter tend to settle out and so the sediment-associated contaminants transported in surface water may be stored temporarily in bed sediments or semi-permanently in floodplain or lake deposits (Walling *et al.*, 2003; Walling and Owens, 2003). Therefore, as noted above, interaction between surface water and bed sediments through retention, release, erosion, and deposition is a key process in controlling surface water composition.

3.5 FURTHER READING ON SOILS, GROUNDWATER, AND SURFACE WATER

Soil Science
- Gerrard, J., 2000, *Fundamentals of Soils*, (Oxford: Routledge).
- Miller, R.W. and Gardiner, D.T., 2008, *Soils in Our Environment*, 11th edition, (Upper Saddle River NJ: Prentice-Hall).

Hydrology
- Ward, R.C. and Robinson, M., 2000, *Principles of Hydrology*, 4th edition, (London: McGraw-Hill).

- Brutsaert, W., 2005, *Hydrology, An Introduction*, (Cambridge: Cambridge University Press).
- Hendriks, M.R., 2010, Physical Hydrology, (Oxford: Oxford University Press).

Limnology
- Cole, G.A., 1994, *Textbook of Limnology*, 4th edition, (Long Grove IL: Waveland Press).
- Kalff, J., 2002, *Limnology*, (Upper Saddle River NJ: Prentice Hall).

EXERCISES

1. Explain why the hydrological cycle is considered to be the most important force driving the transfer of substances between the environmental compartments.

2. Name the most important sources of diffuse soil contamination. How can these sources also contribute to groundwater and surface water contamination?

3. A soil consists of 30 percent clay, 60 percent sand, and 10 percent silt. What is the texture class according to the system of the US Department of Agriculture?

4. Define the following groundwater-related terms:
   a. groundwater recharge
   b. hydraulic head
   c. aquifer
   d. aquitard

5. Explain why the hyporheic zone is important for surface water quality.

6. Describe the evolution of the hypolimnion throughout the seasons.

7. Describe the three steps in wastewater treatment.

8. Name four chemical controls on water composition and indicate the environmental compartment where they have most effect.

9. Name at least three relevant differences and similarities between soil water, groundwater, and surface water.

Part II
Sources, role, and behaviour of substances in soil and water

How persistent and toxic pollutants are in natural environments depends largely on their physical and chemical behaviour. Pollutants can dissolve and precipitate, be adsorbed onto particulate matter, or decompose; these biogeochemical processes are influenced by the presence of other substances or organisms, generally referred to as environmental conditions. When assessing the hazards and risks associated with the toxicity and dispersal of pollutants in soil, groundwater, and surface waters, it is essential to understand the biogeochemical behaviour of pollutants (e.g. their solubility, affinity to particulate matter, and decay rates) under varying environmental conditions. Part II therefore provides a summarising overview of the natural occurrence and anthropogenic sources, environmental role, physico-chemical characteristics, and toxicity of the major polluting substances. These substances are described one by one, according to a common classification based on their natural occurrence and role, or their physical and chemical characteristics.

4

Solid phase constituents

4.1 INTRODUCTION

The solid phase constituents in soil and water are comprised of a variety of components. Based on their chemical composition, they can be subdivided into inorganic minerals (specific density of about 2.6 g cm^{-3}) and organic compounds (specific density of somewhat more than 1.0 g cm^{-3}). Both types of constituents can also be classified according to the size of the particles. Figure 4.1 gives an overview of the particle size of all major components present in soil and water. Note that the size of the particles forms a continuum, from molecules to large inorganic particles. For convenience, when distinguishing between dissolved and particulate matter, a particle size of 0.45 μm is usually adopted; this size is based on the pore size of a certain type of membrane filter. **Colloids** constitute a special intermediate case between dissolved and particulate matter. They consist of particles dispersed in water and ranging in size from 1 nm (the size at which the particles approach molecular dimensions) to about 10 μm. A size of 10 μm is still small enough for Brownian motion (i.e. the apparently random motion experienced by any molecule or small particle immersed in a fluid) to prevent them from settling out. Examples of environmental colloidal particles are clay minerals and humic substances. **Nanomaterials** are natural and man-made particles ranging from 1 nm to 100 nm in size. Like colloidal particles, they form a special group of extremely small solids, and when dispersed in water, nanomaterials include colloidal particles. The types and properties of man-made nanoparticles and their health and environmental risks are discussed in section 4.2.5 In this chapter, both particulate matter and colloidal particles are considered to be solid phase constituents. Note, however, that some

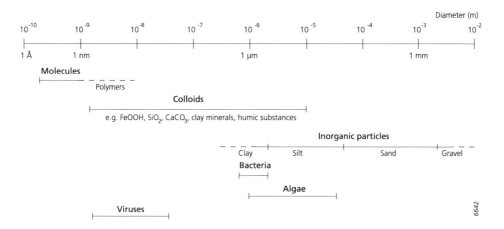

Figure 4.1 Size ranges of common particles in soil and water.

of the colloidal particles are smaller than 0.45 μm and so may also be classed as dissolved matter.

Depending on their physical and chemical properties, solids (including colloidal particles) interact with the dissolved phase via sorption and precipitation–dissolution reactions. Almost all pollutants and other chemicals are subject to sorption or precipitation–dissolution reactions, or both; the rate and extent of these reactions is often referred to as the *reactivity* of the solid phase (e.g. sediment, organic matter). A key factor that determines the reactivity of the solid phase is the specific surface area, expressed in $m^2 \ g^{-1}$. Coarse materials with a small specific surface area are much less reactive than fine pariceles or nanomaterials with a large specific surface area; this is particularly the case for sorption reactions. Usually a specific surface area of $10 \ m^2 \ g^{-1}$ is adopted as the criterion to divide materials that are of no importance for sorption reactions from materials that interact significantly with the liquid phase (Bolt and Bruggenwert, 1978). With regard to dissolution reactions, inorganic materials with a large specific surface area almost always have a low solubility, since very small particles of a readily soluble mineral would normally quickly dissolve and disappear.

In the subsurface environment, i.e. soil and groundwater, the solid phase does not generally change position and so may retain nutrients, metals, and pollutants that are being carried in solution; alternatively, if these compounds are present in the solid phase, they may be released into the flowing groundwater. In surface waters, solid components may be transported by the flowing water as bed load or suspended sediment (synonyms: suspended matter, suspended solids (SS)). Solid particles in surface water are often transported at a slower rate than dissolved constituents, since during transport the particulate matter may settle due to gravity. The rate of settling is positively related to the specific density and the diameter of the particles. Suspended particulate matter may also partly consist of colloidal matter with particle sizes of less than 10 μm, which hardly settles in surface water. Accordingly, colloidal matter is relatively mobile and in combination with its large reactivity, may be an important carrier of pollutants – in both soil and groundwater

Note that suspended solids in surface water are not only a carrier of pollutants; sometimes they may also be considered as pollutants themselves. When concentrations of suspended solids (including living algae) are high, the water clarity is considerably reduced, which affects the photosynthesis and therefore the growth of aquatic plants and algae. Moreover, if suspended solids contain considerable amounts of easily degradable organic matter, the oxidation of organic matter depletes the dissolved oxygen concentration in surface water, which, in turn, may lead to fish mortality. Given the role of solid phase constituents as pollutants and carriers of pollutants, it is clearly important to understand the physical and chemical properties and processes involved. This will shed light on the environmental transport and fate phenomena of pollutants. Below, therefore, the composition and main properties of solid components will be discussed first, prior to other chemical constituents in soil and water.

4.2 INORGANIC COMPONENTS

4.2.1 Composition and formation

Inorganic components occur mainly in a limited number of compounds with a definite crystalline structure, called minerals. Table 4.1 lists some common types of minerals found in soil and grouped according to the anionic constituents. The inorganic components originate from physical and chemical weathering of bedrock materials and can be grouped into primary and secondary minerals. Primary minerals may only have undergone physical weathering and have not changed chemically since their formation in the Earth's crust.

Secondary minerals have been formed by the precipitation or recrystallisation of elements that have been released by the chemical weathering of primary minerals and normally have particle sizes smaller than 2 µm (also referred to as the clay fraction, see also Figure 4.1). The inorganic components may also be classified according to their solubility. Very soluble minerals, such as nitrates, halides, and some sulphates, are usually only present as secondary precipitates in soils under arid conditions (i.e. low soil moisture contents). Carbonates and gypsum are intermediate soluble minerals and are found in soil as primary minerals in parent sedimentary bedrock material or as secondary precipitates. Under natural conditions, gypsum only precipitates out under the influence of evaporation. The dissolution and precipitation of carbonates depend on the pH and CO_2 pressure in the dissolved phase (see Section 3.6). Sulphides (e.g. pyrite) are secondary minerals, which are formed under reducing conditions when sulphate is reduced to sulphide (see Section 5.7). Sulphides are practically insoluble, but under oxic conditions sulphide may oxidise to sulphate and dissolve. Many silicates, such as quartz (note that quartz is also a silicate!), tectosilicates, nesosilicates, inosilicates, micas, and serpentine, are primary minerals and can be considered as the remains of igneous and metamorphic bedrock material. They have a very low reactivity, since they combine very low solubility with a small specific surface area. The remainder of the minerals listed in Table 4.1 are iron (Fe) and aluminium (Al) oxides/hydroxides, and clay minerals. These secondary

Table 4.1 Some major common minerals in soil (adapted from Bolt and Bruggenwert, 1978).

Mineral group	Examples of major minerals	Chemical formula
Oxides/Hydroxides		
Si oxides	Quartz	SiO_2
Fe oxides/hydroxides	Goethite	$FeOOH$
	Haematite	Fe_2O_3
Al oxides/hydroxides	Gibbsite	$Al(OH)_3$
Silicates		
Tectosilicates	Feldspars (e.g. orthoclase, plagioclase, albite, zeolite)	Silicates are arrangements of the elements Si and O with a wide range of other elements (e.g. Al, Fe, Mn, Ca, Mg, Na, K, F, Ba, Ti, Cr)
Nesosilicates	Olivine, garnet, tourmaline, zircon	
Inosilicates	Pyroxene, hornblende, amphibole	
Phyllosilcates	Micas (e.g. biotite, muscovite), serpentine, clay minerals (e.g. kaolonite, illite, montmorillonite, vermiculite)	
Carbonates	Calcite	$CaCO_3$
	Dolomite	$MgCa(CO_3)_2$
	Siderite	$FeCO_3$
Sulphates	Gypsum	$CaSO_4 \cdot 2H_2O$
Sulphides	Pyrite	FeS_2
Halides	Halite	$NaCl$
	Sylvine	KCl
	Carnalite	$KMgCl_3 \cdot 6H_2O$, $CaCl_2 \cdot nH_2O$
Phosphates	Apatite	$Ca_5(F,Cl,OH)(PO_4)_3$
	Vivianite	$Fe_3(PO_4)_2 \cdot 8H_2O$
Nitrates	Soda nitre	$NaNO_3$
	Nitre	KNO_3

minerals may have a large reactivity and are particularly relevant for interactions between the solid and the liquid phase; for this reason, the Fe and Al hydroxides and clay minerals will be discussed separately.

4.2.2 Aluminium and iron oxides/hydroxides

In temperate climates, Al and Fe oxides/hydroxides occur mainly in the form of gibbsite (γ-Al(OH)$_3$) and goethite (α-FeOOH), respectively. The major form of iron oxide in tropical soils is generally haematite (α-Fe$_2$O$_3$). The mineral structure of these Al and Fe oxides/hydroxides, commonly referred to as ***sesquioxides***, is often relatively simple. They consist of a dense packing of oxygen (O^{2-}) and/or hydroxyl (OH$^-$) anions held together in a specific configuration of Fe^{3+} or Al^{3+} cations. The internal configuration may range from perfectly regular, resulting in a crystalline structure, to rather irregular if impurities are present, resulting in amorphous sesquioxides. The specific surface area of sesquioxides depends on the conditions during formation and may vary from several tens of square metres per gram to very low values for macroscopic crystals and concretions. In the interior of the crystals, electroneutrality occurs, i.e. the charge of the Fe^{3+}/Al^{3+} cations and of the O^{2-}/OH$^-$ anions balance each other out. However, at the surface the O^{2-}/OH$^-$ anions are not balanced against the metal cations, because this is the terminal layer. Consequently, there is an excess of anions, which causes the surfaces of sesquioxides to be generally negatively charged. At the surface of the dry crystal, electroneutrality is maintained by adsorption of an appropriate number of cations, generally H$^+$ ions (protons). In contact with the aqueous phase, these protons may dissociate and be exchanged for other cations, depending on the pH of the solution. In this case, the surface acts like an acid. Consequently, the negative charge of the surfaces of the sesquioxides increases with increasing pH and so does the capacity to sorb cations other than protons. At low pH values, the surface may adsorb a larger number of protons than needed for neutralisation. In this case, the surface thus acts like a base and attains a positive charge. These deprotonation and protonation reactions can be written as:

$$\equiv MO^- + 2H^+ \quad \leftrightarrow \quad \equiv MOH + H^+ \quad \leftrightarrow \quad \equiv MOH_2^+ \qquad (4.1)$$

where $\equiv MO$ = the surface oxygen. The pH at which the charge of the mineral surface is neutral (zero charge) is called the ***Point of Zero Charge*** (*PZC*). The *PZC* of gibbsite varies between 5.0 and 6.5, the *PZC* of goethite is about 7.3, and the *PZC* of haematite is about 8.5 (Appelo and Postma, 1996). This implies that under acidic to neutral conditions (pH < 7), goethite and haematite have a positive surface charge and, accordingly, are able to adsorb anions. Note that sesquioxides also have affinity for specific anions, especially phosphate, through specific adsorption. The adsorption phenomena specific to phosphate will be discussed further in Section 4.3.

4.2.3 Clay minerals

At the geological time scale (i.e. thousands to millions of years), clay minerals are formed as the chemical weathering products of relatively readily weatherable minerals such as micas, olivine, pyroxenes, amphiboles, and calcium-rich plagioclases. Examples of abundant clay minerals are kaolinite, illite, montmorillonite, vermiculite, and chlorite. Figure 4.2 gives an overview of the possible pathways of formation of various clay minerals. The clay minerals in soil originate from either the parent sedimentary bedrock material or in-situ formation, and their ultimate form depends on the composition of the parent material and on the climate. Kaolinite is a residue of extensive weathering under humid and acid conditions.

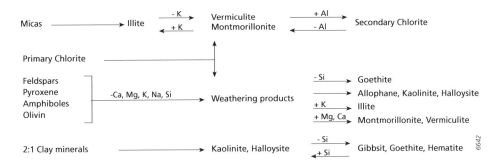

Figure 4.2 Schematic overview of formation of clay minerals from primary minerals. Adapted from Scheffer and Schachtschabel (1989).

It is present in the clay fraction of nearly all soil types. In tropical and subtropical soils, kaolinites constitute the primary part of the clay minerals, but in soils in the temperate climate region they exist only in small amounts, mainly derived from parent sedimentary bedrock. Montmorillonite belongs to the smectites, which are swelling, sticky clays. It occurs mainly in soils formed from basic igneous rocks (e.g. basalt, gabbro). In the tropics and subtropics, montmorillonite can make up the primary clay mineral in soils formed in calcium- and magnesium-rich parent material under poor drainage conditions. Vermiculite occurs in slightly acid soils, but is seldom the dominant clay mineral in soils. Illite is the most abundant clay mineral in soils of the humid temperate climate region, especially in fluvial sediments.

Clay minerals are alumino-silicates with a so-called lattice structure consisting of alumina octahedra joined to one or two layers of silica tetrahedra (Figure 4.3) A silica tetrahedron consists of a relatively small silicon (Si) atom bonded with four larger O^{2-} atoms, three in the base layer and one in the apex layer. In an alumina octahedron, an Al atom bonds with six OH^- ions, three in the base layer and three in the top layer. The spacing of both the OH^- in the alumina lattice and the O^{2-} apex layer of the silica lattice favours a bonding of these two lattices. Either a two-layer (1:1 clay mineral) or a three-layer structure (2:1 clay mineral) is formed, leading to a variety of clay minerals. As examples, Figure 4.4 shows the lattice structures of kaolinite, illite, and montmorillonite. The clay mineral kaolinite consists of stacked sheets of the two-layer unit, and is thus a 1:1 clay mineral. The minerals illite and montmorillonite are both 2:1 clay minerals and consist of stacks of three-layer units.

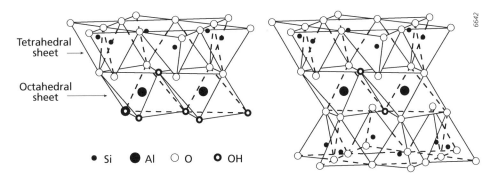

Figure 4.3 Arrangement of tetrahedral and octahedral sheets in a 1:1 clay mineral (left) and a 2:1 clay mineral (right).

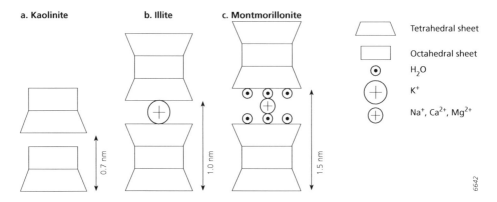

Figure 4.4 Lattice structures of a) kaolinite (1:1 clay mineral), b) illite (2:1 clay mineral), and c) montmorillonite (2:1 clay mineral).

The layers of illite are bonded with dehydrated potassium (K^+) ions (no surrounding water molecules) partly sunken into the hexagonal holes of the tetrahedral layers. The layers of montmorillonite are separated by water molecules containing cations such as sodium (Na^+), calcium (Ca^{2+}), and magnesium (Mg^{2+}).

Aluminium and silicon atoms are similar in size to each other and to other metal ions. It is therefore possible for the Si^{4+} or Al^{3+} in the basic crystal structure to be replaced by Al^{3+}, Fe^{2+}, Mg^{2+}, or Ca^{2+} without disturbances to the crystal structure. This replacement is called ***isomorphous substitution*** and results in a charge imbalance that leaves most of the clay particles negatively charged. The process of isomorphous substitution is rather slow and, therefore, the resulting negative charge is fairly stable and is neutralised by cation adsorption at the clay mineral surface. Like sesquioxides, clay minerals also have an additional pH-dependent charge due to the protonation as described by Equation (4.1). These protonation reactions occur wherever charge imbalances at 'broken bonds' of the tetrahedral and octahedral layers occur, i.e. at the edges of the clay minerals. The faces of the clay minerals do not generally have any free hydroxyl groups for protonation reactions, so the pH-dependent surface charge occurs primarily on the clay edges. The Si-OH groups at the edges of the tetrahedral layer have a very low *PZC* and, therefore, they are negatively charged or near neutral, even at low pH values. In contrast, the behaviour of the edges of the octahedral layer is to some extent like gibbsite, which has a relatively high *PZC* value (about 7.0). Consequently, the edges of the octahedral layer are likely to have a positive charge at low pH values.

Kaolinite and other 1:1 clay minerals are barely susceptible to isomorphous substitution and develop only a pH-dependent charge. Kaolinite has a *PZC* of 4.6, thus at low pH it has a near-zero surface or slightly positive charge, but the surface becomes slightly negative with increasing pH as protons dissociate from surface hydroxyls. The fixed charge on 2:1 clay minerals always contributes more to total surface charge than the variable charge. Illite has a moderately large fixed negative charge. The negative charge of montmorillonite is larger than that of illite.

Because clay minerals are primarily negatively charged, they will basically be sorbents for cations. Cation adsorption can occur via various mechanisms, which results in a different degree of bonding between the negatively charged mineral surface and the cations. These sorption mechanisms include the formation of an inner-sphere complex, formation of an outer-sphere complex, adsorption in the so-called diffuse layer, and specific adsorption. The formation of an inner-sphere surface complex implies a surface complex formed

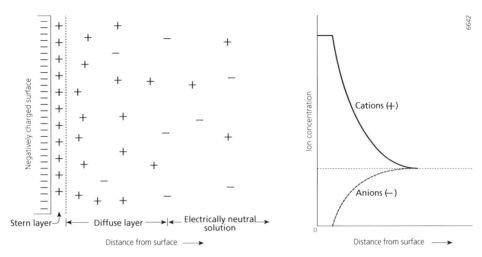

Figure 4.5 Ion distribution and in the double layer. Adapted from Scheffer and Schachtschabel (1989).

directly between a dehydrated cation and the siloxane cavity (i.e. the hexagonal holes of the tetrahedral layers, which serve as reactive sites for the formation of surface complexes with cations). The bond thus formed is very strong. An outer-sphere surface complex involves the formation of a surface complex between a hydrated cation and the siloxane cavity. This bond formed here is much weaker than in the inner-sphere case; outer-sphere complex ions exchange more readily with ions in solution. The inner-sphere and outer-sphere complexes are thus formed directly at the mineral surface, also referred to as the Stern layer. These complexes are usually not sufficient to neutralise the negative charge at the clay mineral surfaces. To balance this residual charge, cation adsorption also occurs in the soil solution near the mineral surface, the so-called ***diffuse layer***. In this diffuse layer, the cations (counter ions) are more abundant than diffuse anions (co-ions). These cations are thus not bonded to the surface, but are in solution. The name of diffuse layer is derived from the tendency of the counter ions to diffuse away from the accumulation zone near the mineral surface towards the region of lower concentrations in the bulk solution (see also Section 11.3.1). It can thus be concluded that the interactions between cations and mineral surfaces in contact with aqueous solutions occur in two layers, together named the diffuse double layer: the layer consisting of inner- and outer-sphere surface complexes and the diffuse ion layer. Figure 4.5 illustrates the distribution of ions in this diffuse double layer.

The greater the valence of the cation and the smaller the hydrated cation, the stronger the cation adsorption is. Therefore, cation adsorption occurs approximately in the following order of preference:

$$H^+ > Al(OH)_2^+ > Ca^{2+} \approx Mg^{2+} > NH_4^+ \approx K^+ > Na^+$$

The adsorbed cations are also referred to as exchangeable cations because when the free water flowing around the particle contains different cations, they may be exchanged with those in solution. The extent to which clay minerals are able to exchange cations depends on their specific surface area and the negative surface charge and is thus partly dependent on the pH. The ***cation exchange capacity*** (CEC) of clay minerals is usually expressed in milliequivalents per unit mass of dry soil, mostly per 100 g. The CEC differ for the various clay minerals (some typical values of the CEC for common clay minerals are shown in Table 4.2). Thus, the content of the various types of clay minerals determines the CEC of soils and sediments.

Table 4.2 Typical values of the cation exchange capacity for some common clay minerals.

Clay mineral	CEC (meq/100 g)
Kaolinite	1–10
Illite	20–40
Montmorillonite	80–120
Vermiculite	120–150

Furthermore, the CEC of soils is determined by the content of sesquioxides (see Section 4.2.2) and organic matter (see Section 4.3), both of which are also able to sorb cations. Section 4.4 gives more details on the CEC of soils.

Besides the interactions between cations at and near the clay mineral surface and in solution, there are some specific adsorption phenomena at the edges of clay minerals. Illitic clay minerals contain considerable amounts of potassium (K^+) which bonds the octahedral and tetrahedral lattices. The amount of K may correspond to twice to four times the CEC associated with the illite surfaces. These K^+ ions are held extremely preferentially and are largely non-exchangeable under natural conditions. However, at the illite edges some K^+ ions may be exchanged, causing the interlattice space to open partially, especially if the illite is brought into contact with a solution containing very small concentrations of K^+. This opening of the interlattice space results in ***frayed edge sites***: open lattice structure at the clay mineral edges. If the K^+ concentration in solution increases, K^+ ions move rapidly into the open edges, subsequently causing the open structure to collapse. The K adsorbed in this manner is trapped irreversibly inside the clay lattice structure. This phenomenon is called K fixation and is specific to K^+ ions and also to ammonium (NH_4^+), caesium (Cs^+), and rubidium (Rb^+) ions, which are of similar size.

A second sorption phenomenon at clay mineral edges involves the adsorption of anions. As mentioned above, depending on the pH of the solution due to protonation reactions at the broken bonds of the octahedral and tetrahedral layers, the clay mineral edges may be slightly positively charged. These positively charged sites give rise to the adsorption of anions to neutralise the positive charge. Because anion adsorption is only limited to the clay mineral edges which usually have a limited surface area, the ***anion exchange capacity*** (AEC) is considerably smaller that the cation exchange capacity. The sorption of anions is very selective, and depends on their valence and size. A larger ion size results in a lower degree of hydration, which in turn favours sorption. For common anions, experimental evidence indicates the following order of preference for anion adsorption at clay mineral edges (Bolt and Bruggenwert, 1978):

$$SiO_4^{4-} > PO_4^{3-} \gg SO_4^{2-} > NO_3^- \approx Cl^-$$

Accordingly, if only very small concentrations of phosphate (PO_4^{3-}) anions are present, sulphate (SO_4^{2-}) and chloride (Cl^-) anions are not adsorbed.

The capacity of clay minerals to adsorb ions is a significant feature for the retention and transport of substances (pollutants) through soil and water. In addition, clay minerals have the ability to stick to each other, which can also be attributed to the charge imbalances at the clay mineral surfaces. This property of clay minerals is crucial for the settling characteristics of clay minerals suspended in surface waters and, therefore, for the transport behaviour of clay minerals and their associated pollutants. The formation of flocs or aggregates is called ***flocculation*** or coagulation; it results in apparently larger particles, which settle much faster than smaller particles, even though the specific density of the flocs is usually less than that of

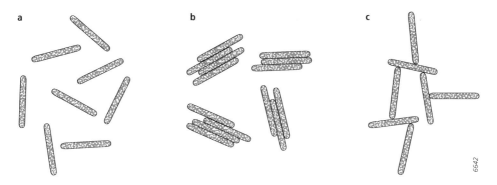

Figure 4.6 Flocculation of clay minerals: a. peptisation; b. face to face coagulation; c. edge to face coagulation.

the individual clay particles. The reverse process of forming a colloidal suspension or sol is called peptisation (see Figure 4.6a).

The tendency of clay minerals to flocculate is determined by the balance of the attractive and repulsive forces between the clay particles. Clay particles are attracted to each other by Van der Waals forces, but these weak forces work only in a close range of less than about 1.5 nm. The particles repel each other because the counter ions in the diffuse double layers around the clay minerals have the same charge. Because the ion concentrations are higher in the diffuse layer than in the surrounding solution, the resulting osmotic pressure also contributes to the repulsion of the particles. To initiate flocculation, the clay particles should thus be permitted to approach each other close enough for Van der Waals attraction, i.e. closer than 1.5 nm. The distance to which the particles are able to approach each other is governed by the thickness of the diffuse double layer, since this layer is the primary cause of repulsion of the particles. The thickness of the diffuse double layer is inversely related to the average valence of the counter ions in the diffuse layer and the total ion concentration in the bulk solution. The occurrence of divalent cations (e.g. Mg^{2+}, Ca^{2+}) results in a thinner double layer than monovalent cations (e.g. Na^+, K^+), because twice as many monovalent ions are needed for electrical neutrality. A larger total ion concentration in solution causes a thinner double layer, because a larger concentration in the bulk solution suppresses the tendency of the counter ions to diffuse away from the clay mineral because the concentration gradient between the diffuse layer and the bulk solution decreases (see the definition of the diffuse layer given above). Consequently, clay minerals tend to flocculate rapidly in brackish or salt water where the total ion concentrations are sufficiently large, even if the cations in the bulk solution are predominantly monovalent (Na^+).

The above described mechanism of flocculation due to Van der Waals forces brings about a face to face aggregation (see Figure 4.6b). Besides Van der Waals forces, clay particles may also be attracted to each another by electrostatic forces. Where conditions promote a thin adsorbed water layer, the positive edges of the clay particle may approach close enough to the negative face to from an effective electrostatic bond. This permits an open 'house of cards' structure, based on edge to face attraction (Figure 4.6c).

4.2.4 Asbestos

Asbestos is a group of six fibrous silicate minerals that occur naturally in metamorphic rocks:
- chrysotile, or white (serpentine) asbestos; its idealised chemical formula is $Mg_3(Si_2O_5)(OH)_4)$

- amosite, or brown (amphibole) asbestos, also called cummingtonite/grunerite: $Fe_7Si_8O_{22}(OH)_2$
- crocidolite, or blue (amphibole) asbestos, also called riebeckite: $Na_2Fe^{2+}{}_3Fe^{3+}{}_2Si_8O_{22}(OH)_2$
- tremolite: $Ca_2Mg_5Si_8O_{22}(OH)_2$
- actinolite: $Ca_2(Mg, Fe)_5(Si_8O_{22})(OH)_2$
- anthophyllite: $(Mg, Fe)_7Si_8O_{22}(OH)_2$.

Asbestos minerals are chemically inert and have a high tensile strength. They can be woven, are resistant to heat and fire and do not conduct electricity. Because of these properties, asbestos has been used in a variety of industrial products and commercial goods, including roofing shingles, ceiling and floor tiles, coatings, electrical insulation wire, textiles, and automobile brakes and clutches.

In its natural form, asbestos is known to be a non-toxic, harmless material. Its health hazard to humans is associated with its fibrous nature. When asbestos fibres become airborne, they can be inhaled deep into the lungs. Prolonged inhalation of asbestos fibres can cause a build-up of scar-like tissue in the lungs called asbestosis, which often results in loss of lung function that may progress to disability or death. Asbestos may also cause lung cancer or other diseases such as mesothelioma, a rare form of cancer that develops in the membrane lining of many of the internal organs. Although the amphibole asbestos types (amosite and crocidolite) are considered to be the most hazardous, exposure to all types of asbestos fibres is known to be a serious health hazard in humans. Heavy exposure tends to occur in asbestos mining, the construction industry and in ship repair, particularly during the removal of asbestos materials due to renovation, repairs, or demolition. In general, the greater the exposure to asbestos, the greater the chance of developing harmful health effects, but disease symptoms may take several years to develop. The European Union and Australia have banned all use of asbestos and extraction, manufacture and processing of asbestos products. In the United States, however, many consumer products may still legally contain trace amounts of asbestos.

Although human exposure to asbestos is mostly associated with indoor environments and construction sites where old material containing asbestos is present or is being processed, soil contaminated by asbestos may also pose a threat to human health. Abandoned industrial premises and former waste disposal sites are particularly likely to be contaminated by asbestos. At such sites, asbestos may be typically found on the surface or buried at shallow depth. Asbestos fibres in soil are not inherently hazardous to humans when left undisturbed. However, asbestos in soil may become airborne as a result of intense activity like digging, ploughing or vehicle movement. The fibres tend to be released more readily into the air from dry, coarse-grained, well-drained soil. Fibres are particularly likely to become airborne if the asbestos is present in the form of high concentrations of friable material. In general, hazardous concentrations of airborne asbestos fibres only occur in a limited radius (about 100 m) around the point of activity.

4.2.5 Nanomaterials

As noted in section 4.1, nanomaterials are particles ranging from 1 nm to 100 nm in size, and comprise nanoparticles, nanotubes, nanocapsules, and nanofibres. As a result of their extremely small sizes, nanomaterials have a large surface to volume ratio and their surface chemistry differs from that of their parent compounds. They also have distinctive optical and electronic properties. Based on their composition, they can be classified into carbon-based materials, metal-based materials, dendrimers and composites. As their name implies, carbon-based nanomaterials are composed mostly of carbon, usually taking the form of a hollow sphere, ellipsoid, or tube. Spherical and ellipsoidal carbon nanomaterials are called fullerenes, whereas cylindrical carbon nanomaterials are called nanotubes. Spherical fullerenes are also called buckyballs. Carbon-based nanomaterials have a variety of potential applications, including improved films and coatings, stronger and lighter materials, and electronic and medical applications. Metal-based nanomaterials include quantum dots, nanosilver, nanogold, and metal oxides such as titanium dioxide and zinc oxide. A quantum dot is a closely packed semiconductor crystal comprising hundreds or thousands of atoms with varying optical properties that depend on the size of the quantum dots. Examples of the application fields of metal-based nanomaterials are catalysts, solar cells, batteries, fuel cells, sunscreens, coatings, and nanopharmaceutical agents for chemotherapy. Dendrimers are nanosized polymers whose surface has numerous chain ends and some contain interior cavities. These forms can be tailored to carry out specific chemical functions. Current and potential applications can be found in catalysis and drug delivery. Nanoparticles can be combined with other nanoparticles or with larger materials to form composites, in order to enhance mechanical, thermal and electrical properties.

Despite their potential beneficial uses, nanoparticles also present potential hazards to human health and the environment. Most of these are associated with their large surface to volume ratio, which can make the particles very reactive. In addition, because they are so small, nanomaterials can pass through biological membranes: however, their interactions with biological systems are still poorly understood. The reactive properties of nanomaterials increase the probability that they could produce unanticipated toxicological effects. Some nanomaterials, such as carbon fullerenes, carbon nanotubes and nanoparticle metal oxides, are known to increase the production of free radicals, particularly of reactive oxygen species (ROS). This is one of the main mechanisms of nanoparticle toxicity to human tissue and cell cultures. The presence of reactive oxygen species may result in oxidative stress, inflammation, and consequent damage to proteins, membranes and DNA, which may ultimately result in cell death.

At present, the health risks and environmental impacts associated with nanomaterials remain largely unknown. That is why nanomaterials are often classified as so-called emerging substances of concern (ESOCs). Other types of ESOCs are described and discussed in Section 9.7.

4.3 ORGANIC COMPONENTS

4.3.1 Composition and formation

Organic matter in soil and water consists of living biomass (plants, animals, bacteria, algae, fungi, and viruses), undecayed dead plant and animal tissues (litter or *detritus*), and their organic transformation products. In soils, the living biomass makes up about 10 percent of the soil organic matter (SOM) on average. However, some scientists do not classify living biomass as organic matter. As soon as the living biomass dies, the detritus is decomposed by microorganisms (bacteria, fungi) to form *humic substances*, which are temporary but relatively resistant intermediate products remaining after considerable degradation. Humic substances are amorphous, colloidal, polymeric, dark-brown organic compounds with a high molecular weight. They are smaller than 2 μm and therefore have a large specific surface area. In general, humic substances make up 60 to 70 percent of the total soil organic matter. The term humus is frequently used as a synonym for humic substances. Unfortunately, here too the terminology is not consistent, since many soil and environmental scientists also use the term humus as a synonym for soil organic matter, i.e. all organic material in the soil including humic substances. The process of formation of humic substances is called *humification*, whereas *mineralisation* refers to the complete breakdown to inorganic compounds, mainly water and carbon dioxide, but also other minerals and nutrients that constitute an essential part of organic matter. Under natural conditions, the mineralisation of organic matter is a major source of nutrients in soil and water. The degree to which detritus is broken down depends on environmental factors such as the pH and redox potential and the accompanying microorganisms.

The composition of organic matter is extremely heterogeneous and consists of numerous organic compounds. For this reason there is no general chemical formula for organic matter. Nonetheless, all organic matter consists of between 45 and 55 percent carbon (C), with smaller amounts of oxygen (O) and hydrogen (H) plus small quantities of nitrogen (N), phosphorus (P), sulphur (S), chlorine (Cl), and several other elements. The carbon in organic matter is referred to as ***organic carbon***. The basic skeleton of organic compounds consists of carbon atoms bonded together into unbranched and branched chains or rings of various sizes with bonds to hydrogen atoms. Together with the other elements attached, they make up a wide range of organic-matter substances, such as lignins, phenols, carbohydrates (cellulose and sugars), proteins, lipids (fats), oils, and waxes. Because of their complex composition, humic substances are often classified according a classical method involving treatment with a sodium hydroxide solution. The portion of humus that is insoluble in the dilute sodium hydroxide is called humin. The part that dissolves in the solution consists of ***humic acids*** and ***fulvic acids***, but humic acids precipitate when the solution is made acidic. Fulvic acids are distinguished from humic acids by their lower molecular weight.

In surface waters, a general distinction is made between dissolved organic matter (DOM[1]) and particulate organic matter (POM). Naturally occurring POM consists largely of recalcitrant remains of woody terrestrial and aquatic plants; the POM concentration generally ranges between 10 and 20 mg l^{-1}. DOM is usually assumed to be approximately similar to the soluble organic matter present in soils. In general, the DOM concentration in rivers varies from about 5 mg l^{-1} in temperate, arid, and semi-arid regions, via about 10 mg l^{-1} in tropical regions, to about 40 mg l^{-1} in subarctic regions (Hem, 1989). Obviously, there is a wide range in DOM concentrations in any given stream, as well as substantial differences amongst streams. DOM concentrations in groundwater are usually smaller than those in surface waters. The presence of coloured DOM in surface water and groundwater may tinge the water from pale yellow to dark brown, although the colour has no direct link with the actual concentration of DOM. Coloured DOM consists largely of colloidal fulvic and humic acids and may significantly affect the water clarity. Among the environments in which intensively coloured water occurs are swamps and bogs.

Similar to clay minerals, the particulate organic matter in soil and water also has a negatively charged surface. Because of the small size and large specific surface area of the humic substances, organic matter is very reactive and is able to interact intensively with both the liquid phase and other solid phase constituents. Before discussing these interaction mechanisms, it is useful to make a few comments on the common chemical structure of organic compounds in order to better understand the nature of these interactions. As mentioned above, the carbon atoms are bonded together to make carbon skeletons consisting of chains and/or rings. The spatial structure of atoms in the carbon skeletons of organic molecules will be further elaborated upon in Chapter 7. In the context of interaction of organic matter with the liquid phase and other solid matter, the so-called functional groups bonded to the skeleton are important. These chemical subunits display a more or less similar behaviour in the variety of carbon skeletons to which they are attached. Figure 4.7 shows some common functional groups present in environmental organic compounds. The next

---

[1]   Some researchers in carbon cycling use DOM as an abbreviation for dead organic matter in or on top of the mineral soil.

Chemical structure	Name	Alternative names
R – OH	Hydroxy group	Alcohol, phenol if R=
R – SH	Mercapto group	Thiol, mercaptan
$R_1 – O – R_2$	Ether	
$R_1 – S – R_2$	Sulphide	Thioether
$R_1 – N$ R_2 R_3	Amino group	Primary amine if $R_2=R_3$ =H, secondary amine if R_3=H, tertiary amine, aniline if R_1=
$R_1 – C – R_2$ (=O)	Carbonyl group	Ketone, aldehyde if R_2=H
$R_1 – C – OH$ (=O)	Carboxy group	Carboxylic acid
$R_1 – C – O – R_2$ (=O)	Ester	Carboxylic acid ester
$R_1 – C – S – H$ (=O)	Thioester	Carboxylic acid thioester
$R_1 – C – N$ (=O) R_2 R_3	Amide	
$R – C \equiv N$	Cyano group	Nitrile
$R_1 – N^{\oplus}$ (=O) $O^{\ominus}$	Nitro group	

Figure 4.7 Some common functional groups of environmental organic chemicals. If not otherwise indicated, R represents a carbon-centred substituent (modified from Schwarzenbach et al., 1993).

section discusses the relevance of these groups for the interaction between organic matter and the liquid phase.

4.3.2 Interaction with the aqueous phase

A number of the characteristic groups of the polymer chains depicted in Figure 4.7 contain OH, notably the hydroxyl (-OH) and carboxyl (-COOH) groups. These functional groups, which are weakly to moderately acidic, are the most active. Most of them may dissociate depending on the pH and the total ion concentration in the bulk solution. Thus, the surface charge of organic matter constituents is negative and variable, very much as in the case of the variable surface charge of sesquioxides and clay minerals. Consequently, cation exchange may also occur and organic matter may contribute considerably to the total CEC of the soil or sediment. The CEC of colloidal organic matter typically ranges between 100 and 300 meq/100 g (Miller and Gardiner, 2004).

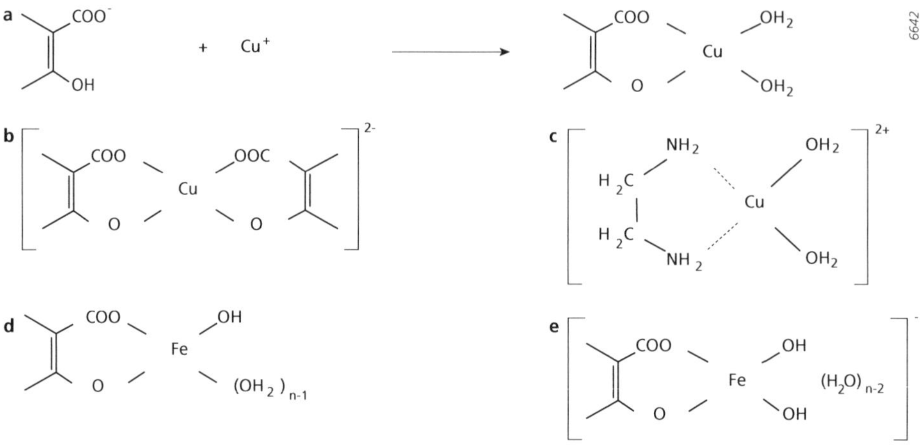

Figure 4.8 Chelate complexes of copper and iron with carboxyl and phenol groups of fulvic acids and amino acids. Adapted from Scheffer and Schachtschabel (1989).

Another important property of the chemical structure of the polymer chains of humic substances is the particular combination of different active groups that are able to form complexes with certain metals. When an organic substance bonds to a metal by two or more contacts, these complexing organic substances are usually referred to as ***ligands*** and the ligand plus complexed metal is called ***chelate***. Examples of natural ligands are amino acids, citric acid, polyphenols, and fulvic and humic acids. One well-known synthetic ligand is EDTA (ethylenediaminetetraacetic acid). Ligands form cyclic structures by bonding to the metal (see Figure 4.8), which makes the chelate very stable; the accordingly complexed metals are not exchangeable like metals regularly adsorbed to the surfaces of organic matter, sesquioxides, or clay minerals. The stability of a chelate is determined by the chelating ligand, the total ion concentration and the pH in the bulk solution, and the metal ion. In general, the chelate stability increases with increasing pH and decreases with increasing ion concentration in the bulk solution. In addition, the chelate stability is greater with metals in approximately the following order (Scheffer and Schachtschabel, 1989):

$$Fe^{3+} > Al^{3+} > Cu^{2+} > Pb^{2+} > Fe^{2+} > Ni^{2+} > Co^{2+} > Cd^{2+} > Zn^{2+} > Mn^{2+} > Ca^{2+} > Mg^{2+}$$

Some of the metal–organic complexes are very soluble in water. The solubility and, accordingly, the environmental mobility of Fe and Al, trace metals (e.g. manganese (Mn), copper (Cu), and zinc (Zn)) and other potentially toxic heavy metals (e.g. cadmium (Cd), lead (Pb), and mercury (Hg)) may be considerably enhanced by the formation of such complexes. Therefore, complexing organic substances are crucial in the supply of micronutrients to plants, the bioavailability of heavy metals, and the leaching of metals in soils.

4.3.3 Interaction with mineral surfaces

Besides interaction with the liquid phase through cation exchange and formation of chelates, organic matter also tends to interact strongly with mineral surfaces such as those of quartz grains, sesquioxides, and clay minerals. This causes almost all mineral particle surfaces to be covered with an organic coating. Because of their large specific surface area, clay minerals sorb most organic matter. Accordingly, the organic matter content in soil or sediment is usually proportional to the clay content.

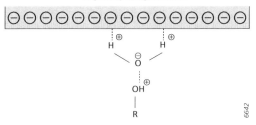

Figure 4.9 Schematic diagram of a hydrogen bond between a hydroxyl group and a mineral surface.

The sorption of organic molecules to mineral surfaces occurs via various mechanisms. Positively charged organic solutes, such as for example amino acids (at pH values below *PZC*) and polypeptides, are readily removed from the dissolved phase by cation exchange. The active functional hydroxyl or carboxyl groups have a dipole character like water molecules, which means that the centres of positive and negative charge do not coincide. The H atoms of the functional groups and water molecules possess a significant amount of positive charge and the O atoms a negative charge. This leads to the formation of so-called hydrogen bonds between the H atoms and the negatively charged mineral surface, often via H_2O molecules (see Figure 4.9). Large humic molecules are often attached to mineral surfaces by multiple hydrogen bonds.

The sorption of organic compounds changes the sorption properties of the mineral surface. If the organic molecule sorbs in a position that leaves its active functional groups or chelating sites exposed, it will tend to take cations from solution to the particulate phase. If it sorbs to the mineral phase in such a way that its reactive sites are obstructed, it becomes less effective at capturing cations. Besides the polar regions as explained above, large organic molecules often also have non-polar regions. If these large molecules bind to minerals, the mineral surface becomes less polar. This makes the mineral surface a better sorbent for non-polar or hydrophobic organic pollutants. This sorption mechanism will be discussed further in Chapter 9.

The presence of these organic coatings is particularly significant because it enhances the cohesion between mineral particles. This leads to the formation of stable aggregates in soil. In surface water, the organic–mineral complexes promote the flocculation process (Johnson *et al.*, 1994). Note too that living microorganisms also contribute actively to the process of coagulation of soil or sediment through feeding. The aggregates thus formed take the form of faecal pellets. In groundwater and on lake and river beds, microorganisms also form so-called **biofilms** that consist of a consortium of bacteria, algae, and fungi, embedded in an extracellular polysaccharide matrix that acts as glue (Lock, 1994). Biofilms form slimy coatings on coarse materials such as sand, gravel, and even pebbles, and may contain inclusions of inorganic particles. The biofilm may be eroded from the lake or river bed and so form fragile organic aggregates. While these flocs are settling, they continually collide with and capture smaller particles and DOM, and therefore grow bigger (Van Leussen, 1988).

4.3.4 Decomposition of organic matter

Organic matter is decomposed through oxidation, which involves numerous microorganisms that break down the organic molecules through multiple enzymatic reactions (Van Cappellen and Wang, 1995). It was shown above that these reactions produce a number of intermediate organic compounds (e.g. humic and fulvic acids) but ultimately, the organic matter is fully decomposed or mineralised to carbon dioxide and water, plus some other inorganic minerals.

However, the decay rates of the various organic compounds may differ considerably and the complete breakdown may be a very slow process taking several decades or – in the case of recalcitrant humic substances – thousands of years. Because of its complex composition, there is no general chemical formula for organic matter. However, it is possible to write a general simplified equation of the reaction, which only involves the initial reactants and the final products:

$$CH_2O + e.a. \quad \rightarrow \quad CO_2 + H_2O + \text{inorganic N, P, S} \tag{4.2}$$

where *CH₂O* refers to organic matter and *e.a.* to the various electron acceptors (oxidants) available in the system (e.g. oxygen (O_2), nitrate (NO_3^-), and sulphate (SO_4^{2-})). In well-aerated soils and surface waters, oxygen is the predominant electron acceptor:

$$CH_2O + O_2. \quad \rightarrow \quad CO_2 + H_2O \quad \leftrightarrow \quad HCO_3^- + H^+ \tag{4.3}$$

In many environments, however, the amount of oxygen is limited, and the other oxidants are used consecutively. Figure 4.10 shows the sequence of redox reactions which operate at successively lower redox potentials. Firstly, if oxygen is depleted, nitrate-reducing microbe species take over the decomposition of organic matter, using nitrate as oxidant. Nitrate is reduced via nitrite (NO_2^-), nitric oxide (NO_x), and nitrous oxide (N_2O) to free nitrogen (N_2). This process is called ***denitrification*** (see also Section 6.2) and this overall transformation can be written as:

$$CH_2O + 0.8\,NO_3^- \quad \rightarrow \quad HCO_3^- + 0.4\,N_2 + 0.4\,H_2O + 0.2\,H^+ \tag{4.4}$$

Not all nitrate is transformed into free nitrogen gas, because both nitric oxide) and nitrous oxide) are also gases, which can escape prematurely from the system. If nitrate has been depleted, manganese-reducing bacteria use Mn(IV) as the principal electron acceptor (oxidant):

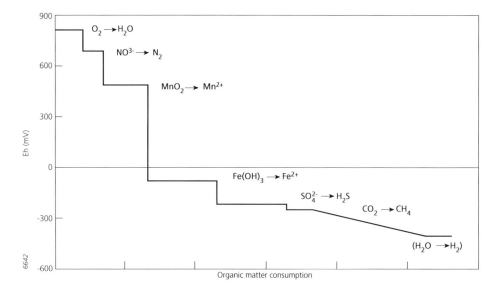

Figure 4.10 Evolution of the redox potential (Eh) as function of organic matter consumption.

$$CH_2O + 3\,H^+ + 2\,MnO_2 \quad \rightarrow \quad 2\,Mn^{2+} + HCO_3^- + 2\,H_2O \tag{4.5}$$

Next, iron-reducing bacteria reduce Fe(III) to Fe(II):

$$CH_2O + 7\,H^+ + 4\,Fe(OH)_3 \quad \rightarrow \quad 4\,Fe^{2+} + HCO_3^- + 10\,H_2O \tag{4.6}$$

Subsequently, other bacterial species use sulphate as oxidant and reduce sulphate to sulphide:

$$2CH_2O + SO_4^{2-} \quad \rightarrow \quad 2HCO_3^- + H_2S \quad \leftrightarrow \quad 2HCO_3^- + 2\,H^+ + S^{2-} \tag{4.7}$$

The hydrogen sulphide, a colourless gas with a rotten-egg odour, may volatilise from the system or dissociate. Many sulphides are barely soluble in water, so will precipitate soon as a metal sulphide (see Section 5.11). Ultimately, if sulphate is also used, methane-producing bacteria continue the decomposition under anaerobic conditions:

$$2CH_2O \quad \rightarrow \quad CO_2 + CH_4 \tag{4.8}$$

The methane thus produced is also known as swamp gas. The above sequence of reactions is accompanied by a decrease of the redox potential (Figure 4.10), which can be observed as step-wise gradients in redox potential, reactants, and reaction products, in, for instance, water-saturated organic soils, bed sediments of rivers and lakes, and along groundwater flow paths.

It was mentioned above that the decomposition rate of organic matter may vary widely, depending on the nature of the organic compounds. The breakdown of organic matter adsorbed to mineral surfaces is slower. Moreover, the decomposition rate of organic matter is governed by environmental conditions such as the nature and concentration of the oxidant, temperature, and pH. The decomposition of organic matter under oxic conditions proceeds faster and produces more energy for the bacteria than decomposition in anoxic environments. In general, decomposition rates increase with increasing temperature and increasing pH, though the pH effect may be small at pH above 5 (Scheffer and Schachtschabel, 1989). The rate of organic matter decomposition is often largely controlled by the nitrogen content. Microbes need nitrogen to build proteins, and nitrogen is often a limiting element for their growth. In this context, the mass ratio of the total nitrogen to organic carbon or **C:N ratio** is an important parameter that determines the decomposition rate. Bacteria require a C:N ratio of about 6:1; detritus with C:N ratios of 20:1 or less (e.g. young green leaves, algal detritus) has sufficient nitrogen for a relatively fast decomposition. Detritus with C:N ratios of more than 30:1 (e.g. straw, pine needles) decomposes slowly.

In surface waters, the decomposition of organic matter causes the dissolved oxygen concentration to decrease. This may adversely affect the aquatic ecosystem, especially if the organic matter is from an anthropogenic source (e.g. effluents from wastewater treatment plants or discharges of untreated sewage water). The depletion of oxygen is usually measured using a standard **biochemical oxygen demand** (BOD) test. The result of this test indicates the amount of dissolved oxygen expressed in g m^{-3} used up by a water sample when incubated in darkness at 20 °C for five days, and is a measure of the concentration of easily biodegradable organic matter (both dissolved and particulate). In fact, the BOD cannot be fully attributed to the decay of organic matter, because other oxidisable substances also contribute to the BOD: oxidisable nitrogen (ammonium NH_4^+; see Section 6.2) in particular. Therefore, a distinction is often made between the biochemical oxygen demand of the carbonaceous matter (CBOD) and that of the nitrogenous matter (NBOD). A BOD test may be supplemented by a **chemical oxygen demand** (COD) test, which measures the amount of oxygen consumed in the complete oxidation of carbonaceous matter, thus also the non-biodegradable matter. In rivers and lakes, a significant part of the oxygen consumption is due to the decomposition of organic matter deposited in the bed sediments. The ***sediment oxygen demand*** (SOD) is the rate of the dissolved oxygen consumption in a water body (river, lake or ocean) due to the decomposition of sediment organic matter. The SOD is expressed in g O_2 m^{-2} d^{-1}. The BOD, COD, and SOD are often important parameters of the dissolved oxygen budget of surface waters and their determination provides crucial information for adequate water quality control.

4.4 SORPTION BY SOILS AND SEDIMENTS

In the previous sections, it was observed that both the inorganic and the organic components in soils and sediments have or can have negatively charged surface sites and so are able to adsorb cations. The CEC of clay minerals varies from less than 10 meq/100 g (kaolinite) to more than 100 meq/100 g (montmorillonite and vermiculite) (see Table 4.2) and the CEC of organic matter ranges between 100 and 300 meq/100 g. The CEC of sesquioxides typically has a much smaller value and is commonly less than 3 meq/100 g soil. All these constituents thus contribute to the CEC of the bulk soil. Obviously, the CEC of a soil is closely related to its organic matter and clay content. Table 4.3 lists some typical values of the CEC for various soils.

The cation exchange capacity is an important property of soils and sediments in governing the cycling and retention of nutrients and pollutants transported by water. The properties and occurrence of the various cations adsorbed to the cation exchange sites are discussed in Chapters 5 through 7, but in general, the concentrations of the cations in the bulk solution tends to be in equilibrium with the amounts on the CEC. The exchangeable potassium (K^+), calcium (Ca^{2+}), magnesium (Mg^{2+}), and ammonium (NH_4^+) are major nutrient sources for plants. The losses of these nutrients by leaching are substantially retarded because they are retained on the cation exchange sites. Moreover, metals (e.g. cadmium (Cd^{2+}), zinc (Zn^{2+}), nickel (Ni^{2+}), lead (Pb^{2+}), and copper (Cu^{2+})) are largely removed from the dissolved phase, as they are adsorbed to the cation exchange sites. The actual cation composition that is adsorbed to the exchange complex depends on the chemical composition

Table 4.3 Cation exchange capacity and base saturation in the topsoil of various soil types in the Netherlands and other parts of the world (source: De Bakker, 1979 (Dutch soils); Scheffer and Schachtschabel, 1989 (other soils)).

Soil*	pH	Clay	Organic C	CEC	Saturation (%)				
		(%)	(%)	(meq/100 g)	Na$^+$	K$^+$	Mg^{2+}	Ca^{2+}	H$^+$/Al^{3+}
Histosol (Netherlands)	5.2	54	11.3	61.5	1.5	1.6	6.0	67.6	23.3
Entisol (Netherlands)	7.5	24	0.5	18.1	3.3	1.7	3.9	87.8	3.3
Entisol (Netherlands)	5.9	68	1.3	46.5	0.9	1.1	9.5	82.5	5.6
Inceptisol (Netherlands)	7.7	24	0.9	20.7	1.5	0.5	5.4	90.8	1.8
Alfisol (Netherlands)	3.8	9	2.0	8.3	0.0	0.0	1.2	9.6	89.2
Spodosol (Germany)	2.6	n.d.	11.7	6.8	2.6	4.6	6	22	65
Vertisol (Sudan)	6.8	n.d.	0.9	47	3.8	0.4	25	71	0
Andisol (Hawaii)	4.5	n.d.	11.7	13.3	2.2	3.8	20	71	3.7
Oxisol (Brazil)	3.5	n.d.	2.8	2.6	1.2	3.1	3.5	2.7	89
Ultisol (Puerto Rico)	3.5	n.d.	3.3	7.2	1.4	2.8	8.3	15	72
Aridisol (Arizona, USA)	9.9	n.d.	0.4	36.4	47	2.5	5.5	45	0

* Classification according to the orders of the Soil Taxonomy (USDA, 1999).

of the bulk solution. In general, calcium (Ca^{2+}) ions dominate the CEC, except in acidic and saline soils. In acidic soils, the exchange complex is largely occupied by protons (H^+) and aluminium species ($Al(OH)_2^+$ and/or Al^{3+}). In saline soils, sodium (Na^+) occupies a considerable part of the CEC. The exchangeable Ca^{2+}, Mg^{2+}, Na^+, and K^+ ions are often called the exchangeable bases, because soils with the CEC fully saturated with these cations react as a base by accepting H^+ and so neutralise the pH of the bulk solution. The proportion of these cations on the CEC is called the ***base saturation***. The base saturation of typical soils in the Netherlands and other parts of the world is indicated in Table 4.3.

EXERCISES

1. Define the following terms:
   a. Colloidal particles
   b. Nanomaterials
   c. Asbestos
   d. Sesquioxides
   e. Diffuse layer
   f. CEC
   g. Base saturation
   h Humic substance
   i. DOM
   j. Chelate
   k. Biofilm
   l. BOD

2. Name the two principal mechanisms responsible for the surface charge of solid particles.

3. Given the *PZC* of the following materials:

Material	PZC
Al_2O_3	9.1
$Al(OH)_3$	5.0
Fe_2O_3	6.7
$Fe(OH)_3$	8.5
Quartz (SiO_2), feldspars	2.0–2.4
Clay minerals: kaolinite	4.6
montmorillonite	2.5

   a. Indicate whether the surface charge of the above minerals is negative or positive at pH = 5.4.
   b. What will happen if the pH increases to 6.8?

4. Why is the CEC of quartz and feldspars much smaller than the CEC of sesquioxides and clay minerals?

5. Describe the differences in structure and physico-chemical properties of 1:1 and 2:1 clay minerals.

6. Briefly describe the process of K fixation in illitic clay minerals.

7. Explain why clay particles tend to flocculate as the salinity of the water increases.

8. What is the difference between decomposition and mineralisation of organic matter?

9. The decomposition of organic matter in lake bed sediments leads to stratified layers that are characterised by the dominant redox reaction. The thickness of these layers is in the order of a few centimetres in organic-rich bed sediments.
    a. Explain why organic matter decomposition leads to the formation of such layers.
    b. Which substances act as oxidants in the successive layers?
    c. How can these layers be recognised in terms of the presence or oxidants of reductants or reaction products?

10. Describe the role of the C:N ratio for the decomposition of organic matter.

11. Describe how the base saturation changes as the pH of a soil decreases.

5

Major dissolved phase constituents

5.1 INTRODUCTION

The major dissolved phase constituents include those substances that are abundant as dissolved ions in natural waters, namely calcium (Ca^{2+}), magnesium (Mg^{2+}), sodium (Na^+), potassium (K^+), iron (Fe^{2+}), manganese (Mn^{2+}), carbonate species (H_2CO_3, HCO_3^-, and CO_3^{2-}), chloride (Cl^-), and sulphate (SO_4^{2-}). Together with dissolved silica (SiO_2), these ions generally account for most of the total dissolved solids (TDS) in subsurface and surface waters. Aluminium (Al^{3+}) may also occur as a dissolved ion, but only under exceptionally acid conditions. Since they occur in natural waters (see Section 1.2.2) these major dissolved constituents are usually not considered as pollutants. Nevertheless, they may cause contamination of soil and water, because they are often released into the environment together with other polluting substances. Consequently, the major dissolved constituents may be an important indicator of pollution when they occur in abnormal concentrations, so knowledge of their behaviour and patterns may help when interpreting the occurrence and dispersal of other, more harmful, pollutants.

Rainwater contains very few dissolved ions, most of which originate from atmospheric gases (e.g. CO_2, SO_2, and HCl) and from aerosols originating from volcanic emissions and biochemical emissions from soil and water. Dissolution of these gases and aerosols enhances the concentrations of Na, carbonate species, SO_4, and Cl, and results in a lowering of the pH of rainwater. Moreover, salt spray from seas and oceans brings substantial amounts of Na and Cl into the atmosphere. These substances may also reach the Earth's surface independent of rainfall, by dry deposition of particulate aerosols and adsorption of atmospheric gases (see also Section 6.2.3). As soon as rainwater infiltrates into soil, its composition is affected by exchange processes between water and soil material through dissolution, sorption, plant uptake, and concentration due to evapotranspiration. In general, the main source of ions in groundwater and river water is the weathering and dissolution of rocks and minerals. Some minerals, such as carbonates (e.g. chalk, limestone, dolomite, and marl) and evaporite minerals (e.g. rock salt or halite, gypsum) dissolve readily in water; others, such as silicates (e.g. quartz, feldspars, clay minerals) dissolve much more slowly. Therefore, the composition of water reflects both the geological setting in the drainage basin and the residence time. Table 5.1 gives an overview of the main sources and typical concentration ranges of the major dissolved constituents in unpolluted rainwater and fresh water.

Another factor in addition to the geology of the drainage basin that governs the water composition is the total amount of annual precipitation, since it determines the degree of dilution. Figure 5.1 shows this effect for typical water composition as function of mean annual runoff from US drainage basins with different geology (Walling, 1980). This figure illustrates that the TDS concentrations decrease sharply with mean annual runoff in catchments dominated by granite, sandstone, shale, schist, and gneiss bedrock. This indicates that the weathering rates of these rocks are in the order of the residence time of the water within the catchment. In catchments dominated by sand and gravel, or volcanic

Table 5.1 Typical concentration ranges of major dissolved constituents in unpolluted fresh water and their sources (source: Appelo and Postma, 1996).

Dissolved constituent	Typical concentration range mmol l⁻¹	mg l⁻¹	Source
Na^+	0.1–2	0.2–46	Feldspar, rock salt, zeolite, atmosphere
K^+	0.01– 0.2	0.4–8	Feldspar, micas
Ca^{2+}	0.05–5	2–200	Limestone, gypsum feldspar, pyroxene, amphibole
Mg^{2+}	0.05–2	1.2–48	Dolomite, serpentine, pyroxene, amphibole, olivine, micas
Cl^-	0.05–2	1.8–70	Rock salt, atmosphere
HCO_3^-	0–5	0–305	Limestone, organic matter
SO_4^-	0.01–5	1–480	Gypsum, sulphides, atmosphere
Fe^{2+}	0–0.5	0–28	Silicates, goethite, haematite, siderite, pyrite
SiO_2	0.02–1	1.2– 60	Silicates

rocks, or limestone, the maximum TDS concentration depends very little on mean annual runoff. Sand and gravel are very resistant to weathering, so the TDS concentrations remain small. In this case, the runoff water composition is largely regulated by rainwater chemistry, evapotranspiration, and plant uptake. In contrast, volcanic rocks and limestone (especially the latter) are much more prone to weathering and the weathering rates are fast relative to the residence time of water; the TDS concentrations are therefore larger than for sand and gravel, and almost independent of mean annual runoff.

Since the constituents listed in Table 5.1 comprise the majority of the dissolved ions in water they should – by approximation – obey the principle of electroneutrality, which means that the total charges of the cations and anions balance each other. This principle thus

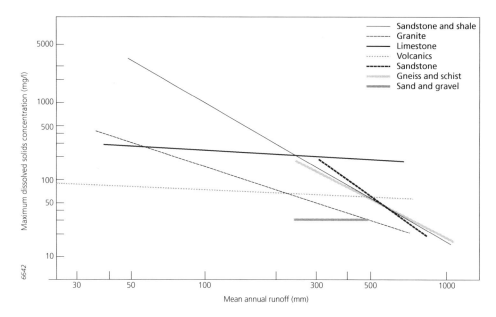

Figure 5.1 Typical dissolved constituents concentrations as function of mean annual runoff from US drainage basins. Adapted from Walling (1980).

provides an opportunity for a first check of the analysis of water composition, by calculating the charge imbalance using:

$$IB = (Cations - Anions)/(Cations + Anions) \tag{5.1}$$

where IB = ionic charge imbalance [–] (usually expressed as percent); *Cations* and *Anions* refer to the respective concentrations of cations and anions expressed in milliequivalents per unit volume (meq l⁻¹). Due to analytical errors in laboratories, ionic charge imbalances of up to ± 2 percent are usual. However, water samples with an absolute charge imbalance greater than 5 percent are considered to be inaccurate. Note that some ions other than the cations and anions mentioned above may contribute significantly, such as NO_3^- in oxic groundwater, NH_4^+ in reduced groundwater, or H^+ and Al^{3+} in acid water. Obviously, if this is the case, these ions should also be considered when calculating the charge imbalance.

The tabulated analysis results are often insufficient for further interpretation of the water composition of different samples, because it is difficult to obtain an immediate overview over the data. Therefore, to facilitate the interpretation of water composition, the concentrations in water are often displayed graphically. Common useful diagrams for the presentation of analysis results of water composition are Stiff and Piper diagrams. Figures 5.2 and 5.3 show some examples of these diagrams displaying the composition of various European bottled mineral waters as tabulated in Table 5.2. Stiff diagrams are constructed by expressing the concentrations of cations and anions in milliequivalents per litre, which are subsequently plotted on three axes as shown in Figure 5.3. The advantage of Stiff diagrams is that the different water types become manifest in different shapes of the diagram, which can easily be recognised. Furthermore, the absolute concentrations are visualised by the width of the graph. The Piper diagram consists of two triangular graphs displaying the relative contribution of various cations and anions to the respective total positive and negative

Table 5.2 Composition of ten bottled mineral waters from Europe (source: Van der Perk and De Groot, 2013).

No.	Brand	Location	Ca^{2+} (mg l⁻¹)	Mg^{2+} (mg l⁻¹)	Na^+ (mg l⁻¹)	K^+ (mg l⁻¹)	HCO_3^- (mg l⁻¹)	Cl^- (mg l⁻¹)	SO_4^{2-} (mg l⁻¹)	NO_3^- (mg l⁻¹)
1	Contrex	Contrexeville, France	467	84	7	3	377	7	1192	n/a
2	Gerolsteiner Sprudel	Gerolstein, Germany	347	108	119	11	1817	40	36	n/a
3	Kaiserbrunnen	Aachen, Germany	62	9	1295	69	876	1486	277	n/a
4	Lete	Pratella, Italy	330	11.1	5.1	2.4	1047	12.2	7.1	3.9
5	Monchique	Monchique, Portugal	1.2	0.1	111	1.9	87.8	39.7	53.2	n/a
6	Parot	St-Romain-le-Puy, France	100	71	635	82	2507	82	47	n/a
7	Ramlösa	Helsingborg, Sweden	2.2	0.5	220	1.5	548	23	7.3	n/a
8	Rogaska	Slatina, Slovenia	398	859	1170	22	6605	50.8	1574	0.5
9	Sourcy	Bunnik, Netherlands	40	3.1	10	0.8	135	12	0.7	0.3
10	Spa Barisart	Spa, Belgium	5.5	1.5	5	0.5	18	5.5	7.5	1.5

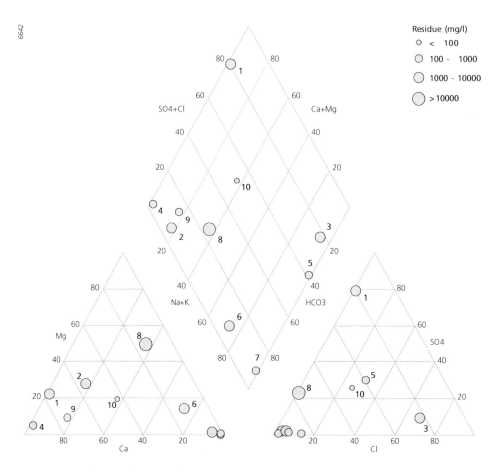

Figure 5.2 Piper diagram of ten bottled mineral waters from Europe (see Table 5.2) (source: Van der Perk and De Groot, 2013).

charge, and a diamond graph combining these contributions of cations and anions. A Piper diagram is constructed by first calculating the relative contributions to the total charge of both cations and anions and plotting them in the triangular graphs. Subsequently, the data points from both triangles are extrapolated to the diamond by following lines parallel to the outer boundary until they intersect. The size of the point in the triangles and diamond represents the total amount of dissolved solids or the electrical conductivity of the sample. The position of the point in the diamond reveals the general water type.

In the subsequent sections of this chapter, the sources, environmental role, and some important features of the chemical behaviour of the major solutes plus aluminium in fresh water are discussed.

5.2 SODIUM

Sodium (Na) is the most abundant of the alkali metal group (see periodic table of elements; Appendix I). Like all alkali metals, all Na occurs in the 1+ oxidation state (Na^+) under natural conditions. Sodium is one of the useful and necessary elements for plants and animals; it

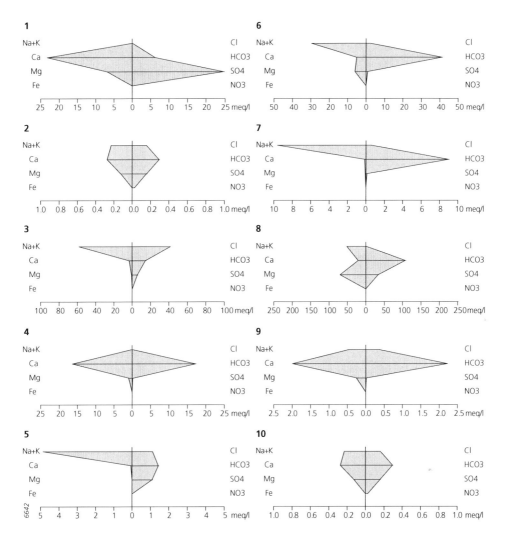

Figure 5.3 Stiff diagrams of ten bottled mineral waters from Europe (see Table 5.2) (source: Van der Perk and De Groot, 2013).

is particularly important for maintaining osmotic pressure in the cells. It is essential for the exchange of water between the cells and the intercellular medium. Although Na is an essential element, it is toxic in large concentrations or amounts. In animals and humans, chronic excess intake of Na leads to high blood pressure, which can result in serious health issues in the long term. The tolerance of different organisms to Na displays a broad range, but the tolerance of individual species may be limited. This implies that a structural increase in Na concentrations will bring about changes in species composition. Drinking water containing elevated Na concentrations of more than about 200 mg l^{-1} has an objectionable salty taste.

Sodium is derived from weathering of Na-bearing feldspars present in igneous rocks (e.g. albite, $NaAlSi_3O_8$) or dissolution of evaporite sediments. In addition, in coastal areas, seawater and brackish water may particularly be a significant source of Na, either via direct input as a result of flooding or via atmospheric input (salt spray). Because the major Na source is sea salt (NaCl), either in seawater or in evaporite minerals, the Na concentration is often strongly proportional to the chloride concentration.

In soils and closed surface water bodies (i.e. lakes without an outlet) enrichment of Na in solution may occur through evaporation. This process of salt build-up (called salinisation) occurs most intensively under natural conditions in arid and semi-arid regions. Salinisation can be amplified on irrigated arable land when the irrigation water is reused. Other human activities can also alter the Na concentrations in surface water and groundwater. The most widespread dispersal of Na in the environment occurs through the application of road salt for de-icing roads during winter periods. The release of municipal and industrial wastewater generally causes elevated Na concentrations downstream of the discharges. Furthermore, releases of mining wastewater and brines pumped or flowing from oil wells can have noticeable regional effects on Na concentrations in surface water or groundwater. In coastal areas, the pumping of water destined for drinking water or for maintaining groundwater levels in low-lying polder regions may cause intrusion of salt or brackish groundwater into freshwater aquifers.

The environmental reactivity of Na is limited. Sodium is readily soluble in water and there are no important precipitation reactions that control the Na concentrations in natural waters. Furthermore, it does not participate in redox reactions, since – as mentioned above – all Na is in the 1+ oxidation state. Because Na is a cation, it participates in sorption reactions at the surfaces of minerals and organic matter, especially of minerals with a large cation exchange capacity such as clay minerals. However, the affinity of the mineral surfaces for Na, a monovalent ion, is much less than for divalent ions. Accordingly, Na is quite mobile in soil and water.

5.3 POTASSIUM

Like Na, potassium (K) is an alkali metal and occurs only in the 1+ oxidation state under natural conditions. Potassium is an essential nutrient for both plants and animals: it is needed for cell division and is involved in various enzyme reactions. Like Na, K plays an important role in the regulation of water in the cells. It maintains the osmotic pressure and so helps to prevent dehydration and excess fluid retention. In animals, K is also responsible for the transmission of nerve impulses and for muscle contractions. Compared to other nutrients (especially nitrogen and phosphorus; see Chapter 6), K shows the largest concentrations in living tissues.

The principal sources of K in soil and water are weathering of K-bearing minerals, decomposition of organic detritus, and fertiliser application. Potassium is slightly less common than sodium in igneous rocks but more abundant in sedimentary rocks. The main K-containing minerals of silicate rocks are the K-bearing feldspars orthoclase and microcline ($KAlSi_3O_8$) and various micas (muscovite and biotite). The weathering rate of the K-feldspars is, however, low compared to other feldspars (Hem, 1989). In sedimentary rocks, K is present in unaltered grains of K-feldspars and micas and in clay minerals, especially illite. The total K content in soils mostly ranges between 0.2 and 3.3 percent (Scheffer and Schachtschabel, 1989). As demonstrated in Section 4.2.3, K exhibits a strong tendency to be incorporated in the lattice structure of illite. Therefore, deposition of fine sediments including illitic clay minerals is an important source of K in floodplain areas. The major anthropogenic source of K is the application of fertilisers. Because K is an essential nutrient, fertilisers contain a considerable amount of K. On arable land, potassium fertilisers are applied to up to 300 kg K ha^{-1} depending on the crop type and the soil characteristics, especially the contents of exchangeable K and clay.

In contrast to the K-containing minerals, potassium salts are readily soluble in water, so are not subject to precipitation–dissolution reactions, except when water evaporates under semi-arid and arid conditions and salts are left behind. The major mechanisms that control K concentrations in soil and water are sorption to clay minerals and plant uptake. Adsorption of K to organic matter plays only a minor role in K retention. Due to its strong affinity with illitic clay minerals, K is partly irreversibly retained in sediments (K fixation). In general, there is equilibrium between the following pools of K:

$$\text{Dissolved } K^+ \quad \leftrightarrow \quad \text{Exchangable } K^+ \quad \leftrightarrow \quad \text{Fixed } K \tag{5.2}$$

The equilibrium between dissolved and exchangeable (adsorbed) K is reached relatively fast, whereas the equilibrium between K adsorbed to the clay mineral surfaces and the K fixed between the clay lattices is reached much more slowly. The exchangeable K content in soil generally increases with increasing clay content. In acid, sandy soils the exchangeable K content only amounts to a few mg K per 100 g soil; in basic, clayey soils it is up to about 100 mg K per 100 g soil (= 0.1 percent). Because of the fixation of K in clay minerals, K seldom occurs in very large concentrations in natural waters and its concentration is usually much lower than the concentration of Na. Because of the relatively strong K sorption in soil, K generally moves slowly through the soil.

The second important factor (after sorption and fixation of K to clay minerals) that controls the availability of K for solution in subsurface water is the biological cycling of K. The potassium taken up by plants comes from the exchangeable and dissolved K pools. The fixed K pools are generally not available for uptake by plants. When exchangeable and dissolved K are available in sufficiently large amounts, luxury consumption by plants may occur, which means that plants take up K in excess amounts. The average K content of living plant material is about 0.3 percent, but concentrations in dry plant material and ash are considerably greater (1.5–4 percent). Potassium assimilated by plants becomes available immediately after the plants die or when leaves and other parts are shed at the end of the growing season. The K present in the litter or detritus is readily released in dissolved form during the dormant season. The released K leaches into the soil where it is adsorbed to cation exchange sites of clay minerals and recycled during the subsequent growing season. A small part of the K may be lost from the soil due to leaching to groundwater or runoff to surface water. In agricultural soils, part of the K is exported by crop harvesting and removal of crop residues.

5.4 CALCIUM

Calcium (Ca) is the most abundant of the alkaline-earth metals; under natural conditions it occurs only in the 2+ oxidation state. Calcium is an essential element for plant and animal life forms: it is needed for the physical integrity and normal functioning of cell membranes and it plays a vital role in cells that are dividing. In animals and humans, Ca is also used to build bones and teeth and to control muscle activity and transmission of nerve signals. In addition, it plays a role in the coagulation of the blood and the regulation heart activity. Calcium is non-toxic, but large concentrations in drinking water are considered as undesirable, because it causes the formation of boiler scale (i.e. calcareous deposits) in much household equipment such as boilers, washing machines, and kettles. The sum of the ions which can precipitate as 'hard particles' from water is called the ***hardness***. The hardness is usually reported as mg $CaCO_3$ per litre or meq/l or in hardness degrees. One German degree corresponds to 17.8 mg $CaCO_3$/l and 1 French degree corresponds to 10 mg $CaCO_3$/l. The temporary hardness is the part of Ca^{2+} and Mg^{2+} ions which are balanced by HCO_3^- and can thus precipitate as carbonate; the permanent hardness is the part of Ca^{2+} and Mg^{2+} ions in excess of HCO_3^-. To avoid the formation of boiler scale, the hardness of drinking water is usually kept below 150–200 mg l^{-1}.

Calcium is an important constituent of many igneous rock minerals, especially plagioclase, pyroxene, and amphibole. The weathering of igneous rocks is slow, so the Ca concentrations in water that has been in contact with igneous rocks are generally low. Furthermore, calcium is commonly found in many sedimentary rocks in the form of calcite and aragonite (both $CaCO_3$), dolomite (Ca, $MgCO_3$), gypsum ($CaSO_4 \cdot 2H_2O$), or anhydrite ($CaSO_4$). Limestone and chalk consist mostly of calcium carbonate (calcite) and calcium carbonate is also a major cementing agent between the particles of sandstone and other sedimentary rocks. Calcium is also a component of montmorillonite. Because of the broad range of Ca-bearing minerals, Ca is plentiful in soil and water, except in pure sands (largely consisting of quartz minerals) and in acidic leached soils. The Ca content of soils normally ranges between 0.1 and 1.2 percent, but in $CaCO_3$-bearing or $CaSO_4$-bearing soils, the Ca contents are often considerably larger and in sandy soils and acidic leached soils (laterite soils) often smaller (Scheffer and Schachtschabel, 1989). In these latter soils, Ca deficiency may occur for plant growth, although Ca deficiency is rare. Therefore, Ca fertilisers are seldom applied. Nevertheless, a major anthropogenic Ca source is lime (calcium carbonate), a fertiliser applied to increase the pH and/or to eliminate toxic aluminium and manganese in agricultural soils, forest soils, or lake waters. In this case, the carbonate is the effective agent.

Two major mechanisms control the Ca concentrations in water: calcite precipitation and dissolution, and cation exchange at the negatively charged surfaces of the soil particles. Furthermore, Ca ions form complexes with some organic anions, but these complexes barely affect the Ca concentrations in natural water (Hem, 1989). Because the Ca concentrations are for the greater part controlled by calcite precipitation and dissolution, Ca concentrations are often correlated with the most abundant carbonate species in natural water: bicarbonate (HCO_3^-). The solubility of calcite is regulated not only by the Ca concentration but also by the carbonate equilibria, which, in turn, are controlled by the CO_2 partial pressure (pCO_2) and pH. Section 5.10 gives more detail about the relation between Ca and HCO_3^- and also the solubility of calcite. Since Ca ions are divalent, they adhere more strongly to surface charge sites than monovalent ions. Because Ca is generally the dominant divalent ion in surface water and groundwater, Ca^{2+} ions occupy most surface sites. Cation exchange may cause changes in the ratio between calcium and other dissolved cations. However, in systems in which water is in contact with an extensive area of solid surface, the Ca:Na ratio tends to remain relatively constant (Hem, 1989).

5.5 MAGNESIUM

Magnesium (Mg) is an alkaline-earth metal and has only one oxidation state of significance: Mg^{2+}. It is an essential constituent of plant and animal nutrition. About one-fifth of the Mg taken up by plants is used for the production of chlorophyll. A deficiency of Mg leads to interveinal chlorosis, i.e. yellowish coloration of leave tissue between the darker veins, especially of older leaves, because Mg is mobilised from older leaves and transported to younger tissues. Furthermore, Mg is an activator for many critical enzymes in both plants and animals. In animals and humans, it is also used for building bones and tendons, and together with calcium it plays a role in the regulation of heart activity. Magnesium also contributes to the total water hardness and is thus undesirable in large concentrations in drinking water (see above).

Magnesium occurs typically in dark-coloured minerals present in igneous rocks such as plagioclase, pyroxenes, amphiboles, and the dark-coloured micas. In metamorphous rocks, magnesium also occurs as a constituent of chlorite and serpentine. The major sedimentary form of magnesium is dolomite, in which calcium and magnesium are present in equal amounts. Furthermore, in most carbonate rocks Mg occurs in significant amounts as dolomite, magnesite ($MgCO_3$), hydrated species of Mg carbonate (nesquehonite ($MgCO_3 \cdot 3H_2O$), and lansfordite ($MgCO_3 \cdot 5H_2O$)), and brucite ($Mg(OH)_2$). Obviously, the dissolution of these carbonate rocks also brings Mg into solution, but the process is not readily reversible. As all Mg carbonate species are more soluble than calcite, a considerable degree of supersaturation may be required before they precipitate. Consequently, the precipitate that forms from a solution that has attacked an Mg-bearing limestone may be nearly pure calcite and, in turn, in Mg-bearing rock or sediment both the Mg concentration and the Mg:Ca ratio tends to increase along a groundwater flow path (Hem, 1989). Magnesium is also present between the lattices of montmorillonite. The cation exchange behaviour of Mg is similar to that of Ca. Both ions are strongly adsorbed to negatively charged sites of clay minerals and organic particles. In general, the content of exchangeable Mg increases with increasing clay and silt content. Podsolic sandy soils and acidic leached soils contain very little Mg.

Anthropogenic sources of Mg are generally rare. In the case of Mg deficiency in agricultural crops, the Mg content of the soil is adjusted by applying dolomitic lime, potassium magnesium sulphate, or magnesium sulphate.

5.6 IRON

Iron is a transition metal and, after aluminium, the second most abundant metallic element in the Earth's outer crust. Nevertheless, the Fe concentrations in natural water are generally small. In plants, about three-quarters of the iron (Fe) is associated with chloroplasts (i.e. the green globules inside plant cells responsible for photosynthesis). It plays an important role in the plants' redox reactions. Plants that suffer Fe deficiency exhibit the typical interveinal chlorosis. In animals, Fe is part of haemoglobin, the carrier of the oxygen in red blood cells. Although Fe is non-toxic, it is an undesirable constituent of drinking water, because it has an adverse effect on the taste, it may clog distribution systems and may produce stubborn, rusty brown stains on textile and ceramics. Therefore, during drinking water preparation it is usually removed by aeration (see below).

Relatively large amounts of iron occur in a wide range of igneous rock minerals, including the pyroxenes, amphiboles, biotite, magnetite, and olivine. In these minerals, Fe is largely present in the ferrous (Fe^{2+} or Fe(II)) oxidation state, but it may also be present in the ferric (Fe^{3+} or Fe(III)) oxidation state. When these primary minerals are weathered, the Fe released generally reprecipitates quickly. Under reducing conditions, Fe is in the ferrous oxidation state; if sulphur is available, pyrite or marcasite (both FeS_2) is formed. When sulphur is less available, siderite ($FeCO_3$) or vivianite ($Fe_3(PO_4)_2 \cdot 8H_2O$) may precipitate. Under oxidising conditions, Fe is in the ferric oxidation state and the precipitates consist of ferric oxides, oxyhydroxides, or hydroxides. Fresh precipitates consist mostly of amorphous ferric hydroxide ($Fe(OH)_3$). In the course of time, this $Fe(OH)_3$ undergoes ageing which results in the formation of more stable, crystalline minerals of Fe(III) oxides/hydroxides, such as haematite (α-Fe_2O_3) or goethite (α-FeOOH) (see Section 4.2.2).

Fe^{2+} is much more soluble than Fe^{3+}. Thus, the most obvious control on the solubility of Fe is the oxidation reaction from Fe^{2+} to Fe^{3+} and the reverse reduction reaction. The oxidation reaction occurs particularly in zones of groundwater discharge, where groundwater containing Fe^{2+} encounters dissolved oxygen from the free atmosphere, and proceeds very rapidly. The Fe^{3+} precipitates primarily as amorphous ferric hydroxide. Consequently, when

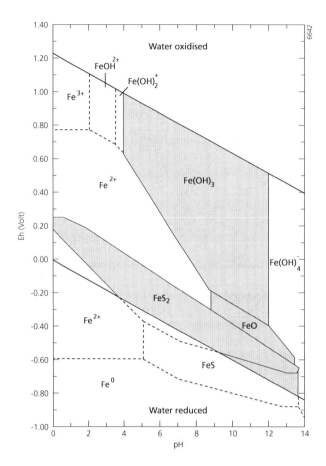

Figure 5.4 pH–Eh diagram for the stability of dissolved and solid Fe species at 25 °C and 1 atmosphere pressure. Activity of sulphur species 96 mg l^{-1} as SO_4^{2-} and carbon dioxide species 61 mg l^{-1} as HCO_3. Boundaries have been drawn for a dissolved Fe activity of 10^{-3} mmol l^{-1} (= 56 μg l^{-1}) (source: Hem, 1989).

groundwater rich in Fe^{2+} is sampled from a well, it is first clear but soon becomes cloudy and brown from the precipitating $Fe(OH)_3$. The reaction equation can be written as:

$$4\,Fe^{2+} + O_2 + 10\,H_2O \quad \rightarrow \quad 4\,Fe(OH)_3 + 8\,H^+ \tag{5.3}$$

The process of oxidation of Fe(II) results in the formation of bands of Fe(III)-rich precipitates in the zone around the water table. In oxic groundwater, oxygen is continuously consumed by many oxidation reactions, especially the decomposition of organic matter. If the oxygen consumption rate exceeds the diffusion rate of oxygen from the atmosphere into groundwater, the dissolved oxygen becomes depleted. The decomposition of organic matter proceeds using other oxidants, such as nitrate and manganese. If these are depleted too, the ferric iron also serves as an oxidant for the decomposition of organic matter and is so reduced to Fe^{2+}, which is released into solution (see Section 4.3.4, Equation 4.6):

As well as depending on the redox potential, the solubility of Fe depends strongly on the pH. In acid solutions, the ferric iron can occur as Fe^{3+}, $FeOH^{2+}$, and $Fe(OH)_2^+$ and in polymeric hydroxide forms. Above a pH of 4.8 however, the total activity of these ions is less than 10 μg l^{-1} (Hem, 1989). Under reducing conditions, the most common form of dissolved iron is ferrous Fe^{2+}, which is much more soluble than the ferric Fe^{3+}. The concentrations can range up to 50 mg l^{-1}, but are controlled by the solubility of pyrite, marcasite, and siderite, as noted above.

Because the solubility of Fe depends on both the pH and the redox potential, the stability of the solid and dissolved forms of Fe can conveniently be depicted in a pH–Eh diagram. The construction, use, and limitations of these types of diagrams were outlined in Section 2.10. Figure 5.4 shows the fields of stability of Fe species in a system containing a constant total amount of dissolved sulphur and carbon dioxide (inorganic carbon) species both of 1 mmol l^{-1} (equivalent to 96 mg SO$_4^{2-}$ l^{-1} and 61 mg HCO$_3^-$ l^{-1}). The solid species indicated by the shaded areas would be thermodynamically stable in their designated domains. The Fe activity does not affect the boundaries between the dissolved species. The boundaries between the solid and dissolved species have been drawn for a dissolved Fe activity of 56 μg l^{-1}, but if more Fe is present, the domains of the solid species will increase.

Figure 5.4 confirms that there is one general Ph–Eh domain in which Fe is relatively soluble: under moderate reducing conditions, especially at low pH. Comparison with Figure 2.3 shows that these conditions may prevail in groundwater, acidic podsolic soils and peat soils, and in the hypolimnion of lakes. The high-solubility domain at high pH is outside the range that is common in natural waters. Under the conditions specified for Figure 5.4, siderite (FeCO$_3$) saturation is not reached. To attain siderite saturation and an accompanying stable solid-phase domain of siderite in the pH–Eh diagram requires a higher activity of inorganic carbon species.

Although Fe is very insoluble under oxidising conditions in soil, the presence of organic ligands can keep considerable amounts of Fe(III) in a mobile form (see also Section 4.3.2). This is a major pathway of the plant uptake of Fe and some plants even produce ligands that chelate soil Fe. Therefore, most plants are not deficient in Fe. The formation of chelates is also an important process of Fe leaching in podsolic soils.

5.7 MANGANESE

Like Fe, manganese (Mn) is a transition metal and its chemistry is similar to that of Fe. However, Mn can occur in three possible oxidation states: 2+, 3+, and 4+, or Mn(II), Mn(III), and Mn(IV). In organisms, Mn is involved in many enzyme systems and helps with the protein, carbohydrate and fat metabolisms, with building of bone and connective tissues, and also with the clotting of the blood. In broad-leaved plants, Mn deficiency cause interveinal chlorosis of the younger leaves. Manganese may become noticeable in drinking water at concentrations greater than 0.05 mg l^{-1} by imparting an undesirable colour, odour, or taste to the water. However, the effects on human health are not a concern until the concentration of Mn is approximately 0.5 mg l^{-1}. Long-term exposure to high concentrations of Mn may be toxic to the nervous system. Like Fe, Mn is usually removed by aeration during the preparation of drinking water.

Because of its three possible oxidation states, manganese can form a wide range of mixed-valence oxides. Divalent Mn is present as a minor constituent in many igneous rock minerals, such as olivines, pyroxenes, and amphiboles, and in sedimentary limestone and dolomite. Divalent Mn is released into solution during weathering and is more stable than Fe^{2+}. Figure 5.6 shows the pH–Eh diagram of Mn species in a system containing a constant total amount of 1 mmol l^{-1} dissolved sulphur and 1 mmol l^{-1} inorganic carbon species. From Figure 5.5 it can be seen that the concentration of dissolved Mn^{2+} is largely controlled by the pH. Therefore, acidic groundwater may contain more than 1 mg l^{-1} of Mn. Comparison with Figure 5.4 demonstrates that Mn^{2+} has a greater stability domain than Fe^{2+}. This implies that

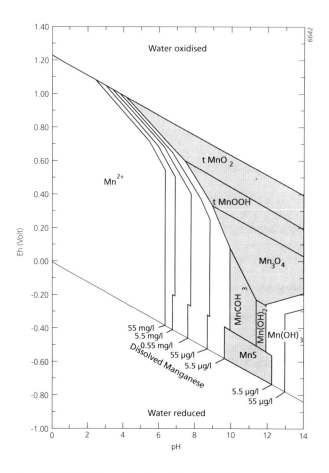

Figure 5.5 pH–Eh diagram for the stability of solid Mn species and equilibrium dissolved Mn activity at 25 °C and 1 atmosphere pressure. Activity of sulphur species 96 mg l^{-1} as SO_4^{2-} and activity of carbon dioxide species 61 mg l^{-1} as HCO_3^- (source: Hem, 1989).

if the redox potential decreases, the solid species of Mn are reduced to dissolved Mn^{2+} at a higher redox potential than Fe hydroxide is reduced to dissolved Fe^{2+}. If the redox potential increases again, the Fe^{2+} is oxidised first and the Mn^{2+} later. Therefore, Mn is more mobile than Fe and is more prone to leaching and transport. According to Figure 5.6, Mn(II) is oxidised to species whose oxidation state is 3+ (MnOOH) or below 3+ (Mn_3O_4) if the pH is sufficiently high. However, such species are relatively uncommon in natural waters. Under naturally occurring conditions, the formation of MnO_2 prevails, in which Mn has an oxidation state of 4+. The Mn(III) species are generally unstable and their formation usually leads to a subsequent disproportionation reaction, which means that two Mn(III) ions may interact and form one Mn(II) and one Mn(IV) ion. For example: Mn_3O_4 contains two Mn(III) and one Mn(II) atoms; thus, the disproportionation reaction of Mn_3O_4 results in one Mn(IV) atom that forms crystalline MnO_2 and two Mn(II) ions that become available again for oxidation by O_2. The reaction equation reads:

$$Mn_3O_4 + 4\,H^+ \quad \rightarrow \quad MnO_2 + 2\,H_2O + 2Mn^{2+} \tag{5.4}$$

Mn oxide is preferentially precipitated on existing Mn oxide surfaces, which results in nodules being formed around a central nucleus. Coatings or small discrete particles of Mn oxide are very common in soils and bed sediments. Subsurface water in contact with these Mn precipitates under moderately reduced conditions may exhibit enhanced dissolved Mn^{2+} concentrations up to about 1 mg l^{-1}, even at neutral pH. Accumulations of Mn oxides are also found in acidic leached laterite soils in tropical regions and in the dark stains on rocks in arid regions. The manganese nodules and coatings usually contain significant amounts of coprecipitated iron and sometimes other metals as well, especially cobalt, lead, zinc, copper, nickel, and barium. Because of the different redox potentials at which Fe and Mn are oxidised (compare Figures 5.4 and 5.5), the zones in which Fe and Mn precipitate may occur spatially separated.

The Mn taken up by plants accumulates mostly in the leaves. As soon as the plant parts die back or the leaves are shed, the Mn is released into solution in surface runoff or soil moisture. Accordingly, the released Mn may cause temporary increases in the Mn concentrations in river water during autumn. Similar to Fe, chelates may play an important role in the environmental transport of Mn in some situations. Note, however, that the Mn^{2+} ion is considerably more stable than Fe^{2+} under oxic conditions, and can also be transported in larger concentrations without organic complexation.

A possible anthropogenic source of Mn in streamwater is the discharge of acidic mine water. These mine water discharges may also contain considerable amounts of Fe and other metals. Manganese usually persists in the water for greater distances downstream from the discharge location than Fe. Downstream from the inflow, the acidity is buffered, so the pH increases gradually. The Fe is removed from the water first, due to precipitation of ferric hydroxide. Later (and thus after more distance has been travelled) Mn also disappears from solution.

5.8 ALUMINIUM

Aluminium (Al), the third most abundant element after oxygen and silicon, is by far the most common metal in the Earth's crust. In nature it is almost always found in the 3+ oxidation state combined with other elements and there are many different compounds which contain Al. Whilst Al is abundant in the environment, the naturally occurring forms are usually stable under near-neutral pH, so Al rarely occurs in natural waters in concentrations greater than a few tens or hundreds of milligrams per litre. Because of this, Al is not usually considered to be a major dissolved phase constituent. However, under extreme pH conditions, Al may become soluble and because of its toxicity it may be a relevant parameter in environmental assessment studies. Therefore, the most important chemical properties of Al are discussed below.

Aluminium does not appear to be an essential trace element and it has no known biological role. Under acidic conditions, Al occurs in a soluble form which can be absorbed by plants and animals. Some plants (ferns and tea, for example) naturally accumulate high amounts of aluminium compounds in their leaves. However, Al in large concentrations is toxic for most plants because it inhibits the root growth. Dissolved Al in surface waters of low pH has a deleterious effect on fish and other aquatic organisms. Humans have highly effective barriers to exclude aluminium. Only a very small fraction of aluminium in the diet is taken up from the stomach and intestines, and in healthy individuals most of this absorbed aluminium is excreted by the kidneys. When the natural barriers that limit the absorption of aluminium are bypassed, or if the ability of the kidneys to excrete aluminium is impaired, this metal accumulates in the body and may cause adverse health effects.

Aluminium occurs in substantial amounts in many silicate, igneous, and sedimentary rock minerals, such as feldspars, amphiboles, micas, and clay minerals (see Section 4.2.3). On weathering of these minerals, Al is usually retained in newly formed solid clay minerals (see also Figure 4.2), some of which may be considerably enriched in Al. Gibbsite $(Al(OH)_3)$ is a common mineral in acidic podsolic soils in temperate and subarctic regions and in highly weathered soils in tropical regions. Gibbsite is also the major constituent of bauxite, the ore from which aluminium is produced.

The solubility of Al is strongly controlled by the solution pH. This pH control on the different forms of Al species can be elucidated by the following equilibrium equation:

$$
\begin{aligned}
\left[Al(H_2O)_6\right]^{3+} \leftrightarrow\ & \left[Al(H_2O)_5(OH)\right]^{2+} + \ H^+ \\
\leftrightarrow\ & \left[Al(H_2O)_4(OH)_2\right]^+ + \ 2H^+ \\
\leftrightarrow\ & \left[Al(H_2O)_3(OH)_3\right]^0 + \ 3H^+ \\
\leftrightarrow\ & \left[Al(H_2O)_2(OH)_4\right]^- + \ 4H^+ \\
\leftrightarrow\ & \left[Al(H_2O)(OH)_3\right]^{2-} + \ 5H^+
\end{aligned}
\tag{5.5}
$$

With decreasing pH the reaction equilibrium shifts to the left. Figure 5.6 shows the solubility of Al and the dominant Al species as function of pH. At pH less than about 4, the cation Al^{3+} is the predominant form of Al in solution and on the cation exchange sites of clay minerals and organic matter. The Al^{3+} ion always forms an octahedral configuration with six water molecules $(Al(H_2O)_6^{3+})$, comparable with six oxygens or hydroxyls in clay minerals. If the pH increases, one or more water molecules dissociate and become hydroxyl (OH^-) ions. In the pH range of 4.5 to 7.5, the dominant soluble form of Al is $Al(H_2O)_4(OH)_2^+$ or, if the hydration water molecules are ignored, $Al(OH)_2^+$. In the pH range of 4.5 to 6.5, the almost insoluble gibbsite is formed. The minimum solubility is about 27 µg l^{-1} for natural

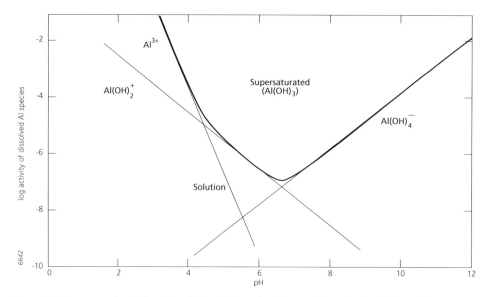

Figure 5.6 Percentage of total aluminium for the different aluminium species and total solubility of aluminium as function of pH (after Drever, 1982).

gibbsite near a pH of 6. Above neutral pH, the predominant dissolved form of Al is the anion $Al(OH)_4^-$.

In addition to the pH, the presence of other dissolved species may control the solubility of Al. In solutions with a concentration above about 0.10 mg l^{-1} fluoride, the solubility of Al increases (Hem, 1989). In solutions with a considerable amount of dissolved silica, the Al solubility is substantially decreased due to the formation of clay-mineral species. Although the solubility of Al at near-neutral pH is very low, in such solutions Al may be detected in concentrations of up to about 1 mg l^{-1}. These Al concentrations can be attributed to the presence of colloidal or particulate Al in either Al hydroxides or aluminosilicates (clay minerals). This Al is thus not in the dissolved phase.

The main anthropogenic source of aluminium in surface water is the discharge of acidic mine water. These waters may contain several hundreds or even thousands milligrams of Al per litre. If the pH of the stream water increases downstream from the point of discharge, the Al is gradually removed from solution through the precipitation of gibbsite. Furthermore, atmospheric deposition of acidifying compounds, which are largely anthropogenic origin, may bring about low pH values in poorly buffered catchments. In the soils of these acidified catchments, Al is mobilised into the soil solution and some Al enters the drainage network. Acid deposition and precipitation may thus contribute significantly to the soil and water pollution by aluminium.

5.9 CHLORIDE

Chlorine (Cl) is the second most abundant halogen in the Earth's crust (after fluorine), although more than 75 percent of the total amount is present as dissolved chloride (Cl^-) in the oceans. Chlorine can exist in various oxidation states, but in soil and water it exists almost entirely as the chloride ion (1- oxidation state). Like Na, Cl is used to maintain osmotic pressure in the cells and to balance the cell cationic charges. The tolerance of plants to Cl can vary widely. In animals and humans, Cl is a substantial constituent of the digestive acids in the stomach and, consequently, is important in digestion.

Igneous rock minerals in which chloride is an important constituent are not very common. Chloride is often only present as an impurity. Sedimentary rocks, particularly the evaporite halite (NaCl), are far more important as a source of Cl. In coastal regions, Cl may be supplied by surface intrusion of estuarine water or seawater or by subsurface intrusion of saline groundwater. Another important source of Cl in these areas is the input from the atmosphere. Salt spray is formed along the shoreline when waves break; small salt particles may be suspended as the small seawater droplets evaporate, leaving aerosols behind. These sea salt aerosols are transported inland by wind and are deposited on the land surface either by dry deposition or by precipitation (wet deposition). The Cl deposition declines sharply with increasing distance from the coast. Near the coast, the Cl deposition rates may be up to 100 kg ha^{-1} y^{-1}, but 200 km further inland the rate will have decreased to about 20 kg ha^{-1} y^{-1}. In areas remote from the coast, for example in the Midwestern USA or central Australia, the atmospheric input is no more than about 1 kg ha^{-1} y^{-1} (Whitehead, 2000). The major anthropogenic sources of Cl in soil are road salt and, more important, agricultural fertilisers. Some fertilisers (especially potassium chloride (KCl) and sewage sludge) contain large amounts of Cl, but Cl is not usually added as an essential soil amendment, since chloride deficiency in soils is very rare. In KCl fertiliser, Cl is the balancing anion for the essential potassium; in sewage sludge and slurries, Cl is commonly present as a contaminant. In surface water, the most important anthropogenic sources are leaching of the Cl of anthropogenic origin from soils in the catchment, irrigation waters, and the discharge of domestic, industrial, and mine wastewater.

Box 5.I Chloride as a natural tracer to calculate evapotranspiration

In infiltration areas, the evapotranspiration rate relative to the precipitation rate can be relatively easily determined by comparing the Cl^- concentration in rainfall with the Cl^- concentration in groundwater. The evapotranspiration rate (mm y^{-1}) can be calculated as follows:

$$E = P \frac{\{Cl\}_{gw} - \{Cl\}_{rain}}{\{Cl\}_{gw}} \qquad (5.Ia)$$

where E = evapotranspiration rate (mm y^{-1}), P = annual precipitation (mm y^{-1}), $\{Cl\}_{gw}$ = chloride concentration in groundwater (mg l^{-1}), and $\{Cl\}_{rain}$ = chloride concentration in rainwater (mg l^{-1}). This calculation only applies for situations where anthropogenic inputs of Cl and dry Cl deposition are negligible, thus in natural, non-coastal areas. In areas where such additional inputs are substantial, they should be accounted for in the calculation:

$$E = \frac{P\{Cl\}_{gw} - \{Cl\}_{rain} - 100\ I}{\{Cl\}_{gw}} \qquad (5.Ib)$$

where I = additional Cl inputs from anthropogenic sources (e.g. fertiliser application) and dry deposition (kg ha^{-1} y^{-1}). The factor 100 is included to correct for the different units (kg ha^{-1} y^{-1} → mg m^{-2} y^{-1}). The additional inputs are difficult to measure and are often unknown, which makes the calculation of evapotranspiration rates from Cl^- concentrations in groundwater very unreliable.

Chloride is a very soluble and mobile ion and is practically inert (i.e. non-reactive or conservative), which means that it has very little tendency to react with anything in soil and water. It does not participate in redox reactions, does not form barely soluble salts, does not form important complexes with other ions unless the Cl^- concentrations are very large, is barely adsorbed onto mineral surfaces, and does not play a vital role in the biochemical cycles (Hem, 1989). Therefore, in soils, most of the Cl is dissolved in the soil solution. However, in acid soils with kaolinitic clays the pH may be below the *PZC* (see Section 4.2.3) and thus these soils may show some adsorption of Cl^-, though the amounts are mostly small. In arid regions, the Cl may precipitate and accumulate in the topsoil due to excess evaporation.

Since Cl^- is a conservative ion, it is frequently used as a natural tracer for studying the rates of evapotranspiration, water flow, or dispersion. For example, in infiltration areas the evapotranspiration rates (mm y^{-1}) can be relatively easily determined by measuring Cl^- concentrations in groundwater (see Box 5.I). Furthermore, Cl^- is sometimes injected into groundwater or surface water to measure the water flow rates. Other commonly used tracers are other conservative ions that occur naturally in a water system, such as iodide (I^-) and bromide (Br^-), environmental isotopes such as tritium (T = ^3H), other contaminants of all kinds, and dyes that are deliberately added to the water, such as rhodamine and uranine. For more information about the use of Cl^- as a tracer and tracer experiments, see hydrological or water quality textbooks (e.g. Thomann and Mueler, 1987; Domenico and Schwarz, 1996).

5.10 INORGANIC CARBON

The inorganic carbon consists of dissolved carbon dioxide species, which include carbonic acid (H_2CO_3), hydrogen carbonate or bicarbonate (HCO_3^-), and carbonate (CO_3^{2-}). These three species together are also often referred to as ***total inorganic carbon (TIC)***. Inorganic carbon is an essential nutrient for green plants that consume it to construct organic molecules through photosynthesis. Terrestrial plants take up carbon dioxide from the free atmosphere, but aquatic plants and phytoplankton (including algae, diatoms, and photosynthetic bacteria) extract carbon dioxide species dissolved in the water to use in their photosynthesis process. Foraminiferans take up dissolved carbon dioxide species to build their calcareous shells. In animals and humans, dissolved inorganic carbon helps to maintain the acid balance in the stomach and the intestines.

The principal sources of the dissolved carbon dioxide species in surface water or groundwater are the CO_2 in the free atmosphere and the CO_2 produced from the respiration and decomposition of organic matter in soil, sediment, or water. The CO_2 content of the atmosphere is about 0.03 volume percent. The air in the unsaturated zone of the soil can be substantially enriched in CO_2 due to respiration by plants and the oxidation reaction in soil. In volcanic areas, CO_2 can also be derived from degassing of the Earth's mantle or metamorphic decarbonation of carbonate rocks. Another important source of inorganic carbon is the dissolution of carbonaceous rocks by acid compounds. The major direct anthropogenic source of inorganic carbon is the application of lime (calcite ($CaCO_3$) or dolomite ($Ca,MgCO_3$)) fertilisers to increase the pH of agricultural or forest soils or lake water. The rate at which the lime is applied to soils depends on the cation exchange capacity and the soil pH, but the dissolved inorganic carbon concentrations in groundwater and surface water may be greatly boosted by anthropogenic emissions of organic carbon or acidic compounds. The organic compounds are decomposed to CO_2, which dissolves readily in water. Immissions of acid compounds into soil or water via direct anthropogenic emissions or through atmospheric deposition ('acid rain') may lead to an increase in dissolved inorganic carbon if these acids react with and dissolve carbonate solids. If these carbonates are not available for reaction, a fall in the pH in response to acid immissions causes dissolved carbon dioxide species to be removed from solution (see below).

Carbon dioxide species are important reactants that control the pH of natural waters. When water is in equilibrium with a gas phase containing CO_2, such as the atmosphere, carbon dioxide dissolves up to a specific solubility limit that depends on pressure and temperature. At 1 atmosphere pressure and 25 °C, the following mass-law equation applies:

$$\frac{pCO_2}{[H_2CO_3]} = K_H = 10^{1.43} \tag{5.6}$$

where K_H = equilibrium (Henry's law) constant (atm · l mol^{-1}) for dissolution of CO_2; pCO_2 = the partial pressure of carbon dioxide in the gas phase (i.e. the volume percent of CO_2 multiplied by the total pressure expressed in atmospheres and divided by 100) (atm); $[H_2CO_3]$ = the activity of carbonic acid (mol l^{-1}). Since this is a dilute solution, the activity of H_2CO_3 is virtually the same as the H_2CO_3 concentration. The dissociation of dissolved carbonic acid proceeds in two steps:

$$H_2CO_3 \underset{1}{\leftrightarrow} H^+ + HCO_3^- \underset{2}{\leftrightarrow} 2H^+ + CO_3^{2-} \tag{5.7}$$

Clearly, the equilibrium of these dissociation reactions is dependent on pH. With increasing H^+ concentrations (decreasing pH) the equilibrium shifts to the left. Figure 5.7 summarises the relationships between the pH and the dissolved carbon dioxide species. In natural

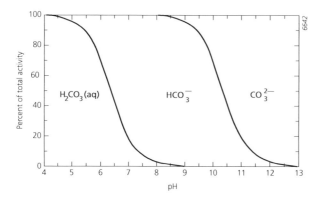

Figure 5.7 Percentages of dissolved carbon dioxide species activities at 25 °C and 1 atmosphere pressure as function of pH (source: Hem, 1989).

waters the dissolved carbon dioxide species are almost solely responsible for the ***alkalinity***, i.e. the capacity to neutralise acid. Although dissolved phosphates and OH⁻ may contribute to a limited extent, in practice only dissociated carbon dioxide species (HCO_3^- and CO_3^{2-}) are important for the alkalinity. At pH values less than 8.3, less than 1 percent of the carbon dioxide species is present as CO_3^{2-} (see Figure 5.7), so the alkalinity approximately corresponds to the HCO_3^- concentration.

The mass-law equations for the two coupled equilibria in Equation (5.7) at 1 atmosphere and 25 °C are:

$$\frac{[H^+][HCO_3^-]}{[H_2CO_3]} \quad = \quad K_1 \quad = \quad 10^{-6.35} \tag{5.8}$$

$$\frac{[H^+][CO_3^{2-}]}{[HCO_3^-]} \quad = \quad K_2 \quad = \quad 10^{-10.33} \tag{5.9}$$

Besides the mass-law Equations (5.7), (5.8), and (5.9), the mass-law equation for the dissociation of water (see Equation 2.49) and the constraint of electrical neutrality applies. Thus, for carbon dioxide dissolving in pure water the following applies:

$$C_{H^+} \quad = \quad C_{HCO_3^-} \quad + \quad 2\,C_{CO_3^{2-}} \quad + \quad C_{OH^-} \tag{5.10}$$

where the C terms represent the ion concentrations (mol l⁻¹), which are approximately equal to the ion activities. Given this constraint and the mass-law equations mentioned above, there are five equations with six variables (the pCO_2 and five dissolved species). So, if one variable is given, the concentrations of the different carbon dioxide species are fixed as well. As mentioned above, the average CO_2 content of the free atmosphere is about 0.03 volume percent; the value of pCO_2 is therefore equal to $10^{-3.53}$. Hence, pure water that is in contact with the free atmosphere has a pH of 5.65. Obviously, other ions affect the pH of the solution, but if these concentrations are insignificant, the pH of water in contact with air is around this value; this happens in rainwater in the absence of atmospheric pollution.

If the water is in contact with an excess of solid calcite, the above mass-law equations are extended with the equation representing the dissolution of calcite:

$$CaCO_3(s) \quad + \quad H^+ \quad \leftrightarrow \quad Ca^{2+} \quad + \quad HCO_3^- \tag{5.11}$$

$$\frac{[Ca^{2+}][HCO_3^-]}{[H^+]} \ = \ K_s \ = \ 10^{2.0} \tag{5.12}$$

Now the electroneutrality principle imposes that

$$2\,C_{Ca^{2+}} \ + \ C_{H^+} \ = \ C_{HCO_3^-} \ + \ 2\,C_{CO_3^{2-}} \ + \ C_{OH^-} \tag{5.13}$$

Together with the above equations, these additional equations allow us to calculate the concentrations of the carbon dioxide species in a solution in equilibrium with calcite. The H⁺ needed for the dissolution of calcite is almost entirely derived from carbonic acid, i.e. dissolved carbon dioxide. The overall reaction equation is then:

$$CaCO_3(s) \ + \ CO_2 \ + \ H_2O \ \leftrightarrow \ Ca^{2+} \ + \ 2HCO_3^- \tag{5.14}$$

In water bodies that are in contact with solid calcite and the free atmosphere, the partial CO_2 pressure will remain constant at $10^{-3.5}$ atmosphere and the corresponding HCO_3^- concentration at pH = 8.3 will be 1 mmol l^{-1} (= 61 mg l^{-1}). Note, however, that equilibration between the gas and liquid phase is a relatively slow process, so water bodies exposed to the free atmosphere may not always be in equilibrium with the atmospheric partial CO_2 pressure; this is especially likely if the CO_2 in the water is produced or consumed chronically in substantial amounts. In surface waters with abundant aquatic vegetation or algae, the photosynthesis process consumes lots of dissolved CO_2, so the equilibrium in Equation (5.14) shifts to the left. The dissolved CO_2 is then partly replenished from the atmosphere. The CO_2 consumption in the water may lead to supersaturation, followed by precipitation of calcite. This process occurs in particular in isolated lakes and leads to the formation of calcareous deposits (gyttja) on the lake bed. In the unsaturated zone or in volcanic areas, where the pCO_2 is higher, the equilibrium in Equation (5.14) shifts to the right and more calcite dissolves. If this water comes into contact with the free atmosphere, the reduction of the pCO_2 pushes the equilibrium in Equation (5.14) to the left again and calcite is precipitated. Accordingly, the degassing of CO_2 from the exfiltrating water brings about the formation of stalactites and stalagmites in caves and massive travertine (calcite) formations near springs.

In the case of a water body isolated from the free atmosphere, for instance in deep groundwater, the pCO_2 drops as calcite dissolves, since the CO_2 is not replenished. The dissolution of calcite consumes H⁺ ions (see Equation 5.11); as a result, the pH increases and the carbonate equilibrium in Equation (5.11) shifts to the right. The total TIC concentration in such isolated groundwater bodies is given by the concentration of TIC in the zone where the water became isolated from the gas phase plus the amount of TIC derived from additional dissolution of calcite. In the case of deep groundwater, the zone where the water becomes isolated from the atmosphere is usually the root zone. Because the calcite dissolution results in an increase of the Ca^{2+} ions that equals the increase of the HCO_3^- ions, the TIC concentration can be calculated from the following mass balance:

$$TIC \ = \ TIC_{root} \ + \ \Delta Ca^{2+} \tag{5.15}$$

where *TIC* = the concentration of total inorganic carbon in the isolated groundwater body (mol l^{-1}), *TIC_{root}* = the TIC concentration in the root zone (mol l^{-1}), and ΔCa^{2+} = the increase of the Ca^{2+} concentration resulting from calcite dissolution (mol l^{-1}). If there is no supply of CO_2 from deeper layers or from the decomposition of organic matter, as is usually the case in deep aquifers, the initial CO_2 from the root zone is rapidly depleted. In such environments, a low pCO_2 occurs together with low Ca^{2+} concentrations and a high pH.

It is worth mentioning that calcite dissolution by carbonic acid yields a Ca^{2+}:HCO_3^- molar ratio of 0.5, i.e. for every mole $CaCO_3$ that dissolves, two HCO_3^- ions are produced (see Equation 5.14). If calcite is dissolved by another acid (e.g. sulphuric acid (H_2SO_4) derived from atmospheric pollution, acidic mine wastewater, or pyrite oxidation (see also Section 5.11), the Ca^{2+}:HCO_3^- molar ratio amounts to 1.0 (see Equation 5.11). This implies that a Ca^{2+}:HCO_3^- molar ratio greater than 0.5 may be an indication of acidification in waters in which the Ca^{2+} concentration is controlled by calcite dissolution and precipitation. Obviously, in the case of dissolution of dolomite ($Ca,MgCO_3$) or gypsum ($CaSO_4 \cdot 2H_2O$) also contributing to the dissolved Ca^{2+} concentration, this relationship becomes more complex.

5.11 SULPHATE AND SULPHIDE

Sulphate (SO_4^{2-}) is the dominant dissolved species and most highly oxidised form of sulphur (S). As sulphur can occur in oxidation states varying from 2- to 6+ sulphur, its chemical behaviour is strongly governed by the redox conditions. As well as occurring as sulphate, sulphur also occurs naturally as sulphide (S^{2-}, which is the 2- oxidation state) and, to a lesser extent, in the neutral oxidation state (S^0) and in the sulphite form (SO_3^{2-}, the 4+ oxidation state). Sulphur is a constituent of some amino acids and so forms an essential part of proteins: the sulphur content of protein materials can be as high as 8 percent. It is therefore an important nutrient for both plants and animals. Sulphate is also an oxidant in the microbial decomposition of organic matter, if oxygen is depleted under anaerobic conditions. In this process, sulphate is reduced to sulphide, which either precipitates or forms H_2S gas which is toxic to many plants (e.g. the *Akiochi* disease in some paddy rice).

Sulphate-containing minerals occur most extensively in evaporite sediments. The most important of these are gypsum ($CaSO_4 \cdot 2H_2O$) and anhydrite ($CaSO_4$; without crystalline water). In other sedimentary rocks and igneous rocks, sulphur occurs commonly as metal sulphides. The most abundant and widely distributed form of metal sulphide is pyrite (FeS_2), which is common in coal and slate. If the metal sulphides come into contact with oxygenated water, they oxidise to SO_4^{2-} that readily dissolves. In soils, the decomposition of soil organic matter including proteins constitutes an important supply of S to plants.

Another substantial source of S in soil and water is the atmospheric deposition of SO_4^{2-} from volcanic or anthropogenic sources. Volcanic eruptions, fuel combustion, and ore smelting lead to the emission of sulphur dioxide (SO_2) that readily oxidises to sulphuric acid (H_2SO_4), the main compound of sulphate aerosols, in the atmosphere. Therefore, these aerosols also contribute significantly to acid deposition and have led to the widespread degeneration of forest ecosystems in Europe and North America. In Central Europe in particular, where sulphur-rich brown coal (lignite) has for long been used for electricity generation and heating, acid deposition has caused massive die-back of forest trees. The rate of S deposition in the mountainous areas most affected by forest die-back reached 150 kg S ha^{-1} y^{-1}. Atmospheric deposition of SO_4^{2-} has also greatly enhanced the SO_4^{2-} concentrations in runoff water from catchments in industrialised countries. Other anthropogenic sources of SO_4^{2-} involve the enhanced oxidation of sulphides or organic S in, for example, mine heaps and domestic waste disposal sites.

In contrast to sulphides, sulphates are generally soluble, having solubility products greater than 5. The exceptions are barium sulphate ($K_s = 10^{-9.97}$), lead sulphate ($K_s = 10^{-7.79}$), strontium sulphate ($K_s = 10^{-6.5}$) and calcium sulphate ($K_s = 10^{-4.58}$). Sulphate does not adsorb to mineral and organic surfaces and is rather mobile in solution. It tends to form complexes with cations. In natural waters, the most important sulphate complexes are $NaSO_4^-$ and $CaSO_4^0$ (aq), and, under acid conditions (pH less than 4), HSO_4^-. The strongest complexes are formed with divalent or trivalent cations. For aqueous calcium sulphate, the equilibrium constant is:

$$\frac{[CaSO_4^0]}{[Ca^{2+}][SO_4^{2-}]} = 10^{2.31} \tag{5.16}$$

This implies that solutions containing 10^{-3}–10^{-2} mol l^{-1} of SO_4^{2-} ($\approx$ 100–1000 mg l^{-1}) contain considerable amounts of this complex, and so the solubility of $CaSO_4$ is usually much greater (up to more than three times) than can be expected from the solubility product of gypsum alone. In the absence of Na^+, the SO_4^{2-} concentration in equilibrium with gypsum is about 1480 mg l^{-1} and increases with increasing Na^+ concentrations (Hem, 1989).

Under reducing conditions, sulphate is reduced to sulphide (see Equation 4.7). The reduction of sulphate to sulphide is often mediated by bacteria that use sulphate as an energy source in anoxic environments. Hydrogen sulphide causes water to have a rotten egg odour; this is noticeable in waters having only a few tenths of milligram H_2S per litre in solution. Figure 5.8 shows the fields of dominance of sulphur species at equilibrium as function of pH and Eh at 25 °C and 1 atmosphere pressure. Note that the redox reactions involving the sulphur species are generally slow, unless governed by microorganisms, so the solution may not necessarily be in equilibrium.

Figure 5.8 shows the sulphur species in absence of other constituents. As mentioned above, metal sulphides are barely soluble, so they tend to precipitate rapidly if metal cations are present in solution. Because iron is the most common metal in solution under reduced conditions, the most abundant metal sulphide formed is iron(II) disulphide (pyrite). In sediments, pyrite is mostly formed in two steps: first metastable FeS is formed, which is subsequently transformed to FeS_2 by the overall reaction:

$$FeS + S^0 \rightarrow FeS_2 \tag{5.17}$$

The formation of metal sulphides also governs the solubility of many heavy metals under reducing conditions (see Chapter 7). As the redox potential increases, the metal sulphides are reoxidised and dissolved. The overall oxidation reaction of pyrite can be summarised by the following equation:

$$FeS_2 + 15/4\,O_2 + 7/2\,H_2O \leftrightarrow Fe(OH)_3 + 4H^+ + 2SO_4^{2-} \tag{5.18}$$

This equation illustrates the strong acidification that occurs as pyrite is oxidised. In deep groundwater, the dissolved oxygen concentration is limited because oxygen cannot be replenished by exchange with the free atmosphere. Since aerated groundwater in equilibrium with the atmospheric partial O_2 pressure contains about 0.33 mmol l^{-1} (= 10.6 mg l^{-1}) O_2, the maximum increase in sulphate caused by the complete pyrite oxidation according to Equation (5.18) is 0.18 mmol l^{-1} (= 17.3 mg l^{-1}) SO_4^{2-} (Appelo and Postma, 1996). Obviously, the oxygen is also consumed by the decomposition of sediment organic matter, so these figures are maximum values. The pH usually remains unchanged because of sediment buffering, i.e. cation exchange and dissolution of carbonates. Therefore, pyrite oxidation and the subsequent dissolution of calcite often cause an increase in both Ca^{2+} and HCO_3^- concentrations (see Equation 5.11). If the pyrite-containing sediments are drained and/ or exposed to air, the pyrite oxidation is not hampered by oxygen limitation. The slow oxidation of mineral sulphides in these sediments is non-biological until the pH reaches a value of about 4. Below this pH the *Thiobacillus ferrooxidans* bacteria are the most active

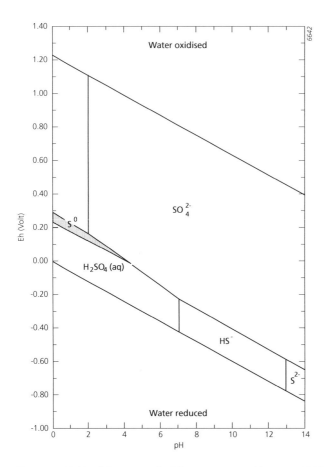

Figure 5.8 Fields of dominance of sulphur species at equilibrium at 25 °C and 1 atmosphere pressure. The total dissolved sulphur activity is 1 mmol l^{-1} (= 96 mg SO$_4^{2-}$ l^{-1}) (source: Hem, 1989).

oxidisers. This gives rise to a rapid increase of the acidity, causing the pH to plummet to values around 2 and the SO$_4^{2-}$ concentrations to soar. Soils that are rich in sulphides and have become strongly acidic after being drained and aerated are called acid sulphate soils. In Dutch these soils are referred to as *Katteklei* (cat clay). Furthermore, oxidation of pyrite and other metal sulphides is primarily responsible for the extreme acidity of acid mine drainage. The strong acidity resulting from oxidation of sulphides may result in toxic concentrations of aluminium, iron, manganese, and hydrogen sulphide (H$_2$S).

EXERCISES

1. Consider the samples of mineral water presented in Figure 5.2.
   a. Calculate the charge imbalance of these samples.
   b. Do you think the analyses are reliable?
   c. Calculate the hardness of these samples in mg l^{-1} and in German degrees of hardness.

2. What is the advantage of depicting analysis results in Piper and Stiff diagrams?

3. What is the natural origin of dissolved sodium, potassium, magnesium, and sulphate in unpolluted groundwater?

4. Explain why K fixation affects the availability of K for plant uptake.

5. At which pH is aluminium least soluble? What could be the cause of detectable amounts of Al being present in water?

6. The Cl^- concentration in precipitation is measured as 4.7 mg l^{-1} and the annual precipitation is 800 mm.
   a. Calculate the local evapotranspiration rate if the Cl^- concentration in groundwater is 15 mg l^{-1}.
   b. What are the assumptions in the above calculation?

7. What is the difference between alkalinity and TIC? Both alkalinity and TIC are usually reported as mg l^{-1} or mmol l^{-1} of HCO_3^-. Why?

8. Calcite ($CaCO_3$) in soil and sediment dissolves as a result of acids that are present in the percolating water. These acids (e.g. carbonic acid (CO_2) and nitric acid (HNO_3)) may be naturally present or anthropogenic in origin. The supply of additional acids of anthropogenic origin and its consequences for soil and water is called acidification.
   a. Give the reaction equation for the dissolution of calcite by carbonic acid (CO_2).
   b. Give the reaction equation for the dissolution of calcite by nitric acid (HNO_3).
   c. Will acidification lead to a decrease in pH in soils containing calcite?
   The table below lists the analysis results for Ca^{2+} and HCO_3^- concentrations in shallow groundwater (< 10 m below the surface) in a sandy, unconfined aquifer below fields of grass.

Sample no.	Ca^{2+} (mg l^{-1})	HCO_3^- (mg l^{-1})
GW1999-Gr03	61	182
GW1999-Gr06	204	488
GW2000-Gr02	98	234
GW2001-Gr05	105	236
GW2001-Gr08	305	753
GW2002-Gr11	124	298

   d. What is the average Ca^{2+}:HCO_3^- molar ratio for the above samples?
   e. What can you conclude about the contribution of acidification to calcite dissolution in the area?

9. Give the half reaction equation for the reduction of sulphate to sulphide. What will happen to the sulphide ion in reduced environments?

6

Nutrients

6.1 INTRODUCTION

Nutrients are defined as the raw materials that are assimilated by living organisms to promote growth, development, and reproduction. The types of nutrients required and the amounts in which they are consumed vary for the different plant and animal species. In general, nutrients include proteins, carbohydrates, fats, inorganic salts and minerals, and water. However, the term 'nutrients' is more commonly applied to essential elements such as nitrogen and phosphorus. Plants need at least sixteen essential elements to grow. They utilise oxygen, carbon, and water from air and soil. The primary **macronutrients** are nitrogen (N), phosphorus (P), and potassium (K). The secondary nutrients include calcium (Ca), magnesium (Mg), and sulphur (S). Micronutrients are absorbed in lesser amounts and include chloride (Cl), boron (B), iron (Fe), manganese (Mn), molybdenum (Mo), copper (Cu), and zinc (Zn). Some plants need some other elements, often called beneficial elements, such as cobalt (Co), nickel (Ni), silicon (Si), sodium (Na), and vanadium (V). Most of these elements are described in other chapters (Chapters 5 and 7); this chapter focuses mainly on nitrogen and phosphorus.

The reserves of nutrients in soil and water often deviate from those needed by plants; often, therefore, optimal plant growth is hampered because not enough of one or more of these nutrients is available. Because the primary nutrients N, P, and K are so important for plant growth, they are most often the limiting nutrients and are the main constituents of fertilisers. In most cases, the most limiting nutrient is N or P. The 'optimal' molar **N:P ratio** for plants, in which N and P are in balance and neither N nor P is limiting, is about 16:1 (Redfield *et al.*, 1963). This corresponds to a mass ratio of 7.2:1, although some experts use a mass N:P ratio of 10:1 (Thomann and Mueller, 1987). This implies that a mass N:P ratio greater than 10 indicates limitation of P. In this case, plant growth is largely controlled by the concentration of P. An N:P ratio of less than 7.2 indicates N limitation, which implies that N controls plant growth. Deficiencies of the secondary nutrients Ca, Mg, Fe, and S are less common, although plants may absorb large amounts of these elements.

An important property of nutrients is that they undergo a relatively rapid cycling within ecosystems. **Nutrient cycling** includes all the processes by which nutrients are transferred from one organism to another (see Figure 6.1). Nutrients are taken up by primary producers (e.g. green plants, algae, and mosses) from the soil solution or surface water, whereby they enter the food web and become part of living organic materials. The nutrients are further transferred to higher **trophic levels** (i.e. the position of organisms in the food chain or food web) by grazing and predation, until the organic material (the whole or parts of an organism, litter, or faeces) dies and becomes detritus. The dead organic materials are decomposed by microorganisms (bacteria and fungi), which use the organic matter for nutrition. Eventually, the organic matter is mineralised, which makes the nutrients available again for primary producers. Additional local inputs of nutrients come from atmospheric deposition, fertilisers, and human wastes. In addition, N in soil may come from the fixation of atmospheric N_2 by

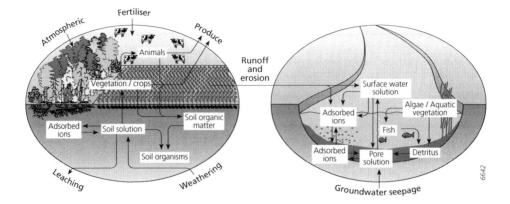

Figure 6.1 Schematic overview of nutrient cycling in terrestrial and aquatic ecosystems.

some soil bacteria, and P may be released by slightly soluble inorganic minerals. Nutrients are withdrawn from local cycling due to crop harvest, leaching to deeper soil horizons beyond the reach of plant roots, release of gaseous compounds to the atmosphere (only nitrogen), storage in peat or other organic deposits, or lateral runoff and erosion to downstream areas. Most of these local nutrient sources and sinks are part of the nutrient cycle at the regional or global scale.

Nutrient cycling is a natural process, but it may be enhanced or accelerated by external nutrient inputs of anthropogenic origin, such as fertilisers and wastes. Obviously, the enhancement is intentional in farming, to increase agricultural production, but nutrients of anthropogenic origin may also enter natural ecosystems via various hydrological or atmospheric pathways (e.g. surface runoff, atmospheric deposition, groundwater discharge). The resulting enhanced nutrient cycling is also referred to as ***eutrophication***.

The condition of an ecosystem can be described in terms of its trophic state, i.e. its degree of eutrophication or the lack thereof. Three designations are used: 1) oligotrophic, i.e. low productivity, ecosystems; 2) mesotrophic, i.e. intermediate productivity ecosystems, and 3) eutrophic, i.e. high productivity ecosystems. In both terrestrial and aquatic ecosystems, eutrophication causes an increase of biomass production and a loss of biodiversity. Furthermore, in surface waters, the increased nutrient levels may result in an excessive growth of aquatic weeds and algae (phytoplankton). The excessive growth of phytoplankton is also known as algal bloom. This increased production of aquatic plants has several consequences for water uses (Thomann and Mueller, 1987):

1. Large diurnal variations in dissolved oxygen (DO): by day phytoplankton produce oxygen through photosynthesis, but at night they consume oxygen through respiration. The excess occurrence of algae then causes a rapid depletion of DO, which, in turn, can result in fish dying.
2. Phytoplankton and weeds settle to the bottom of the water system and create a sediment oxygen demand (SOD). In deep lakes and reservoirs, this results in low values of DO in the hypolimnion.
3. Extensive growth of rooted aquatic macrophytes interferes with channel carrying capacity, aeration, and navigation.
4. Aesthetic and recreational drawback: algal mats, decaying algal clumps, odours, and discolouration may occur.
5. Public health drawback: some blue algae (cyanobacteria) species secrete toxins that can be harmful to the liver. Blooms of these algae in summer may make surface water bodies unsuitable for swimming.

Nutrient cycling causes the nutrient concentrations in the dissolved phase in the soil solution or surface water to vary greatly over the seasons. In the temperate climate zone, the nutrients are absorbed in large amounts by plants at the beginning of the growing season during springtime, so that the concentrations are considerably reduced. Therefore, nutrient concentrations in surface water and the soil solution are generally lower in summer than in winter. During autumn, when plants produce much detritus, the concentrations increase again. The chemistry of nitrogen and phosphorus is quite different. This causes notable spatial differences between nitrogen and phosphorus with respect to their occurrence, speciation, mobility, and bioavailability. In the following sections the chemical properties of nitrogen and phosphorus will be explained in more detail.

6.2 NITROGEN

6.2.1 Environmental role and occurrence of nitrogen

Nitrogen may occur in the +5, +3, 0, and -3 oxidation states, and occurs in the environment as the gaseous compounds free nitrogen (N_2), nitric oxide (NO_x)), nitrous oxide (N_2O)), and ammonia (NH_3), as nitrate (NO_3^-), nitrite (NO_2^-), ammonium (NH_4^+), or as organic nitrogen (in the form of plant material or other organic compounds). It is a fundamental constituent of nucleic acids, proteins, and plant chlorophyll. The major dissolved forms of nitrogen are ammonium and nitrate. Table 6.1 shows some typical ranges of concentrations of nitrogen species in soil, groundwater, and surface water, which includes variations in both space and time in unpolluted and polluted situations. This table demonstrates that there are huge variations (several orders of magnitude) in the concentrations.

Ammonium is very soluble in water, but is also readily adsorbed to the negatively charged cation exchange sites of colloidal particles (clay minerals and organic matter). ***Ammonium fixation*** is the process by which clay minerals (especially illite, vermiculite, and montmorillonite) capture ammonium tightly between the mineral lattices (see potassium fixation, Sections 4.2.3 and 5.3). Most of this fixed ammonium is irreversibly bonded and cannot be exchanged with the bulk solution. The amount of ammonium fixation is much less compared to the quantities adsorbed to cation exchange sites. Ammonium is a weak acid and under basic conditions is converted into volatile ammonia:

$$NH_4^+ \quad \rightarrow \quad NH_3 \quad + \quad H^+ \quad (pK_a = 9.24) \tag{6.1}$$

Like ammonium, nitrate is very soluble in water, but because nitrate does not interact with soil particles, it is rather mobile and readily moves downstream with soil moisture,

Table 6.1 Typical concentration ranges of nitrogen species in soil, groundwater, and surface water.

	Soil[1]	Soil solution[2]	Groundwater[3]	Surface water[3]	Wastewater[4]
	g N kg^{-1}	mg N l^{-1}	mg N l^{-1}	mg N l^{-1}	mg N l^{-1}
Total N	0.8 – 9.7	-	-	0.56–17	-
NO$_3^-$	-	1.3–280	trace–35	0.04–10.5	0.04–2.5
NH$_4$	-	1.4–28	trace–9	0.004–7.5	1.5–45

1 Source: De Bakker (1979)
2 Source: Miller and Gardiner (2004)
3 Source: EEA (2006a)
4 Source: Van der Perk (1996)

groundwater, or surface water until it is taken up by biota or transformed through redox reactions (see next section).

Although nitrogen is a key nutrient, inorganic aqueous nitrogen, like any substance, is toxic to plants, animals, and humans at high concentrations. Elevated levels of nitrate or nitrite in drinking water have been known to cause a potentially fatal blood disorder in infants below six months of age, called methemoglobinemia or 'blue-baby syndrome', in which the oxygen-carrying capacity of blood is reduced. High concentrations of ammonium in the soil solution are toxic to plants. Furthermore, free ammonia formed from ammonium under basic conditions in surface water is highly toxic to fish. The role of nitrogen as a key factor in the process of eutrophication and the development of algal blooms was discussed above.

6.2.2 Nitrogen cycle

Primary producers and some bacteria absorb nitrogen mainly in the form of aqueous nitrate or ammonium. Because ammonium is essentially toxic inside cells, it is assimilated into organic nitrogen very quickly. Some plants (papilionaceous legumes, such as lupine, lucerne, and clover, and some other plants) host *Rhizobium* bacteria or cyanobacteria, which are able to fix nitrogen (N_2) from the atmosphere. This process is known as **nitrogen fixation**. For this reason, legumes are commonly included in agricultural crop rotations on poorly fertile soils to increase the amount of soil nitrogen and to reduce the need for fertiliser applications. Organic nitrogen is made available by the decomposition and subsequent mineralisation of organic matter by heterotrophic bacteria (i.e. bacteria that derive their cell carbon from organic carbon) and fungi. Organic nitrogen is first oxidised to amino acids and finally to ammonium (**ammonification**).

Under oxic conditions, autotrophic bacteria oxidise ammonium (3- oxidation state) to nitrate (5+ oxidation state) in a multiple-step **nitrification** process (see Figure 6.2). The overall nitrification process can be summarised as a two-step process:

$$2NH_4^+ \; + \; 3O_2 \; \rightarrow \; 2NO_2^- \; + \; 4H^+ \; + \; 2H_2O \tag{6.2a}$$

$$2NO_2^- \; + \; O_2 \; \rightarrow \; 2NO_3^- \tag{6.2b}$$

Nitrosomonas bacteria carry out the first step, the oxidation of ammonium to nitrite, and *Nitrobacter* bacteria carry out the second step, the final oxidation to nitrate. Nitrite is very reactive, and the second reaction step (Equation 6.2b) proceeds very fast. Therefore, nitrite generally only occurs in small amounts in soil and the primary nitrogen species in aqueous solutions are nitrate and ammonium. The nitrification reaction requires oxygen, so the reaction only takes place in well-aerated soils or surface waters. Furthermore, because the nitrification reaction is carried out by live bacteria, the reaction rate depends greatly on environmental factors such as temperature and pH. At temperatures below 10 °C the reaction is inhibited. Between 10 °C and about 32 °C the reaction rate increases with temperature. The optimal pH for nitrification is between 6.6 and 8. The reaction is slowed at pH less than 6 and comes to a standstill at pH less than 4.5.

Under slightly reducing conditions, for example due to the biologically mediated decomposition of organic matter (see Section 4.3.4) or pyrite oxidation (see Section 5.11), the process of **denitrification** breaks down nitrate (5+ oxidation state) to N_2 (0 oxidation state) through intermediates including nitrite, nitric oxide, and nitrous oxide:

$$2\,NO_3^- \;+\; 10\,e^- \;+\; 12\,H^+ \;\rightarrow$$
$$2\,NO_2^- \;+\; 6\,e^- \;+\; 8\,H^+ \;+\; H_2O \;\rightarrow$$
$$2\,NO \;+\; 4\,e^- \;+\; 6\,H^+ \;+\; 4\,H_2O \;\rightarrow \qquad\qquad (6.3)$$
$$N_2O \;+\; 2\,e^- \;+\; 2\,H^+ \;+\; 5\,H_2O \;\rightarrow$$
$$N_2 \;+\; 6\,H_2O$$

This reduction reaction is mediated by bacteria as well and is thus temperature dependent. Note that the denitrification process is not the opposite to the nitrification process. The intermediate products NO and N_2O are gases that can volatilise from the system before the denitrification reaction has been completed. Both NO and N_2O are greenhouse gases and so their emission contributes to global warming.

Microorganisms utilise substantial quantities of inorganic nitrogen that thereby become part of organic matter. This process is called ***immobilisation*** and is the opposite reaction to mineralisation. Together with plant uptake, immobilisation is the process of assimilatory reduction. Immobilisation proceeds faster than mineralisation when nitrogen is limiting relative to organic carbon. This means that the C:N ratio (see Section 4.3.4) determines whether mineralisation or immobilisation dominates. For soil microbial biomass the typical C:N ratio is 20:1. Thus, if the C:N ratio is less than 20, mineralisation prevails and

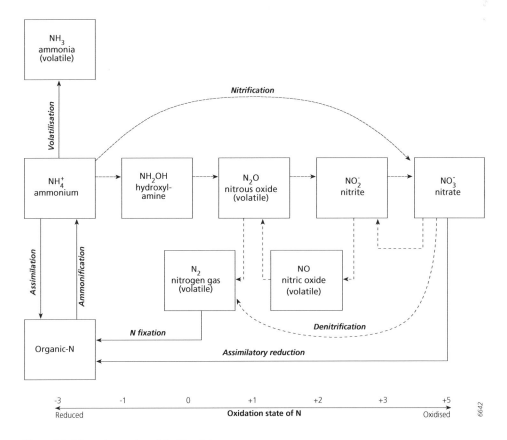

Figure 6.2 Schematic overview of the N-cycle in ecosystems.

a C:N ratio of greater than 20 brings about a net immobilisation. Immobilised nitrogen is temporarily unavailable for plant uptake. Ammonification, nitrification, denitrification, and assimilation of nitrate and ammonium are the major process of the nitrogen cycle. In general, nitrogen fixation and ammonium volatilisation are relatively insignificant in comparison with the other nitrogen fluxes. Figure 6.2 is a schematic diagram of the entire N-cycle including N-fixation, ammonification, nitrification, and denitrification.

Human activities have considerably enhanced the natural global nitrogen cycle in a number of ways. First, the combustion of fossil fuels is currently adding 20-30 Tg yr^{-1} of reactive N species to the atmosphere. Second, legume crops are responsible for an additional 40 Tg yr^{-1} of N$_2$ being fixed naturally from the atmosphere. Third, the invention of the Haber-Bosch chemical process in 1913 has enabled the artificial (i.e. non-biological) fixation of atmospheric N$_2$ to ammonium (NH$_4^+$) for the production of artificial fertilisers. As a result, the global artificial nitrogen fixation rate has increased from about 5 Tg yr^{-1} just after the second World War to about 80 Tg yr^{-1} at present (Galloway *et al.*, 1995; Vitousek *et al.*, 1997a; Galloway, 1998). This means that nowadays more nitrogen is fixed artificially by humans than by natural fixation processes in terrestrial ecosystems (Vitousek *et al.*, 1997b; Crutzen and Steffen, 2003).

6.2.3 External sources and sinks

Figure 6.2 demonstrates that volatilisation of NH$_3$, NO, and N$_2$O are sinks in the N-cycle of ecosystems. Other sinks include the harvest of N-containing biomass and leaching or runoff of dissolved nitrogen to downstream areas. Leaching occurs mainly in the form of nitrate, because, as mentioned above, nitrate is very mobile. In addition to nitrogen fixation by soil bacteria, other important external sources of nitrogen sources are fertiliser inputs, inflow of contaminated water from upstream areas, and atmospheric deposition. Fertiliser inputs depend on initial soil fertility, crop type and intended crop yield, and other possible sinks in the N-cycle. In intensive cropping systems, the N input by fertilisers may range from 50 to more than 250 kg N ha^{-1} y^{-1}. Inputs by atmospheric deposition are derived from nitrogen emissions by industry and traffic (mainly nitrogen oxides (NO$_x$)) and agriculture (mainly ammonia (NH$_3$)). Current rates of atmospheric N deposition in Europe and North America have increased by a factor of 5–20 compared to pre-industrial times (25–100 kg N ha^{-1} y^{-1} compared to 5–10 kg N ha^{-1} y^{-1}) (Hatch *et al.*, 2002). In areas with intensive agriculture, the dominant form of atmospheric N deposition is the deposition of ammonia-N. In 2001, the average total N deposition in the Netherlands was 32 kg N ha^{-1} y^{-1} of which about 70 percent was ammonia-N (RIVM, 2002). Locally, deposition rates may depart considerably from these average values.

By far the greatest emission of ammonia (83 percent) occurs during the decomposition of manure. The volatilisation of ammonia from manure is positively related to the pH, temperature, and moisture content of the manure. Once in the atmosphere, ammonia is one of the most important gases that neutralise atmospheric nitric acid and sulphuric acid originating respectively from the oxidation of nitrogen oxides (NO$_x$) and sulphur dioxide (SO$_2$). This process causes the formation of ammonium aerosols ((NH$_4$)$_2$SO$_4$ and NH$_4$NO$_3$). Under normal atmospheric conditions in Europe, sulphuric acid occurs in hydrated form, which accelerates the neutralisation reaction considerably. The formation of the (NH$_4$)$_2$SO$_4$ aerosol is irreversible. The formation of the NH$_4$NO$_3$ is reversible and slows down with increasing temperature. Both ammonia and ammonium aerosols may be dissolved in the droplets of moisture in clouds.

Ammonia and its derivatives reach the Earth's surface via different mechanisms:
* Dry deposition: removal of gases and aerosols from the atmosphere without the interference of precipitation (rain, snow, hail).

- Wet deposition: wash-out of gases and aerosols from the atmosphere in solution by precipitation.
- Occult deposition: removal of gases and aerosols from the atmosphere through diffusion across the interface between air and wet surface, in fog and dew drops.

In contrast to nitrogen dioxide, the transport range of ammonia in the atmosphere is limited and dry deposition is mainly concentrated around local emission sources: about 20 percent of the emitted ammonia is deposited within 5 km of the source (Erisman *et al.*, 1988). Atmospheric deposition is locally affected by aerodynamic surface roughness controlled by local surface topography and land cover. The presence of hills and transitions in the height and structure of vegetation cover causes airflow perturbations, which alter the rates of dry and wet atmospheric deposition. Because forest canopies have a large roughness and deposition surface, the deposition rate is higher on forests than on other vegetation surfaces. The deposition rate is especially enhanced where sharp transitions occur in surface roughness: for example, in the zone from 50 to 100 m from the forest edge the rate of atmospheric deposition of N may be 50 percent higher (Ivens *et al.*, 1988; Draaijers, 1993) (see Section 16.3.3). The nitrogen deposited by dry and occult deposition reaches the soil surface as a consequence of rainfall via throughfall (i.e. the precipitation that falls through or drips off of the plant canopy) and stemflow (i.e. the precipitation intercepted by plants that flows down the stem to the ground). Furthermore, atmospheric deposition rates follow local weather patterns. Obviously, the deposition rates increase with increasing amounts of precipitation. Due to more stable atmospheric conditions and lower wind speeds, the deposition rates are lower at night than during the day. During springtime, when large amounts of manure are applied, the ammonia deposition is largest. During autumn and winter, as deciduous trees shed their leaves, the deposition surface is reduced and, accordingly, so are the atmospheric deposition rates.

As soon as ammonium reaches the soil, it is nitrified to nitrate, producing nitric acid. For every ammonium ion that is nitrified, 4 protons are released (see Equation 6.2). As a consequence, the process of deposition of ammonium aerosols contributes substantially to the acidification of natural ecosystems. In poorly buffered ecosystems, such as coniferous forests, heaths, and oligotrophic lakes and peat bogs, the processes of eutrophication and acidification are closely interrelated due to the atmospheric nitrogen inputs.

6.3 PHOSPHORUS

6.3.1 Environmental role and occurrence of phosphorus

Phosphorus is the second key plant nutrient. In living organisms, phosphorus makes up part of critical proteins, such as DNA. Phosphate to phosphate ester bonds in ATP (adenosine triphosphate) and ADP (adenosine diphosphate) are the major energy storage and energy transfer bonds in cells. ATP is a high-energy nucleotide that has a ribose sugar (adenosine) and three phosphate groups. The breakdown of ATP releases a great deal of energy which the cell uses for its various activities. ADP has a ribose sugar and two phosphate groups. ADP is used to synthesise ATP with the energy released in cell respiration. When ATP is used for cellular activities, ADP is re-formed. In plant cells, ATP is produced in the mitochondria and chloroplasts. Besides the energy transformation within cells, phosphorus plays a role in cell growth, the stimulation of early root growth, and in fruiting and seed production. In animals and humans, phosphorus is essential for the growth of bones and teeth, which are primarily made from calcium phosphate.

In the environment, phosphorus occurs as organic phosphorus, i.e. as part of living or dead organic materials, or as inorganic orthophosphates, i.e. phosphoric acid (H_3PO_4) and

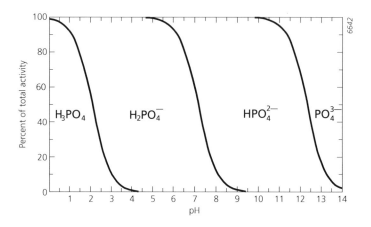

Figure 6.3 Percentages of phosphate species activities as a function of pH in a 10 mM phosphoric acid solution.

its conjugate bases ($H_2PO_4^-$, HPO_4^{2-}, and PO_4^{3-}). Orthophosphate is often abbreviated to phosphate. Similar to carbonic acid, the distribution of phosphate species is a function of pH (see Figure 6.3).

The most common method for the determination of orthophosphate is the molybdate blue reaction (Murphy and Riley, 1962). However, the sample preparation procedure that precedes the measurement can readily transform other phosphorus species into the reactive orthophosphate, resulting in overestimations of orthophosphate concentrations. Moreover, the Murphy–Riley molybdate blue procedure can also determine loosely bound inorganic and organic forms of phosphorus. Thus, the molybdate blue reaction is not specific to inorganic orthophosphate, but also determines other reactive forms of phosphorus. Therefore, Haygarth and Sharpley (2000) have proposed the more appropriate term 'reactive phosphorus' (RP) for phosphorus species measured using the molybdate blue reaction. Besides RP, the concentration of total phosphorus (TP) is also often reported. Total phosphorus includes all phosphorus species (inorganic and organic) and is determined through the molybdate blue reaction after digesting the water sample or through direct measurement (e.g. using inductively coupled plasma atomic emission spectrometry (ICP–AES) or inductively coupled plasma mass spectrometry (ICP–MS)). Water samples are commonly filtered through a 0.45 μm membrane filter to distinguish between dissolved and particulate fractions. But as discussed in Section 4.1, the filtered samples may contain < 0.45 μm sized particles or colloidal particles, which also may contain phosphorus. For this reason, Haygarth and Sharpley (2000) also recommend reporting the analytical results for phosphorus concentrations in water samples according to the filter size instead of using terms such as 'dissolved' or 'soluble'. The phosphorus species (RP or TP) is followed by a suffix (in parentheses) to indicate whether the sample was filtered (<0.45) or not (unf).

The major natural source of dissolved inorganic phosphorus is the mineral apatite ($Ca_5(F,Cl,OH)(PO_4)_3$). Phosphates are barely soluble in water and are not readily available for plant uptake. Under acid conditions (pH < 4.5), phosphate ions react with dissolved iron and aluminium ions and precipitate as ferric iron and aluminium phosphates. In addition, phosphates adsorb preferentially to surfaces of sesquioxides and edges of clay minerals. The adsorption capacity of clay minerals for phosphate is much less than the adsorption capacity of sesquioxides, because of the small edge surface of clay minerals. The specific phosphate adsorption to sesquioxides causes the OH and OH_2^+ ligands to be exchanged with phosphate (ligand exchange), as shown in the following reactions (Scheffer and Schachtschabel, 1989):

Table 6.2 Typical concentration ranges of phosphorus species in soil, groundwater, and surface water.

	Soil[1]	Soil solution[2]	Groundwater[3]	Surface water[4]	Wastewater[5]
	g P kg^{-1}	mg P l^{-1}	mg P l^{-1}	mg P l^{-1}	mg P l^{-1}
Total P	0.01–5.6	-	-	-	-
TP (< 0.45)	-	0.02–7.8	0.005–6.6	0.01–2.1	3–8
RP (< 0.45)	-	0.004–7.8	trace–4.2	0.003–1.9	3–6

1 Source: Whitehead (2000); Willems et al. (2002)
2 Source: Whitehead (2000); Sims et al. (1998)
3 Source: Willems et al. (2002); Kedziora et al. (1995)
4 Source: EEA (2006a)
5 Source: Van der Perk (1996); Corbett et al. (2002)

$$(Al, Fe) - OH_2^+ \quad + \quad H_2PO_4 \quad \leftrightarrow \quad (Al, Fe) - O - PO(OH)_2 \quad + \quad H_2O \qquad (6.4)$$

$$O \overset{\displaystyle (Al, Fe) - OH}{\underset{\displaystyle (Al, Fe) - OH}{\bigg\langle}} \quad + \quad H_2PO_4^- \quad \leftrightarrow \quad O \overset{\displaystyle (Al, Fe) - O}{\underset{\displaystyle (Al, Fe) - O}{\bigg\langle}} P \overset{\displaystyle O}{\underset{\displaystyle OH}{\big\langle}} \qquad (6.5)$$

$$+ \quad OH^- \quad + \quad H_2O$$

In reaction (6.5), a ring structure with two O-bridges is formed, in which the phosphate is strongly bonded. These type of specific adsorption can also occur at a pH above the point of zero charge, and so in a wide pH range. As a consequence, phosphate ions are bonded strongly in soils containing sesquioxides. Under reducing conditions, ferric iron is reduced to ferrous iron, which is much more soluble (see Section 5.6). This reductive dissolution of ferric hydroxides causes phosphate adsorbed to the ferric hydroxides to be released into solution too. Under these conditions, the phosphate concentration is controlled by the solubility of ferrous iron phosphate (vivianite; $Fe_3(PO_4)_2 \cdot 8H_2O$). Since the solubility product of vivianite is much higher than that of ferric iron phosphates, phosphate is much more soluble and is therefore mobile in acid to neutral anaerobic environments.

In the higher pH range (pH > 7), phosphates also precipitate as calcium phosphates, mainly in the form of hydroxyapatite ($Ca_5OH(PO_4)_3$) and fluor apatite ($Ca_5F(PO_4)_3$). The interactions of phosphates with iron, aluminium, sesquioxides, and calcium are collectively referred to as ***phosphate fixation***. As a result of phosphate fixation, the concentration of dissolved phosphates is often only one-twentieth or less than the concentration of nitrogen or potassium. Table 6.2 shows some typical ranges of concentrations of phosphorus species in soil, groundwater, and surface water, including variations in space and time in unpolluted and polluted situations.

6.3.2 Phosphorus cycle

The phosphorus cycle is roughly similar to the cycle of nitrogen through the food chain. However, it is much less complex than the nitrogen cycle, because phosphorus occurs in fewer forms and there are no volatile gaseous compounds.

Dissolved phosphate is used by primary producers and some heterotrophic bacteria. These organisms convert inorganic phosphate into organic phosphorus. Subsequently, part of the organic phosphorus is transferred into the food chain and the remains are left behind in detritus (e.g. phosphorus in plant roots). Ultimately, all living organic materials die and become detritus. The microbial decomposition of organic matter mineralises organically bound phosphorus to inorganic phosphates. The turnover rate of organic phosphorus is rapid in conditions favourable for microorganisms. If the C: organic P ratio is about 200:1 or smaller, phosphorus is readily mineralised and released into solution and becomes available for plant uptake. If the C: organic P ratio is larger than 300:1 (less than about 2 g kg^{-1} P in organic matter), the microorganisms use most of the phosphorus and immobilise it in their cells instead of releasing it for plant uptake (Miller and Gardiner, 2008). Part of the phosphorus, however, remains in organic form. Some of the organic phosphorus is present in complex humus polymers, but most (about 60 percent) is present in the form of small molecular compounds, such as inositol phosphates. Inositol phosphates are primarily bacterial in origin and occur mainly in the form of insoluble Ca, Fe, and Al salts. They are only slowly mineralised. In general, the soil organic phosphorus correlates well with organic matter content and organic nitrogen, but the C: organic P ratio displays more variation than the C:N ratio, because organic phosphorus is less associated with large humus polymers than organic nitrogen (Whitehead, 2000).

As mentioned above, the availability of phosphorus is generally low, because of its affinity with iron, aluminium, sesquioxides, and calcium. Phosphorus availability depends on soil pH, redox conditions, and organic matter content. Phosphorus available for plant uptake is greatest at pH values between 6 and 7, or under anaerobic conditions.

6.3.3 External sources and sinks

The major external input of phosphorus in agricultural soils is phosphorus in fertilisers (artificial fertilisers, manure, or sewage sludge) and may amount up to 1000 kg P ha^{-1} y^{-1}. Table 6.2 lists the main fertilisers and fertiliser constituents that contain phosphorus. Artificial fertilisers are initially very soluble, but the phosphates are increasingly adsorbed, precipitated, or immobilised by microorganisms. Fertiliser application may cause phosphorus in the topsoil to build up substantially over many years, to up to 2–3 times the initial phosphorus content.

Atmospheric deposition of phosphorus is usually negligible. Wet and dry phosphorus deposition generally vary between 0.2 and 1.5 kg P ha^{-1} y^{-1}. Atmospheric inputs of phosphorus are mainly derived from suspended soil particulates eroded by wind, though there may be a small contribution from the burning of plant materials and fossil fuels (Whitehead, 2000).

The main phosphorus losses are due to crop harvest and grazing, which lead to a removal of 5 to 40 kg P ha^{-1} y^{-1}. However, if grazing is accompanied by supplementary feed, there is usually a net phosphorus input from the excreta of the grazing animals. Because phosphorus has a great affinity for soil minerals, leaching of phosphorus rarely occurs. Nevertheless, phosphate leaching may occur on agricultural areas where fertiliser applications are so high that the phosphate binding capacity becomes saturated. This phenomenon has been reported in intensively used agricultural areas of north-western Europe. In the case of a shallow water

Table 6.3 Some main types of phosphate fertilisers (source: Whitehead, 2000; Miller and Gardiner, 2004).

Fertiliser	Chemical formula	%P	Solubility
Superphosphate	$Ca(H_2PO_4)_2 + CaSO_4 \cdot 2H_2O$	8–9	High
Triple superphosphate	$Ca(H_2PO_4)_2$	20	High
Monoammonium phosphate	$NH_4H_2PO_4$	26	High
Diammonium phosphate	$(NH_4)_2HPO_4$	23	High
Dicalcium phosphate dehydrate	$CaHPO_4 \cdot 2H_2O$	18	Low
Tricalcium phosphate	$Ca_3(PO_4)_2$	12–16	Low

table, phosphate breakthrough to the groundwater may occur. In anaerobic groundwater, phosphate is much more mobile, because of the absence of ferric hydroxides.

Another phosphorus sink on agricultural land is soil loss due to soil erosion; this may be up to 60 kg P ha^{-1} y^{-1}. Soil erosion is greater on arable land than on grassland and forested land and increases with increasing slope gradient. Most of the eroded soil particles are redeposited at the foot of slopes and may ultimately enter a river network.

Besides inputs of eroded soil, the principal external sources of phosphorus in surface water consist of effluents from wastewater treatment plants, untreated wastewater discharges, and industrial releases. Phosphates, primarily sodium tripolyphosphate, are used as a laundry detergent additive to eliminate free Ca^{2+} during the washing process. Phosphate detergents may contribute up to one third of the total phosphorus load of effluent from wastewater treatment plants. The rest is primarily from human excrement and wasted foodstuffs. Since the introduction of phosphate-free detergents during the 1980s and the implementation of tertiary wastewater treatment (see Section 3.4.3 and Box 3.I), the phosphorus loads from wastewater treatment plants have been reduced substantially. In surface water, too, phosphates are mainly associated with particulate matter, and the deposition of these suspended particles promotes the removal of phosphates from the water column. As a consequence, phosphorus may build up in the bottom sediments of rivers and lakes. If, as is generally the case, the redox potential in bed sediments is low, the iron hydroxides that bond phosphate dissolve. Therefore, the phosphate concentrations in the pore water of bed sediments are much larger than the concentrations in the overlying water column. The resulting concentration gradient between the pore water of bed sediments and surface water may cause a net phosphorus flux from the bed sediment to the surface water. This process is called ***internal loading***, as it refers to the recycling of previously present phosphorus in the water system. Internal loading may continue to cause phosphorus releases into surface waters many years after the direct releases have ceased. The effect of internal loading is greater in shallow lakes than in rivers or deep lakes. This is because the length of the water column over which the released phosphorus is distributed is greater and the residence time is longer in deep lakes than in shallow lakes and rivers. The process of internal phosphorus loading may seriously hamper the ecological rehabilitation of shallow lakes that have had a long history of external phosphorus loading (Van der Molen and Boers, 1994).

EXERCISES

1. Define the following terms
   a. Micronutrients
   b. Eutrophication
   c. Nitrogen immobilisation
   d. Nitrogen fixation
   e. Ammonium fixation
   f. Phosphate fixation
   g. Ammonification
   h. Nitrification
   i. Denitrification
   j. Internal loading

2. The water in a lake contains 0.63 mg N l^{-1} and 0.041 mg P l^{-1}. Is this lake N or P limited?

3. Name four consequences of eutrophication of aquatic ecosystems.

4. Describe the fate of nitrogen that has been added as fertiliser on arable land.

5. Explain why atmospheric deposition of aerosols is greater near the forest edge than further away.

6. Is nitrate leaching from the root zone greater:
   a. During summer or winter?
   b. In dry sandy soils or moist loamy soils?
   c. In soils with a shallow or deep water table?
   d. In soils rich or poor in organic matter?
   Give reasons for your answers.

7. Is ammonium volatilisation greater:
   a. During summer or winter?
   b. From acid or neutral soils?
   c. In soils rich or poor in organic matter?
   Give reasons for your answers.

8. What is the principal natural source of phosphorus in soil?

9. Indicate in a pH–Eh diagram (see Figure 2.3) the region where phosphate is most mobile.

7

Heavy metals

7.1 INTRODUCTION

The term heavy metals (or trace metals) is applied to the group of metals and semimetals (metalloids) that have been associated with contamination and potential toxicity or ecotoxicity; it usually refers to common metals such as copper, lead, or zinc. However, the term is only loosely defined and there is no single authoritative definition (see Duffus, 2002). Some define a heavy metal as a metal with an atomic mass greater than that of sodium, whereas others define it as a metal with a density above 3.5–6 g cm^{-3}. As mentioned above, the term is also applied to semimetals (elements that have the physical appearance and properties of a metal but behave chemically like a non-metal), such as arsenic, or non-metals, such as selenium, presumably because of the hidden assumption that 'heaviness' and 'toxicity' are in some way identical. Despite the fact that the term heavy metals has no sound terminological or scientific basis, it is used here in the way it has been used in much of the scientific environmental literature, namely to refer to metals or semi-metals which meet the definitions given above. Common heavy metals include zinc (Zn), copper (Cu), lead (Pb), cadmium (Cd), mercury (Hg), chromium (Cr), nickel (Ni), tin (Sn), arsenic (As), and silver (Ag).

At their natural concentrations, many metals play an essential role in biochemical processes and are thus required in small amounts by most organisms for normally healthy growth (e.g. Zn, Cu, Se, Cr). Other metals, however, are not essential and do not cause deficiency disorders if absent (e.g. Cd, Pb, Hg, Sn, and the semi-metal As). Figure 7.1 shows typical dose–response curves for these two types of trace metals. If ingested in excessive quantities, virtually all heavy metals are toxic – especially to animals and humans – although organisms are also able to adapt themselves, at least partly, to increased levels of metals. Most heavy metals accumulate in organism tissues (bioaccumulation) and as they are transferred through the food chain (biomagnification). Metals generally produce their toxicity by forming complexes with organic compounds (ligands). The modified molecules lose their ability to function properly, causing the affected cells to malfunction or die. Metals commonly bind to biological compounds containing oxygen, sulphur, and nitrogen, which may inactivate certain enzyme systems. This is especially true for enzymes that are directly or indirectly involved in ATP production. In acute poisoning, large excesses of metal ions can disrupt membrane and mitochondrial function and the generation of free radicals. In most cases this leads to general weakness and malaise.

In the environment, most heavy metals are present as cations, though some semi-metals may occur as oxyanions (e.g. arsenate AsO_4^{3-}). Heavy metals occur naturally in the Earth's crust as impurities isomorphously substituted for various macroelement constituents in the lattices of many primary and secondary minerals. The heavy metal content varies greatly both within and between different types of rocks. Typical ranges of heavy metal concentrations in major igneous and sedimentary rock types, as well as in fresh water, are given in Table 7.1. The maximum concentrations of trace elements are commonly found in areas near ore

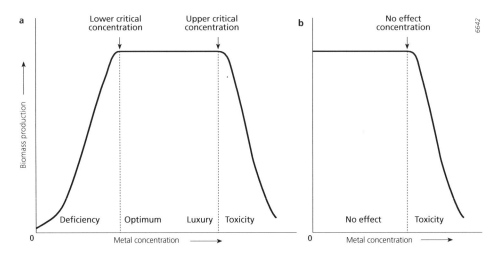

Figure 7.1 Typical dose–response curves for a) essential trace metals, and b) non-essential trace metals. Adapted from Alloway and Ayres (1997).

deposits, which are often associated with past or present volcanic activity. This may give rise to anomalously large natural concentrations in soil, groundwater, stream water and stream sediments. Nevertheless, such natural enrichment of metals may still be harmful to living organisms.

Enhanced environmental concentrations of heavy metals are often associated with mining and smelting. These activities cause air pollution and associated atmospheric deposition of contaminated dust. Most mine tailing ponds and heaps are potentially hazardous, because pyrite contained in the ores oxidises to form sulphuric acid. As a consequence, drainage waters from mine tailings are often very acid (pH less than 3) and carry large amounts of mobilised heavy metals and aluminium. Moreover, until the 1980s, slag containing high levels of leachable heavy metals was widely used for road metalling, and in this way the metals were dispersed over large areas. Other important potential anthropogenic sources of heavy metals include sewage sludge (when spread on the land), phosphate fertilisers, manure, and atmospheric fallout (from smelting, or from burning coal and gasoline), leaching

Table 7.1 Typical ranges of heavy metal concentrations in major igneous and sedimentary rock types, and fresh water (source: Alloway and Ayres, 1997; Smedley and Kinniburgh, 2001).

	Igneous rocks			**Sedimentary rocks**			
	Ultramafic mg kg^{-1}	Mafic mg kg^{-1}	Granitic mg kg^{-1}	Limestone mg kg^{-1}	Sandstone mg kg^{-1}	Shales/Clays mg kg^{-1}	Fresh water µg l^{-1}
As	1	1.5	1.5	2.6	4.1	13 (< 900)	0.2–230
Cd	0.12	0.13	0.09	0.028	0.05	0.22 (<240)	0.01–3
Cr	2980	200	4	11	35	39	0.1–6
Cu	42	90	13	5.5	30	39	2–30
Hg	0.004	0.01	0.08	0.16	0.29	0.18	0.0001–2.8
Ni	2000	150	0.5	7	9	68	0.02–27
Pb	14	3	24	5.7	10	23	0.06–120
Sn	0.5	1.5	3.5	0.5	0.5	6	0.0004–0.09
Zn	58	100	52	20	30	120	0.2–100

from building materials (roofs, gutters, pipes, lead slabs), deposition of contaminated river sediments, and direct domestic or industrial discharges and disposals. Computers, televisions, and other electronic equipment contain an array of trace materials, including lead, mercury, cadmium, and arsenic. In the past twenty years, the releases of heavy metals to the environment has been considerably reduced as a result of improved waste air and water purification techniques, waste recycling, and the implementation of more stringent environmental regulations. Figure 7.2 shows the declining trends in heavy metal releases from wastewater treatment plants (effluents) in the Netherlands in the period 1981–2011.

The principal geochemical processes controlling the retention of heavy metals in soil and water are adsorption and precipitation. For these processes the redox potential and pH are the key variables governing the distribution of metals between the solid and dissolved phases and, consequently, their dispersal in the environment and their bioavailability. In general, many solids control the fixation of heavy metals, namely clay minerals, organic matter, iron, manganese, and aluminium oxides and hydroxides for adsorption, and poorly soluble sulphide, carbonate, and phosphate minerals for precipitation (Bourg and Loch, 1995).

Under oxidised conditions, the major process controlling the speciation of heavy metals is adsorption to the negatively charged exchange sites of clay minerals and organic matter. In general, adsorption causes the heavy metals to be relatively immobile in soils. Many metals show specific adsorption and compete actively with protons for surface sites. They may even be adsorbed on mineral and organic matter surfaces that are positively charged. Nevertheless, the amount of adsorbed metals decreases with decreasing pH. Another reason why the pH is often found to be the most important factor determining the distribution coefficient of heavy metals in soil and sediment is the specificity of heavy metals for surfaces that can deprotonate (Appelo and Postma, 1996). At a given pH, the concentration in the dissolved phase is approximately proportional to the concentration adsorbed to the solid phase (see Figure 7.3a). Some metals (e.g. copper and lead) also tend to form complexes with dissolved and sediment organic matter, some of which are mobile. This process of ligand formation increases with decreasing pH. At high pH values, heavy metals may also precipitate as carbonates or hydroxides. Furthermore, heavy metals may be removed from an aqueous solution due to ***coprecipitation*** (i.e. the inclusion of additional species within or on the surface of a precipitate as it is formed) with calcite or iron, aluminium,

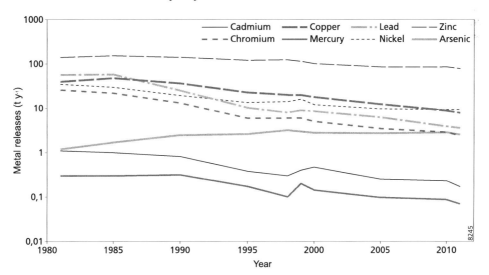

Figure 7.2 Heavy metal releases by effluent discharges from wastewater treatment plants in the period 1981–2011. (source: RIVM (2003) and Rijkswaterstaat (2013)).

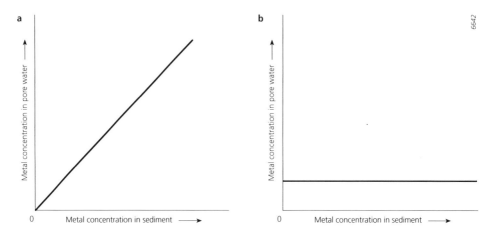

Figure 7.3 Relation between metal concentrations in sediment and pore water under oxidised (a) and reduced (b) conditions.

and manganese oxyhydroxides. It should be clear from the above that the pH is the master variable determining the mobility of heavy metals under oxidising conditions, as it controls adsorption, complexation, and precipitation. All these processes bring about a decrease in the mobility of heavy metals with increasing pH.

Under reduced conditions, the mobility of most metals is further decreased due to the formation of barely soluble sulphide minerals. In this case, the concentration of heavy metals in the dissolved phase is controlled by the solubility product of the sulphide minerals, which means that the total concentration of heavy metals barely influences the concentration of dissolved heavy metals (Salomons and Förstner, 1984) (see Figure 7.3b).

7.2 ZINC

Zinc is found mainly in the 2+ oxidation state and occurs in metalliferous ores as zinc sulphide (ZnS; sphalerite). In general, zinc has about the same abundance in crustal rocks as copper and nickel (see Tables 1.1 and 7.1), but tends to be more soluble in most types of natural water (Hem, 1989). The concentrations of zinc in soil are usually the highest of all heavy metals. Zinc is subject to specific adsorption in soils and sediments, and the strength of the bond tends to increase with the ageing of zinc (i.e. how long ago the soluble zinc was released).

Zinc is an essential trace nutrient and forms part of many enzymes needed for growth and development and DNA synthesis. Zinc metalloenzymes play an important role in many aspects of cellular metabolism, including DNA replication, repair and transcription protein synthesis, and energy metabolism. In animals and humans, it also plays important roles in neurosensory functions, insulin synthesis, and cell-mediated immunity, and is required in sufficient levels to ensure and maintain the body's immunity, strength, and general well-being. Normal average zinc intake by adult humans through the diet ranges from 7 mg d^{-1} to 16 mg d^{-1}. Food may contain levels of zinc ranging from 6 mg kg^{-1} dry matter (e.g. potatoes) to 24 mg kg^{-1} dry matter (e.g. meat, fish, and poultry). In humans and animals, long-term exposure to excess levels of zinc may result in copper deficiency, reduced immune function, reduced levels of high-density lipoproteins (the good cholesterol), anaemia, death of foetuses, and damage to the liver, pancreas, and kidneys (ATSDR, 2013). However, in the context of pollution, zinc is more a cause of phytotoxicity rather than being toxic to animals and humans. If the zinc concentration in the soil exceeds about 300 mg kg^{-1}, which is frequently the case if the soil is contaminated, zinc may be toxic to several species of soil microorganisms. This affects the decomposition of plant litter and hence nutrient cycling in contaminated ecosystems. Nitrogen-fixing bacteria in the root nodules of legumes such as clover are especially susceptible to zinc toxicity (Alloway and Ayres, 1997). As a consequence, soil pollution by zinc may cause a noticeable shift in plant species composition.

The most severe pollution by zinc is usually found in mining and smelting areas. Fragments of sulphides in mine tailings oxidise on weathering. The oxidation reaction of sphalerite can be written in simplified form as:

$$ZnS \ + \ 2O_2 \ \rightarrow \ Zn^{2+} \ + \ SO_4^{2-} \tag{7.1}$$

This reaction is accompanied by the oxidation of pyrite and creates acidic solutions which tend to decrease adsorption and hence increase the environmental mobility of zinc and other metals.

Other anthropogenic sources of zinc are galvanised steel, sewage sludge, waste disposals, and industrial releases. Galvanised steel is used in roofs, gutters, drainpipes, and wire fences. Exposure to acid rain slowly dissolves these materials, and the zinc ends up in soil or runoff water. Zinc is also widely used in domestic products such as skin care products (cosmetics, baby creams, shampoos). Together with drainage water from galvanised surfaces, these are the main sources of zinc pollution of sewage water, effluent from wastewater treatment plants, and sewage sludge. Hence, spreading sewage sludge on land as a fertiliser progressively increases the zinc concentrations in agricultural soils. The other major sources of zinc in domestic waste in addition to discarded galvanised materials and domestic products containing zinc are batteries, pigments, and paints. Waste disposal can therefore lead to local soil and groundwater pollution around landfills.

7.3 COPPER

Copper may occur in the Earth's crust as the free native metal or in the 1+ or 2+ oxidation states. Copper may be in solution as either Cu^{2+} or Cu^+, but the more oxidised Cu^{2+} predominates due to redox reactions in aerated water and the tendency of Cu^+ ions to disproportionate ($2Cu^+ \rightarrow Cu^0 + Cu^{2+}$). Copper ions are prone to complexation, especially with hydroxide and carbonate ligands. In water above neutral pH, the $Cu(OH)_3^-$ complex is formed and a strong $CuCO_3$ (aq) ion pair predominates in aerated natural waters containing dissolved carbonate species. These complexes are slightly soluble and may keep the copper concentration in water below 10 μg l^{-1}. Moreover, adsorption and coprecipitation with ferric oxyhydroxides may cause even smaller copper concentrations (Hem, 1989). Under reduced conditions and in bedrock minerals, copper is contained in rather stable sulphide minerals. Some of the common copper sulphide minerals that are important as ores also contain iron, such as chalcopyrite ($CuFeS_2$).

Like zinc, copper is an essential micronutrient that is a constituent of many enzyme systems. Copper deficiency in soils may depress crop yields, especially in cereals. Animals and humans need copper to be able to use iron properly. Herbage with less than 5 mg kg^{-1} can cause deficiency in sheep and cattle. The most commonly observed results of copper deficiency in cattle are diarrhoea, broken bones, infertility, anaemia, poor weight gains, and reduced immune response. Humans need between 1 mg and 3 mg each day. It takes several days for copper to leave the body.

Copper contamination of soils may cause phytotoxicity. The bioavailability of copper decreases with increasing pH, so the toxicity to plants can be mitigated by liming soils contaminated with copper, to bring the pH up to 7. In addition, copper is highly toxic to soil microorganisms. Like zinc, it may affect nitrogen fixation and the mineralisation of plant litter (Obbard and Jones, 2000). Although sheep are prone to copper deficiency, they are also most sensitive to copper toxicity (Alloway and Ayres, 1997). When copper is fed in excess of requirements it tends to accumulate in the sheep's liver. With minimal stress, the stored copper can cause cattle or sheep to die within a few hours to a few days. Copper toxicity to humans is rare, usually occurring after prolonged exposure to copper. Very young children may be more sensitive to copper, and long-term exposure to high levels of copper in food or water may cause liver damage and death.

Pollution of soil and water by copper may arise from copper mining and smelting. Copper is primarily used as the metal or alloy in sheet metal, wires, pipes, and other metal products. It is widely used for water pipes. Corrosion of these metal products as well as disposal of their scrap allows the copper to enter the environment either directly or indirectly via effluent from wastewater treatment plants, sewage sludge, compost, or landfills. Copper compounds are most commonly used in agriculture to treat plant diseases or for water treatment and as preservatives for wood, leather, and fabrics. An example of a commonly used copper-based fungicide is Bordeaux mixture, which contains copper sulphate. It is widely used in viniculture, where the vines are frequently sprayed after rainfall to prevent mildew. The copper applied can accumulate to large concentrations in the topsoil. Manure, especially swine slurry from intensive pig farming, may also contain significant amounts of Cu, which is added to the livestock's diet as copper sulphate at up to 250 mg Cu kg^{-1} to promote growth.

7.4 LEAD

Lead is present in moderate amounts in igneous and sedimentary rocks, mostly in the form of lead sulphide (galena; PbS). Nevertheless, its natural mobility is low, because of the low solubility of lead hydroxide, carbonate, and phosphate. The adsorption of lead on mineral and organic sediment surfaces and its coprecipitation with manganese oxide also tend to maintain low concentrations in natural waters. The major dissolved inorganic forms of lead are the free Pb^{2+} ion, hydroxide complexes, and, the carbonate and sulphate ion pairs. Equilibria involving basic lead carbonates maintain the dissolved lead concentration below about 50 µg l^{-1} in water having 61 mg l^{-1} HCO$_3^-$ and a pH between 7.5 and 8.5. Water having lower alkalinity and pH can contain larger concentrations of dissolved lead (Hem, 1989).

Lead is a non-essential element and it is not as bioavailable as other metals. However, it is toxic to mammals, including humans. Exposure to high levels of lead can cause damage to the brain and kidneys and may affect haemoglobin production and male fertility (ATSDR, 2013). Children under the age of seven are particularly sensitive to even small amounts of lead. Furthermore, lead exposure may increase blood pressure in middle-aged men, but it is unknown if lead increases blood pressure in women. Lead can enter your body when you inhale air with lead-containing dust or particles of lead. Almost all of the lead in the lungs enters the blood and moves to other parts of the body. Most of the lead is stored in bone and the levels of lead in bone and teeth increase as a person ages. In adults, only a small part of the lead ingested from food, beverages, water, soil, or dust enters the blood from the gastrointestinal tract and moves to other parts of the body. However, when children swallow food or materials containing lead, such as polluted soil found at a hazardous waste site or chips of lead-containing paint, much more of the lead enters their blood. In the first half of the 20th century, many malnourished inner-city children satisfied their cravings by eating old paint flakes from dilapidated buildings, which caused many cases of lead poisoning. For this reason, lead was phased out of paint pigments (EPA, 2013).

The lead used by industry comes from mined ores or from recycled scrap metal. Lead has a wide range of uses. Its main use is in the manufacture of storage batteries. Other uses include the production of chemicals, including petrol additives, paint pigments, pigments for glazing ceramics, various metal products (for example, sheet lead, solder, pipes, and fisherman's weights), and ammunition. Contamination of air, water, and soil by lead is due to dispersal of these chemicals in the environment via similar pathways as has been described for zinc and copper (see previous sections). However, the pathway via the air used to be more significant, especially in areas of heavy automobile traffic. Before the introduction of lead-free petrol, tetraethyl lead was added to the petrol to promote more efficient combustion. The lead content of the fuel was emitted as aerosol particulates in the exhaust gases, which are deposited within a relatively limited range from the source. The use of lead in ammunition for clay pigeon shooting, game hunting, and warfare is another 'special case' causing lead pollution. Military training grounds and clay pigeon shooting club sites can therefore contain particularly large concentrations of lead. Although lead ammunition contains lead as native metal, the weathering of the shot pellets, bullets, and shells in the soil results in relatively large quantities of lead being dispersed within the soil (Alloway and Ayres, 1997).

7.5 CADMIUM

Cadmium has some chemical similarities with zinc and occurs only in the 2+ oxidation state. However, it is much less abundant than zinc. In general, cadmium is present in zinc ore minerals such as sphalerite, and also in some copper ores. The solubility of cadmium in water is mainly controlled by adsorption on cation exchange sites and coprecipitation with manganese oxides. However, cadmium tends to be less strongly adsorbed than other divalent metals and is, therefore, more labile in soil and sediments and more bioavailable (Hem, 1989).

Cadmium is a highly toxic metal not known to have any beneficial effects for plants and animals. High human exposures to cadmium are rare today, but long-term, low-level exposure may cause adverse chronic health effects. The main toxic effect in humans from chronic exposure to the metal is kidney damage and, ultimately, kidney failure. Many cadmium compounds are also believed to be carcinogenic (ATSDR, 2013). The normal intake of cadmium by humans amounts to about 1-3 $\mu g\ d^{-1}$, which does not appear to cause health problems. Most cadmium enters the body via the gastrointestinal tract by eating food products grown on contaminated soil, although smokers may receive a considerable part of their cadmium intake by inhaling cigarette smoke. A notorious case of cadmium poisoning occurred in the Jintsu Valley in Japan during the 1940s, when it was found that more than 200 elderly women who had had several children had developed kidney damage and skeletal deformities. This disease was called the 'itai-itai' disease, which literally means 'ouch-ouch' because of the pain caused by the deformed bones (Copius Peereboom, 1976).

Cadmium used in industry is a byproduct of zinc, lead, and copper refining. It is used for metal plating, rechargeable and non-rechargeable batteries, and for pigments used in paint, printer ink, and plastics. The principal source of cadmium release to the general environment is the burning of fossil fuels (in power stations, furnaces, stoves, automobiles, etc.) and the incineration of municipal waste materials. Cadmium may also be released from zinc, lead, or copper smelters. Cadmium becomes very volatile at temperatures above 400 °C (Alloway and Ayres, 1997). Therefore, if materials containing cadmium are heated, the cadmium is readily released as aerosols. These cadmium aerosols are usually deposited within several tens of kilometres from the source. The largest source of cadmium in the human food chain is, however, the application of phosphate fertilisers, in which cadmium is present as impurity. In addition, the spreading of contaminated sewage sludge on agricultural land contributes to the human cadmium intake.

7.6 NICKEL

Nickel is a ferromagnetic transition metal sharing many chemical properties with iron. It occurs naturally in the Earth's crust in various forms such as nickel sulphides and oxides. Nickel may occur in the oxidation states 1-, 0, 2+, 3+, and 4+. However, the aqueous chemistry of nickel is primarily concerned with the 2+ oxidation state. Other oxidation states occur in special complexes and oxides. In alkaline solutions, Ni(II) hydroxide can be oxidised to a hydrated Ni(IV) oxide. Like zinc, nickel is specifically adsorbed in soils and sediments and the binding strength increases with ageing of nickel (Hem, 1989).

Nickel is an essential nutrient that is only required in very small amounts. In plants, symptoms of nickel toxicity are generally induced by iron deficiency: chlorosis and foliar necrosis. Excess nickel affects nutrient absorption by roots, root development, and metabolism, and it inhibits photosynthesis and transpiration. Nickel can replace other metals located at active sites in metallo-enzymes and disrupt their functioning. Its absorption into the human body is affected by factors of consumption, the acidity of the gut, and various binding or competing substances, including other metals such as iron, magnesium, zinc, and calcium. An excess of nickel in tissues can cause damage to chromosomes and other cell structures, alter hormone and enzyme activities, and affect membrane permeability and immune function. As a result, changes in glucose tolerance, blood pressure, response to stress, growth rate, bone development, and resistance to infection are possible. Nickel has also been associated with possible carcinogenic effects (ATSDR, 2013).

Since the body's requirement for nickel is so low, excessive exposure from the environment, leading to toxicity is very common, e.g. from tobacco, dental implants, stainless steel kitchen utensils, coins, and inexpensive jewellery. The most common trouble is allergy of the skin to nickel due to direct contact. On the other hand, toxicity by consumption of food products grown on contaminated soil is very rare and would require 1000 times the amount normally consumed in food.

The nickel used by industries comes from mined ores or from recycled scrap metal and has a wide range of industrial uses. When alloyed with other elements, nickel imparts toughness, strength, resistance to corrosion and various other electrical, magnetic, and heat-resisting properties. It is primarily used in making various steels and alloys (e.g. stainless steel) and in electroplating. Minor applications include the use in ceramics, permanent magnet materials, and nickel–cadmium batteries.

7.7 CHROMIUM

Chromium is found in the environment in three major states: chromium(0), chromium(III), and chromium(VI). The predominant oxidation state in the environment is trivalent chromium. Ultramafic igneous rocks contain the most chromium. Chromite ($FeCr_2O_4$), which is highly resistant to weathering, may be concentrated in the lateric soils overlying such rocks. In general, chromium (III) is largely present in soil as relatively unavailable, insoluble chromium or chromium–iron oxyhydroxides. It can also exist as a substitute for Al(III) in the octahedral groups of aluminosilicates. The solubility of Cr(III) in soil is dependent on pH and decreases dramatically at pH values greater than 4.5. In aqueous solutions the reduced forms of chromium are Cr^{3+}, $CrOH^{2+}$, $Cr(OH)_2^+$, $Cr(OH)_3^0$, and $Cr(OH)_4^-$. Chromium hydroxide ($Cr(OH)_3$) has a minimum solubility of about 5 μg l^{-1} at about pH 9.0. Chromium(III) can form stable complexes with organic ligands at pH values as high as 8.5. These complexes form slowly, and once formed, are difficult to break. The presence of organic complexes can significantly influence the concentration of total dissolved Cr(III) in the soil solution. Chromium(VI) is not thermodynamically stable in soils except in alkaline, oxidising environments, and is readily reduced to Cr(III). Hexavalent chromium may be present as the anions dichromate $Cr_2O_7^{2-}$ and chromate CrO_4^{2-} (Hem, 1989). Chromium(VI) is considerably more soluble than chromium(III). Concentrations of chromium in natural waters that have not been affected by waste disposal are commonly less than 6 μg l^{-1} (see Table 7.1), although under exceptional conditions they may increase over 1 to 2 orders of magnitude. Dissolution and oxidation of Cr^{3+} under basic conditions may cause the groundwater to be enriched in chromium(VI) up to concentrations greater than 100 μg l^{-1}.

There is no evidence that chromium has any physiological function in plants. Chromium is toxic to agronomic plants at about 5 to 100 μg g^{-1} of available chromium in soil (Hossner *et al.*, 1998). Plants absorb chromium(VI) better than chromium(III), whereas chromium(VI) is more toxic to plants than chromium(III). In animals and humans, chromium(III) is considered to be an essential nutrient that helps to maintain normal metabolism of glucose, cholesterol, and fat (ATSDR, 2013). Signs of chromium deficiency in humans include weight loss and impairment of the body's ability to remove glucose from the blood. The minimum human daily requirement of chromium for optimal health is not known, but a daily ingestion of 50–200 μg d^{-1} has been estimated to be safe and adequate. Only very large doses of chromium(III) are harmful. Long-term exposure to hexavalent chromium may cause adverse effects in the kidney and liver. Chromium(VI) and some chromium compounds are also known carcinogens. Because small amounts of trivalent chromium occur in many foods, most of it enters the body from dietary intake. In the case of excess intake of chromium (VI), the chromium originates mainly from the use of contaminated groundwater as drinking water.

The metal (chromium(0)) is a steel-grey solid with a high melting point (1857.0 °C). It is mainly used for making steel and other alloys. Chromite is used by the refractory industry to make bricks for metallurgical furnaces. Chromium compounds produced by the chemical industry are used for chrome plating, the manufacture of pigments, leather tanning, wood treatment, and water treatment. Soil and groundwater pollution by chromium is primarily caused by leaching from waste disposal sites and industrial zones.

7.8 MERCURY

Mercury is a metal that occurs naturally in the environment in several forms. Elemental mercury has a melting point of –38.9 °C and is thus a liquid at normal environmental temperatures near the Earth's surface. However, it has a relatively high vapour pressure at these temperatures, so it is also slightly volatile. This free metallic form, which has an equilibrium solubility of 25 μg l^{-1} in a closed system without a gas phase over a considerable pH–Eh range (Hem, 1989) is the stable form in most natural water systems. However, the concentration that would be present in water in contact with the free atmosphere is likely to be much lower, because of mercury's tendency to vaporise. Mercury can also combine with other chemicals, such as chlorine, carbon, or oxygen to form either inorganic or organic mercury compounds. Organic complexes such as methyl mercury ($HgCH_3^+$) may be produced by methane-producing bacteria in contact with metallic mercury in lake or stream sediments. In this form, mercury is in the 2+ oxidation state. Like other trace metals, the mobility of mercury(II) in soil and water is largely controlled by adsorption reactions with fixed or mobile adsorbents.

Soil mercury is poorly available to plants and there is a tendency for mercury to accumulate in the roots, indicating that the roots serve as a barrier to mercury uptake. The mercury concentration in aboveground parts of plants appears to largely depend on foliar uptake of elementary mercury volatilised from the soil (Lindqvist *et al.*, 1991). Uptake of mercury is often plant-specific; organic mercury compounds are taken up in larger amounts than elemental mercury. Factors affecting plant uptake include soil or sediment organic content, cation exchange capacity, oxide and carbonate content, redox potential, and the total metal content. Methyl mercury is lipophyllic and can build up in the fat of certain fish. For this reason, low levels of mercury in lakes and rivers can contaminate these fish.

Mercury can easily enter the human body if its vapour is breathed in or if it is eaten in organic forms in contaminated fish or other foods. Mercury can also enter the body when food or water contaminated with inorganic mercury is eaten or drunk. Mercury in all forms may also enter the body directly through the skin. Once mercury has entered the body, it may be months before all of it leaves. Long-term exposure to organic or inorganic mercury can permanently damage the brain, kidneys, and developing foetuses (ATSDR, 2013). The form of mercury and the way humans are exposed to it determine which of these health effects will be more severe. For example, organic methyl mercury is a neurotoxin and may cause greater harm to the brain and developing foetuses than to the kidney, whereas inorganic mercury ingested with contaminated food or water may cause greater harm to the kidneys. Metallic mercury vapour that enters the body via the lungs may cause greater harm to the brain (ATSDR, 2013).

In the mid 1950s, severe neurological disorders were diagnosed in the fishing population living around the Minamata Bay in Japan. Mercury containing liquid wastes were discharged into this bay, and the inorganic mercury was transformed into methyl mercury through the action of bacteria in the bay sediments. The methyl mercury accumulated in the food chain of fish in the bay, which resulted in fish catches containing high mercury concentrations. The large daily fish consumption of the local population resulted in a daily intake of approximately 2 mg mercury per day. This led to ataxia, limited range of vision, loss of hearing, speaking disorders, trembling, stiffness, and psychological disorders in the affected population (Copius Peereboom, 1976). According to the Japanese government, 2955 people contracted this so-called Minamata disease, of whom 1784 died.

Mercury is introduced to the environment by natural and anthropogenic emissions. Natural emissions of mercury form two-thirds of the input while man-made releases form about one-third, although the amounts released from anthropogenic sources have fallen greatly since 1970. Because mercury is rather volatile, it can be dispersed over great distances in the atmosphere. The amounts escaping to the atmosphere in smelting and fossil fuel combustion have probably enhanced the mercury levels in the environment above pre-industrial background levels. In addition to the burning of fossil fuel, a major present-day source of mercury in the western world is the production, consumption, and final waste disposal of materials containing mercury. Metallic mercury is used in thermometers, barometers, cell batteries, and other common consumer products. One of the most important sources of contamination in agricultural soils used to be the use of organomercuric compounds as a seed coating to prevent fungal diseases in germinating seeds. Together with several other agricultural applications of mercury, this was banned in the 1960s. Important sources of water and sediment pollution by mercury are its uses in electrolysis cells for the production of chlorine and sodium hydroxide from sodium chloride brine, and as a catalyst in the production of some plastics. Another source of mercury pollution of surface waters in its use in small-scale artisanal gold mining in East Africa, the Philippines, and the Amazon region.

7.9 ARSENIC

Arsenic is a semi-metal which may form metal arsenides in which the oxidation state is negative. In addition, it may form sulphides that can be found in sulphide ore minerals. In aqueous solutions, arsenate (5+ oxidation state) and arsenite (3+ oxidation state) are the thermodynamically stable forms. Under mildly reduced conditions, the arsenite uncharged ion $HAsO_2$(aq) is the predominant form. Under oxidising conditions, the monovalent arsenate anion $H_2AsO_4^-$ predominates in the pH range between 3 and 7, whereas the divalent species $HAsO_4^{2-}$ predominates between pH 7 and pH 11 (Hem, 1989). Arsenic compounds can adsorb to oxides and hydroxides of Fe(III), Al(III), Mn(III/IV), to humic substances, and clay minerals (Bissen and Frimmel, 2003). Like phosphate, arsenate is strongly adsorbed by ferric oxyhydroxides. Furthermore, ferric and other metal arsenates are poorly soluble. This allows arsenates to accumulate in zones of iron precipitation when reduced groundwater seeping upwards crosses the oxidation–reduction interface. In aerated surface water, adsorption by hydrous iron oxides, or precipitation or coprecipitation with sulphide in reduced bottom sediments maintain the arsenic concentrations in water at very low levels (usually below 10 μg l^{-1}). Arsenic is also involved in biochemical processes. Biologically mediated methylation produces organic complexes, such as dimethyl arsenic acid ((CH_3)$_2$AsOOH) and methyl arsonic acid (CH_3AsO(OH)$_2$). Dimethyl arsenic acid is difficult to oxidise and may constitute the main arsenic species in surface water. The importance of the redox state and the formation of soluble organic complexes of arsenic result in the arsenic concentrations in natural water varying over several orders of magnitude (see Table 7.1). Some arsenic compounds are relatively volatile, which is an important factor in the natural circulation of arsenic in the environment.

Arsenic is a toxic, non-essential element. As arsenic's toxicological importance is partly attributable to its chemical similarity to phosphorus (though the oxidation state of phosphorus is not sensitive to the redox potential), it can readily disrupt the metabolic pathways involving phosphorus (Alloway and Ayres, 1997). Organic forms of arsenic are usually less toxic than the inorganic forms. The largest source of arsenic intake for humans is food (generally about 25 to 50 μg is ingested per day), with smaller amounts coming from drinking water and air. However, in areas with naturally elevated arsenic concentrations in groundwater used for drinking water, such as in southern Bangladesh and West Bengal, India, the principal source of arsenic intake is drinking water. Some edible fish and shellfish contain elevated levels of arsenic, but this is predominantly in less toxic organic forms. Small children who swallow small amounts of soil while playing, may be exposed to arsenic if they ingest soil containing elevated levels of arsenic. Most arsenic that is absorbed into the body is converted by the liver to a less toxic form that is efficiently excreted in the urine. Consequently, arsenic does not have a strong tendency to accumulate in the body except at high exposure levels. Inorganic arsenic has been recognised as a human poison since ancient times. Large doses can produce death: a daily intake of 5 to 50 mg d^{-1} is considered to be toxic and above 50 mg d^{-1} as lethal. Long-term exposure to lower levels of arsenic may cause decreased production of red and white blood cells, abnormal heart function, blood vessel damage, liver and/or kidney injury, and impaired nerve function. Arsenic has also been associated with skin cancers (ATSDR, 2013).

Arsenic is present in volcanic gases and is a common constituent of geothermal waters. In some parts of the world, natural mineral deposits contain large quantities of arsenic; this may result in elevated levels of inorganic arsenic in soil and water. Enhanced arsenic concentrations of 50 μg l^{-1} or more have been reported in groundwater from a number of large aquifers in various parts of the world, including Bangladesh, South-east Asia, inner Mongolia, Hungary, south-west USA, central Mexico, and Argentina. However, these enhanced concentrations have different causes: as well as occurring in geothermal waters (as mentioned above), they may be the result of evaporative concentration or mobilisation from the aquifer material. Nevertheless, high arsenic concentrations in groundwater are not necessarily related to high arsenic concentrations in the source rocks. In sedimentary aquifers in arid inland basins, the elevated concentrations are probably linked to the desorption of arsenate from iron and other oxides under oxic, alkaline conditions, whereas in young alluvial and deltaic aquifers, such as in Bangladesh, arsenic mobilisation occurs under strongly reducing conditions (Smedley and Kinniburgh, 2001). Such conditions are produced by the microbially mediated decomposition of buried peat deposits, which induces both the reduction of arsenate to arsenite (less strongly adsorbed by the ferric oxyhydroxides) and the reductive dissolution of ferric oxyhydroxides, thereby releasing the adsorbed load of arsenic to groundwater (Smedley and Kinniburgh, 2001; McArthur *et al.*, 2001).

Anthropogenic sources of arsenic include mining activities, metal smelters, combustion of fossil fuels, pesticide applications, household products, and waste disposal. About 40 percent of the anthropogenic emissions are derived from the smelting of copper and other metals that releases inorganic arsenic into the atmosphere. Low levels of arsenic are found in most fossil fuels (oil, coal, gasoline), so the burning of these materials results in inorganic arsenic emissions into the air. Fuel combustion accounts for approximately 20 percent of the anthropogenic emissions. Arsenic has been used as a component of several pesticides. Some products, mostly weed killers, contain organic arsenic as the active ingredient, whereas other pesticides used to control weeds, insects, or rodents, or for wood preservation, contain inorganic arsenic. In the past, inorganic arsenic was also contained in household products such as paints, dyes, medicines, and rat poisons. These products are, however, no longer in general use. As a result of the disposal of these agricultural and domestic products, some waste disposal sites contain large quantities of arsenic and may be an important local source of soil and water pollution by arsenic.

7.10 SELENIUM

Selenium is a member of the sulphur group of non-metallic elements. Despite officially being a non-metal, selenium is sometimes considered to be a metalloid because it shares physical, chemical, biological, and toxic properties with heavy metals. In its pure form, selenium occurs as metallic grey to black hexagonal crystals, but in nature it is commonly found in sulphide minerals (e.g. pyrite), where it partly replaces sulphur, or combined with silver, copper, lead, and nickel minerals. Selenium occurs naturally in five oxidation states: 2-, 0, 2+, 4+, and 6+. The primary species are selenate (SeO_4^{2-} or Se[VI]), selenite (SeO_3^{2-}, or Se[IV]) and organo-selenide (Se(II); e.g. selenomethionine). The chemical properties of selenium are similar to sulphur. Consequently, selenium plays a role analogous to that of sulphur in organic compounds. Selenium combines with metals and many non-metals directly or in aqueous solution. It reacts with oxygen to form a number of oxides, the most stable of which is selenium dioxide.

Selenium is found naturally in igneous rocks, volcanic deposits, ore deposits, and in sedimentary rocks such as sandstone, carbonaceous siltstones and shales. The average occurrence of selenium in the Earth's continental crust is 0.12 mg kg^{-1} (see Table 1.1). It is typically found in marine, carbonaceous (organic-rich) shale formations, in which selenium is primarily hosted by organic matter and pyrite. Se-rich shales can contain bulk Se concentrations up to 9.1%. Weathering usually transforms most inorganic Se into more oxidised species (i.e. selenite and selenate), which are very soluble in water. Therefore, the release of selenium during weathering is largely controlled by the oxidation of pyrite. Furthermore, pyrite oxidation promotes selenium release from other pools by the associated release of acidity (Matamoros-Veloza *et al.*, 2011).

In most soils, selenium concentrations have been estimated to vary between 0.01 and 0.2 mg kg^{-1}. Selenium-enriched natural soils can be found in areas with Se-bearing rocks. Soils may also become naturally enriched in selenium as a result of seepage and subsequent evaporation of groundwater originating from Se-bearing aquifers. This typically occurs in semi-arid regions and results in alkaline and saline soils. In North America, selenium is found in high concentrations across the Great Plains (in particular Wyoming and South Dakota) and Canadian Prairies, westward to the Pacific and south into Mexico. Selenium also occurs in alkaline soils in certain localities in Colombia, Ireland, Israel, South Africa, and China. High soil salinity and pH promote the adsorption of selenite and selenate onto clays and sequioxides. Selenite is adsorbed much more strongly than selenate, so selenate is the major bioavailable form of selenium. Some soil anions, such as phosphate, compete with selenium anions for adsorption site and so increase selenium concentrations in the soil solution and bioavailability.

In pristine surface waters, typical concentrations of dissolved selenium range from 0.07 to 0.19 µg l^{-1} (Luoma and Rainbow, 2008). As a result of its geochemical behaviour in rocks and soils, in natural waters, selenium is primarily present in an anionic speciation. In surface waters, selenate can be taken up by plants –including phytoplankton– or can be reduced by microorganisms to particulate forms in sediments. In plants, selenite is transformed to organo-selenide. When organo-selenium is released it can be oxidised to selenite. Further oxidation back to selenate is extremely slow and therefore only little selenite is reconverted into selenate. Dissolved selenite and organo-selenium are bioavailable to animals, but the uptake rate is very slow. The biogeochemical cycling in aquatic ecosystems leads to the predominance of organo-selenium and selenite, which would not be expected based upon chemical-thermodynamic predictions alone (Luoma and Rainbow, 2008).

Selenium bioaccumulates in aquatic habitats, which leads to concentrations being much higher in organisms than in the surrounding water. Zooplankton can concentrate organo-selenium compounds over 200 000 times. Inorganic selenium bioaccumulates more readily in phytoplankton than in zooplankton. Phytoplankton can concentrate inorganic selenium by a factor of 3000. Further concentration through biomagnification may occur along the food chain.

Trace amounts of selenium are necessary for cellular function in animals and humans. Unlike animals, plants do not appear to require selenium for survival. The toxicity of elemental selenium and most metallic selenides is relatively low, because of their relatively low solubility and bioavailability. By contrast, the more oxidised species of selenates and selenites are very toxic, having an oxidant mode of action similar to that of arsenite. In humans, acute and fatal toxicities may occur after accidental or deliberate ingestion of high doses of selenium. Poisoning due to excessive intake of selenium is also referred to as selenosis. Long-term exposure to smaller doses of selenium in food and water may result in chronic selenosis. The most common symptoms of selenosis are hair and nail brittleness and loss. Other symptoms may include gastrointestinal disorders, fatigue, skin rashes, a garlic breath odour, and neurological damage.

The major pathway to animals is ingestion of plants or sediments. Organic forms of selenium found in grain, cereals, and forage crops constitute a major source of dietary intake of selenium. In general, there is a wide variation in the selenium concentration in plants and grains. Because plants do not appear to require selenium, the incorporation of selenium into plants depends largely on the selenium concentration in soil. Other, non-vegetable food sources of selenium include organ meats, seafood, and muscle meats.

Anthropogenic sources of selenium include the mining and smelting of sulphide ores and coal burning. After coal burning, selenium is concentrated in the remaining ash and so ash disposal sites may be an important local source of selenium contamination. Mining activities often cause a massively increased exposure of selenium-bearing pyrite to the atmosphere.

The consequent pyrite oxidation may cause and substantially enhance leaching and runoff of selenium from mining sites.

The importance of selenium as an environmental contaminant has been recognised by scientists and, especially in the USA, also by policy makers and the general public. Nevertheless, there are wide differences in environmental regulations with respect to selenium between countries across the world (Luoma and Presser, 2009). In Europe, selenium is not considered an ecological threat. It is not listed as a chemical of concern, whereas in the USA selenium is ranked 141 in the ATSDR 2011 Priority List of Hazardous Substances (ATSDR, 2013).

EXERCISES

1. Where can one expect relatively high background concentrations of heavy metals in soil or water?

2. What are the principal anthropogenic sources of heavy metals
   a. in general?
   b. in agricultural soils?
   c. in urban areas?
   d. in forests?
   e. in river water?

3. Explain why acid mine drainage often contains high concentrations of heavy metals.

4. a. Name the two most important chemical processes that determine the environmental mobility of heavy metals.
   b. Which other substances are involved in these processes?

5. a. Discuss the mobility of cadmium, zinc, and lead compared to that of iron and manganese as a function of redox conditions.
   b. Under which conditions (pH, redox, aquifer material) are cadmium, zinc, and lead most mobile in groundwater?

6. Which metal is more mobile under oxidising conditions: zinc or cadmium?

7. Name a specific peculiarity of each of the following elements, which plays a part in their environmental behaviour.
   a. Nickel
   b. Chromium
   c. Mercury
   d. Arsenic
   e. Selenium

8. The heavy metal contamination of the river Geul, a tributary of the river Meuse, is largely due to historic erosion of mine tailings near the zinc and lead mines in Plombières and Kelmis, Belgium. Near this source the coarse particle size fraction (sand) of the river sediments is heavily contaminated by heavy metals (zinc, lead, cadmium). The river Geul flows into the river Meuse just north of Maastricht, the Netherlands. The Geul valley between Plombières and its confluence with the river Geul is about 35 kilometres long.

About 100 km further downstream, heavy metal contamination of the sand fraction of Meuse sediments is low, whereas the contamination in the bulk sediment remains high.

a. What was the speciation of the heavy metals as they entered the river system?
b. What is the main cause of the decrease in contamination in the coarse particle size fraction with increasing distance downstream of the contamination source?
c. What has happened to the metals associated to the coarse particle size fraction?
d. What is the main difference in the relationship between heavy metal concentrations and clay and organic matter content for Geul sediments just downstream of Plombières and for Meuse sediments 100 km downstream of the Geul–Meuse confluence?

8

Radionuclides

8.1 INTRODUCTION

Radionuclides are elements that have unstable nuclei which disintegrate or change spontaneously with a loss of energy in the form of ***ionising radiation*** (that is any radiation which displaces orbital electrons from atoms, so producing ions). The instability of an atomic nucleus is caused by an imbalance of the number of protons (Z) and neutrons (N) in the nuclei. Stable nuclei have neutron and proton numbers which are closely related. This is illustrated in Figure 8.1, which shows that stable nuclei occur only within a narrow band of increasing neutron and proton numbers. Unstable nuclei break down, ultimately forming stable nuclei. The most unstable nuclei disintegrate rapidly and do not now exist in measurable quantities in the environment. Other unstable nuclides, however, have a slow decay rate and still exist in significant amounts.

Atoms that have the same number of protons but different numbers of neutrons are known as ***isotopes*** of an element. Isotopes are usually denoted by their mass number A (i.e. the sum of the number of protons and neutrons in a nucleus: $A = Z + N$), e.g. ^{1}H or H-1 for hydrogen. For example, lead has four stable isotopes ^{204}Pb (1.4 %), ^{206}Pb (24.1 %), ^{207}Pb (22.1 %), and ^{208}Pb (52.3 %), and eight instable isotopes ^{202}Pb, ^{203}Pb, ^{205}Pb, ^{209}Pb, ^{210}Pb, ^{211}Pb, ^{212}Pb and ^{214}Pb. The percentages in brackets refer to the average relative abundance of each isotope in naturally occurring lead. The unstable lead isotopes are found in only trace amounts. Note that the relative abundance of the isotopes varies for the different ores, giving each region its own lead isotope signature.

Radioactive decay is a first-order kinetic process, which implies that the number of nuclei that disintegrate per second (dN/dt) is proportional to the number of nuclei N at any time t:

$$\frac{dN}{dt} = -\lambda N \tag{8.1}$$

where λ is the disintegration or decay constant. Radioactive decay is a stochastic process and the decay rate constant is an invariable property of each radioisotope and is independent of factors such as pressure, temperature, chemical form, and time. The number of disintegrations per second is measured in ***Becquerel*** (Bq), named after the discoverer of radioactivity in 1896. One Bq is equal to one disintegration per second. It replaces the older unit Curie (1 Ci = $3.7 \cdot 10^{10}$ Bq) named after Marie and Pierre Curie, the discoverers of the radioactive element radium.

If N_0 is the number of nuclei present at time $t = 0$, then by integration of Equation (8.1) we obtain:

$$N = N_0\, e^{-\lambda t} \tag{8.2}$$

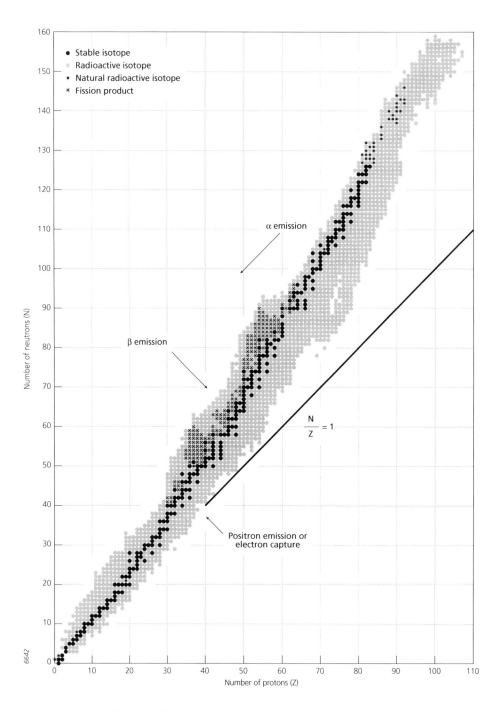

Figure 8.1 Nuclei stability band (after Pattenden, 2001).

Thus, the number of nuclei present decreases exponentially with time. Usually, for radionuclides the decay rate is expressed as a half-life T, that is the time required for half the amount initially present at time $t = 0$ to disintegrate. It can easily be shown that

$$T = \frac{\ln(2)}{\lambda} \approx \frac{0.693}{\lambda} \qquad (8.3)$$

Half-lives for various elements can range from milliseconds to billions of years. Elements with a short half-life are more radioactive than elements with a long half-life. On the other hand, elements with a long half-life remain radioactive for a long period. Besides the half-life and associated decay rate, another factor that determines the environmental hazard of radionuclides is the nature and intensity of the ionising radiation emitted during decay of the atomic nuclei. The three forms of ionising radiation of principal interest in environmental pollution are 1) *alpha radiation*, consisting of positively charged helium nuclei, 2) *beta radiation*, consisting of electrons or positrons (i.e. positively charged electrons), and 3) *gamma radiation*, consisting of a discrete quantity of electromagnetic energy without mass or charge. The energy associated with radiation is usually expressed in mega-electron-volts (MeV). One eV is the energy acquired when a particle carrying unit electronic charge is accelerated through a potential difference of 1 volt (1 eV = $1.6022 \cdot 10^{-19}$ J).

Alpha particles are the nuclei of stable ^{4}He atoms and are emitted in a process called alpha decay, which only occurs in the nuclei of heavy elements. In this process, the α emission from an initial nucleus (Z, A) leads to the formation of a product nucleus $(Z - 2, A - 4)$. In general, the emission energy of most alpha particles from natural radioactivity is in the range from 4 to 11 MeV. Along their travel path, alpha particles interact with matter by the transfer of kinetic energy to atomic orbital electrons during collision. As a consequence, one or more electrons are ejected from their orbit, leaving the atoms as cations. The energy transfer during each collision is relatively small, so the alpha particle remains in a straight line until its energy is dissipated. The range or travel distance of an alpha particle is very short and depends on its initial energy. In gases, the distance travelled is in the order of several centimetres, whereas in high-density solid materials the range amounts to hundredths of millimetres.

Beta particles are electrons and positrons emitted from nuclei in a process known as beta decay, which occurs both in heavy and light elements. Electron emission (β^- emission) is initiated by the conversion of a neutron into a proton inside the nucleus and yields a product nucleus $(Z + 1, A)$. On the other hand, positron emission (β^+ emission) is produced by the conversion of a proton into a neutron inside the nucleus and gives a product nucleus $(Z - 1, A)$. The emission energy of beta particles is approximately between 0.05 and 4 MeV. The principal mechanisms of interaction of beta particles with materials are basically the same as those of alpha particles. Beta particles also eject electrons from their atomic shield. Their range depends on the emission energy. In air, the range is about 1 m for particles having an emission energy of 0.5 MeV and about 10 m for those of 3.0 MeV. Air is, however, an inconvenient absorbing medium for beta particles. Solids absorb beta particles much more effectively. The range of beta particles in solid material is in the order of a few millimetres and does not depend strongly on the absorbing material.

Gamma radiation (or photons) consists of electromagnetic waves with very short wavelengths and, therefore, a high energy. Gamma radiation is emitted from the nuclei together with the emission of an alpha or beta particle, or due to the transfer from a higher to a lower energy state. It can also be produced as a consequence of the collision and subsequent annihilation of an electron–positron pair. The typical energy range for most photons resulting from radioactive decay is about 1 keV to 2 MeV, but may be larger. Gamma rays interact with matter in various ways. The three main mechanisms are photoelectric absorption, Compton scattering, and electron–positron pair production. Photoelectric absorption involves the energy transfer from the photon to an orbital electron, which is ejected from its atom. This process tends to predominate at low energies below about 500 keV. In Compton scattering, the gamma ray collides with an electron, transferring part of its energy to the electron, while itself being scattered at a reduced energy. This process tends to predominate at higher energies above 500 keV. In electron–positron pair production, a gamma ray of sufficient energy (equal or greater than 1.02 MeV, which is the residual mass energy equivalent of the pair) disappears, resulting in the creation of an electron and a positron. The penetration of gamma rays is much greater than that of alpha or beta particles. In solid materials, the depth over which about 99 percent of the initial gamma-ray energy is dissipated is in the order of a decimetre and is inversely related to the density of the absorbing material and the initial energy of the gamma rays.

Radioactivity is both natural and man-made. At present, the radioactive elements which occur near the Earth's surface include radionuclides having half-lives in the order of the age of the Earth (about 5 billion years) or longer (e.g. uranium-238 (half-life $4.5 \cdot 10^9$ y), thorium-232 (half-life $1.4 \cdot 10^{10}$ y), potassium-40 (half-life $1.3 \cdot 10^9$ y)) together with their decay products. These are commonly known as the terrestrial sources of natural radioactivity. In addition, radionuclides may also be derived from the effects of cosmic radiation: for example, the formation of naturally occurring radiocarbon, ^{14}C. These are known as cosmogenic radiation. The sum of terrestrial radiation, cosmogenic radiation, and the portion cosmic radiation that reaches the Earth's surface is usually referred to as ***background radiation***. Naturally occurring radiation levels have been anthropogenically enhanced due to mining of natural uranium and phosphate rocks (used as fertilisers), fossil fuel combustion, and the production of man-made radionuclides through nuclear reactions in particle accelerators, nuclear reactors, and nuclear weapons. The most important source of man-made environmental radioactivity is the process of ***nuclear fission*** of certain elements. The fission process implies the break-up of heavy elements such as uranium, plutonium, or americium, into smaller nuclei (e.g. strontium-90, caesium-137, and iodine-131) through the bombardment with neutrons in a nuclear fission reactor. Nuclear fission is primarily used in nuclear power plants to generate electricity, but also in nuclear weapons to generate an enormous explosive power. Another source is the production of radioisotopes through ***neutron activation***, i.e. the process of capture of neutrons in the nuclei of elements. Some of these radioisotopes are produced in nuclear reactors for medical, industrial, scientific, or military purposes; others arise as radioactive waste. In addition, radionuclides may be produced by bombarding nuclei with charged particles, such as protons and alpha particles. Man-made radioactivity is dispersed in the environment due to discharges under normal operation of nuclear reactors, radioactive waste disposal, accidental releases, and nuclear explosions.

Radioactive material is considered to be an environmental hazard because its ionising radiation affects the cells of biota, especially animals and humans, mainly because of the damage it can cause to vital enzymes, chromosomes, and hormones. Typically, the cells most easily harmed are those that divide most rapidly (e.g. white blood cells, bone marrow, hair follicles, and foetuses). The magnitude of the damage caused by ionising radiation depends on the type of radiation, its energy, and whether the exposure is internal or external. The hazard to humans from exposure to alpha particles is insignificant if the exposure is external. In this case the particles are absorbed by the outer layer of the skin or clothing. However, if the exposure is internal due to inhalation or ingestion, alpha radiation may severely damage sensitive internal organs. In contrast to alpha particles, when considering the potential hazard of beta particles and gamma radiation to humans it is necessary to take into account both the internal and external exposure. The exposure to radiation can be measured as the ***absorbed dose***, i.e. the energy imparted by ionising radiation per unit mass of irradiated material. The unit for absorbed dose is gray (Gy): one gray is equal to an absorbed dose of 1 J kg^{-1}

Table 8.1 Average annual dose from background radiation in Europe (source: UKAEA, 2006).

Country	Average dose rate from background radiation (mSv y⁻¹)
Austria	2.8
Belgium	3.3
Finland	7.6
France	5.1
Germany	3.3
Ireland	3.8
Netherlands	2.1
Spain	4.9
Sweden	6.0
UK	2.2

and 100 rads (the old unit for absorbed dose). The current usage is to refer to exposure in terms of ***dose equivalents***, which takes account of a quality factor, which is 1 for beta and gamma radiation and 10 for the more dangerous alpha radiation. The dose equivalent is expressed in sieverts (Sv), where one sievert equals the absorbed dose in grays times the quality factor. One sievert is also equal to 100 rems (the old unit for dose equivalent). The dose rate (i.e. dose equivalents per unit time) from background radiation varies from region to region, depending on the composition of the bedrock. Table 8.1 shows the average annual doses to humans for several European countries. Higher background radiation levels have been reported in other parts of the world: for example, in India, large numbers of people are exposed to an annual dose rate of 15 mSv y^{-1}. In other places in Europe, Africa, and South America, background radiation produces levels as high as 50 mSv y^{-1}.

At high radioactive doses of more than 1 Sv, many cells of vital organs may be killed, seriously injuring the body; doses above 5 Sv are likely to be lethal within a few weeks. Lower doses (50 mSv – 1 Sv) do not instantly cause obvious injury, but a number of the cells that survive may carry mutations, which implies that damage to the DNA has been incorrectly repaired. Some specific mutations leave the cell at greater risk of being triggered to become cancerous in the future. There is usually a 5- to 20-year lag before cancer due to radiation exposure develops. About 10 percent of cancer is estimated to be attributable to exposure to background radiation.

8.2 NATURAL RADIONUCLIDES

8.2.1 Terrestrial radionuclides

The terrestrial radionuclides can be assigned to two groups. The first group consists of heavy radionuclides that occur in three radioactive decay series, which decay stepwise until a stable lead isotope is formed. Most of the naturally occurring radioactive isotopes are members of one of these three decay series. The other group consists of lighter nuclides that do not belong to such a series (Pattenden, 2001).

The three radioactive decay series comprising the radionuclides of the first group are: 1) uranium series in which uranium-238 decays to lead-206; 2) actinium series in which uranium-235 decays to lead-207; and 3) thorium series in which thorium-232 decays to lead-208. Figure 8.2 is a schematic diagram of these three radioactive decay series. Uranium (U) and thorium (Th), which constitute the heads of these three radioactive decay series, are predominantly found in alkali feldspathoidal rocks and acid rocks rich in silica (see Table 8.2).

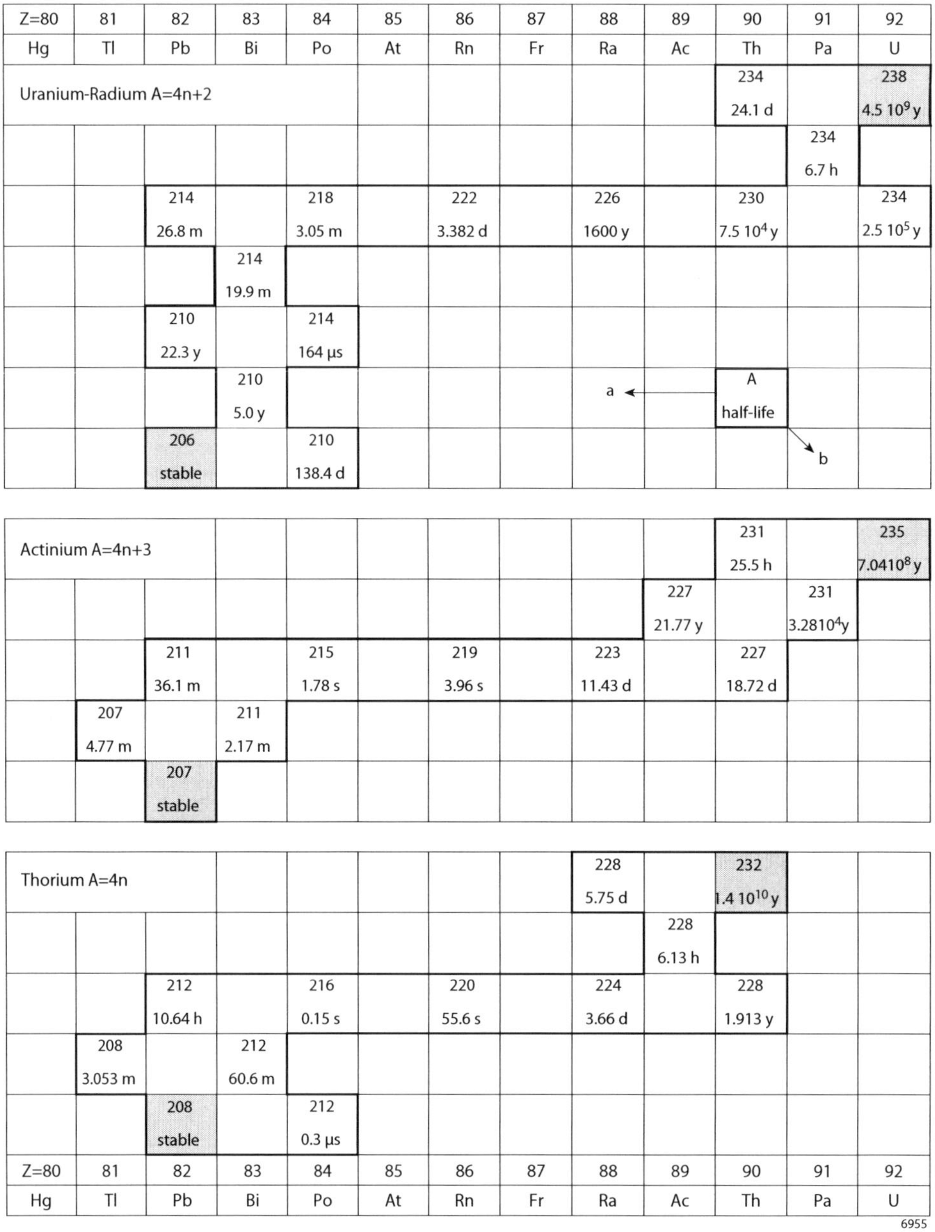

Z=80	81	82	83	84	85	86	87	88	89	90	91	92
Hg	Tl	Pb	Bi	Po	At	Rn	Fr	Ra	Ac	Th	Pa	U
Uranium-Radium A=4n+2										234 24.1 d		238 4.5 10^9 y
											234 6.7 h	
		214 26.8 m		218 3.05 m		222 3.382 d		226 1600 y		230 7.5 10^4 y		234 2.5 10^5 y
			214 19.9 m									
		210 22.3 y		214 164 µs								
			210 5.0 y					a ←		A half-life		
		206 stable		210 138.4 d								b

Z=80	81	82	83	84	85	86	87	88	89	90	91	92
Actinium A=4n+3										231 25.5 h		235 7.0410^8 y
									227 21.77 y		231 3.2810^4 y	
		211 36.1 m		215 1.78 s		219 3.96 s		223 11.43 d		227 18.72 d		
	207 4.77 m		211 2.17 m									
		207 stable										

Z=80	81	82	83	84	85	86	87	88	89	90	91	92
Thorium A=4n								228 5.75 d		232 1.4 10^{10} y		
									228 6.13 h			
		212 10.64 h		216 0.15 s		220 55.6 s		224 3.66 d		228 1.913 y		
	208 3.053 m		212 60.6 m									
		208 stable		212 0.3 µs								
Z=80	81	82	83	84	85	86	87	88	89	90	91	92
Hg	Tl	Pb	Bi	Po	At	Rn	Fr	Ra	Ac	Th	Pa	U

6955

Figure 8.2 Natural radioactive decay series (minor branching not shown) (after Pattenden, 2001).

Both U and Th are actinides, i.e. the group of elements (heavy metals) in the periodic table (see Appendix 1) from actinium (Ac) with atomic number 89 to lawrencium (Lr) with atomic number 103.

Uranium occurs naturally in the 3+, 4+, 5+, and 6+ oxidation states, but it is most commonly found in the 4+ and 6+ oxidation states present in reducing and oxidising conditions, respectively. In aqueous solutions without carbonates, the soluble U(VI) species

UO_2^{2+}, UO_2H^+, $(UO_2)_3(OH)_5^+$, and $(UO_2)_3(OH)_7^-$ occur in proportions depending on the pH. In water containing carbonate, dissolved U(VI)-carbonate species predominate. These U(VI)-carbonate species include $UO_2CO_3^0$, $UO_2(CO_3)_2^-$, and $UO_2(CO_3)_3^{2-}$. The anionic U-carbonate species dominate at and above neutral pH and tend to cause the desorption of U(VI) from mineral surfaces and the dissolution of U(VI) solids (Zhang *et al.*, 2002). As well as forming complexes with hydroxyl and carbonate, U(VI) also forms complexes with sulphate, fluoride, and phosphate. Despite complexation, the natural concentrations in soil and water are low (see Table 8.2). The major pathway leading to human exposure to uranium is soil U taken up by plants. Uranium tends to accumulate in bones and bone marrow. The chemical toxicity of U is, however, more significant than its radiotoxicity.

Thorium can be found in the 2+, 3+, and 4+ oxidation states, but occurs predominantly as hydroxides of Th(IV) in soil and water, although carbonate complexes may also occur (Zhang *et al.*, 2002). Further data on the heads of the decay series ^{238}U, ^{235}U, and ^{232}Th, and some important decay products are listed in Table 8.3. Each of the series includes both alpha and beta emitters with half-lives ranging from less than a millisecond to thousands of years. The greatest part of the radiation from the nuclides in the three decay series is emitted from the short-lived isotopes. Provided that the radionuclides in a series are not separated due to transport in the gaseous or aqueous phase, they exist in state of radioactive equilibrium in which the activity, i.e. decay rate, is the same as that of the radionuclide preceding it. As a result, the molar concentrations of the series members are proportional to their half-lives. Due to chemical and physical separation processes, full equilibrium with the heads of the series is rarely found.

Table 8.2 Uranium and thorium concentrations in different rock types (source: Wollenberg and Smith, 1990).

Rock type	Uranium		Thorium	
	range mg kg^{-1}	mean mg kg^{-1}	range mg kg^{-1}	mean mg kg^{-1}
Acid extrusive	0.8–23	5.7	1.1–116	22.4
Acid intrusive	0.1–30	6.3	0.1–253	27.3
Intermediate extrusive	0.2–5.2	2.1	0.4–28	6.7
Intermediate intrusive	0.1–23	3.2	0.4–106	12.2
Basic extrusive	0.03–3.3	0.9	0.05–8.8	2.5
Basic intrusive	0.01–5.7	0.8	0.03–15	2.3
Ultrabasic	0–1.6	0.3	0–7.5	1.4
Alkali Feldspathoidal intermediate extrusive	1.9–62	29.7	9.5–265	134
Alkali Feldspathoidal intermediate intrusive	0.3–720	55.8	0.4–880	133
Alkalic basic extrusive	0.5–12	2.3	2.1–60	8.9
Alkalic basic intrusive	0.4–5.4	2.3	2.8–20	8.4
Chemical sedimentary rocks	0.03–27	3.6	0.03–132	14.9
Carbonates	0.03–18	2.0	0–11	1.8
Detrital sedimentary rocks	0.1–80	4.8	0.2–362	12.4
Clay	1.1–16	4.0	1.9–55	8.6
Shale	0.9–80	5.9	5.3–39	16.3
Sandstone and conglomerates	0.1–62	4.1	0.7–227	9.7
Metamorphosed igneous rocks	0.1–148	4.0	0.1–104	14.8
Metamorphosed sedimentary rocks	0.1–53	3.0	0.1–91	12.0

Table 8.3 Data on the heads of the radioactive decay series (^{238}U, ^{235}U, and ^{232}Th) and some important members of these series.

Nuclide	Half-life	Isotopic abundance	Specific activity	Mode of decay	Decay energy	Typical concentrations**	
						Fresh water	Soils
		(% weight)	(Bq g^{-1})	(-)	(MeV)	(Bq m^{-3})	(Bq kg^{-1})
Uranium (U)							
^{238}U	4.47·10^9 y	99.28	12.4·10^3	Alpha	4.270	4.8	8–110
^{235}U	7.04·10^8 y	0.71	7.99·10^4	Alpha	4.679	0.22	-
^{234}U	2.45·10^5 y	0.0054	2.31·10^8	Alpha	4.859	5.2	9–120
Thorium (Th)							
^{232}Th	4.5·10^{10} y	~100	4.1·10^3	Alpha	4.083	0.12	4–73
^{230}Th	7.5·10^4 y	Nil	7.6·10^8	Alpha	4.770	-	100
Radium (Ra)							
^{226}Ra	1600 y	> 99	3.7·10^{10}	Alpha	4.871	4–400	7–180
Radon (Rn)							
^{222}Rn	3.825 d	-	5.95·10^{15}	Alpha	5.590	9	-
Polonium (Po)							
^{210}Po	138.4 d	-	1.66·10^{14}	Alpha	5.407	0.5–2.6	8–220
Lead (Pb)							
^{210}Pb	22.3 y	Nil	2.8·10^{12}	Beta	0.064	3–8	75

* Specific activity = the activity of 1 g pure radionuclide. This can be calculated from:
  Specific activity = Avogadro's number (6.022·10^{23}) × ln(2)/half-life (in s)/atomic weight
** Source: Bowen (1979).

Further inspection of the radioactive decay series shows that each series includes an isotope of the inert gas radon. The uranium series contains radon-222, the actinium series contains radon-219, and the thorium series contain radon-220 (also called thoron). These isotopes are particularly of interest because they and their decay products provide the largest single source of radiation dose to humans. Since they are gases and chemically inert, they tend to diffuse away from the rock in which they have been formed, to surface waters and the atmosphere. Radon-222 is most likely to escape since it has the longest half-life (3.825 d), but the rate depends on surface conditions and local weather conditions. Water, whether liquid or in the form of an ice or snow cover, reduces the escape of radon gas from the bedrock to the atmosphere, whereas wind and high temperatures enhance it. In the atmosphere, the gas is rapidly dispersed. The activity concentrations of radon-222 and its short-lived products in air vary greatly spatially, being low in marine air (about 0.01 Bq m^{-3}) and high in areas with granitic rock (up to about 9 Bq m^{-3}). In poorly ventilated buildings, radon can build up rapidly, since atmospheric dispersal is inhibited. In addition, radon also escapes from inorganic building materials, such as brick, concrete, and gypsum board. Typical average indoor activity concentrations of radon-22 and its short-lived products vary between 5 and 25 Bq m^{-3}. The longer-lived and stable decay products (lead, bismuth, and polonium isotopes) formed due to the decay of atmospheric radon gas are metals and tend to be deposited as solid aerosols on the Earth's surface. This atmospheric deposition adds to the activity concentrations in water and soil due to the decay of radon-222 already present.

Table 8.4 Non-series terrestrial radionuclides (source: Pattenden, 2001).

Nuclide	Half-life	Isotopic abundance	Element abundance in crustal rock	Activity concentration in crustal rock
	(y)	(% atoms)	(mg kg^{-1})	(Bq kg^{-1})
Potassium-40 (^{40}K)	$1.28 \cdot 10^{9}$	0.0118	$2.59 \cdot 10^{4}$	789
Rubidium-87 (^{87}Rb)	$4.8 \cdot 10^{10}$	27.83	90	79
Samarium-147 (^{147}Sm)	$1.06 \cdot 10^{11}$	15.0	6.0	0.73
Lutetium-176 (^{176}Lu)	$3.8 \cdot 10^{10}$	2.59	0.50	$3.07 \cdot 10^{-2}$
Lanthanum-138 (^{138}La)	$1.05 \cdot 10^{11}$	0.0902	30	$2.47 \cdot 10^{-2}$

In addition to the radionuclides in the abovementioned decay series there are also a number of terrestrial radionuclides that are not part of a series because they disintegrate to stable isotopes. Like the heads of the decay series, they are characterised by long half-lives. Table 8.4 lists some of these radionuclides; others exist, but they have such long half-lives, or such small abundances that they are insignificant. The most significant non-series radionuclides are potassium-40 and rubidium-87. Potassium emits both gamma radiation and beta particles, whereas rubidium-87 is only a beta emitter. Like uranium and thorium, their concentrations in rock and soil vary widely depending on rock type. Rubidium behaves chemically very much like the more common potassium; both are mainly associated with K-bearing feldspars, micas and clays (see also section 5.3). Since potassium is an essential nutrient, it is taken up by plants, animals, and humans. A 70 kg human contains about 140 g of potassium, mostly in the muscle. The potassium-40 component emits about 3700 Bq and delivers about 0.15 mSv to the bone (Alloway and Ayres, 1997).

8.2.2 Cosmogenic radionuclides

Cosmogenic radionuclides are produced as a consequence of cosmic irradiation of the Earth's atmosphere. The major part of cosmic radiation consists of nuclear particles with very high energy: approximately 70 percent protons, 20 percent alpha-particles, 0.7 % lithium, beryllium, and boron ions, 1.7 percent carbon, nitrogen, and oxygen, and 0.6 percent other ions with $Z > 10$ (Choppin *et al.*, 1995). These ions are bare nuclei, because their kinetic energies exceed the binding energies of all of the orbital electrons. When the high-energy cosmic particles enter the atmosphere, they interact with atmospheric gases such as nitrogen (N_2), oxygen (O_2), and argon (Ar), and are annihilated. This results in the production of a large number of secondary particles and gamma radiation. The particles include cosmogenic radionuclides and neutrons. These neutrons may be captured by the nuclei of atmospheric gases to produce other radionuclides. This process results in the formation of many radionuclides with half-lives ranging from very short to very long; examples are tritium (^{3}H), beryllium-7, beryllium-10, radiocarbon (^{14}C), and sodium-22. Table 8.5 lists some properties of these radionuclides. The production rate of cosmogenic radionuclides in the atmosphere is fairly constant in time but increases with altitude (by approximately four orders of magnitude from sea level to 20 km) and latitude (by more than tripling from the equator to 60° north).

Although the radionuclides are formed in extremely small concentrations, the global inventory is considerable and measurable. The radionuclides formed in the atmosphere are generally oxidised and they attach to aerosols, which reach the Earth's surface by wet or dry deposition. Obviously, cosmogenic radionuclides with short half-lives compared with the atmospheric residence time are likely to decay before they reach the Earth's surface.

Table 8.5 Properties of some cosmogenic radionuclides (source: Pattenden, 2001; Choppin *et al.*, 1995)

Nuclide	Half-life	Main radiation	Decay energy	Main target	Average rate of atmospheric production (atoms m^{-2} s^{-1})
Tritium (^{3}H)	12.323 y	Beta	18.6 keV	N, O	2500
Berrylium-7 (^{7}Be)	53.29 d	Gamma	478 keV	N, O	81
Berrylium-10 (^{10}Be)	1.6·10^6 y	Beta	555 keV	N, O	300
Carbon-14 (^{14}C)	5730 y	Beta	156 keV	N, O	17000–25000
Sodium-22 (^{22}Na)	3.603 y	Beta+ Gamma	0.545 MeV, 1.82 MeV 0.511 MeV, 1.275 MeV	Ar	0.5

Tritium and radiocarbon form a special case. They are also oxidised to form water (known as tritiated water, HTO) and carbon dioxide, respectively, after which they follow the global water and carbon cycles, including the uptake by plants, animals, and humans. The average natural activity concentration in continental surface water is in the range 200–900 Bq m^{-3} (values measured before nuclear bomb testing; see also section 8.3) (Samuelsson, 1994). The activity concentrations in other parts of the ecosystems, including plants and human bodies, do not deviate much from these values, since significant enrichment of tritium does not occur. Tritium at natural concentrations does not contribute relevantly to the radioactive dose to humans. The average natural concentration of radiocarbon in biota is 227 Bq kg^{-1} of carbon. The annual average equivalent dose from radiocarbon is estimated at about 10 μSv y^{-1} for the entire human body.

8.3 MAN-MADE RADIONUCLIDES

Large-scale production of man-made radionuclides started with the operation of the first nuclear reactor in Chigaco, USA, in 1942, less than four years after the discovery of nuclear fission. Three years later, in 1945, the first radioactive contamination of the environment by man-made radionuclides occurred when the first nuclear bombs were tested in New Mexico, USA, and subsequently deployed in Hiroshima and Nagasaki in Japan. Since then, there have been many instances of environmental contamination by man-made radionuclides. They include liquid discharges from nuclear fuel reprocessing plants under normal operation, accidental releases from nuclear reactors, and nuclear weapons testing.

8.3.1 Production and releases

About 430 nuclear reactors are currently connected to the world's electricity grids, supplying 16 percent of the world's electricity demand. The majority of these reactors are located in the industrialised countries of the world. In these reactors, a wide range of radionuclides is produced as a result of nuclear fission of fissionable heavy nuclei such as uranium-235 and plutonium-239. The fuel for nuclear reactors consists of rods of enriched uranium oxide (UO_2). Uranium oxide concentrate from mining is refined to form so-called yellowcake (U_3O_8). Subsequently, it is converted to uranium hexafluoride gas (UF_6), in which state it undergoes enrichment to increase the uranium-235 content from 0.7 percent to about 3.5 percent. It is then turned into a hard ceramic oxide (UO2) for assembly as reactor fuel elements. The main byproduct of enrichment is depleted uranium, principally the uranium-238 isotope, which is stored, either as UF_6 or as U_3O_8. Depleted uranium is less radioactive than natural uranium. Because of the very large specific density of metallic

Table 8.6 Properties of some radioecologically important radionuclides produced by nuclear fission (source: Aarkrog, 2001).

Nuclide	Half-life	Decay	Decay energy (MeV)	Production rate* (PBq/GWy)
Radioiodine (^{131}I)	8.02 d	Beta, Gamma to ^{131}Xe (stable)	0.606 0.364	640
Radiocaesium (^{137}Cs)	30.17 y	Beta to ^{137m}Ba (2.554 m) Gamma to ^{137m}Ba (stable)	0.514 0.662	45
Radiostrontium (^{90}Sr)	28.64 y	Beta to ^{90}Y (T = 64.1 h) Beta to ^{90}Zr (stable)	0.546 2.24	38

* Production rate at Chernobyl per gigawatt thermic year (continuous production).

uranium (19.0 g cm^{-3}), depleted uranium has been used in boat keels, aeroplane wings, and penetrators of artillery shells.

Fission products arise from the splitting of uranium-235 or plutonium-239 nuclei by bombarding them with slow-moving neutrons. In the process of nuclear fission, the nucleus of a heavy fuel element absorbs a slow-moving free neutron, becomes unstable, and then splits into two smaller atoms. The fission process for uranium atoms yields two smaller atoms, one to three fast-moving free neutrons, plus an amount of energy. Because more free neutrons are released from a uranium fission event than are required to initiate the event, under controlled conditions a chain reaction starts, resulting in the release of an enormous amount of energy in the form of radiation and heat. The newly released fast neutrons must be slowed down (moderated) before they can be absorbed by the next fuel atom. This slowing-down process is caused by the neutrons colliding with atoms from an introduced moderator, mostly water, which is introduced between the nuclear fuel rods. The fission products include radioecologically important isotopes such as radioiodine (^{131}I), radiocaesium (^{137}Cs), and radiostrontium (^{90}Sr). Table 8.6 lists some characteristics of these isotopes. Their chemical and radioecological behaviour is discussed in section 8.3.5.

Besides the production of fission products, nuclear fission also yields a number of byproducts as a consequence of neutron reactions with ^{238}U and its fission products, the construction materials of the nuclear reactor wall, reactor coolants, and fuel impurities. Neutron capture in ^{238}U results in the formation of plutonium-239:

$$_{92}^{238}\text{U} \; + \; _{0}^{1}\text{n} \; \rightarrow \; _{92}^{239}\text{U} \; \overset{\beta^-}{\rightarrow} \; _{93}^{239}\text{Np} \; \overset{\beta^-}{\rightarrow} \; _{94}^{239}\text{Pu} \tag{8.4}$$

Plutonium-239 is a so-called transuranic element, i.e. an element with an atomic number greater that 92 (uranium). Other examples of transuranic elements occurring in nuclear reactors are plutonium-238, plutonium-240, plutonium-241, americium-241, curium-242, and curium-244. Typical examples of corrosion products that are generated through neutron activation of construction materials in nuclear reactors are cobalt-60 (half-life 5.27 y) and zinc-65 (half-life 0.67 y). Another radioisotope of caesium, ^{134}Cs, is produced through neutron activation of ^{133}Cs, which is the stable end-product of the fission product decay chain with mass number 133. The half-life of ^{134}Cs is 2.06 y and the activity in reactors is approximately half that of ^{137}Cs. Radiocarbon (^{14}C) and tritium are also generated in large quantities in nuclear reactors. Carbon-14 is produced by neutron capture by stable nitrogen (^{14}N) followed by the emission of a proton:

$$_{7}^{14}\text{N} \; + \; _{0}^{1}\text{n} \; \rightarrow \; _{6}^{14}\text{C} \; + \; _{1}^{1}\text{H} \tag{8.5}$$

Carbon-14 can also be generated by neutron capture and subsequent alpha emission by oxygen-17:

$$^{17}_{8}O + ^{1}_{0}n \rightarrow ^{14}_{6}C + ^{4}_{2}He \tag{8.6}$$

Tritium is formed due to neutron activation reactions with lithium and boron isotopes dissolved in or in contact with the reactor coolant (Aarkrog, 2001).

Radionuclides are also produced in cyclotrons, i.e. circular particle accelerators. In these devices the nuclei of atoms are bombarded with accelerated and focused beams of charged atomic particles (protons, deuterons ($^2H^+$), or alpha particles). They are primarily used for scientific research in nuclear physics and for the production of short-lived isotopes for medical treatment and diagnosis (e.g. technecium-99m (half-life 6.0 h) and iodine-131 (half-life 8.04 d)). In general, the quantities produced are small.

During normal operation of nuclear reactors, only minor amounts of radioactivity are released into the atmosphere and water. Airborne discharges involve tritium, radiocarbon, radioiodine, and noble gases (e.g Xe). The airborne releases of ^{131}I are in the order of 2.5–14 GBq per GWy (GigaWattyear). Liquid discharges into water bodies are radioecologically more important. For example, the liquid discharges of ^{137}Cs from reactors vary between 15 and 2500 GBq GWy^{-1}, depending on the type of reactor. Uranium and plutonium in spent fuel from reactors are recovered in reprocessing plants like those in Sellafield, UK, or La Hague, France. During the fuel reprocessing process the fuel is dissolved by strong acids, from which the uranium and plutonium are recovered. If economic and institutional

Table 8.7 Major nuclear accidents (sources: Aarkrog, 1995; Aarkrog, 2001; ATSDR, 2013; Stohl *et al.*, 2012).

Location	Date	Description	Total radioactivity released	Major radionuclides of radioecological concern
Chernobyl, USSR (Ukraine)	26 April 1986	Explosion of nuclear reactor 4 after engineering test; followed by fire	1–2 EBq	^{131}I, ^{137}Cs, ^{134}Cs, ^{90}Sr
Fukushima (Japan)	11 March 2011	Equipment failures and nuclear meltdowns at the Fukushima I Nuclear Power Plant, following the Tōhoku earthquake and tsunami	15.3 EBq 36.6 PBq	^{133}Xe, ^{131}I, ^{137}Cs, ^{134}Cs
Windscale (now Sellafield), UK	10 October 1957	Fire in nuclear reactor	2 PBq	^{131}I, (^{137}Cs, ^{90}Sr)
Three Miles Island, Harrisburg, USA	28 March 1979	Core of the reactor became overheated due to drainage of primary coolant	378 PBq	^{133}Xe, ^{131}I
Kyshtym (Mayak/ Chelyabinsk-40), USSR (Russia)	29 September 1957	Chemical explosion in a tank at the nuclear weapons plutonium production site	1 EBq	^{95}Zr, ^{144}Ce, (^{90}Sr)
Kyshtym	-	Groundwater pollution and airborne resuspension from open reservoirs for waste disposal of high-level radioactive waste	4 EBq (total inventory)	^{137}Cs, ^{90}Sr
Indian Ocean	21 April 1964	US satellite SNAP 9-A powered by a radioisotope generator re-entered the atmosphere	0.6 PBq	^{238}Pu
Northwest Territories, Canada	24 January 1978	Soviet satellite Cosmos 954 powered by a nuclear reactor re-entered the atmosphere	0.2 PBq	^{131}I, ^{137}Cs, ^{90}Sr

conditions permit, the recovered uranium and plutonium can be recycled for use as nuclear fuel. The residual liquid is still highly radioactive and part of this radioactive waste is released as effluent. During the 1970s and 1980s, these liquid discharges (excluding tritium) from reprocessing plants normalised per TBq GWy^{-1} were more than a thousand times greater than liquid discharges from nuclear reactors (UNSCEAR, 2000). However, the discharges have been greatly reduced since the mid-1980s, thanks to the introduction of improved waste-treatment practices.

The spent fuel from reactors or, if the fuel has been reprocessed, wastes from reprocessing plants are high-level radioactive wastes. The spent fuel is mostly stored within the reactor basins or at aboveground waste storage facilities. Releases of radionuclides from these storage facilities are negligible. However, in the long term these materials must also be safely disposed of and isolated from the biosphere until the radioactivity they contain has diminished to a safe level. Currently a preferred option is for the ultimate disposal of the wastes in solid form in licensed deep, stable geological structures.

8.3.2 Accidental releases

The exploitation of nuclear energy has been accompanied by a number of accidents, resulting in radioactive contamination of the environment. Most of these accidents have only been

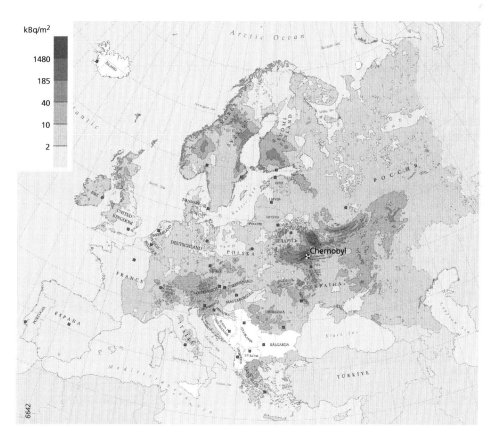

Figure 8.3 Cs-137 deposition in Europe after the Chernobyl accident in April 1986 (source: DeCort *et al.*, 1998).

of local or regional concern. However, some of them have led to continental or even global contamination by various radionuclides. Table 8.7 lists some of the major nuclear accidents.

The most serious nuclear accident to date was at Chernobyl. During the explosion in reactor 4 and the subsequent fire, which lasted for 10 days, radioactive contamination spread over large area in Ukraine, Belarus, Russia, and the rest of Europe. From a radioecological perspective, the 1760 PBq of the short-lived ^{131}I was of most concern in the first weeks following the accident. Later, attention shifted to ^{137}Cs and ^{90}Sr, which are longer lived and are the main contributors to internal and external doses to humans in the long term. Figure 8.3 shows the spatial distribution of soil contamination in soil by ^{137}Cs in Europe.

On 11 March 2011, a magnitude 9.0 undersea earthquake off the Pacific coast of Tōhoku triggered a tsunami that hit the coast of north-eastern Japan. The waves, which were up to about 14 m high, caused the Fukushima I Nuclear Power Plant (also known as the Fukushima Daiichi Nuclear Power Plant) to become disconnected from the power grid

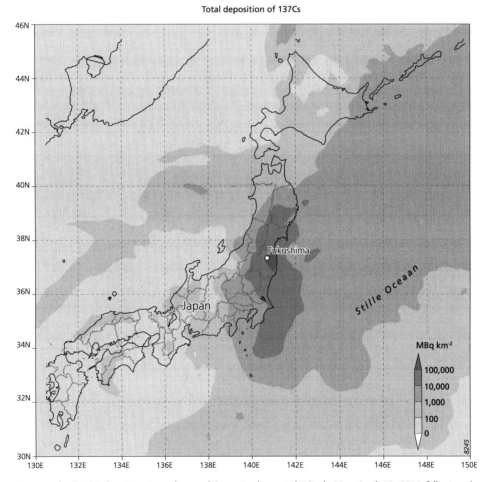

Figure 8.4 Cs-137 deposition in and around Japan in the period March 20 – April 19, 2011 following the Fukushima accident. Adapted from Yasunari *et al.* (2011).

and destroyed the emergency power back-up systems. Severe cooling problems developed in the nuclear reactors of Units 1-3, and these led to explosions in the three reactors in the subsequent days. Temperatures inside the reactor became so high that meltdown occurred in units 1 and 2. As a consequence, radioactive materials were released to the atmosphere, groundwater, and sea. Most of the releases were uncontrolled, but some materials were released deliberately: for example, to reduce gas pressure or to discharge contaminated cooling water. Most of the ^{137}Cs released into the atmosphere was deposited on the ocean northeast of Fukushima. Terrestrial deposition of ^{137}Cs occurred primarily northwest of Fukushima (Yasunari *et al.*, 2011; Figure 8.4).

8.3.3 Nuclear weapons

Nuclear fission weapons work similar to nuclear fission in nuclear reactors, except that the process is not moderated, so the chain reaction can proceed unhindered and rapidly expands. As the atoms that undergo fission with each step of the chain increase exponentially, the energy release rapidly becomes sufficient to vaporise the warhead, causing the explosion. Typically, 1–3 percent of the fissile material actually splits; once vaporised, the atoms are too far apart to sustain the chain reaction. Another process that is used in nuclear weapons is nuclear fusion. Fusion bombs (also called hydrogen bombs or thermonuclear bombs) explode with enormous power using uncontrolled self-sustaining chain fusion reactions. Under extremely high temperatures, deuterium (D = ^{2}H) and tritium (T = ^{3}H) fuse to helium providing the energy:

$$D \; + \; T \; \rightarrow \; {}_{2}^{4}He \; + \; n \tag{8.7}$$

In principle, a mixture of D, T, and lithium-6 heated to very high temperature and confined to a high density will start a chain fusion reaction. The explosive process begins with a fission chain reaction of ^{238}U, which produces a temperature of several million degrees. When the temperature of the mixture reaches 10 000 000 K, fusion reactions take place. Fusion causes the temperature to rise, and neutrons released in fusion cause further fission of ^{238}U that releases more energy and radioactive fallout.

Nuclear weapons have only been deployed twice in war. In August 1945 the USA detonated two fission bombs over Hiroshima and Nagasaki in Japan. Although these weapons killed a large number of people, the principal source of contamination by man-made radionuclides at a global scale has been the releases during nuclear weapons testing in the open atmosphere. In total, 541 atmospheric tests have been performed by the USA, USSR, UK, France, and China. In 1962, USA, USSR, and UK ended atmospheric testing; no atmospheric test has been performed since 1980. Since 1980 only underground nuclear tests have been performed. The major test sites for the atmospheric testing of nuclear weapons include Novaya Zemlya (Russia), Semipalatinsk-21 (Kazakhstan), Nevada test site (USA), Lopnor (China), and several test sites in the Pacific Ocean (Bikini and Eniwetok atolls (USA), Mururoa (France), Christmas Island (UK)). Fission bombs were tested at Novaya Zemlya (USSR) and the Bikini and Eniwetok atolls (USA). Explosions of thermonuclear bombs caused substantial releases of tritium into the atmosphere and atmospheric explosions of fission bombs released significant amounts of fissile materials (^{238}U, ^{238}U, ^{239}Pu) and fission products (e.g. ^{131}I, ^{137}Cs, ^{90}Sr, and many others). The released radioactive materials have been dispersed over large areas and have contributed to an overall background contamination level of long-lived fission products and transuranic radionuclides in the environment. It has been estimated that about 12 percent of the fallout was deposited close to the test site, about 10 percent was tropospheric fallout deposited in a relatively narrow band around the latitude of the test site, and 78 percent was global

Table 8.8 Deposition density of selected radionuclides in the 40°-50° N latitude band (Values not corrected for radioactive decay)(source: Aarkrog, 2001).

Radionuclide	Total deposition density to 1990 (Bq m^{-2})
^{89}Sr	20000
^{90}Sr	3230
^{131}I	19000
^{137}Cs	5200
^{140}Ba	23000
^{238}Pu	1.5
^{239}Pu	35
^{240}Pu	23
^{241}Pu	730
^{241}Am	25

fallout mainly deposited in the same hemisphere as the test site (UNSCEAR, 1993). The total global fallout amounted to 604 PBq of ^{90}Sr and 940 PBq of ^{137}Cs (value not corrected for radioactive decay). Since most of the tests were carried out in the northern hemisphere, about three quarters was deposited there. Table 8.8 lists the deposition density of some selected radionuclides resulting from weapons fallout in the 40°–50° N latitude band, which represents the global maximum belt of fallout from weapons testing.

On two occasions, nuclear weapons have led to environmental contamination by transuranic elements due to crashes of aircraft carrying nuclear weapons. This happened on uncultivated farmland near Palomares, Spain, in January 1966, and in the Atlantic Ocean near Thule, Greenland, in January 1968. Furthermore, eight nuclear submarine losses have been reported. Two were lost by the USA and six by the USSR. The submarines contained nuclear reactors and probably also nuclear weapons. Most losses have occurred in the North Atlantic.

Recently, concern has been growing about another potential source of radioactive contamination by radiological weapons, also called dirty bombs. These weapons have been suggested as a possible terrorist weapon to create panic in densely populated areas. They do not require weapons-grade materials: common materials such as ^{137}Cs used in radiological medical equipment could be used. So far, such weapons have never been deployed.

As mentioned above, metallic depleted uranium (DU) is also used in conventional weapons, such as penetrators of shells and bombs to increase the penetration depth during impact. Ammunition containing DU is known to have been used in Iraq during the first Gulf war in 1991, in Bosnia–Herzegovina in 1995, and in Kosovo, Serbia, and Montenegro in 1999. The forming of aerosols containing DU during impact on hard surfaces and the corrosion of the penetrators in soil may give rise to localised spots of DU contamination. Nevertheless, the hazards related to such contamination points in terms of possible contamination of water and plants are considered to be negligible (UNEP, 2002).

8.3.4 Environmental behaviour and effects of selected man-made radionuclides

The most important man-made radionuclides from a radioecological perspective are ^{131}I, ^{137}Cs, ^{90}Sr, and ^{239}Pu. A brief overview of their behaviour in soils, plants and ingestion pathways is given below.

Iodine-131

Iodine-131 is only short-lived (half-life 8.04 d) and is therefore highly radioactive, but only significantly so in the first weeks after a release of radionuclides into the environment. Iodide (I^-) is the primary species of I in soil except under exceptional arid and alkaline conditions. Iodide sorbs poorly to most soil materials, but can be taken up by some clays, organic matter, and metal sulphides (Zhang *et al.*, 2002). The main exposure pathway of ^{131}I is from cow's milk. Cows that have ingested ^{131}I deposited on the leaves of grass transfer it rapidly to their milk. Radioiodine is easily accumulated in the thyroid gland, especially in humans who suffer iodine shortage, and so may induce thyroid cancer. The biological half-life of ^{131}I is approximately 100 days in the thyroid, 14 days in bone, and 7 days in the kidney and reproductive organs.

Caesium-137

Caesium is chemically similar to potassium and occurs as Cs^+ cation in soil and water. It is very soluble but also readily adsorbed by clay minerals. Directly after atmospheric deposition of radiocaesium, it is very mobile in runoff water, but the bulk of ^{137}Cs is adsorbed by soil materials in the topsoil. Fixation of caesium by illitic clay minerals causes the mobility and bioavailability to decrease by a factor of ten during the first five years after initial deposition. After this period, the decrease of bioavailability follows radioactive decay (half-life 30.17 y) (Smith *et al.*, 2000). The gamma radiation emitted by decay product ^{137m}Ba (see Table 8.6) is of most concern from a health perspective. Caesium-137 uptake by crops is the main pathway of Cs transfer into the animal and human food chains. Since Cs competes with potassium, ^{137}Cs uptake is greater in soils with low potassium content. Like potassium, Cs tends to be absorbed in neural and muscle tissues. The average biological half-life of ^{137}Cs is about 45 days. Furthermore, radiation from soil contaminated by ^{137}Cs contributes significantly to external radiation exposure.

Strontium-90

Like ^{137}Cs, ^{90}Sr has a relative long radioactive half-life (28.64 y) and is readily absorbed by plants in the form of Sr^{2+}. It is chemically similar to calcium and therefore accumulates in bone and bone marrow. Since Sr is not specifically adsorbed by clay minerals, it is more mobile than ^{137}Cs in soil and water.

Plutonium

Plutonium and other transuranic elements do not exist naturally, with the exception of the small amounts of Pu generated by natural fission. Natural fission is very rare and is only found in a few locations: for example, the Oklo natural reactor in Gabon. In general, plutonium exhibits a complex and diverse geochemistry. It may change its oxidation state due to slight changes in redox potential and often two or more of its oxidation states co-exist (Runde, 2002). At the low concentrations in which they occur under natural conditions in soil and water, the major process that determines their environmental behaviour is sorption onto particulates and clay mineral surfaces. Surface contamination, rather than uptake by plant roots, is the main pathway of Pu accumulation in plants and crops (Zhang *et al.*, 2002). Although there is some controversy about whether Pu is one of the most dangerous substances known, there is no doubt that it is very toxic. Therefore, the toxicity of Pu as a heavy metal accumulating in the bone and liver is likely to be greater than its radiotoxicity, which is relatively low due to the relatively long physical half-lives. Nevertheless, most Pu isotopes are alpha emitters and may induce lung cancer when inhaled: for example, due to airborne resuspension of contaminated soil particles by wind erosion.

EXERCISES

1. Define the following terms:
   a. Isotope
   b. Alpha radiation
   c. Beta radiation
   d. Gamma radiation
   e. Nuclear fission
   f. Neutron activation
   g. Absorbed dose
   h. Cosmogenic radionuclides
   i. Man-made radionuclides

2. a. What does background radiation consist of?
   b. Explain why background radiation varies from location to location.

3. Name the three natural radioactive decay series of heavy radionuclides.

4. a. Name a terrestrial radionuclide that is not a member of these decay series.
   b. Name a cosmogenic radionuclide.
   c. Explain why the heads of the three decay series and other non-series terrestrial radionuclides have very long half-lives in the order of the Earth's age.

5. Radon-222 gas has a half-life of 3.825 days.
   a. Given an initial amount of 1 g of ^{222}Rn, how much will be left after one week?
   b. Which decay product predominates then? (Hint: consider the half-lives of the successive decay products (see Figure 8.2)).

6. Name the two major events that have caused global fallout of fission products.

7. On 26 April 1986 the Chernobyl accident happened in Ukraine. From a radioecological perspective, the most important radionuclides released were ^{131}I, ^{137}Cs, and ^{90}Sr. Which of these radionuclides was of most concern for humans in the contaminated area in the following situations?
   a. Drinking water preparation from lake water in a peat area in 1995.
   b. Potato cultivation on a sandy soil in 1990.
   c. Milk production on 5 May 1986.
   Give reasons for your answers

8. Does plutonium-239 occur naturally in the environment?

9. Describe the differences between the possible radioecological effects of the Chernobyl nuclear accident and those of the Fukushima nuclear accident.

9

Organic pollutants

9.1 INTRODUCTION

In Section 4.3 we saw that organic substances consist of a variety of compounds made up of carbon, oxygen, hydrogen, and small amounts of nitrogen, phosphorus, sulphur, chlorine and several other elements. The chemical bonds between the carbon atoms and between these atoms and hydrogen or other elements are established by covalent bonding, which implies that a carbon atom shares one or more electrons with its neighbouring atoms. In the case of one shared electron, a single bond occurs, indicated by a single line (e.g. ethane: $CH_3 - CH_3$). If two atoms share two electrons, a double bond is formed (e.g. ethene: $CH_2 = CH_2$), and, logically, a triple bond is formed when two atoms share three electrons (e.g. ethyne: $CH \equiv CH$). A chain of carbon molecules can thus have single, double, or triple bonds at any arbitrary location within the chain. If a chain with two or more substituents (i.e. atoms or groups other than hydrogen) consists of four or more carbon atoms, the atoms forming a substance can be structured spatially in more than one manner, resulting in so-called *isomers*. Obviously, if organic matter molecules consist of many carbon atoms, multiple branches of short or long chains may occur. Although isomers are built up from the same elements, their physico-chemical properties (e.g. melting point, vapour pressure, aqueous solubility, and chemical reactivity) may differ substantially.

Instead of chains, carbon atoms may also form ring structures that sometimes have double bonds. Such rings are usually composed predominantly of carbon atoms, but they may also contain elements other than carbon, such as oxygen or nitrogen. Organic molecules and their isomers are named according to the systematic nomenclature that has been briefly summarised in Box 9.I, which also gives some examples of the structure of organic molecules and isomers. Further details about the nomenclature can be found in any organic chemistry textbook. Apart from the strictly structural approach adopted in this nomenclature, organic chemicals are commonly classified on the basis of their source (e.g. petroleum products) or use (e.g. pesticides (insecticides, herbicides, fungicides), plasticisers, solvents) or some of their physico-chemical properties (e.g. volatile organic compounds (VOCs), persistent organic pollutants (POPs), or absorbable organic halogens (AOX)). The latter classification often reflects the analytical procedure applied for determining the compounds.

Most organic substances are potential pollutants as they enter groundwater or surface water, since they may cause a depletion of dissolved oxygen. However, some organic compounds are directly toxic, thereby potentially causing direct harmful effects on living organisms and ecosystems. These organic compounds include petroleum and its derivatives, polycyclic aromatic hydrocarbons (PAHs), and chlorinated hydrocarbons. The possible effects of organic pollutants on human health include cancer, allergies, disruption of the immune system, damage to the nervous system, and reproductive disorders. Most polluting organic compounds are man-made and industrially produced, although a number of these chemicals also occur naturally, produced by biochemical synthesis, incomplete decomposition of organic matter, volcanic eruptions, forest fires, or lightning. The uses made

Box 9.1 Nomenclature for organic compounds

The root name for an organic compound indicates the number of carbon atoms in the longest continuous chain of carbon atoms containing the functional group (meth-, eth-, prop-, but-, pent-, hex-, hept, oct-, non- dec- for 1 to 10 carbons, respectively). A prefix and/or suffix indicates the family to which the compound belongs. The family name is based on the functional groups; for example, the suffix -ane refers to single carbon bonds, -ene to double carbon bonds, -yne to triple carbon bonds, -anol to an alcohol or hydroxyl group (-OH), and -anone to an oxide group (= O). Functional groups consisting of nitrogen atoms or halogens are indicated by a prefix (e.g. nitro- ($-NO_2$), chloro- (-Cl), fluoro (-F)). If a functional group occurs more than once, the number is indicated by the Greek name for the number (di-, tri-, tetra-, penta-, etc.). The position of a functional group is indicated by a prefix with the appropriate number as a position identifier. The numbering of the main chain is done in the direction which gives the lowest number to the first branching point. Examples are given in Figure 9.Ia.

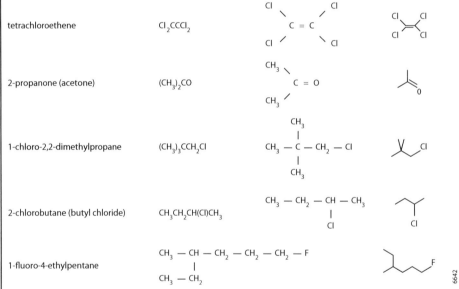

Figure 9.Ia Examples of simple open chain organic molecules. The compounds are symbolised using three different conventions.

Ring structures are prefixed by cyclo-, for example in dichlorocyclohexane. Figure 9.Ib depicts some examples of ring structures. The ring structure of cyclohexa-1,3,5-triene (C_6H_6) is a special case and is commonly called benzene. Benzene rings are extremely stable and this quality is referred to as aromaticity. Consequently, compounds that include benzene ring structures are referred to as **_aromatic_**. Hexagonal ring structures, in which a carbon is replaced by nitrogen (as in quinoline; see Figure 9.Ib), or pentagonal ring structures where one or more carbons are replaced by a NH group, sulphur, or oxygen (as in furan) are also aromatic. Hydrocarbons that are not aromatic (open chains and their cyclic isomers) are called **_aliphatic_**.

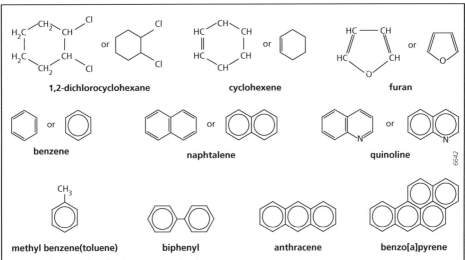

Figure 9.Ib Examples of cyclic organic molecules.

Because of the symmetry of the planar ring, there are only three isomers for any particular disubstituted benzene: those identified as 1,2; 1,3; and 1,4. However, another commonly adopted system uses the prefixes ortho-, or o- for the 1,2-isomer; meta-, or m- for the 1,3-isomer; and para-, or p- for the 1,4-isomer.

The prefixes cis- and trans- describe the relative position of two substituents relative to a double (or triple) carbon bond or the ring plane. The term cis is used is used if the two substituents are located on the same side of the double bond or plane and the term trans is used if they are on opposite sides (see Figure 9.Ic).

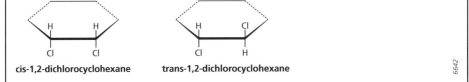

Figure 9.Ic Cis and trans isomers of 1,2-dichlorocyclohexane.

In ring systems with more than two substituents, more isomers are possible. For example, Figure 9.Id shows three possible isomers of 1,2,3,4,5,6-hexachlorocyclohexane (HCH). The γ-isomer is known as the insecticide lindane. Note that, like lindane, many organic compounds have another 'trivial' name, which is used more frequently than the systematic name.

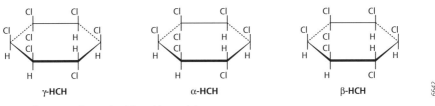

Figure 9.Id Isomers of 1,2,3,4,5,6-hexachlorocyclohexane.

of organic compounds include fuel, pesticides, plasticisers, surfactants, solvents, pigments, or coolants. They may cause environmental pollution as they enter the environment as a result of industrial processes, product use and applications, spills and leaks, combustion of fuel, dumping or incineration of organic wastes or wastes contaminated with organic compounds, and application of pesticides.

The dispersal and persistence of organic compounds in the environment are principally controlled by their physico-chemical properties. Organic pollutants may occur in the solid, liquid, and gas phases. Some organic compounds evaporate readily and completely and remain in the gas phase when exposed to the atmosphere at normal environmental temperatures and pressures. These – mostly liquid – compounds are often referred to as *volatile organic compounds* (VOCs). VOCs include pure hydrocarbons (e.g. benzene, toluene), partially oxidised hydrocarbons (e.g. acetone), and organic compounds containing chlorine, sulphur or nitrogen (e.g. methylene chloride, methyl chloroform). *Semi-volatile organic compounds* (SVOCs) comprise another set of organic compounds, which evaporate slowly and partially when exposed to the atmosphere.

Some groups of organic compounds, for example carbohydrates, alcohols, and organic acids, contain polar functional groups that allow the formation of hydrogen bonds (e.g. hydroxyls). Consequently, they are hydrophilic and soluble in water. More polar organic compounds, particularly those containing oxygen and nitrogen, tend to be more water-soluble than less polar organic compounds. However, most organic pollutants are non-polar and are hydrophobic. The inability to form hydrogen bonds prevents water from solubilising these compounds and mixing with them, and so they are excluded from the aqueous phase. Liquid hydrophobic organic compounds, such as petroleum and benzene, are therefore often referred to as *non-aqueous phase liquids* (NAPLs). Given similar structures of the organic molecules, compounds with higher molar masses are generally more hydrophobic and less volatile. Hydrophobic organic compounds tend to occur both attached to solid surfaces and as liquid droplets. They interact strongly with the organic matter associated with the solid phase. To describe the affinity of an organic compound for water and organic matter, the octanol–water partition coefficient K_{ow} is commonly used (see Section 2.5.3).

The behaviour of immiscible NAPLs in the subsurface depends on their density. Compounds with densities less than the aqueous phase will float on the capillary fringe if a sufficient amount of the compound is present. The NAPLs are called *light non-aqueous phase liquids* (LNAPLs). *Dense non-aqueous phase liquids* (DNAPLs) have densities greater than the aqueous phase and will sink until an impermeable barrier is reached. This flow behaviour of NAPLs will be discussed further in Section 11.4.

The persistence of organic compounds in the environment is largely controlled by their degradation rate, which is promoted by microorganisms (biodegradation) or light (photochemical degradation). Organic chemicals are distributed in a continuum, from very reactive to extremely persistent ones. The latter, referred to as *persistent organic pollutants* (POPs), degrade very slowly, with long environmental half-lives ranging from two months to several decades in water, soil, and sediments. POPs include anthropogenic chemicals such as pesticides (for example DDT, dieldrin, aldrin, and hexachlorobenzene) and industrial chemicals (for example PCBs, dioxins, and furans). In general, POPs are hydrophobic, and, because they persist in the environment for long periods, are prone to bioaccumulation as they are transferred through the food chain. Furthermore, most POPs are semi-volatile, and therefore tend to be transported long distances in the atmosphere and deposited long range. This has resulted in ubiquitous contamination by POPs of the global environment; they are found even in remote regions where they have never been used or produced (e.g. the Arctic and Antarctic).

Many organic compounds occur only in very small concentrations in soil and water, but even at these low levels can cause severe ecotoxicological effects. Therefore, these compounds are often referred to specifically by the generic term ***organic micro-pollutants***. Although this term is poorly defined, it is widely used in the environmental literature. Nevertheless, it should be realised that it does not cover all polluting organic compounds: for example, some organic compounds, such as petroleum constituents or trichloroethene (TCE), may occur in large quantities in soil or groundwater near massive and localised releases.

The following sections cover the sources, use, environmental behaviour, and potential toxic effects of a selection of environmentally relevant classes of organic pollutants. Given the vast number of polluting organic compounds, it is impossible, and perhaps even undesirable, to cover them all in this text. For further reading on environmental organic chemistry and toxicology, see Schwarzenbach *et al.* (1993), Alloway and Ayres (1997), Fawell and Hunt (1988), or Hayes and Laws (1991). For more information on individual substances or groups of substances, see the detailed toxicological profiles issued by the U.S. Agency for Toxic Substances and Disease Registry (ATSDR, 2013). Much of the information given in the following section has been excerpted from these toxicological profiles. Other sources of information, if any, are also referenced at appropriate places.

9.2 PETROLEUM AND DERIVATIVES

Petroleum or crude oil is a thick, dark brown inflammable liquid formed by the anaerobic decay of organic matter in conditions of increased temperature and pressure in enclosing sedimentary rocks. The organic matter breaks down into liquid petroleum and natural gas. Both the liquid and gas tend to migrate from the source rock (usually shale) through porous rocks and permeable rock (usually sandstone), until they encounter impermeable layers, underneath which they tend to collect. Crude oil consists of a complex mixture of many different chemical compounds ranging from very volatile, light compounds like pentane and benzene, to heavy compounds such as bitumens and asphaltenes. It contains mainly hydrocarbons (i.e. organic compounds made up of only carbon and hydrogen) of the alkane series, which usually make up about 95 percent of the crude oil. These aliphatic components are barely soluble in water. Crude oil also includes impurities such as monocyclic and polycyclic aromatic hydrocarbons. The remaining part is comprised of small amounts of oxygen, nitrogen, and sulphur, and traces of other elements, such as heavy metals.

Petroleum is extracted from reservoirs by drilling and pumping, after which it is refined by distillation. The products include kerosene, benzene, petrol, paraffin wax, asphalt, etc. The four lightest hydrocarbons methane (CH_4), ethane (C_2H_6), propane (C_3H_8), and butane (C_4H_{10}) are all gases used as fuel. The alkane chains with five to seven carbons are all light and volatile and are used as solvents and dry cleaning fluids. The alkane chains with six to twelve carbons are mixed together and used for petrol. Kerosene is made up of chains in the range between 10 to 15 carbons, followed by diesel fuel (10 to 20 carbons), and heavier fuel oils, such as those used in ship engines. All these petroleum compounds are liquid at room temperature and are less dense than water. Because oil does not mix with water, it belongs to the group of LNAPLs. Alkane chains with more than 20 carbons (paraffin wax, tar, and asphaltic bitumen, respectively) are solid.

Environmental pollution by petroleum and oil products may occur during extraction, refinement, transport, storage, and use. Spills and leaks are the principal causes of oil pollution of soil, groundwater, and surface water, and pose a threat to soil and water quality, plant and animal life, and human health. Traces of oil in water can substantially affect odour and taste and can make the water unfit for use as drinking water. On sandy soils, phytotoxicological effects may already occur at concentrations of 0.5 mg kg^{-1} (Scheffer and Schachtschabel, 1989). Possible hazards are also related to impurities in petroleum and derivatives, such as benzene (see Section 9.3) and PAHs (see Section 9.4), or additives to petroleum products, such as MTBE (methyl tertiary-butyl ether) that has replaced lead as petrol additive. In humans, ingestion of mineral oil, the major by-product in the distillation of petroleum to produce petrol, can hamper the absorption of vitamin A. Mineral oils used as lubricants for metal workers have been associated with enhanced risk of occupational skin cancer (Irwin *et al.*, 1998). Because of the broad variety of constituents in petroleum and petroleum products, which vary in density, mobility, degradability, and toxicity, the composition of each individual oil or oil product must be taken into consideration in order to determine the environmental impact of the oil in question.

The environmental fate of spilled oil is controlled by a variety of natural processes that reduce the amount and toxicity of oil in soil or water and, consequently, mitigate the severity of an oil spill. These processes include evaporation, emulsification, oxidation, adsorption, and biodegradation (EPA, 1999a). Evaporation occurs when the lighter substances within the oil mixture volatilise. This process leaves behind a residue consisting of the heavier oil components, which may undergo further weathering. Some of these components are denser than water, so, in surface waters, they may sink to the bottom. Lighter refined petroleum-based products such as kerosene and petrol may volatilise completely within a number of hours, thereby reducing the toxic effects to the environment. The rate of volatilisation increases with increasing wind speed. In surface water, this effect is further promoted by waves and currents. The action of waves also causes emulsification, i.e. the formation of water-in-oil emulsions. The uptake of water in oil can increase the volume of a floating oil layer up to four times. Heavy, very viscous oils tend to take up water more slowly than light, more liquid oils. The formation of emulsions reduces the rate of other weathering processes, making water-in-oil emulsions very persistent oil slicks. Oxidation of the oil leads to the formation of water-soluble compounds or persistent tar. Oxidation is promoted by sunlight, but compared to other weathering processes its overall effect on dissipation is small. Even under intense sunlight, oil oxidises relatively very slowly: usually less than 0.1 percent per day (ITOPF, 2011). Thick slicks may only partially oxidise, which results in the formation of tar balls. These dense, sticky, black spheres can collect in the bed sediments of lakes or slow moving streams and may persist in the environment for a long time. In sediments and soils, oil is particularly adsorbed by organic matter. As the ability of sediments to bind oil increases with increasing organic matter content, the phytotoxicity of oil is less in organic soils than in soil with low organic matter content. Biodegradation occurs when microorganisms feed on oil. The biodegradation rate of oil is optimal under warm conditions with sufficient supply of oxygen and nutrients (particularly nitrogen and phosphorus).

9.3 MONOCYCLIC AROMATIC HYDROCARBONS

Monocyclic aromatic hydrocarbons are compounds consisting solely of carbon and hydrogen and containing one benzene ring. The most common monocyclic aromatic hydrocarbons are benzene, toluene, ethylbenzene, and the three isomers of xylene (or dimethylbenzene): ortho-xylene, meta-xylene, and para-xylene (see Box 9.I), which all exist as clear, colourless, non-corrosive, volatile liquids with a sweet odour. These compounds are collectively referred to by the acronym BTEX. BTEXs can make up a significant percentage of petroleum products: about 18 percent on a weight basis in a standard petrol blend. Benzene is used in the production of synthetic materials and consumer products, such as synthetic rubber, plastics, nylon, insecticides, and paints. Toluene is used as a solvent for paints, coatings, glues, oils, and resins. Ethylbenzene may be present in consumer products such as paints, inks, plastics, and pesticides, and as an additive in petrol and aviation fuel. Xylene is used as a solvent in the printing, rubber, and leather industries. It is also used as a cleaning agent and as a paint thinner.

BTEXs are among the most hazardous constituents of petrol and solvents. Except for short-term hazards from concentrated spills, BTEX compounds have been more frequently associated with risk to human health than with risk to plants and animals. This is partly due to the fact that plants and animals take up BTEX compounds in very small amounts and BTEXs tend to volatilise relatively rapidly into the atmosphere rather than persisting in surface waters and soils. However, BTEX compounds may pose a threat to drinking water quality when they accumulate in groundwater. Short-term hazards of BTEX include potential acute toxicity to aquatic life in surface water and potential inhalation hazards. Acute exposures to high levels of BTEX may cause irritation of the skin, eyes, and respiratory tract, and depression of the central nervous system. Long-term potential hazards of BTEX compounds include chronic inhalation and contamination of groundwater. Prolonged exposure to BTEXs may cause changes in the liver and adverse effects on the kidneys, heart, lungs, and the central nervous system. Benzene is the most dangerous and carcinogenic to humans. Its chronic effects encompass the destruction of bone marrow, leading to a reduction of red and white blood cells. Workers exposed to high levels of benzene in occupational settings have been found to have an increased incidence of leukaemia. Toluene may also adversely affect reproduction.

The acronym BTEX suggests that benzene, toluene, ethylbenzene, and xylenes are often found together at contaminated sites. The principal source of BTEX contamination is the leakage of petrol from faulty underground storage tanks. Other sources of BTEX contamination are releases from large bulk facilities, surface spills, and pipeline leaks. Once released to the environment, BTEXs are liable to volatilisation, dissolution, adsorption, and biodegradation. Although BTEXs can behave as a LNAPL, which means that the bulk of the BTEX floats on water, it can also dissolve in water. Compared to the other components in petrol, such as the aliphatic components, BTEXs are very soluble in water. BTEXs can be adsorbed by organic matter in soil, but are not sorbed to soil particles as strongly as the aliphatic components. The bacterial flora naturally present in soil is capable of breaking down BTEXs under aerobic conditions (Brady *et al.*, 1998). Like the decomposition of the other mineral oil components, the biodegradation rate depends on temperature and the supply of oxygen and nutrients. Toluene and – to some extent – xylene can also be biodegraded under anaerobic conditions (Edwards *et al.*, 1992). In anaerobic sediments, the natural biodegradation is largely controlled by the amount of bioavailable Fe(III), which can act as an electron acceptor. Typically, bioavailable Fe(III) constitutes between 10% and 30% of the total amount of iron present in aquifer material (Manshoven *et al.*, 2010).

9.4 POLYCYCLIC AROMATIC HYDROCARBONS

Polycyclic aromatic hydrocarbons (PAHs) are a class of over 100 different, very stable organic molecules that are made up of only carbon and hydrogen and contain two or more connected benzene rings (see examples in Figure 9.Ib). These molecules are formed during the incomplete burning of coal, oil and gas, garbage, or other organic substances like tobacco or charbroiled meat. PAHs are found in coal tar, crude oil, creosote, and roofing tar. Some PAHs are manufactured and are used in medicines or to make dyes, plastics, and pesticides. Pure PAHs usually exist as colourless, white, or pale yellow-green solids. PAHs are commonly divided into two groups, depending upon their physical and chemical properties: low-molecular-weight PAHs, containing three or fewer aromatic rings, and high-molecular-weight PAHs, containing more than three aromatic rings. Most PAHs are semi-volatile and the volatility generally increases with decreasing molecular weight. PAHs containing 6 or more benzene rings are barely volatile. The US Environmental Protection Agency has defined a subset of 16 PAHs that are most important from an environmental perspective. These 16

PAHs include (the number between brackets indicates the number of benzene rings, but note that PAHs may also contain additional non-benzene rings): acenaphtene (2), acenaphthalene (2), anthracene (6), benz[a]anthracene (4), benzo[b]fluoranthene (4), benzo[k]fluoranthene (4), benzo[ghi]perylene (6), benzo[a]pyrene (5), chrysene (4), dibenz[a.h]anthracene (5), fluorantene (3), fluorine (2), indeno[1.2.3-cd]pyrene (5), naphthalene (2), phenanthrene (3), and pyrene (4).

PAHs enter the atmosphere mostly as releases from volcanoes, forest and peat fires, fossil fuels combustion, coke and asphalt production, waste incineration, and aluminium smelting. In air, PAHs occur mostly attached to dust particles smaller than 1–2 μm. Over a period of days to weeks, PAHs can break down by reacting with sunlight and other airborne chemicals. They are also removed from the atmosphere by deposition, as a result of which they reach the Earth's surface. The lighter molecular weight PAHs can be revolatilised and redistributed before they are redeposited. Although PAH levels in air, soil, and water are particularly enhanced in urbanised regions, PAHs are ubiquitous in the environment, even in remote areas, due to long-range atmospheric transport. In addition to atmospheric deposition, the sources of PAHs to terrestrial environments include releases from creosote-treated products, spills of petroleum products, and application of compost.

The PAH contents of terrestrial plants and animals may be substantially greater than the PAH contents of soil on which they live, albeit that the PAH contents in terrestrial plants are often independent from those in soil because PAH compounds are poorly soluble and poorly available in soils (ATSDR, 2013). A major source of PAHs in plants is the accumulation of airborne PAHs on plant leaves; plant uptake of PAHs via the root system is often negligible. The process of accumulation is affected by a variety of factors. It increases with decreasing ambient temperature, and with increasing leaf surface area and lipid concentration in plant tissues. Volatile PAHs with a lower molecular weight are primarily subject to dry gaseous deposition, whereas non-volatile PAHs mainly accumulate on plant surfaces in the form of dry particulates (Bakker, 2000). The volatile PAHs enter plants primarily through gaseous diffusion via open stomata, although absorption by the waxy leaf surface accounts for a portion of the total PAHs in tissues. The waxy surface of leaves intercepts both gaseous and particle-bound PAHs. Because of their ability to intercept PAHs from the atmosphere, a wide range of plant species, including lichens, mosses, algae, and trees have been used to evaluate environmental accumulation of PAHs (e.g. Franzaring *et al.*, 1992; Wegener *et al.*, 1992). Mosses are particularly effective filters due to their large surface area, their capacity to absorb large organic molecules, and their ability to obtain water and nutrients from the air.

Surface water contamination by PAHs can occur through runoff from contaminated soil surfaces (particularly urban runoff), and direct discharges from industrial and wastewater treatment plants. In general, most PAHs are not very mobile in soil and water, due to their relatively low solubilities and strong affinity for organic particulate matter. As a consequence, PAHs tend to partition into sediments and soils. PAH levels are usually much higher in sediments than in surface water, i.e. in the range of $\mu g\ kg^{-1}$ (ppb) rather than $ng\ kg^{-1}$ (ppt).

PAHs are potentially hazardous to plants due to photo-enhanced toxicity in the presence of ultraviolet (UV) or other types of solar radiation (Irwin *et al.*, 1998). PAHs are good photosensitisers, forming biologically harmful excited or singlet-state oxygen radicals. Their toxic effects include chlorosis, inhibition of photosynthesis, and diminished biomass accumulation. Soil contamination by PAHs may also inhibit root growth. PAH toxicity in aquatic environments is mainly associated with the more soluble compounds, particularly two-ring compounds such as the naphthalenes, and some heavier PAHs that affect aquatic macrophytes, benthic aquatic invertebrates (i.e. invertebrates that live on the bed of a water body), and fish. As a consequence, aquatic animals in the water will have a decreased individual fitness and will not be able to develop through their life stages successfully and reproduce effectively. Some PAH compounds, however, appear to stimulate reproduction, which suggest they have an oestrogen-like action. Microorganisms can break down PAHs in soil or water after a period of weeks to months, though some PAHs are resistant to decay. Microorganisms degrade low molecular weight PAHs more easily than higher molecular weight PAHs. Non-substituted PAHs will also be degraded faster than alkyl-substituted PAHs (Lundstedt *et al.*, 2003).

With respect to humans, PAHs are generally associated with chronic risks, which are often the result of exposure to complex mixtures of aromatic compounds rather than to low levels of a single compound. Toxic effects include DNA adducts and cancer. In general, the heavier (4-, 5-, and 6-ring) PAHs have greater carcinogenic potential than the lighter (2- and 3-ring) PAHs (ATSDR, 2013). Human exposure to PAHs is mainly via the inhalation of contaminated air in the work environment and outdoors, and the ingestion of grilled or charred meats or contaminated water, cow's milk, or other foodstuffs (e.g. cereals, vegetables, fruits).

9.5 CHLORINATED HYDROCARBONS

Chlorinated hydrocarbons, also known as organochlorins, are hydrocarbons with one or more chlorine substituents. These compounds can consist of aliphatic as well as aromatic structures, and include a wide range of compounds. The carbon–chlorine bond is strong and makes organochlorins very stable. Because of this property, many chlorinated hydrocarbon compounds have been synthesised and used extensively in agricultural and industrial applications. In particular, the use of organochlorins as pesticides has resulted in a widespread contamination of aquatic and soil systems. Chlorinated hydrocarbons that are pesticides include DDT, aldrin, dieldrin, heptachlor, chlordane, lindane, endrin, and hexachloride. The chemical structures of these compounds are depicted in Figure 9.1. Other sources of environmental contamination by chlorinated hydrocarbons include spills from chemical facilities, improper disposal or leakage from storage containers or waste sites, and the burning of waste containing plastics, especially polyvinylchloride (PVC).

Because organochlorins are very stable, these molecules tend to persist in the environment for a long time and biomagnify in the food chain. Chlorinated hydrocarbon compounds are very toxic to fish and invertebrates, as well as to any other animals that may feed on them. The 96-hour LC50 for tested animals ranges from 1–60 μg l^{-1}, depending on the species. To account for their great bioaccumulation and biomagnification potential, the US Environmental Protection Agency has stipulated that the amount of chlorinated hydrocarbon compounds permitted in the water must not exceed 0.001 μg l^{-1}. However, even at low chronic doses, these compounds have the potential to cause problems, including hormone disruption leading to reduced reproductive success and damage to the central nervous system. In humans, they may also cause liver and kidney damage. Because of their persistence, tendency to bioaccumulate, and toxicity to non-target species, most chlorinated hydrocarbon pesticide uses have been phased out in the developed countries.

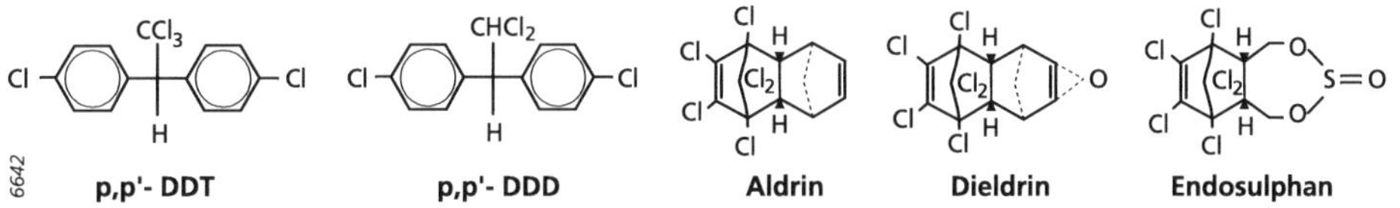

Figure 9.1 Examples of some organochlorine pesticides.

9.5.1 Aliphatic chlorinated hydrocarbons

The most widely used chemicals from the class of aliphatic chlorinated hydrocarbons consist of 1 or 2 carbons with one or more chlorine substituents. Examples include dichloromethane (methylene chloride), trichloromethane (chloroform), tetrachloromethane (carbon tetrachloride or tetra), 1,2-dichloroethane (ethylene dichloride), 1,1,1-trichloroethane (methyl chloroform), 1,1,1,2,2,2-hexachloroethane (perchloroethane), chloroethene (vinyl chloride), 1,2,2-trichloroethene (TCE or tri), 1,1,2,2-tetrachloroethene (perchloroethylene, PCE, or per). These compounds are used as organic solvents in the metal and electronic industries and as dry cleaning liquids in laundries. They are non-flammable, colourless, very volatile, organic liquids with a sweet odour. In the presence of ultraviolet light or excess heat they decompose to form the poisonous gas phosgene ($COCl_2$). Their half-lives in air are in the order of a few days to a week. In surface water, the half-lives range from days to weeks. In soil and groundwater they break down much more slowly because of the much slower evaporation rate. All of the chemicals listed above are denser than water, and since their aqueous solubility is limited, they behave like DNAPLs when spilled in substantial amounts.

Large concentrations of the above listed compounds may cause narcosis, lung irritation, and damage to the liver and kidneys. Tetrachloromethane is one of the most toxic chlorinated hydrocarbons. It can be absorbed through the skin; exposure to small amounts can cause severe liver and kidney damage. Chloroform, ethylene dichloride, perchloroethylene, and trichloroethene have been shown to cause liver cancer in rats and mice and it is reasonably likely that they are carcinogens to humans as well. Methyl chloroform is less toxic than other chlorinated hydrocarbons at low concentrations, but may be lethal when inhaled at high concentrations (ATSDR, 2013).

9.5.2 Hexachlorocyclohexane (HCH)

The most common isomer of hexachlorocyclohexane (HCH) is γ-HCH (also known as lindane; see Figure 9.Ib). Lindane is a white solid substance that may volatilise as a colourless vapour with a slightly musty odour. Lindane used to be used as an insecticide on fruit and vegetable crops and forest crops. Its use was phased out in the developed countries during the late 1970s, but it is still a widely used insecticide in Third World countries. The other isomers of HCH do not have the insecticide property. During the production of lindane, these isomers are separated and stored. HCH enters the environment due to application or releases from lindane production facilities. In air, HCH can be present as a vapour or attached to dust particles. Lindane can persist in air for up to 3 months and can be transported long distances in the atmosphere. In soil, sediments, and water, microorganisms break down HCH to less harmful substances. HCH isomers are broken down quickly in water; lindane does not remain in water longer than 30 days. The HCH degradation in soil is much slower, especially under aerobic conditions (Herbst and Van Esch, 1991). Under anaerobic conditions, the degradation may proceed considerably faster. The β-isomer usually predominates, because it is the most stable of the isomers. The HCH in soil is predominantly fixed by organic matter in the top 35 cm of the soil.

Plants may take up considerable amounts of HCH from soil; the HCH concentration in plant tissue is usually positively correlated with the HCH concentration in soil. The relationships vary with plant species and soil type. In the various plant parts, the HCH concentrations are higher in the roots and shoots than in stems, leaves, and fruits (Herbst and Van Esch, 1991). The LD50 value in rats is 90 mg kg^{-1} and the LC50 value in rainbow trout is 0.06 mg kg^{-1} (Alloway and Ayres, 1997).

The effects on human health of high exposures to lindane include muscular weakness, dizziness, and nausea. Animals that were fed with high levels of HCH showed convulsions and some became comatose. At moderate levels, a decreased ability to reproduce, resistance disorders, and effects on the liver and kidneys have been observed. Liver cancer has been observed in laboratory rodents that ate HCH for a long period of time and HCH may reasonably be anticipated to be a carcinogen to humans (ATSDR, 2013).

9.5.3 Hexachlorobenzene (HCB)

Hexachlorobenzene (HCB: C_6Cl_6) consists of a benzene ring with 6 chlorine groups attached to it and occurs as a white crystalline solid. HCB is man-made and does not occur naturally in the environment. It is formed as a byproduct during the manufacture of other chemicals. Small amounts can also be formed during the incineration of municipal waste. HCB used to be widely used as a fungicide to protect the seeds of onions and sorghum, wheat, and other grains. It was also used to make fireworks, ammunition, and synthetic rubber. It has not been used commercially in the developed countries since the 1970s.

HCB is very persistent and can remain in the environment for a long time. Its half-life in soil and surface water is 3–6 years. HCB is barely soluble in water and binds strongly to organic matter. For this reason, plants take up HCB in only very small amounts. Studies in animals show that eating HCB for a long time can damage the liver, thyroid gland, nervous system, bones, kidneys, blood, and immune system (ATSDR, 2013). HCB is classified as a probable human carcinogen.

9.5.4 Dichlorodiphenyl trichloroethane (DDT)

Dichlorodiphenyl trichloroethane (DDT: see Figure 9.1) is a white, crystalline solid with no odour, which does not occur naturally in the environment. It is an organochlorine pesticide developed during World War Two to control insects for agriculture and to assist in the elimination of insects known to spread diseases such as malaria. It was banned in the developed countries in 1972 but it is still used in Third World countries. Although not used for a long time, it is still present in the environment of developed countries due to its persistence. Illegal use of old stock and long-range atmospheric transport from countries in which DDT is still being used also contributes to the ubiquity of this chemical in the environment. Commercial DDT preparations are contaminated by chemicals similar to DDT, namely DDD (dichlorodiphenyldichloroethane) and DDE (dichlorodiphenyldichloroethylene). DDD was also used to kill pests, but this use has also been banned. DDE has no commercial use.

In air, DDT is broken down relatively rapidly under the influence of UV light. The half-life of DDT is about 2 days. In soils and sediments, DDT binds strongly to the organic fraction and does not dissolve easily in water. The biodegradation of DDT by microorganisms in soil proceeds slowly and the half-life of DDT in soil ranges from 5 to 8 years, depending on soil type (WHO, 1989). The microorganisms break down DDT to DDE and DDD. DDT accumulates in plant tissues and fatty parts of fish, birds, and mammals and has a high potential to biomagnify in the food chain. DDT is a hormone disruptor which acts in a similar way to oestrogen and binds to the cell's oestrogen receptors. This results in the impaired reproductive success in many of the higher organisms living in the aquatic environment. It also affects the nervous system. In acute exposure, DDT is highly toxic to aquatic invertebrates at concentrations as low as 0.3 µg l^{-1}. DDT is also highly toxic to fish: the 96 hour LC50s reported range from 1.5 to 56 µg l^{-1} (WHO, 1989). The sensitivity of birds to DDT varies greatly. Predatory birds are especially sensitive, as DDT and its metabolites are responsible for the thinning of eggshells and the consequent increased egg breakage (Alloway and Ayres, 1997).

In humans and mammals, DDT, DDE, or DDD enters the body mainly via ingestion of contaminated food. Studies in animals fed with DDT have shown that DDT may cause liver cancer, but studies in DDT-exposed workers have not shown increased incidences of cancer (ATSDR, 2013). Nevertheless, the US Environmental Protection Agency classifies both DDT and its breakdown products DDE and DDD as probable human carcinogens.

9.5.5 Polychlorinated biphenyls (PCBs)

Polychlorinated biphenyls (PCBs; $C_{12}H_{10-(x+y)}Cl_{(x+y)}$; cf. Figure 9.Ib) are mixtures of up to 209 individual man-made chlorinated compounds. They are either oily liquids or solids that are colourless to light yellow and have no odour. Some PCBs can exist as a vapour in air. The vapour pressure and water solubility of PCBs decreases with an increasing degree of chlorination. PCBs have a very high chemical stability, low flammability, low electrical conductivity, and good heat-conducting properties. For this reason, PCBs were widely used as coolants and lubricants in transformers and capacitors, heat exchangers, and hydraulic systems. They are also used as plastics solvents and paint strippers. Many commercial PCB mixtures are known by the trade name Aroclor. The production of PCBs worldwide had almost ceased by 1977. However, they still occur in old equipment and the destruction of these residues remains an issue. PCBs enter the environment via leaks from and fires in apparatus containing PCBs, leakage from landfill sites, and discharges of sewage effluents

PCBs are very persistent in the environment; the more chlorinated PCBs are more persistent than less chlorinated ones. PCBs can be broken down by microorganisms or photochemically under the influence of UV light. They can travel long distances through the atmosphere and be deposited in areas far from where they were released. In water, PCBs are strongly bound to organic particles and bottom sediments. Only a small amount of PCBs may remain dissolved. In soil, PCBs are fixed strongly to organic matter and to clay minerals. Because of their strong affinity with soil, PCBs are mostly fixed in the topsoil. In sandy soils poor in organic matter, however, PCBs can be transported to deeper horizons. Because the different PCB compounds have different solubility and volatility, they can be spatially separated due to differential volatilisation of PCBs or downward percolation of soil moisture. As a result, the more persistent compounds remain in the topsoil, whereas the relatively mobile compounds are either volatilised or transported downward in the soil profile.

The strong fixation of PCBs to soil particles means that there is very limited plant uptake of PCBs from soil. However, algae and fish can take up PCBs in considerable amounts and via this pathway, PCBs have a high potential to biomagnify in the food chain. In birds and mammals, PCBs accumulate primarily in fat tissues. The toxic effects of PCBs include damage to the liver and skin. Animals that ate food containing large amounts of PCBs for short periods of time had mild liver damage and some died. Rats that were fed with high levels of PCBs for two years developed liver cancer. Animals that ate smaller amounts of PCBs in food over several weeks or months developed various kinds of health effects, including anaemia, acne-like skin conditions, and damage to the liver, stomach, and thyroid gland. Other effects of PCBs in animals comprise changes in the immune system, altered behaviour, and impaired reproduction. At normal levels of exposure, PCBs are not very toxic to humans. The most commonly observed health effects in humans exposed to large amounts of PCBs are skin conditions such as acne and rashes. Studies on exposed workers have shown changes in blood and urine that may indicate liver damage. Few studies of workers have indicated that PCBs are associated with liver cancer in humans (ATSDR, 2013).

Figure 9.2 Examples of dioxins: 2,3,7,8-tetrachlorodibenzo-p-diozin and 1,2,3,7,8-pentachlorodibenzo-p-diozin.

9.6 DIOXINS

Chlorinated dibenzo-*p*-dioxins (CDDs) are a class of 75 chemically related compounds that contain two benzene rings with a varying number of chlorine atoms attached to them and which are linked by two bridging oxygen atoms (Figure 9.2). They are commonly known as chlorinated dioxins. One of the most hazardous dioxins is 2,3,7,8-tetrachlorodibenzo-*p*-diozin (2,3,7,8-TCDD; see Figure 9.2). In pure form, CDDs are crystals or colourless solids. 2,3,7,8-TCDD is odourless and the odours of the other CDDs are not known.

CDDs enter the environment as mixtures containing a number of individual components. They are not intentionally manufactured by industry, except for research purposes. They (2,3,7,8-TCDD in particular) may be formed during the chlorine bleaching process at pulp and paper mills or during chlorination at plants that treat waste or drinking water. CDDs can occur as contaminants in the manufacture of certain organic chemicals and are released into the atmosphere during the incineration of industrial and municipal solid waste. They may also be released during industrial accidents, such as the Séveso accident in Italy in 1976. This accident led to the release of a cloud, which contaminated over 1800 ha of land and affected the vegetation and many birds. CDDs were also an ingredient of Agent Orange, a defoliation herbicide widely used by the US Air Force during the Vietnam War in the 1960s (Alloway and Ayres, 1997).

In the atmosphere, CDDs may be transported over long distances, which is why dioxins are ubiquitous in the global environment. When released in wastewaters, some CDDs are broken down by sunlight, some evaporate to air, but most attach strongly to solid particles and settle to the bottom sediments. CDD has a high potential to biomagnify in the food chain, resulting in measurable levels in animals. In animals, exposure to low levels of 2,3,7,8-TCDD can induce various toxic effects, including damage to the liver, weight loss, disruption of the immune system, reduced reproductive ability, and birth defects including skeletal deformities, kidney defects, and weakened immune responses. Human exposure to large amounts of 2,3,7,8-TCDD can cause chloracne, a severe skin disease with acne-like lesions on the face and upper body. Other skin effects noted in people exposed to large doses of 2,3,7,8-TCDD include skin rashes, discoloration, and excessive body hair. Exposure to high concentrations of CDDs may also cause liver damage and changes in glucose metabolism and hormonal levels (ATSDR, 2013). According to the World Health Organisation, 2,3,7,8-TCDD is a human carcinogen.

9.7 EMERGING SUBSTANCES OF CONCERN

The term 'Emerging Substances of Concern', or ESOC, has recently been introduced for a wide range of chemicals – mostly organic – that are currently causing concern to environmental toxicologists. These substances are termed 'emerging' because they have only recently been introduced into the environment, have only recently been detected in the environment as a result of improved chemical analytical technologies, or their potential toxicity has only recently been recognised (Ternes and Von Gunten, 2010; Stuart *et al.*,

2012). The term 'substances' rather than 'contaminants' is used to recognise the fact that some of these substances are of natural origin.

Considerable uncertainty surrounds the environmental transport, fate and toxicological effects of ESOCs, since they are not often routinely measured in environmental monitoring programmes, or their precise toxic effects are still unknown. The lack or absence of environmentally relevant information of substances is not surprising, given the vast number of new organic compounds. By 2013, the CAS Registry database, the most comprehensive substance database maintained by the Chemical Abstract Service (a division of the American Chemical Society), contained descriptions of more than 71 million organic and inorganic substances and approximately 12 000 new substances were being added every day.

Categories of ESOCs include current-use pesticides, pharmaceuticals and personal care products (PPCPs), and endocrine-modulating compounds (EMCs). As is the case with the classification of organic pollutants in general, the various classes of ESOCs also overlap: for example, some pesticides and pharmaceuticals have endocrine-modulating properties, and personal care products may contain nanoparticles. In addition to the above classes of ESOCs, microplastics, i.e. particles of plastic debris smaller than 1 or 5 mm, have been identified as an emerging global environmental issue, especially in the marine environment. The various categories of ESOCs are discussed below.

9.7.1 Current-use pesticides

Current-use pesticides include modern agricultural and domestic pesticides, such as glyphosate, atrazine, and alachlor. Although widely used, they are less persistent than past pesticides such as DDT, dieldrin and aldrin.

Glyphosate, also known by its trade name Roundup, is a herbicide used to kill broadleaved weeds and grasses in agricultural and urban areas. It is the mostly widely used herbicide in the world. Glyphosate is considered to be relatively low in acute toxicity: in rats, the oral LD50 of pure glyphosate is 4230 mg kg^{-1}. It is not carcinogenic and barely bioaccumulates. Despite being chemically stable in water and not subject to photochemical degradation, glyphosate degrades microbially in soil, aquatic sediments and water. It also binds readily to particles. The sorption mechanisms of glyphosate to soils and sediments are not yet fully understood, but in general, sorption is controlled by available phosphate binding sites, which means that sorption is promoted in the presence of combinations of clay, sesquioxides and organic matter. Given its low mobility in soil, there is minimal potential for glyphosate contamination of groundwater. Glyphosate can, however, enter surface and subsurface waters through direct aquatic applications or through runoff or leaching from terrestrial applications. In a recent study in the Mississippi basin (Coupe *et al.*, 2011), glyphosate was frequently and consistently detected in surface waters, rain and air, which indicates it is transported from its point of use into the broader environment.

Another well-known agricultural herbicide is atrazine. Although it has been banned in European Union since 2004, it is still one of the most widely used herbicides in the world. As is the case of glyphosate, the World Health Organisation classifies atrazine as a pesticide unlikely to present acute hazard in normal use. The acute oral LD50 for rats is 1869-3090 mg kg^{-1}, indicating its relatively low toxicity, although ruminants seem to be much more sensitive to the acute toxic action than rodents. The carcinogenicity of atrazine is rather controversial, but the United States Environmental Protection Agency classifies it as being unlikely to be carcinogenic to humans. Atrazine is an endocrine-modulating compound and may cause low oestrogen levels and menstrual irregularities in women, even when it occurs in drinking water at concentrations far below the drinking water threshold of 3 μg l^{-1}. For aquatic organisms, serious effects have been observed on frogs' sexual development: at levels often found in the environment, atrazine causes demasculinisation of male tadpoles and turns them into hermaphrodites with ovaries in their testes, much smaller vocal organs and testosterone levels one-tenth of normal. Microbial degradation accounts for most of the breakdown of atrazine, although the degradation rate is slow. As a result, atrazine can persist in the environment for a long time, especially under dry or cold conditions. Atrazine is not easily absorbed by soil particles. Despite being moderately soluble in water, it is one of the most significant water pollutants in rain, surface, and groundwater, especially in areas where it is intensively applied.

Alachlor, also called metachlor, is another herbicide, used to control annual grasses and many broadleaved weeds on cropland. It is classified by the United States Environmental Protection Agency as slightly toxic. It may cause slight skin and eye irritation in people exposed to levels above the maximum contaminant level (MCL) in drinking water. Lifetime exposure to levels above the maximum contaminant levels may result in liver and kidney damage. Its carcinogenic effects remain uncertain. In soil, the breakdown of alachlor occurs primarily through biodegradation. Its half-life in soil ranges from about 8 to15 days. Alachlor is highly to moderately mobile in soil and its mobility in soil decreases with increasing soil organic carbon and clay content. Consequently, alachlor is widely detected in surface water and groundwater in the vicinity of farms where it has been applied. In water, alachlor is lost through photodegradation and biodegradation, although photodegradation takes place only in shallow, clear surface water. Alachlor does not bioaccumulate in aquatic organisms.

9.7.2 Pharmaceuticals and Personal Care Products

Pharmaceuticals and Personal Care Products (PPCPs) are a diverse category of ESOCs comprising prescription and non-prescription drugs, diagnostic agents used in medicine, caffeine and nicotine, and the ingredients of dietary supplements, soaps, conditioners, cosmetics, fragrances, and sunscreens. PPCPs also include veterinary antibiotics. The most important pathways of PPCPs to the environment are wastewater discharges, the application of sewage sludge and manure as a soil conditioner and fertiliser, and landfill leachate. Given the variety of compounds present in PPCPs, it is not possible to generalise their environmental fate and impacts. The substances in PPCPs that attract most concern, however, are the compounds that are water-soluble, are not readily sorbed to soil and sediment particles, and are only slowly biodegradable.

PPCPs that are excreted or disposed by humans and washed away in wastewater, such as caffeine, pain relievers (e.g. ibuprofen), antidepressants, birth control pills, drugs used in chemotherapeutic cancer treatment and stimulant drugs (e.g. cocaine) can typically be traced downstream of urban areas. The effectiveness of wastewater treatment in removing these compounds from the effluent water depends on the type of wastewater treatment and the physico-chemical properties of the specific ingredients of the PPCPs. Concentrations of PPCPs in the receiving water typically range from nanograms per litre (ng l^{-1}) to micrograms per litre (μg l^{-1}), with the highest concentrations typically being those of common pain relievers and caffeine.

Veterinary antibiotics, hormones and other animal-growth regulators which are incompletely metabolised and excreted in livestock urine and manure can be traced in soils and groundwater beneath intensive agricultural areas. This may pose a threat to human health if the groundwater is abstracted for drinking water. Although in most cases groundwater contamination by veterinary drugs is not an acute health concern, the long-term health and environmental effects are largely unknown. One of the potential environmental hazards is the development of antibiotic-resistant microorganisms due to the presence of antibiotic contaminants.

9.7.3 Endocrine-Modulating Chemicals

Endocrine-Modulating Chemicals (EMCs) (also called endocrine disruptor chemicals or endocrine toxicants) are substances that interfere with the normal function of the endocrine or hormone system. Consequently, they may cause serious adverse health effects, such as cancer, reproductive and developmental defects, or altered immune function in an organism or its progeny. EMCs include hormones (e.g. natural and synthetic oestrogens), surfactants, pesticides, plasticisers, polychlorinated biphenyls (PCBs), and dioxins/furans. These substances or their ingredients possess a wide range of physical and biochemical properties: some are lipophilic and persistent, while others are hydrophilic and rapidly degraded. Routes of EMC exposure include intake of medication, ingestion of contaminated food products and water, contact with contaminated soil and dust, and contact with plastic household products (e.g. packaging, rainwear, footwear, carpets, and toys). Note that for the majority of chemicals in use today, the long-term effects of exposure on the endocrine system are unknown.

For aquatic ecosystems, the occurrence of all EMCs is of concern, but the most intensely investigated EMCs are oestrogens, which mimic or inhibit the effects of the vertebrate female reproductive hormones (e.g. Campbell *et al.*, 2006). The major source of oestrogens in surface water is the effluent from municipal and industrial wastewater treatment facilities. The actual primary source of oestrogens entering the wastewater is humans who excrete or dispose of natural hormones and pharmaceutical oestrogens (e.g. birth control pills) in toilets. Conjugated oestrogens, which are formed when the body eliminates oestrogens, are not oestrogenically active. However, in wastewater treatment systems, these conjugated oestrogens can be deconjugated, liberating active oestrogenic compounds in the discharge. The largest part of the oestrogens will be removed by degradation or sorption in the wastewater treatment process, but when sewage sludge is applied to land, pharmaceutical oestrogens may enter surface waters through runoff and groundwater flow. Oestrogens are readily degradable and have relatively short environmental half-lives under aerobic conditions. Under anaerobic conditions, however, they degrade slowly. Since many oestrogens have moderately to high log Koc values, the mass that does not remain dissolved is rapidly sorbed to organic matter in bed or suspended sediments, or ends up in organic complexes.

Similarly, steroid oestrogens present in wastewater treatment plant effluents cause feminising effects in fish, i.e. skewed male-to-female ratios and an increase in individuals of indeterminate sex. Disturbance of the development and expression of sexual characteristics in amphibians, reptiles and mammals has also been confirmed in laboratory studies. However, it is still not well understood to what extent EMCs affect the sexual characteristics and reproductive capabilities of natural populations (Wright-Walters, 2009).

9.7.4 Microplastics

As noted above, microplastics are plastic (i.e. synthetic, organic polymers of high molecular mass) particles smaller than 1 or 5 mm. They can be subdivided into three groups according to their source: primary microplastics, secondary microplastics and synthetic textile fibres. Primary microplastics are microplastics produced either for direct use, for example as industrial abrasives or in cosmetics, or for indirect use as raw material for the production of plastic products (pre-production plastic pellets or nurdles). Primary plastics enter the environment mainly due to accidental spills during transport or storage. Secondary microplastics are formed in the environment as a consequence of the breakdown of larger plastic material. As plastics are barely biodegradable, the breakdown of plastics into small fragments is mainly due to mechanical abrasion, oxidative degradation or sunlight-driven photochemical processes. Synthetic textile fibres originate from the abrasion of synthetic textiles (e.g. nylon, polyester, acrylic) during domestic clothes washing. They are transported in the rinsing water, eventually ending up, via the sewage system, in surface waters.

In recent years, microplastics have been found in sediments in inland waters and seas worldwide (e.g. Andrady, 2011; Browne, 2011; Faure *et al.*, 2012). Primary microplastics that enter the aquatic environment can easily be transported by rivers to seas and oceans. Nevertheless, the majority of microplastics found in oceans are secondary microplastics, but 80% the plastic litter from which they originate have terrestrial sources including beach litter (Andrady, 2011).

As well as being pollutants themselves, microplastics are also carriers of two types of micropollutants: native plastic additives and adsorbed hydrophobic pollutants (e.g. POPs). As seen in section 9.7.3, some plastic additives may have endocrine-modulating properties. Although microplastics have been frequently found in the digestive tracts of various aquatic and marine organisms, little is known about their effects on aquatic life and ecosystems. Potential effects of ingestion of microplastics by organisms are reduced food intake due to false satiation, blockage or physical damage of digestive tract or feeding appendages, leaching and bioaccumulation of sorbed pollutants by the organism, and transfer of these pollutants across the food chain.

EXERCISES

1. Define the following terms:
   a. Isomers
   b. Aromatic
   c. Aliphatic
   d. Cis and trans isomers
   e. PAHs
   f. VOCs
   g. POPs
   h. NAPLs
   i. BTEX
   j. CDDs

2. a. Give the chemical formulas of the following substances:
      - 1,2-Dichlorocyclohexane
      - Biphenyl
      - Anthracene
   b. Draw the structure of
      - 3-bromo-2-chloro-2-methylbutane
      - 1,2,4-trichlorocyclohexane
      - cis-1,2-difluoroethene
      - m-dichlorobenzene

3. Describe in brief the fate and toxicity of 1000 litres of oil spilled
   a. into a sandy soil with a shallow water table;
   b. into a peaty soil;
   c. into a lake.

4. a. What can be said about the water solubility of PAHs?
   b. To which sediment components do PAHs adsorb in particular?

   c. If reduced lake bed sediment contaminated by PAHs is exposed to air, what does this mean for the adsorption of PAHs in the sediment?

5. Describe the effect of halogens in organic compounds on:
   a. Water solubility
   b. Environmental persistence

6. Name three POPs that are ubiquitous in the global environment and give their main sources.

7. What is the principal source of environmental contamination by aliphatic chlorinated hydrocarbons?

8. What is the main physical property that determines the difference in environmental behaviour of aliphatic chlorinated hydrocarbons and BTEX in groundwater?

9. Describe in brief the fate of PCBs and dioxins released into stream water.

10. Name three emerging substances of concern and give their main sources.

11. What are the main sources of microplastics in aquatic ecosystems?

Part III
Transport processes of substances in soil and water

The environmental fate of chemical constituents is determined not only by the reactivity of the soil–water environment and of the chemical constituents themselves, but also by physical movement. The flux of matter through river basins is an essential parameter in biogeochemical cycles in riverine and coastal floodplains and wetlands. Various subdisciplines of earth science, such as hydrology, soil science, and geochemistry, have elucidated the mechanisms that govern the flux of dissolved and sediment-associated nutrients and contaminants in soil, groundwater, and river networks. The conveyance of dissolved and sediment-associated contaminants along the surface or underground flow paths connecting each point within a catchment to the catchment outlet is governed by complex, often non-linear, physico-chemical interactions between water, sediment, contaminants, and biological components, which operate at different spatial and temporal scales. These interactions include underground and/or river channel dispersion along and between these flowpaths, adsorption and desorption processes between the dissolved phase and soil and suspended sediment, and biochemical decay. Moreover, the interactions vary in magnitude due to spatial and temporal variation in emission of nutrients and contaminants at the soil surface or river channel, the spatio-temporal distribution of rainfall, and the length, tortuosity, and velocity of the flow paths. The interactions can therefore be seen as a response to hydrological events in the short term, and to climate change, land use change, and changes in nutrient and contaminant emissions in the long-term. So, knowledge of the fundamentals of transport and fate of chemicals in the environment allows us to identify and analyse environmental issues at scales ranging from local to global. This part therefore gives an overview of the relevant transport processes at the landscape or catchment scale, their driving mechanisms, and the ways they can be formalised in mathematical models.

10

Systems and models

10.1 A SYSTEMS APPROACH

Water participates as a reagent in a wide range of adsorption–desorption, dissolution–precipitation, acid–base, and redox reactions. Moreover, water is the major medium conveying dissolved ions, colloidal particles and particulate matter through soil, groundwater and surface water. Given water's role in the dispersal and fate of contaminants in the environment, if we are to understand the direction and rate of dispersal and the spatial and temporal variation of contaminant concentrations in soil and water, we need an understanding of hydrology and hydrological pathways.

A useful way to study hydrological pathways is to consider the Earth or a part of it as a system with clearly defined boundaries that exchanges energy and mass (water) with its surroundings. In this manner, concepts and principles from *systems theory*, the transdisciplinary study of the abstract organisation of phenomena proposed by the biologist Ludwig von Bertalanffy in the 1940s (Von Bertalanffy, 1968), can be applied. Rather than reducing an entity (e.g. soil) to the properties of its parts or elements (e.g. mineral grains or organic matter), systems theory focuses on the quantitative description of the arrangement of and relations between the parts and connects them into a whole. Many disciplines (physics, chemistry, biology, geography, sociology, etc.) base their concepts and principles of organisation on this theory.

Conceptual hydrological systems are based on the notion of *stores* of water or substance, whose state depends on a variable amount of water or substance (volume or mass) in them. These stores or subsystems may be defined in many ways: for example as functional units (e.g. soil water, groundwater, surface water), morphological units (e.g. hill slope unit, river channel, lake, estuary), or discrete spatial units constructed through regular or irregular tessellation of space. Discrete spatial units are often applied in numerical modelling and hydrological modelling using geographical information systems (GIS) (see Burrough and McDonnell, 1998) (Figure 10.1). The relations between the stores are determined by the *fluxes* between them, which have the potential to change the *state* of the store. The stores at the system boundaries may also be influenced by the possible presence of inputs and outputs across these boundaries. Because of the spatial organisation and ordering of the stores distinguished, the hydrological system is often modelled by a cascade of stores (see Figure 10.2). The system should obey fundamental principles of thermodynamics and continuity; in other words, it must obey the conservation laws of energy, momentum, and mass. This implies that the change of mass and energy in each store and the whole system equals the sum of input and output fluxes integrated over time. Section 11.1 goes into more detail on the concepts of mass balance.

Water and substance transport is influenced by numerous feedback mechanisms. A *negative feedback* mechanism is a controlling mechanism that tends to counteract some kind of initial imbalance or perturbation. A good example is a simple system of a reservoir

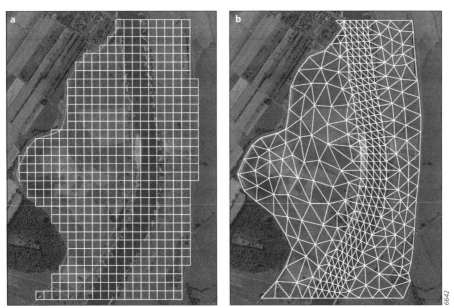

Figure 10.1 Tessellation of a floodplain area along the river Elbe near Bleckede, Germany: a) regular raster of square grid cells, commonly used in a raster GIS and finite difference models; b) triangular irregular network (TIN) commonly used in a vector GIS and finite element models. Background: Digital Orthophoto 1:5000, reprinted with permission from the publisher: LGN – Landesvermessung und Geobasisinformation Niedersachsen – D10390.

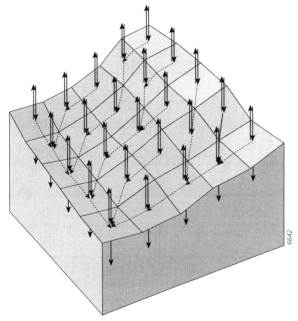

Figure 10.2 Schematic diagram of a two-dimensional cascade system of surface runoff over a landscape: the vertical arrows represent the inputs (precipitation) and outputs (evapotranspiration and infiltration) at the soil surface. The lateral arrows represent the overland flow between the adjacent grid cells.

lake (one store) with a drain that discharges excess water as function of the lake volume: the water discharge (flux) is proportional to the water volume in the lake:

$$Q_{out} = k\,V \tag{10.1}$$

where Q_{out} = discharge from the lake [$L^3\,T^{-1}$], k = rate constant [T^{-1}], and V = water volume in the lake [L^3]. Note that in this feedback mechanism, the rate of change of the store, determined by Q_{out}, is controlled by the state of the store itself (V). Given a constant inflow into the lake, the reservoir system evolves to a condition of ***steady state*** or dynamic equilibrium, which means that the inputs and outputs are in balance, so that the state of the system (water level) does not change. If an additional amount of water is instantaneously added to the reservoir, the system tends to return to a state with a constant inflow and outflow. By definition, the ***response time*** of the system is defined as 1/k, which corresponds to the time needed to reduce the difference between the actual value and the steady state value by 63 percent.

In contrast, a ***positive feedback*** mechanism exacerbates some initial change from the steady state, leading to a 'blow-up' condition. For example, if in the same reservoir a crack or hole occurs in the dam, the eroding power of the outflowing water enlarges the gap, so the water discharge through the hole increases rapidly. It is only a matter of time before the whole dam collapses. Eventually, when the reservoir is nearly empty the negative feedback mechanism takes over, so a new steady state condition is reached for which the outflow rate equals the inflow. The fact that the Earth has had water and an atmosphere for a very long time suggests that our Earth system is dominated by negative feedback mechanisms. Nevertheless, positive feedback mechanisms may be very important, since they have the potential to bring about dramatic changes.

If we consider the hydrological cycle at the global scale, we may assume a ***closed system***, i.e. a system that only exchanges energy with its surroundings; water does not enter or leave the Earth system. The global hydrological cycle (see Figure 3.1) is driven by solar energy that powers the evaporation and translocation by wind. The evaporated water is returned to the Earth's surface by precipitation, some of which falls on land. In most areas, more water enters the land via precipitation than leaves it by evaporation. If the water is not temporarily stored on the soil surface in the form of snow and ice, part of the excess water is discharged directly into streams and lakes via overland flow. The rest infiltrates the soil and percolates to the groundwater, which also ultimately discharges into surface waters. Rivers are the main routes transporting water from land to sea.

Although the global hydrological cycle may be looked upon as a closed system, the water cycle is often investigated at the smaller scale of a local groundwater system or single drainage basin. The drainage basin as a hydrological unit can be envisaged as an ***open system*** receiving mass (in the form of water) and energy from the weather over the basin and losing them by evaporation and water discharge through the basin outlet. Therefore, the drainage basin provides a useful framework for studying the phenomena related to the transport and fate of contaminants.

Groundwater transports chemicals that are produced by internal weathering and dissolution reactions or delivered to the soil surface via atmospheric deposition (e.g. acid rain) or anthropogenic immissions (e.g. application of manure, fertiliser, and pesticides in agriculture). River water transports materials produced by erosion processes on hill slopes and river banks and by internal production of organic matter, as well as dissolved ions originating from natural or anthropogenic sources. The fine fraction of sediments, which consists mainly of clay minerals and organic matter, is particularly able to bind various chemicals; this fraction is therefore considered to be a very important vector for transporting nutrients and a large variety of contaminants, including heavy metals and organic pollutants.

The transport of chemicals associated with sediment is often referred to as ***particulate transport*** and the transport in dissolved form as ***solute transport***.

While travelling along the surface and subsurface flow paths, the transported chemicals may be subjected to a wide range of physical and biogeochemical processes. Key processes are ***transformation*** and ***retardation***, which are commonly referred to as ***retention***. Transformation refers to processes that irreversibly alter a substance to a different substance, such as denitrification, microbial degradation, or physical decay. Retardation means delay due to, for example, sorption to sediments, chemical precipitation, diffusion into stagnant water, and sediment deposition; it leads to a temporary or permanent storage of substances within the catchment system. The ***residence time*** of pollutants in a given store or subsystem is a critical factor in estimating environmental risk. Residence time is the average amount of time a particular substance remains in a given store. For example, the residence time of water in a lake is the amount of water in the lake divided by the rate at which water is lost due to evaporation or runoff through a stream. Likewise, the residence time of nitrate in a groundwater body is the amount of nitrate in that groundwater body divided by the rate of mass loss due to groundwater flow and denitrification.

The rates of the physical and biogeochemical processes are rarely constant but are controlled by the physico-chemical environmental conditions. For example, the denitrification rate is controlled by temperature and redox conditions. The study of the rate of the processes is called ***kinetics***. As we saw above, the process rates are often also subject to manifold feedback mechanisms, which thus means that the process rates are controlled by the state variable itself. For example, the discharge from a lake is controlled by the lake level (see Equation 10.1) and the denitrification rate is controlled by the nitrate concentration.

Studies of the environmental fate of substances usually start by identifying the different natural and man-made sources in the system under study, their magnitude, and spatial and temporal variability. As noted in chapter 1, pollutants can come from point sources or diffuse sources. Pollution around a point source is mostly confined to a plume in the downstream direction from the source with sharp concentration gradients at the fringe of the plume. In contrast, the pollutant concentration gradients in soil and water originating from diffuse sources are usually gradual. An exception is if the discharge rates from the diffuse sources or the physical and biogeochemical processes that can alter the concentrations are spatially or temporally discontinuous or regulated by thresholds. Sharp spatial boundaries occur in groundwater due to discontinuities in diffuse sources: for example, when there are adjacent plots of land with different land use (e.g. arable land and forest), causing a discontinuity in nitrate loading from the soil surface. This is important because landscapes are often patchy and discontinuous and the boundaries between the landscape units encompass pronounced spatial heterogeneities. These sharp boundaries are propagated underground along the groundwater flowpaths (Figure 10.3). In general, such sharp boundaries occur more in groundwater than in surface water, since turbulence caused by currents and waves mixes surface waters more thoroughly. Temporal variation of input from diffuse sources arises due to seasonal changes in climate variables such as temperature and precipitation, and to abrupt disturbance due to extreme weather events, the planting and harvesting of crops, the application of fertilisers and pesticides, and the dumping of wastes.

Because the substances are retained and lost along their transport pathways, the loads from the small subcatchments and the different sources often do not simply add up to the total transport from a river basin. We have also seen that physical and biogeochemical processes result in spatial and temporal patterns of concentrations of substances in soil, groundwater, and surface water, not only because external and internal factors control the process rates but also because the sources vary in space and time. The effects of the whole of inputs and processes make the transport and fate of sediment and chemicals through drainage basins a complex issue. The advantages of a systems approach arise from the fact

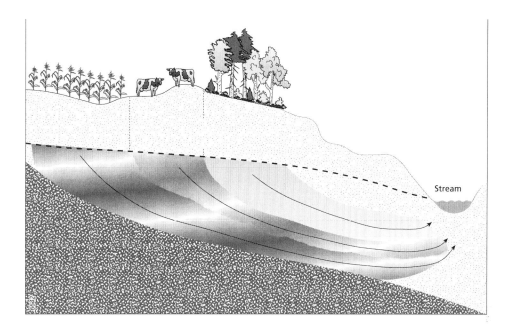

Figure 10.3 Propagation of spatial and temporal variation in groundwater.

that it emphasises relationships between driving processes and resulting patterns. This makes it possible to link observed patterns of contaminants in soil and water to the most important processes causing the patterns.

10.2 THE ROLE OF MATHEMATICAL MODELS

In the past thirty years, the increased availability of computing power has led to hydrological, hydrochemical, and ecological computer models becoming widely used for managing soil and water quality with respect to a variety of environmental impacts. Mathematical models that simulate the transport of contaminating substances allow us to evaluate the fate and persistence of these substances in landscape compartments in space and time and to accurately predict the physical, chemical, and biological processes occurring during transport. The main output from these models is the chemical concentration of one or more chemical substances as a function of space or time, or both. The principles of mathematical modelling of solute and particle transport are based on those of systems theory.

We build and apply such environmental models of contaminant transport for the following reasons:

1. To forecast travel times of pollutants in rivers or groundwater when calamities occur.
2. To assess past, present, or future human exposure to contaminants.
3. To predict future conditions under various scenarios of environmental change and management strategies and to evaluate the effectiveness of possible management actions.
4. To reduce soil and water quality monitoring costs by replacing or supplementing expensive measurements by cheaper model predictions.

5. To gain an improved understanding of the mechanisms controlling the dispersion and fate of chemicals in the environment by quantifying their reactions, speciation, and movement.

Concerning the first reason (to anticipate accidental releases of pollutants to the environment), emergency managers need appropriate tools that can quickly give essential information about the travel and residence times of pollutants in soil, groundwater, and surface water bodies. An example of such an accidental spill was the discharge of organic pollutants to the river Rhine during a fire at the Sandoz chemical manufacturing plant in Basle, Switzerland, on 1 November 1986. Approximately 90 different chemicals were stored at this facility, including organophosphorus pesticides (disulphoton, parathion, thiometon), mercury-based pesticides (ethoxyethylmercury hydroxide, phenylmercury acetate), and other pesticides (captaphol, endosulphan, metoxuron). While the fire was being extinguished, 10 000 to 15 000 m³ of contaminated firefighting water was discharged to the river and transported downstream (Figure 10.4). Not only did this have catastrophic ecological impacts (such as fish mortality in the river Rhine), it also forced the water authorities in the Netherlands, about 1000 km downstream from the chemical plant, to close the water intake for drinking water supply for a number of days. The real-time water quality alarm model available in 1986 was not suitable for forecasting when concentrations of organic pollutants would exceed the intervention limits for intake of drinking water in the short-term. As a result of the Sandoz accident the Rhine alarm model was improved; the improved version is currently deployed by alarm stations along the Rhine (Spreafico and Van Mazijk, 1993; Van Mazijk *et al.*, 1999).

Other recent examples of accidental releases are the dam-burst at the Aznalcóllar mines in Andalucia, Spain, in April 1998, the cyanide and heavy metal spills into the headwaters of the Tisza River in Baia Mare and Baia Borsa, Romania, January/March 2000, and the benzene and nitrobenzene spills into the Songhua and Armur rivers following an explosion in the petrochemical plant in Jilin, China, in November 2005. The dam-burst at the Aznalcóllar zinc mines caused a spill of mine waste sludge contaminated by heavy metals

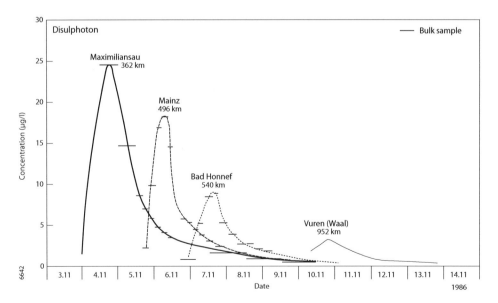

Figure 10.4 Chemographs of disulphoton concentrations in the river Rhine following the Sandoz accident in 1986 (source: IKSR, 1986).

into the Guadiamar river, which polluted the river water and approximately 5000 ha of marshlands including parts of the Doñana national park (see Grimalt *et al.*, 1999). Apart from the immediate effects, these spill also had long-term effects on the river ecosystem, since heavy metals are retained in the natural environment as they are readily adsorbed by fine sediments and deposited on the river bed and floodplains. Despite the extensive clean-up of the flooded areas of the Doñana national park in the months following the Aznalcóllar accident, much of the flooded area still shows elevated concentrations of arsenic and other heavy metals far above tolerable limits (Galán *et al.*, 2002). Long-term environmental models may help us to understand where the contaminants are stored in the river sediments and how they are dispersed further downstream in the river system.

It is clear that the pollution of river water and sediments, whether due to accidental releases as described above or to long-term emissions, have implications far beyond the increased concentrations of chemicals in water and sediment, since the pollutants accumulate in the food chain. If we can quantify the spatial and temporal distribution of the environmental concentrations of pollutants in water and sediment and their transfer rates into the food web via the various pathways, we may also be able to assess the exposure of humans and other organisms to pollutants. As you will recall, that was the second purpose for developing mathematical models.

The third purpose – of predicting future conditions under various scenarios of environmental change and management strategies – is particularly relevant for supporting decisions in the policymaking process. Because there are too many uncertainties in the factors that determine future pollutant emissions and dispersion (for example, future weather and climate conditions, socio-economic development, and occurrence of hazardous releases) it is not realistic to forecast future distributions of pollutant concentrations in the environment. So, instead of forecasting the future, it is usual to use scenarios; these can be defined as a hypothetical, generally intelligible descriptions of sequences of possible future events. The scenarios can then serve as input for mathematical models and, accordingly, help answer 'what if' questions. Examples of such questions are: What are the effects on nutrient runoff from large river basins if temperature and precipitation patterns change due to global warming or if landuse changes due to socio-economic development? What are the effects on regional groundwater quality if urbanisation intensifies? If the what-if question relates to possible management actions, such as reducing groundwater abstraction rates or expanding wastewater treatment facilities, models may also help to evaluate the impact of such actions on water quality.

The fourth purpose – of reducing soil and water quality monitoring costs – is related to the high costs of sampling and laboratory analysis. If we are able to predict environmental concentrations of contaminating substances at unsampled locations or points in time with sufficient accuracy, we do not need to sample at these locations or points in time. These samples therefore become redundant in monitoring strategies and may be omitted without loss of information. In this way, environmental models may aid in the economic optimisation of monitoring networks of soil, groundwater, and surface water quality.

The fifth purpose, to gain an improved understanding of the controlling mechanisms, is especially important in a research and development setting. Traditional models of contaminant transport address biochemical oxygen demand dynamics in river water, and the development of a pollution plume for a single chemical in groundwater. Recently, much effort has been put into detailing the biogeochemical speciation and reactions occurring during transport and into integrating them into water quality models (e.g. Hunter *et al.*, 1998). On the other hand, chemical transport modelling is often very time-consuming if the model includes the interactions amongst several environmental compartments (soil, groundwater, surface water, sediment, and biota) and the area modelled is large. Therefore, methods are being sought for effectively identifying the driving forces and for upscaling the model descriptions to the regional or supra-regional scale (e.g. De Wit, 1999). A better insight into the major governing processes and their rates may lead to improved models; this would improve the prediction of chemical concentrations or loads and therefore benefit the other purposes too. The development of improved models of contaminant transport need not result in more complex models, but may even lead to simpler models from which irrelevant processes have been omitted; this would save computational effort and therefore time and money.

10.3 CLASSIFICATION OF MATHEMATICAL MODELS

In general, mathematical models can be divided into two broad classes. ***Physical models*** use known science to model the processes on the basis of systems behaviour. Examples of physical models are groundwater and surface water transport models based on physical principles from fluid mechanics. ***Empirical models*** are used when insufficient is known about the underlying physics, or when application of the underlying physics would require an unpractically detailed resolution to characterise the system. These models are based on experimental observation and then functionally fitting the models to these observations either by trial and error or by statistical techniques. The fitting procedure is known as model calibration. Examples of empirical models include models describing the temperature dependency of the biodegradation rate of organic chemicals. In practice, there is no sharp boundary between strict physical models and complete empirical models. Complex environmental models, however, are often a mix of models based on physical laws and empirical models, so this book introduces both approaches.

Another important classification of mathematical environmental models involves the distinction between ***dynamic models***, i.e. models that involve change over time and time-varying interactions, and ***static models***, i.e. models that describe a steady state. If the temporal variation in the system state is negligible, the system may be considered to be at steady state and the time dimension may be disregarded and a static model is sufficient to describe the state of the system. Conversely, if the system state changes significantly in time, a dynamic model is necessary. When solving a mathematical model, the main model input besides the model parameters is the so-called ***boundary conditions*** that describe the constant or time-varying values of the state variables at the system boundaries. A dynamic model also requires ***initial conditions***; this implies that the state of a system at the beginning of the model period should be defined. A static model does not require initial conditions, but because chemical transport processes are dynamic, these models are only applied under special conditions.

To summarise, mathematical environmental models are comprised of five conceptual components (Jørgensen, 1988):
- The set of state variables that define the state or condition of the system;
- The set of input variables or forcing functions that influence the state of the system;
- The set of mathematical equations that represent the processes within the system and the relations between the forcing functions and the state variables, including the mathematical solution technique;
- The set of universal physical constants that are not subject to variation, such as Avogadro's number or the gas constant;
- The set of other coefficients or parameters in the mathematical equations, which are either constant within the system or vary in space and time.

Thus, the definition of an environmental model used here comprises more than just an executable computer program.

An essential aspect in environmental modelling is the choice of the number of spatial model dimensions. The number of spatial dimensions (zero to three) chosen to describe an environmental system is likewise based on the spatial variability of the key factors governing the state of the system. If this spatial variation is insignificant or irrelevant, a zero-dimensional model or ***lumped model*** may be chosen, in which the system is spatially averaged. Examples of widely used zero-dimensional models are so-called box models or Mackay models that describe the average partitioning of pollutants among various environmental compartments such as sediment, pore water, surface water, and various trophic levels of the food chain (Mackay, 1991). Such models are commonly used in environmental risk assessment; they are applied to larger areas, such as an entire lake or river basin. In contrast, a ***distributed model*** takes account of the spatial variation of the patterns of environmental entities and processes. One-dimensional models are generally used when the spatial variation predominates in one direction: for example, the vertical distribution of contaminants over a soil profile, or the longitudinal variation of water quality in a river. For some river water quality problems a two-dimensional model may be required: for instance, when modelling the dispersal of contaminants over part of a local floodplain. For more complex surface water bodies a three-dimensional model may be needed: for instance, when stratification of temperature in deep lakes or of salt concentration in estuaries occurs. Because groundwater composition usually varies both vertically and laterally, groundwater quality models usually also require three dimensions.

In recent years, there has been a growing interest in the development of spatially distributed models that simulate or predict the sources, transport, and fate of contaminants at the catchment scale (see e.g. Addiscott and Mirza, 1998; De Wit, 2001; Whelan and Gandolfi, 2002; Mourad and Van der Perk, 2004; Håkanson, 2004; Heathwaite *et al.*, 2005; Destouni *et al.*, 2010; Zanardo *et al.*, 2012). Concurrently, Mackay models to model the partitioning of pollutants over the environmental compartments have also been implemented in a spatially distributed manner (e.g. Coulibaly *et al.*, 2004; Warren *et al.*, 2005). Most of the model studies mentioned have used GIS, because it is advantageous and useful for preparing model input from large spatial data sets on catchment characteristics (e.g. altitude, soil type, land use, hydrology) and for the post-processing of model output. Some models are entirely GIS-based, which implies that these models operate wholly in a GIS environment using an integrated programming language (Burrough, 1996). These developments have opened the door for scientists who want to construct their own environmental models tailored to their specific purposes and available data (see Karssenberg, 2002).

EXERCISES

1. Give examples of at least three feedback mechanisms that control the decay of organic matter in soil. Indicate whether these feedback mechanisms are positive or negative.

2. What is the difference between a closed system and open system?

3. Why is the drainage basin a convenient unit for studying contaminant transport and fate?

4. The nitrate concentration in a lake with a volume of 100 000 m^3 which does not have an outflow is determined by the input via atmospheric deposition and losses due to denitrification. The concentration in the lake is 0.2 mg l^{-1}. The denitrification rate is given by:

Denitrification (mg l^{-1} d^{-1}) = 0.1 (d^{-1}) × nitrate concentration (mg l^{-1})

    a. Calculate the response time of this lake system
    b. Calculate the mass of nitrate in the lake system
    c. Calculate the residence time of nitrate in the lake system

5. Describe the main difference between point and diffuse sources and their impact on regional surface water quality.

6. Describe the differences between physical and empirical models and the differences between lumped and distributed models.

7. Why does a static model not require initial conditions?

8. Give an adequate number of spatial dimensions for modelling soil quality for the following cases:
    a. Leaching of phosphorus into the soil profile
    b. Spatial distribution of radiocaesium in a hilly catchment
    c. Uptake of pesticides by soil organisms in a field
    Explain your answers.

11

Substance transport

11.1 MASS BALANCE

The amounts of chemical constituents are measured in mass quantities (e.g. mg, kg, or tonnes) or related units (moles, or becquerels (Bq) in the case of radioactive elements). These amounts are expressed as such, or as concentration per unit mass of water, rock, soil, or sediment (**mass concentration**, e.g. mg kg^{-1}, % (percent), ppm (parts per million), ppb (parts per billion)), or per unit volume (**volume concentration**, e.g. mg l^{-1}, M (= moles l^{-1})). Since fresh water has a specific density of 1.00 kg l^{-1}, the mass concentrations in fresh water are approximately equal to volume concentrations. Furthermore, the amount of chemicals can also be expressed per unit area (e.g. Bq m^{-2}, tonnes ha^{-1}) or per unit time (e.g. kg s^{-1}, tonnes y^{-1}). The amount of chemicals or sediment transported per unit time is also referred to as **load**.

The variety of options available for expressing the amount of substances stresses the need for a consistent use of physical dimensions and units. A careful check for consistency of units is a simple yet powerful manner to intercept errors in mathematical equations. To test whether a calculation result has the correct dimensions, you should check that the left- and right-hand sides of the equation resolve to the same units of measure. Besides giving insight into the mathematical expression, a check like this will also reveal missing or redundant terms that cause spurious units and erroneous results. For example, the chemical load of a river can be calculated by:

Load = Flow velocity (m s^{-1}) × Cross-sectional area (m^2) × Concentration (g m^{-3}) (11.1)

The units of the load would be (g s^{-1}) with dimensions [M T^{-1}]. Note that the unit for concentration (g m^{-3}) is equal to (mg l^{-1}). If we want to express the load in (tonnes y^{-1}), the load in (g s^{-1}) should be multiplied by (31 536 000 s y^{-1} × 10^{-6} tonnes g^{-1} = 31.536). If the cross-sectional area of the river channel had been omitted from the equation, the load would have had the units of (g m^{-2} s^{-1}) with dimensions [M L^{-2} T^{-1}], which are clearly not the dimensions for chemical load. On the other hand, it is also worth noting that consistency of units and dimensions does not guarantee that the result is correct. For example, in Equation (11.1) if we use the surface area of a part of the river instead of the cross-sectional area, though the units are consistent the result is meaningless. If we calculate the inorganic nitrogen load in a river but in Equation (11.1) we fill in the sum of the nitrate and ammonium concentrations in the river water, the result is also obviously wrong even though the units are again consistent. In this case, we have to convert the nitrate and ammonium concentrations to nitrogen concentrations using stoichiometric constants:

Nitrogen load (g N s^{-1}) = *Flow velocity* (m s^{-1}) × *Cross-sectional area* (m^2) ×
   (14/62 × *Nitrate Concentration* (g NO$_3^-$ m^{-3}) +
   14/18 × *Ammonium Concentration* (g NH$_4^+$ m^{-3})) (11.2)

Example 11.1 Load calculation

The mean annual ortho-phosphate concentration in a river is 0.04 mg l^{-1}. The mean annual discharge of the river is 8.2 m^3 s^{-1}. Calculate the annual load of dissolved inorganic phosphorus in tonnes y^{-1}.

Solution
First, calculate the inorganic phosphorus concentration from the ortho-phosphate concentration (compare Equation 11.2). The molar mass of ortho-phosphate (PO$_4^{3-}$) is 31 (from the P; see Appendix I) + 4 × 16 (from O) = 95. The inorganic phosphorus load is
Dissolved inorganic P load = 31/95 × 0.04 g m^{-3} × 8.2 m^3 s^{-1} = 0.107 g s^{-1}
= 0.107 × 31.536 = 3.37 tonnes y^{-1}

Note that the concentration may fluctuate considerably during the year, often depending on the discharge of the river. If this is the case, multiplying the average concentration by the average discharge gives inaccurate results for the annual load. For more details on this, see Section 18.3.

In Chapter 10 we saw that the conservation law of mass imposes that changes in the total mass in a store, or control volume, should be accounted for in terms of inputs and outputs. A control volume is a soil or water body with clearly defined boundaries and, as mentioned before, can consist of a functional unit, a morphological unit, or an arbitrary box anywhere on Earth. It can also be as small as an infinitesimally thin slice of water in a river. Water can be viewed as a conservative substance with only inputs and outflows from an adjacent water body. Thus, the same principles apply for other conservative substances (i.e non-reactive or inert) transported by water, such as chloride. Changes in the total mass of non-conservative substances may also be due to physical and biogeochemical transformation. A mass balance is simply the accounting of mass inputs, outputs, reactions, and change in storage:

ΔStorage = Inputs – Outflows ± Transformations   (11.3)

If we consider, for example, the simple reservoir lake system (see Section 10.1), the transformation term is zero, because water is a conservative substance. If we assume that the specific density of water remains the same over time, the mass balance for water in the lake could be expressed by the following algebraic difference equation:

$$\Delta V = (Q_{in} - Q_{out})\,\Delta t$$ (11.4)

where ΔV = change in storage volume [L^3], Q_{in} = input from feeding streams, groundwater, and precipitation [L^3 T^{-1}], Q_{out} = output by discharging stream, seepage to groundwater, and evaporation [L^3 T^{-1}], Δt = time increment [T]. If we assume that the surface area of the lake does not change with the water level, the change in lake water depth H [L] can be obtained by dividing both sides of the equation by the surface area A [L^2]:

$$\Delta H = \frac{\Delta V}{A} = \frac{(Q_{in} - Q_{out})}{A}\,\Delta t$$ (11.5)

This difference equation can be made into a differential equation if we divide both sides of the equation by Δt and take the limit as $\Delta t \rightarrow 0$:

$$\frac{dH}{dt} = \frac{(Q_{in} - Q_{out})}{A} \tag{11.6}$$

The output flux through the stream was given by Equation (10.1), so the equation can be rewritten as:

$$\frac{dH}{dt} = \frac{Q_{in}}{A} - \frac{kV}{A} = \frac{Q_{in}}{A} - kH \tag{11.7}$$

where H = the mean water depth [L]. Note that this equation is a first-order differential equation, in which H represents the state variable, Q_{in} the upstream boundary condition, and k a parameter that is constant within the system. This problem has an ***analytical solution***, provided that Q_{in} is either constant over time or described by a continuous function of time. If Q_{in} is constant over time, the analytical solution is (see Box 11.I):

$$H(t) = \frac{Q_{in}}{kA} + \left(H_0 - \frac{Q_{in}}{kA} \right) e^{-kt} \tag{11.8}$$

The system is in steady state if the inputs and outputs are in balance and the water depth does not change in time, thus:

$$\frac{dH}{dt} = 0 \tag{11.9}$$

Hence, the analytical steady state solution of the lake water depth in the reservoir lake reads:

$$H_{eq} = \frac{Q_{in}}{kA} \tag{11.10}$$

Example 11.2 Lake water depth and discharge

A lake with a surface area of 4 ha is fed by a stream with a discharge of 3.8 m³ s⁻¹. The lake water depth is 5 m.
a. Given $k = 2 \cdot 10^{-5}$ s will the lake level rise or fall?
b. Calculate the lake water depth after the response time.
c. Calculate what the discharge of the feeder stream must be to maintain a lake water depth of 5 m.

Solution
a.
Use Equation (11.7) to calculate the change in the lake water depth:

$$\frac{dH}{dt} = \frac{3.8}{4 \cdot 10^4} - 2 \cdot 10^{-5} \times 5 = -5 \cdot 10^{-6} \text{ m s}^{-1}$$

The change is negative, so the lake water level will drop.

Another approach is to calculate the equilibrium lake water depth (Equation 11.10):

$$H_{eq} = \frac{3.8}{2 \cdot 10^{-5} \times 4 \cdot 10^4} = 4.75 \text{ m}$$

The actual lake water depth is greater than this equilibrium depth, which implies that to achieve equilibrium the lake water needs to drop.

b.
The response time (= 1/k = 50 000 s = 13.9 h) gives the time needed to reduce the difference between the actual and equilibrium lake water depths by 63 percent (see Section 10.1). Thus, after 13.9 hours the lake water depth is

$$H(13.9\ h) = 5 - 0.63 \times (5 - 4.75) = 4.84\ \text{m}$$

This can also be calculated by using Equation (11.8):

$$H(50\,000\ s) = \frac{3.8}{2 \cdot 10^{-5} \times 4 \cdot 10^4} + \left(5 - \frac{3.8}{2 \cdot 10^{-5} \times 4 \cdot 10^4}\right) e^{-2 \cdot 10^{-5} \times 50\,000} =$$

$$4.75 + 0.25 \times 0.37 = 4.84\ \text{m}$$

This is indeed the same result as in the above calculation.

c.
Use Equation (11.10) to calculate what the discharge Q_{in} would be if H_{eq} = 5 m:

$$H_{eq} = \frac{Q_{in}}{2 \cdot 10^{-5} \times 4 \cdot 10^4} = 5\ \text{m}$$

$$Q_{in} = 5 \times 2 \cdot 10^{-5} \times 4 \cdot 10^4 = 4\ \text{m}^3\ \text{s}^{-1}$$

In more complex systems, for example when we simulate a cascade of reservoir lakes with different values for parameter *k,* which vary in time, and an upstream boundary condition Q_{in} that varies in time, finding an analytical solution may be very complex or even impossible. In such cases, a ***numerical solution*** of the differential equation must be found by numerical integration, which involves discretisation in space and time (i.e. dividing space and time into discrete units) and local linearisation of the problem at the level of the discrete units. Box 11.II illustrates a simple numerical solution technique for the Equation (11.7). In cases in which the parameter *k* is not constant in time, different values of *k* may be applied for the different time-steps; this is an important advantage of numerical modelling over analytical solutions.

The transport of substances in soil and water is principally governed by two kinds of physical processes: ***advection***, i.e. bulk movement with water from one location to another (also referred to as convection), and ***dispersion***, i.e. random or seemingly random mixing with the water. Non-aqueous phase liquids (NAPLs) can move independently from water flow, due to density flow. Chemicals may also be transported in the gas phase or transported biologically by moving organisms, such as the swimming of contaminated fish or the burrowing of contaminated soil fauna. Furthermore, substances can undergo chemical or physical transformation. A chemical transport model should take account of the whole body of these processes and their complex dependencies on different environmental factors. Because the model parameters vary spatially and temporally, the differential equations describing chemical transport can rarely be solved analytically; instead, the equations are usually solved using numerical techniques.

Box 11.I Analytical solution to differential equation (11.7)

Equation (11.7) can be rewritten as:

$$\frac{dH}{dt} = \frac{Q_{in}}{A} - kH \quad = \quad p - qH \tag{11.Ia}$$

where $p = \dfrac{Q_{in}}{A}$, and $q = k$

To solve this differential equation, we define:

$$H' = He^{qt} \tag{11.Ib}$$

If we differentiate this equation with respect to t (with the help of the chain rule), we obtain:

$$\frac{dH'}{dt} = \frac{dH}{dt} e^{qt} + Hqe^{qt} \quad = \quad e^{qt}(p - qH + qH) \quad = \quad pe^{qt} \tag{11.Ic}$$

Integrating this equation gives:

$$H' = \frac{p}{q} e^{qt} + H'_0 \tag{11.Id}$$

Combining equations (11.Ib) and (11.Id) gives:

$$H = \frac{p}{q} + H'_0 e^{-qt} \tag{11.Ie}$$

If we take H_0 to be the initial value of H at $t = 0$, then:

$$H'_0 = H_0 - \frac{p}{q} \tag{11.If}$$

Combining equations (11.Ie) and (11.If) gives:

$$H(t) = \frac{p}{q} + \left(H_0 - \frac{p}{q} \right) e^{-qt} \tag{11.Ig}$$

If we put back the values for p and q, we obtain:

$$H(t) = \frac{Q_{in}}{kA} + \left(H_0 - \frac{Q_{in}}{kA} \right) e^{-kt} \tag{11.Ih}$$

Note that the steady state lake water level is reached as $t \to \infty$, so $H(\infty) = \dfrac{Q_{in}}{kA}$.

This chapter explores the background and derivation of the governing equations that are widely used in chemical transport models for the different environmental compartments of soil, groundwater, and surface water. In principle, the mathematical descriptions of chemical transport in these compartments are virtually identical: the transport equation that describes the movement of solutes in groundwater can also be adopted for modelling the mixing of industrial effluent into a river. If differences in the equations occur for the specific compartments, they will be indicated. The mathematical models are helpful for analysing

---

Box 11.II Numerical solution to differential equation (11.7): Simple Euler's method

If we discretise time in time steps of Δt, then:

$$t_i = t_{i-1} + \Delta t \qquad (11.\text{IIa})$$

The reservoir lake water level at time step t_i is:

$$H(t_i) = H(t_{i-1}) + \frac{dH}{dt} \Delta t \qquad (11.\text{IIb})$$

Combining Equation (IIb) with Equation 11.7 gives:

$$H(t_i) = H(t_{i-1}) + \left(\frac{Q_{in}}{A} - kH(t_{i-1}) \right) \Delta t \qquad (11.\text{IIc})$$

By means of this simple numerical integration method the lake level at time-step t_i can be explicitly calculated from the lake level at time step t_{i-1} using a linear estimate of the rate at which the lake level is falling at time step t_{i-1}. Note that this Euler's method is only a simple numerical method with a considerable numerical error that can be reduced by using small time-step of size Δt. Too large values for Δt may result in numerical instability: for example, because the fall in the lake level during one time-step becomes larger than possible rise due to water inflow plus the lake water level itself.

For other, more sophisticated numerical methods, see textbooks on numerical methods, such as Press *et al.* (1992).

---

the contrasting chemical behaviour in soil, groundwater, and surface water dealt with in the subsequent chapters.

11.2 ADVECTION

11.2.1 Advection equation

Advection is the process of transporting substances in solution or suspension with the movement of the medium (usually water). Water flows due to gravity forces and is retarded by internal friction (viscosity) and friction at the contact between the moving water and the sediment over or through which it flows. The physics of water movement are dealt with in many hydrological textbooks (e.g. Chow *et al.*, 1988; Domenico and Schwarz, 1998; Ward and Robinson, 2000; Brutsaert, 2005) and are summarised in Boxes 11.III and 11.IV. If a substance is brought into flowing water, it is transported in the same direction and at the same velocity as the water. The change in the mass of a substance in a control volume over time due to advection equals the difference between mass inflow and mass outflow (see Figure 11.1) and may be written as the following difference equation:

$$\Delta(VC) = - \Delta(QC) \, \Delta t \qquad (11.11)$$

where V = the volume of water in the control volume [L^3], C = concentration [M L^{-3}], and Q = discharge [L^3 T^{-1}]. The negative sign on the right-hand side of the equation depicts an increasing concentration within the control volume if the mass inflow is greater than the

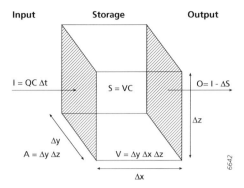

Figure 11.1 Mass balance of a substance in a control volume over time-step Δt.

mass outflow. Dividing Equation (11.11) by **Δ***t* and further division by the incremental volume V = *A* **Δ***x* gives:

$$\frac{\Delta C}{\Delta t} = \frac{-\Delta(QC)}{A\Delta x} \tag{11.12}$$

where *A* = the cross-sectional area of the control volume [L^2]. If we take the limits as **Δ***x* → 0 and **Δ**t → 0, we obtain:

$$\frac{\partial C}{\partial t} = -\frac{1}{A}\frac{\partial(QC)}{\partial x} \tag{11.13}$$

If we assume the discharge *Q* constant over **Δ***x*, we may rewrite Equation (11.13) in:

$$\frac{C}{\partial t} = -\bar{u}_x \frac{C}{\partial x} \tag{11.14}$$

where $\bar{u}_x$ = the area-averaged flow velocity [L T^{-1}]. For the initial condition $C(x,t_0) = C_0(x)$, the analytical solution of Equation (11.14) is:

$$C(x,t) = C_0\left(x - u_x(t - t_0)\right) \tag{11.15}$$

It is important to note that if we use the one-dimensional advection equation (Equation 11.14), we assume that the concentration of the pollutant is homogeneous throughout the river cross-section (laterally and vertically). If we use the advection equation to calculate the transport of a pollution wave downstream in a river, the shape of the wave remains unchanged (see Figure 11.2). Figure 11.3 shows the downward propagation of a continuous input of a substance into groundwater.

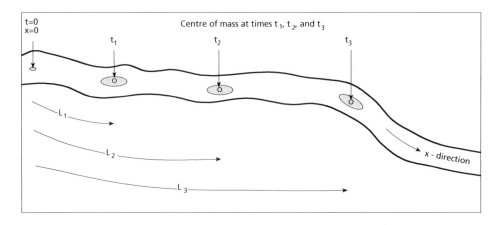

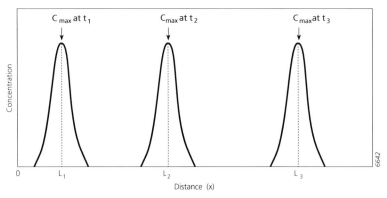

Figure 11.2 Downstream propagation of a pulse injection into a river due to advection. Note that the shape and the size of the chemographs remain unchanged.

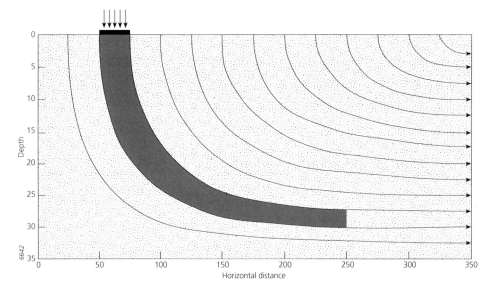

Figure 11.3 Downstream propagation of a continuous input into groundwater due to advection.

Box 11.III Physics of groundwater and soil water flow

The governing equations that describe ground and soil water movement are based on the conservation laws of mass (continuity) and momentum (Newton's second law of motion). Water in the soil occupies the pore space between the soil particles. The porosity n is defined as:

$$n = \frac{volume \ of \ pores}{total \ volume}$$

(11.IIIa)

and the volumetric soil moisture content θ is defined as:

$$\theta = \frac{volume \ of \ water}{total \ volume}$$

(11.IIIb)

Note that both n and θ are dimensionless and θ equals n if the pores are completely saturated with water. The total energy of water is described by the hydraulic head that consists of three components, namely the elevation head, pressure head, and velocity head. Because the flow velocity of ground and soil water is slow, the velocity head can be neglected and the hydraulic head can be expressed as:

$$h = \psi + z = \frac{P}{\rho_w g} + z$$

(11.IIIc)

where h = the hydraulic head [L], ψ = the pressure head [L], z = the elevation [L], P = the pressure [M L^{-1} T^{-2}], ρ_w = the density of water (= 1000 kg m^{-3}) [M L^{-3}], and g = the gravitational acceleration constant (= 9.8 m s^{-2}) [L T^{-2}]. In the case of groundwater, the hydraulic head at a particular location below the water table can be determined by measuring the water level in a tube (piezometer) with the bottom opening at the location where the hydraulic head is being determined and the top opening in contact with the atmosphere. If the soil or rock matrix is not fully saturated, the pores are partially occupied by air. This is responsible for the most important difference compared with saturated conditions, namely that the pressure head ψ is negative because of capillary suction by the soil matrix. The relation between pressure head (or suction head) ψ and the soil moisture content θ is usually depicted in a water retention curve or pF curve (pF = log($-\psi$)).

The three-dimensional continuity equation for subsurface water is:

$$-\left[\frac{\partial q_x}{\partial x} + \frac{\partial q_y}{\partial y} + \frac{\partial q_z}{\partial z} \right] = \frac{1}{\rho_w} \frac{\partial \left(\rho_w n \right)}{\partial t}$$

(11.IIId)

where q = the volumetric flow rate per unit area of soil (= Q/A) [L T^{-1}]. Note that q is not equal to the actual flow velocity of the water between the pores. To derive the actual flow velocity, q must be divided by the effective porosity n_{eff} i.e. the porosity available for flow:

$$u = \frac{q}{n_{eff}}$$

(11.IIIe)

where u = the actual flow velocity of groundwater or soil water [L T^{-1}]. Because groundwater flow is slow, it is laminar and free from turbulence.

Box 11.III (continued) Physics of groundwater and soil water flow

In the case of saturated groundwater flow, the right-hand side of Equation (11.III.4) can be rewritten as:

$$\frac{1}{\rho_w}\frac{\partial\left(\rho_w n\right)}{\partial t} = S_s \frac{\partial h}{\partial t} \qquad (11.\text{III}f)$$

where S_s = the specific storage [L^{-1}], which is a proportionality constant relating the changes in water volume per unit change in hydraulic head. Thus, a change in water flow in space results in a change in the hydraulic head. In the case of an unconfined aquifer, the specific storage is equal to the part of the pore volume that can yet be occupied by water. This is often somewhat less than the porosity, because the soil matrix is rarely completely dry. In the case of a confined aquifer, the specific storage is much smaller and depends on the changes in pore volume as result of changes in hydraulic head, and thus on the compressibility of the rock matrix.

In the case of unsaturated soil water flow, the right-hand side of Equation (11.III.4) equals:

$$\frac{1}{\rho_w}\frac{\partial\left(\rho_w n\right)}{\partial t} = \frac{\partial\theta}{\partial t} \qquad (11.\text{III}g)$$

which implies that a change in water flow in space results in a change in moisture content of the soil in time.

The momentum equation is represented by Darcy's law, which relates the flow rate to the hydraulic gradient. The one-dimensional form of Darcy's law for flow in direction x reads:

$$q_x = -K_x \frac{\partial h}{\partial x} \qquad (11.\text{III}h)$$

where K_x = the hydraulic conductivity [$L\ T^{-1}$], which depends on the grain size distribution of the soil matrix and varies over several orders of magnitude. The saturated hydraulic conductivity is in the order of 10^{-2} m s^{-1} for gravel, 10^{-5} m s^{-1} for sand, and 10^{-9} m s^{-1} for clay. In unsaturated conditions, the air in the soil greatly reduces the connectivity of the pores and hence also greatly reduces the hydraulic conductivity. Accordingly, the hydraulic conductivity is a function of soil moisture (and thus of the pressure head).

The combination of Darcy's law (Equation 11.IIIh) and the continuity equation (Equation 11.IIId) yields:

$$\frac{\partial}{\partial x}\left(K_x\frac{\partial h}{\partial x}\right) + \frac{\partial}{\partial y}\left(K_y\frac{\partial h}{\partial y}\right) + \frac{\partial}{\partial z}\left(K_z\frac{\partial h}{\partial z}\right) = \frac{1}{\rho_w}\frac{\partial\left(\rho_w n\right)}{\partial t} \qquad (11.\text{III}i)$$

This equation has numerous analytical solutions for various simplified situations for both groundwater and unsaturated flow. Because unsaturated flow largely consists of vertical drainage driven by gravity forces, unsaturated models are usually one-dimensional. For more complex distributed models, the equations are mostly solved numerically using available model codes, such as MODFLOW (McDonald and Harbaugh, 1988), a widely used three-dimensional finite-difference groundwater flow model.

Box 11.IV Physics of surface water flow

The governing equations that describe surface water flow are based on the conservation laws of mass (continuity) and momentum (Newton's second law of motion). Working out these basic equations for one-dimensional, unsteady (dynamic) flow in an open channel and neglecting lateral inflow (from precipitation or tributary rivers), wind shear, and energy losses due to eddies yields the so-called Saint-Venant equations:

Continuity equation

$$\frac{\partial Q}{\partial x} + \frac{\partial A}{\partial t} = 0 \tag{11.IVa}$$

Momentum equation

$$\frac{1}{A}\frac{\partial Q}{\partial t} + \frac{1}{A}\frac{\partial}{\partial x}\left(\frac{Q^2}{A}\right) + g\frac{\partial H}{\partial x} - g\left(S_0 - S_f\right) = 0 \tag{11.IVb}$$

| Local acceleration term | Convective acceleration term | Pressure force term | Gravity force term | Friction force term |

Kinematic wave

Diffusion wave

Dynamic wave

where Q = the water discharge [L³ T⁻¹], A = the cross-sectional area of the channel [L²], H = the water depth [L], g = the gravitational acceleration constant (= 9.8 m s⁻²) [L T⁻²], S_0 = the bed slope [-], and S_f = the friction slope [-]. If the water level or flow velocity (= Q/A) changes at a given location in a channel, the effects of this change propagate back upstream. These backwater effects are represented by the first three terms of Equation (IVb). These terms may be neglected for overland flow and flow through channels with a more or less uniform cross-section along their length. The remaining terms comprise the kinematic wave, which assumes $S_0 = S_f$; the friction and gravity forces balance each other. The relation between the friction slope and the flow rate is described by either Chezy's equation or Manning's equation. Chezy's equation reads:

$$u = C\sqrt{R\,S_f} \tag{11.IVc}$$

where u = cross-sectional averaged flow velocity [L T⁻¹], C = Chezy's coefficient [L⁰·⁵ T⁻¹], and R = hydraulic radius [L]. The hydraulic radius equals the wetted perimeter divided by the cross- sectional area. For most rivers that are much wider than they are deep, the hydraulic radius is approximately equal to the water depth. Typical values for C for rivers are in the order of 30–50 m⁰·⁵ s⁻¹. Manning's equation reads:

$$u = \frac{R^{2/3}\,S_f^{1/2}}{n} \tag{11.IVd}$$

where n = Manning roughness coefficient, with u expressed in m s⁻¹ and R expressed in m. Typical values for n range between 0.012 for smooth, concrete channels, via 0.040 for meandering streams, to 0.100 for channels in densely vegetated floodplains. Chezy's and Manning's equations are both only applicable for turbulent flow, which is almost always

---

Box 11.IV (continued) Physics of surface water flow

the case in rivers. Using $Q = u \cdot A$, Equations (11.IVa) and (11.IVb) can be solved for Q. Because these Saint-Venant equations are not amenable to analytical solutions except in a few simple cases, the equations are usually solved numerically.

Example 11.3 Advection

In a river with a mean flow velocity of 0.80 m s^{-1}, the chloride concentration increases due to increasing wastewater discharge from an upstream salt mine at location $x = 0$ km. At $x = 4$ km, the chloride concentration increases linearly from 20 mg l^{-1} to 140 mg l^{-1} between 4 a.m. (04.00 h) until 4 p.m. (16.00 h). Calculate the chloride concentration at $x = 40$ km and 8 p.m. on the same day.

Solution
First, convert the flow velocity from m s^{-1} to km h^{-1}:
$\bar{u}_x = 0.80$ m s^{-1} = 0.80 m s^{-1} × 3600 s h^{-1} = 2.880 km h^{-1}

Next, assume that t_0 is the time that the chloride concentration at location $x = 4$ km is equal to the concentration at $x = 40$ km and 8 p.m. Thus, the initial condition at $x = 4$ km and $t = t_0$ is:

$$C(4,t_0) \quad = \quad C_0(4)$$

Use equation (11.15) to define the concentration at $x = 40$ km at $t = t$ (8 p.m.):

$$C(40,t) \quad = \quad C_0(40 - 2.880 \cdot (t - t_0))$$

Because we have assumed that $C(4,t_0) = C(40,t)$, we may write:

$$C_0(4) \quad = \quad C_0(40 - 2.880 \cdot (t - t_0))$$

$$4 \quad = \quad 40 - 2.880 \cdot (t - t_0)$$

$$2.880 \cdot (t - t_0) \quad = \quad 36$$

$$t - t_0 \quad = \quad 12.5 \text{ h}$$

Thus, it takes 12.5 hours for the water to travel from location $x = 4$ km to location $x = 40$ km. This can also be calculated directly from $\Delta t = \Delta x / \bar{u}_x$. From the calculation it appears that t_0 is 12.5 hours before 8 p.m. = 7.30 a.m., which is 3.5 hours after the chloride concentration started increasing. At location $x = 4$ km, the chloride concentration increases from 20 mg l^{-1} to 140 mg l^{-1} in 12 hours, so 10 mg l^{-1} h^{-1}. This implies that after 3.5 hours, the concentration had increased by 10 mg l^{-1} h^{-1} × 3.5 h = 35 mg l^{-1}. The chloride concentration at location $x = 0$ and $t = t_0$ is

$$C(4,t_0) = 20 + 35 = 55 \text{ mg l}^{-1}.$$

This is also the concentration at $x = 40$ km and 8 p.m.

11.2.2 Load calculation and mixing

In Section 11.1 we saw that the load of a substance in a river is defined as the product of its concentration and the water discharge (see Equation 11.1), and thus represents the amount of substance transported by advection. Actually, this calculation is only valid for a momentaneous load. The average load that has passed a point over a longer period of time can be defined as:

$$L = \frac{\int_0^{t_1} C(t) \cdot Q(t)\, dt}{t_1 - t_0} \qquad (11.16)$$

where L = Load that passes a point in the time interval from $t=0$ to $t=t_1$ [M T^{-1}]. Because concentration and discharge are usually correlated, the average load over a long period, say a year, is not equal to the annual average concentration times the annual average discharge. If there is a positive correlation between discharge and concentration due to increased substance inputs during high flow conditions, for instance as a result of soil erosion, the simple product of annual average concentration and discharge underestimates the annual load. If there is a negative correlation due to dilution during high flow conditions, the simple product of annual average concentration and discharge overestimates the annual load.

The load calculation also enables us to calculate concentrations downstream from a point source, e.g. an effluent discharge or a tributary river. In contrast to concentrations, loads may be added up. If we assume instantaneous mixing of two converging streams, the load downstream of the confluence ($C_3 Q_3$) is the sum of the loads from both streams ($C_1 Q_1$ and $C_2 Q_2$, respectively). The concentration downstream is:

$$C_3 = \frac{C_1 Q_1 + C_2 Q_2}{Q_1 + Q_2} \qquad (11.17)$$

In reality, it takes a certain distance downstream of a confluence to complete mixing over the entire cross-section of the river. This mixing length is treated in more detail in Section 11.3.4.

Advection is one of the most important modes for substance transport in both groundwater and surface water. Note that in all the equations related to advective transport presented in this chapter, the concentration is calculated using water discharge. This emphasises the importance of an accurate water balance, because without a balanced water budget it is impossible to obtain an accurate mass balance for advective transport.

Example 11.4 Load calculation and mixing

A wastewater treatment plant discharges its effluent in a stream at a constant rate of 40 l s^{-1}. The stream has a depth of 0.6 m, a width of 5.0 m, and a flow velocity of 12 cm s^{-1}. The chloride concentration in the stream just upstream from the effluent outfall is 20 mg l^{-1} and the chloride concentration of the effluent water is 180 mg l^{-1}. Calculate the chloride concentration directly downstream from the effluent outfall, assuming instantaneous mixing.

Solution
This is a rather straightforward application of Equation (11.17): First, calculate the discharge of the stream just upstream from the effluent outfall:
Q_{stream} = 0.6 m × 5 m × 0.12 m s^{-1} = 0.36 m^3 s^{-1}

Then, calculate the chloride loads of the effluent and the stream just upstream from the effluent outfall:

$(CQ)_{effluent}$ = 180 g m^3 × 0.040 m^3 s^{-1} = 7.2 g s^{-1}
$(CQ)_{stream}$ = 20 g m^{-3} × 0.36 m^3 s^{-1} = 7.2 g s^{-1}

The total chloride load is 7.2 + 7.2 = 14.4 g s^{-1}; the total discharge is 0.040 + 0.36 = 0.40 m^3 s^{-1}. The chloride concentration downstream from the outfall is

$$C_{downstream} = \frac{(CQ)_{effluent} + (CQ)_{stream}}{Q_{effluent} + Q_{stream}} = \frac{7.2 + 7.2}{0.04 + 0.36} = \frac{14.4}{0.4} = 36 \text{ mg l}^{-1}$$

11.3 DIFFUSION AND DISPERSION

11.3.1 Molecular diffusion

Molecules move randomly through water due to so-called Brownian motion. Even if water seems to be entirely quiescent, chemicals move from regions of high concentrations to regions of low concentrations. This process results in increase of entropy and is known as ***molecular diffusion***. In nature, molecular diffusion occurs primarily through thin, laminar boundary layers, at, for instance, air–water or sediment–water interfaces or in stagnant pore water. In the mid-19th century, Fick determined that the mass transfer by molecular diffusion was linearly related to the cross-sectional area over which the transfer takes place and the concentration gradient. Accordingly, Fick's first law describes the flux density of mass transport as follows:

$$J = -D\frac{dC}{dx} \tag{11.18}$$

where J = the flux density [M L^{-2} T^{-1}], D = molecular diffusion coefficient [L^2 T^{-1}], C = the chemical concentration [M L^{-3}], and x = the distance over which the change in concentration is considered. The negative sign reflects that the direction of the mass transfer is in the same direction as the concentration gradient, i.e. opposite to the direction of positive change in concentration. The mass transfer continues until equilibrium is reached, i.e. the concentration gradient is zero everywhere.

The diffusion coefficient D depends on the chemical and thermodynamic properties of the substance under consideration and the water temperature. Molecular diffusion increases in magnitude with increasing temperatures and with decreasing size of the molecules. At environmental temperatures, the molecular diffusion coefficients of most chemicals are in the order of 10^{-9} m^2 s^{-1}, indicating a very slow transport process. Except for transport through stagnant boundary layers (e.g. water–air interface) and in quiescent sediment pore waters, molecular diffusion is generally not a relevant transport process in natural waters.

11.3.2 Turbulent diffusion and mechanical dispersion

Besides moving through molecular diffusion, molecules in surface water also move due to constantly changing swirls or eddies of different sizes depending on the flow regime. The random mixing caused by this type of turbulence is called ***turbulent diffusion*** or eddy diffusion. In fact, this mixing process is a differential advective process at the microscale. The mass transfer due to turbulent diffusion is several orders in magnitude larger than the mass transfer due to molecular diffusion, and contributes significantly to mixing in rivers and lakes. In groundwater, the typical flow velocities are much slower than in surface water and there are no eddy effects.

Mixing, or dispersion, also occurs as a result of differences in travel times of parcels of water between adjacent water flow paths due to differences in flow velocities or in the length of the flow paths. Although groundwater flow is laminar, at the microscale groundwater does not flow straight on as it must take myriad detours around soil particles, as shown in Figure 11.4. Figure 11.4 also shows that the flow velocity varies within and between the pores. At the macroscale, the same principle applies for flow around regions of less permeable soil. Furthermore, in the unsaturated zone of the soil, inhomogeneities in effective permeability (for instance due to lithological stratification or the presence of macropores and fractures) lead to differences in travel times along parallel flow paths. Accordingly, heterogeneity at the macroscopic scale contributes significantly to dispersion. In surface waters, velocity gradients both laterally and vertically are caused by shear forces at the boundaries of the water body, such as shear stresses at the water–air interface (due to wind) and at the water bottom and the banks (Figure 11.5).

In addition, secondary currents can develop in rivers due to channel morphology: for instance, back currents into stagnant backwaters, eddies induced by ripples, dunes, and rocks on the river bed, and the corkscrew-like helicoidal flow in meandering river channels. These variations from the average water flow velocity (because of the velocity gradients) and travel distances (because of the random variability or tortuosity of flow paths) also cause mixing of parcels of water. This mixing is called ***mechanical dispersion*** and likewise results

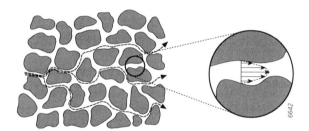

Figure 11.4 Causes of mechanical dispersion in groundwater: flow variation around soil particles.

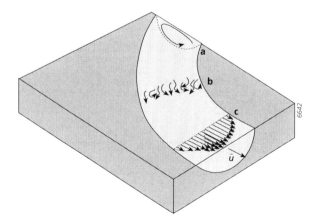

Figure 11.5 Causes of mechanical dispersion in rivers: a. helicoidal flow; b. small-scale flow variations, swirls, and eddies; c. horizontal and vertical flow variations.

Table 11.1 Typical values of dispersion coefficients under various conditions (adapted from Schnoor 1996)

Condition	Dispersion coefficient ($m^2 s^{-1}$)
Molecular diffusion	10^{-9}
Compacted sediment	$10^{-11}-10^{-9}$
Bioturbated sediment	$10^{-9}- 10^{-8}$
Lakes – vertically	$10^{-6}- 10^{-3}$
Rivers – laterally	$10^{-2}- 10^{1}$
Rivers – longitudinally	$10^{1}-10^{3}$
Estuaries – longitudinally	$10^{3}-10^{4}$

in a net mass transport of chemicals from regions of high concentrations to regions of low concentrations. Dispersion thus depends on the average flow velocity and its variability, and varies both in space and time. As a matter of fact, mixing takes place in three dimensions and the magnitude of dispersion may be different for the different directions, i.e. dispersion may be subject to anisotropy. In groundwater, anisotropy in dispersion is largely due to anisotropy in hydraulic conductivity. In river water, anisotropy in dispersion is mainly caused by the differences in velocity gradients parallel and perpendicular to the river channel. In both groundwater and surface water, we distinguish ***longitudinal dispersion***, i.e. mixing in the main flow direction, and ***transverse dispersion***, i.e. mixing perpendicular to the main flow direction.

Although the physical mechanisms that cause the mixing are different, the description of net mass transport by turbulent diffusion and mechanical dispersion is analogous to the description of molecular diffusion. So, in Fick's first law (Equation 11.18) for molecular diffusion, coefficient D is replaced by the respective turbulent diffusion coefficient and dispersion coefficient. The rate of mass transfer is thus proportional to the concentration gradient and the coefficient D. Dispersion coefficients are much larger than the turbulent diffusion coefficients, which are in turn much greater than the molecular diffusion coefficients. These differences reveal an important scale effect: the value of D increases as the scale of the problem increases. This is also the case if we consider mechanical dispersion only: the dispersion coefficient for chemical transport in groundwater is much larger for macro-scale variations in permeability than for micro-scale variations caused by the presence of soil particles. This scale effect also has far-reaching implications for the resolution at which we consider the concentration gradients. For example, in the case of longitudinal mixing resulting from vertical and lateral flow velocity gradients in medium to large river channels downstream from an effluent discharge, we are mostly modelling cross-sectional average concentrations (see Section 11.3.3). In this case, the Fickian analogy for dispersion is not valid at the local scale for the reach just downstream of the effluent, e.g. for a single one litre sample from the river water collected near the river bank. This is due to the difference in sample support and the scale at which turbulent diffusion and differential advection takes place. As already observed in Section 11.2.2, it takes a certain distance downstream from the discharge before the river water is completely mixed over the entire cross-section. Table 11.1 gives a brief summary of the order of magnitude of dispersion coefficients under various conditions.

If we describe mass transport due to dispersion, we may thus apply a Fickian type law. Combining this with the law of conservation of mass gives:

$$\frac{\partial C}{\partial t} = -\frac{\partial J}{\partial x} \qquad\qquad (11.19)$$

This Equation (11.19) says that the sum of the concentration changes in space and time equals zero. Combining this equation with Equation (11.18) yields:

$$\frac{\partial C}{\partial t} = D \frac{\partial^2 C}{\partial x^2} \tag{11.20}$$

This is a second-order partial differential equation and is known as Fick's second law. To solve this Equation (11.20), one initial condition and two boundary conditions (one for each order) are needed and for each set of boundary and initial conditions there is a solution. For a pulse injection of a tracer or pollutant into a water body at $t=0$, we may pose the following initial condition:

$$C(x) = 0 \text{ at } t = 0 \tag{11.21}$$

The boundary conditions are formulated so that the concentration integrated over the semi-infinite length equals the mass M of the substance injected and the concentration at infinite distance remains zero:

$$C(\infty) = 0 \text{ for all } t \tag{11.22}$$

$$M = A \int_{-\infty}^{\infty} C \, dx \tag{11.23}$$

where A = the cross-sectional area [L^2]. The integration of the dispersion Equation (11.20) may be achieved by trial and error methods and the analytical solution reads:

$$C(x,t) = \frac{M}{2 A \sqrt{\pi D_x t}} e^{-x^2/(4 D_x t)} \tag{11.24}$$

where D_x = the longitudinal dispersion coefficient. This Equation (11.24) describes the substance concentration as function of space and time and may be rewritten as:

$$C(x,t) = \frac{M}{A \sqrt{2 D_x t} \sqrt{2\pi}} e^{-x^2/(2 \cdot 2 D_x t)} \tag{11.25}$$

This equation is analogous to the description of the 'bell-shaped' probability density curve of the normal or normalised Gaussian distribution:

$$\phi(x) = \frac{1}{\sigma \sqrt{2\pi}} e^{-x^2/2\sigma^2} \tag{11.26}$$

where the arithmetic mean of $\phi(x)$ equals zero and σ = the standard deviation of $\phi(x)$. Thus, Equation (11.25) describes the concentration distributions as function of time and distance from the point at which the substance was injected and the concentration profile in the water body is described by a Gaussian curve with a mean of $x = 0$ and a standard deviation of

$$\sigma = \sqrt{2 D_x t} \tag{11.27}$$

This implies that the centre of mass remains at the point of injection and the standard deviation increases with time. Accordingly, the Gaussian curve becomes broader with time.

The mass balance for a system in which both advection and dispersion occurs is simply a combination of the advection equation with the dispersion equation, which gives the **advection–dispersion equation**:

$$\frac{\partial C}{\partial t} = -\bar{u}_x \frac{\partial C}{\partial x} + D_x \frac{\partial^2 C}{\partial x^2} \qquad (11.28)$$

Actually, this one-dimensional Equation (11.28) is the simplest form of the advection–dispersion equation and describes longitudinal dispersion (in the direction of the water flow). The dispersion coefficient D_x is therefore also referred to as the longitudinal dispersion coefficient. The full three-dimensional advection–dispersion equation reads:

$$\frac{\partial C}{\partial t} = -\bar{u}_x \frac{\partial C}{\partial x} + D_x \frac{\partial^2 C}{\partial x^2} + D_y \frac{\partial^2 C}{\partial y^2} + D_z \frac{\partial^2 C}{\partial z^2} \qquad (11.29)$$

where x refers to the average direction of water flow and D_y and D_z refer to the respective dispersion laterally and vertically perpendicular to the average flow direction, also referred to as the transverse dispersion coefficients. Advection and dispersion can also be defined in relation to a Cartesian coordinate system:

$$\frac{\partial C}{\partial t} = -\bar{u}_x \frac{\partial C}{\partial x} - \bar{u}_y \frac{\partial C}{\partial y} - \bar{u}_z \frac{\partial C}{\partial z} + D_x \frac{\partial^2 C}{\partial x^2} + D_y \frac{\partial^2 C}{\partial y^2} + D_z \frac{\partial^2 C}{\partial z^2} \qquad (11.30)$$

In this case, the dispersion coefficients D_x, D_y and D_z are defined in relation to a grid system. The three-dimensional version of the advection–dispersion equation is used when a pollutant disperses from a single point into a large three-dimensional water body, such as in groundwater, a deep lake or an estuary. For rivers, the two-dimensional version is usually used at a local scale (for instance, to model the dispersal of sediments and pollutants over floodplain areas) and the one-dimensional version is usually used for long stretches of rivers. In the next section, the one-dimensional advection–dispersion equation is further explored to model longitudinal dispersion in a river or groundwater.

11.3.3 Longitudinal dispersion

Longitudinal dispersion is the process of mixing in the direction of the average water flow, so we need to consider both advection and dispersion. The analytical solution of the one-dimensional advection–dispersion equation for a pulse injection is similar to that of the dispersion equation, but in the right-hand side of Equation (11.24) the x^2 replaced by $(x-u_x t)^2$:

$$C(x,t) = \frac{M}{2 A \sqrt{\pi D_x t}} e^{-(x-u_x t)^2 / 4 D_x t} \qquad (11.31)$$

The centre of the mass travels at a velocity u_x and while the mass is travelling downstream, the Gaussian curve becomes broader as a result of longitudinal dispersion. This can be illustrated by the development of a spreading cloud moving downstream after a pulse release of a tracer into a river (see Figure 11.6).

In groundwater, dispersion occurs in a similar manner, as can be illustrated with a simple column experiment (Figure 11.7). The column is filled with a permeable sediment and the test begins with a continuous inflow of a conservative tracer at a relative concentration of $C/C_0 = 1$ added across the entire cross-section at the inflow end. Figure 11.7 shows how the relative concentration of the tracer varies with time at both ends of the column. The

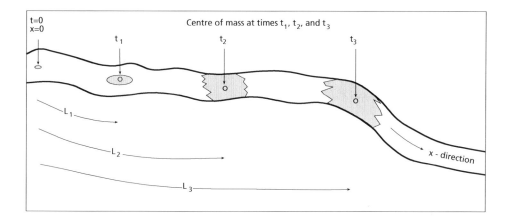

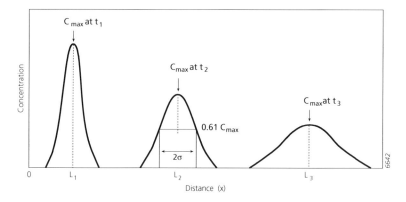

Figure 11.6 Downstream propagation due to advection and dispersion of a pulse injection into a river. Note that the size of the chemographs remains unchanged but their shape becomes broader while travelling downstream. Adapted from Hemond and Fechner-Levy (2000).

so-called ***breakthrough curve*** at the outflow end displays a gradual increase from $C/C_0 = 0$ to $C/C_0 = 1$ and follows a cumulative probability curve of the Gaussian distribution because of the continuous tracer inflow. Just as in the river example, within the column the size of the zone of spread of the tracer increases as the advective front moves further from the source. The position of the advective front at the breakthrough corresponds to a C/C_0 value of 0.5.

The analytical solution to the one-dimensional advection–dispersion equation for a continuous source with an initial condition of $C(x,0) = 0$ and a boundary condition of $C(0,t) = C_0$ reads:

$$C(x,t) = \frac{C_0}{2}\left(erfc\left(\frac{x - u_x t}{2\sqrt{D_x t}}\right) + erfc\left(\frac{x + u_x t}{2\sqrt{D_x t}}\right) e^{u_x x / D_x} \right) \tag{11.32}$$

where C_0 = the continuous inflow concentration [M L^{-3}] and *erfc* is the complementary error function, which equals one minus the error function (*erf*), which is the integral of the normalised Gaussian distribution (Equation 11.26) or the Gaussian cumulative distribution function. The complementary error function *erfc*(β) decreases from 2 at $\beta \to -\infty$ via 1 at $\beta = 0$ to 0 at $\beta \to \infty$ (see Table 11.2 and Figure 11.8). Both the error function

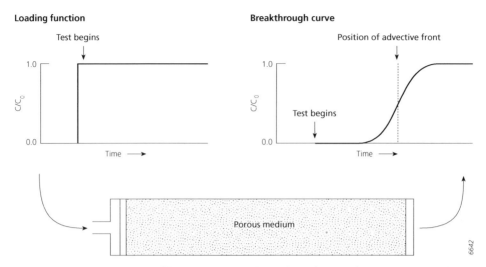

Figure 11.7 Dispersion in a column filled with a porous medium. The test begins with a continuous input of tracer with a relative concentration of $C/C_0 = 1$ at the inflow end. The breakthrough curve at the outflow end characterises longitudinal dispersion in the column. Adapted from Domenico and Schwarz (1998).

Table 11.2 The complementary error function erfc(β)

β	erfc(β)	β	erfc(β)	β	erfc(β)	β	erfc(β)
-3.0	1.999978	-0.95	1.820891	0.05	0.943628	1.1	0.119795
-2.8	1.999925	-0.90	1.796908	0.10	0.887537	1.2	0.089686
-2.7	1.999866	-0.85	1.770668	0.15	0.832004	1.3	0.065992
-2.6	1.999764	-0.80	1.742101	0.20	0.777297	1.4	0.047715
-2.5	1.999593	-0.75	1.711155	0.25	0.723674	1.5	0.033895
-2.4	1.999311	-0.70	1.677801	0.30	0.671373	1.6	0.023652
-2.3	1.998857	-0.65	1.642029	0.35	0.620618	1.7	0.016210
-2.2	1.998137	-0.60	1.603856	0.40	0.571608	1.8	0.010909
-2.1	1.997021	-0.55	1.563323	0.45	0.524518	1.9	0.007210
-2.0	1.995322	-0.50	1.520500	0.50	0.479500	2.0	0.004678
-1.9	1.992790	-0.45	1.475482	0.55	0.436677	2.1	0.002979
-1.8	1.989091	-0.40	1.428392	0.60	0.396144	2.2	0.001863
-1.7	1.983790	-0.35	1.379382	0.65	0.357971	2.3	0.001143
-1.6	1.976348	-0.30	1.328627	0.70	0.322199	2.4	0.000689
-1.5	1.966105	-0.25	1.276326	0.75	0.288845	2.5	0.000407
-1.4	1.952285	-0.20	1.222703	0.80	0.257899	2.6	0.000236
-1.3	1.934008	-0.15	1.167996	0.85	0.229332	2.7	0.000134
-1.2	1.910314	-0.10	1.112463	0.90	0.203092	2.8	0.000075
-1.1	1.880205	-0.05	1.056372	0.95	0.179109	2.9	0.000041
-1.0	1.842701	0	1.000000	1.0	0.157299	3.0	0.000022

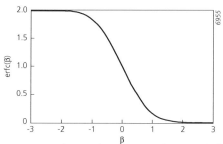

Figure 11.8 The complementary error function erfc(β).

and the complementary error function are well tabulated and can be calculated using most spreadsheets. In many practical situations, the second term of Equation (11.32) is small and can therefore be ignored.

Example 11.5 Dispersion

A cylindrical column is filled with sandy sediment fully saturated with water with a chloride concentration of 0 mg l^{-1}. At time $t = 0$ a continuous inflow of water with a chloride concentration of 500 mg l^{-1} begins, with a flow velocity of 60 cm d^{-1}. Calculate the chloride concentration at x = 35 cm after 12 hours, given the longitudinal dispersion coefficient = 5·10^{-3} m^2 d^{-1}.

Solution
Use Equation (11.32) to calculate the chloride concentration. First, calculate the complementary error function terms:

First term:

$$\frac{x - u_x t}{2\sqrt{D_x t}} = \frac{0.35 - 0.6 \times 0.5}{2\sqrt{0.005 \times 0.5}} = \frac{0.05}{0.1} = 0.5$$

Use the values of *erfc*(β) listed in Table 11.2 to estimate the value of *erfc*(0.5):

$$erfc(0.5) = 0.4795$$

In this case the value of β is listed, but if the value of β is between two values listed, the values of *erfc*(β) can be interpolated.

Second term:

$$\frac{x + u_x t}{2\sqrt{D_x t}} = \frac{0.35 + 0.6 \times 0.5}{2\sqrt{0.005 \times 0.5}} = \frac{0.65}{0.1} = 6.5$$

The value of *erfc*(6.5) is very close to zero (see Table 11.2). However, the exponential term in Equation (11.32) is large ($\approx 1.7 \cdot 10^{18}$). Calculation using a spreadsheet shows that the entire second term of Equation (11.32) (including the error function and exponential terms) is very small and approximates zero.
    Use Equation (11.32) with $C_0 = 500$ mg l^{-1} to calculate the chloride concentration at $x = 0.35$ m and $t = 0.5$ d:

$$C(0.35, 0.5) = \frac{500}{2}(0.4795 + 0) = 120 \ \text{mg l}^{-1}$$

This example shows that in practical situations it is indeed safe to ignore the second term in Equation (11.32).

The longitudinal dispersion coefficient D_x can thus be estimated by fitting a Gaussian curve from tracer experiments in rivers, groundwater column experiments, and soil. From the graph of concentration versus time or distance it is possible to estimate the variance σ and to compute the longitudinal dispersion coefficient using Equation (11.27) (see also Figure 11.7). The longitudinal dispersion coefficient in natural streams can also be estimated empirically from river characteristics. Fischer *et al.* (1979) suggested the following empirical equation:

$$D_x = 0.011 \frac{u_x^2 \ B^2}{H \ u_*} \tag{11.33}$$

where D_x = the longitudinal dispersion coefficient in m² s⁻¹ [L² T⁻¹] B = the width of the river channel [L], H = the depth of the river channel [L], and u_* = the shear velocity [L T⁻¹] given by

$$u_* = \sqrt{gHS} = \frac{\sqrt{g}}{C} u_x \tag{11.34}$$

where g = the gravitational acceleration constant (= 9.8 m s⁻²) [L T⁻²], S = the river slope [-], and C = Chezy's coefficient [L⁰·⁵ T⁻¹] (see Equation 11.IVc). The shear velocity u_* can be approximated by:

$$u_* \approx 0.10 \ u_x \tag{11.35}$$

Equation (11.35) is only an approximation and does not explicitly account for 'dead zones' in the river.

The dispersion coefficient in groundwater is the sum of the diffusion coefficient and the mechanical dispersion coefficient and is generally linearly related to the linear groundwater flow velocity if mixing is dominated by mechanical dispersion:

$$D = \alpha \cdot u \tag{11.36}$$

where D = the dispersion coefficient [L² T⁻¹], α = the dispersivity of the aquifer [L], and u is the groundwater flow velocity [L T⁻¹]. However, at very low groundwater flow velocities, molecular diffusion may contribute significantly to mixing. The relative importance of mass transport by advection compared to dispersion over a certain stretch can be estimated using the **Peclet number**:

$$Pe = \frac{u_x \Delta x}{D_x} \tag{11.37}$$

where Pe = Peclet number [-] and Δx = some characteristic segment length [L]. The Peclet number expresses the ratio of transport by bulk fluid motion or advection to the mass transport by dispersion. If the Peclet number is much larger than 1, advection dominates the mass transfer, and if it is much smaller than 1, dispersion predominates.

Longitudinal mechanical dispersion also occurs in soil, primarily as a consequence of bioturbation. Due to the reworking of the soil particles by burrowing earthworms, arthropoda, or macrofauna (e.g. moles, voles) contaminated soil is displaced and diluted with deeper uncontaminated soil. If an uncontaminated soil is contaminated by deposition

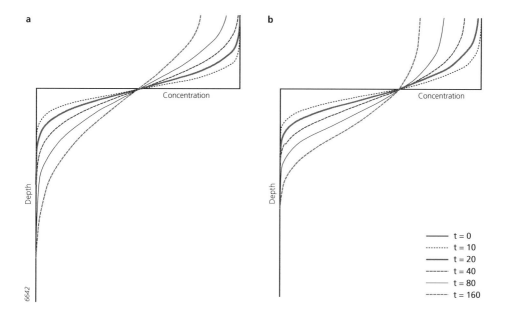

Figure 11.9 Vertical dispersion in soil with a) a constant dispersion coefficient over the vertical soil profile and b) a decreasing dispersion coefficient with soil depth.

of a layer of contaminated sediment, for example after flooding, the initial sharp boundary between the uncontaminated soil and the fresh contaminated sediment becomes blurred in the course of time, as shown in Figure 11.9a. Note that in this example advective transport does not take place and the dispersion coefficient is constant throughout the soil profile. Just as in the groundwater example, the depth profiles of the contaminant concentrations follow a Gaussian cumulative distribution function. Under natural conditions, however, the biological activity responsible for the dispersion process tends to diminish with depth (Middelkoop 1997). Accordingly, the mixing decreases with depth and results in the contamination moving more slowly deeper in the soil profile (Figure 11.9b).

11.3.4 Transverse dispersion

Transverse dispersion is the process of mixing perpendicular to the average flow direction. When mass spreads in two or three dimensions, the distribution of mass sampled perpendicular to the direction of flow is also normally distributed with a standard deviation that increases proportionally to $\sqrt{2t}$ (compare Equation 11.27). The two-dimensional dispersal of a substance in a unidirectional flow field therefore leads to an elliptically shaped distribution of the concentration. Typical transverse dispersion coefficients are smaller than longitudinal dispersion coefficients. In unconfined or semi-confined groundwater systems, transverse dispersion results in a reduction of concentration everywhere beyond the advective front, while longitudinal dispersion only does so at the front of the plume (Figure 11.10). In general, in groundwater dispersion does not occur through the confining layers and in rivers it does not occur through the river banks. In rivers, the transverse dispersion results in lateral mixing when a substance has entered the river from the river bank.

Similar to the estimation of the longitudinal dispersion coefficient, the transverse dispersion coefficient can be estimated by fitting a Gaussian curve on a graph of concentration versus distance perpendicular to the main flow direction. Furthermore, based

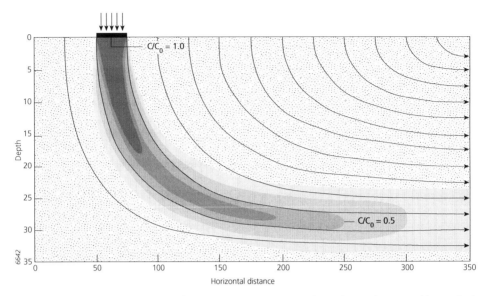

Figure 11.10 Downstream propagation of a continuous input into groundwater due to advection and longitudinal and transverse dispersion.

on a wealth of laboratory and field experiments reported in Fischer *et al.* (1979), the average transverse turbulent diffusion coefficient in a uniform river channel can be estimated by:

$$D_y = 0.15\ H\ u_*$$ (11.38)

where H is the river depth [L] and u_* is the shear velocity (= $\sqrt{gHS}$; see Equation 11.34) [L T^{-1}]. The experiments indicate that the width of the river channel plays some role in transverse mixing. However, it is unclear how that effect should be incorporated (Fischer *et al.*, 1979). In natural streams, transverse mixing deviates from the behaviour in Equation (11.20) primarily due to large helicoidal motions, which are not properties of the turbulent diffusion. Moreover, the cross-section is rarely of uniform depth and the river slope varies. These effects enhance transverse mixing, and for natural streams the transverse dispersion coefficient can be estimated by:

$$D_y = 0.6\ H\ u_*$$ (11.39)

If the stream is slowly meandering and the irregularities in the river banks are moderate, the empirical coefficient of 0.6 in Equation (11.39) usually varies in the range from 0.4 to 0.8 (Fischer *et al.*, 1979).

The mixing length, i.e. the distance over which the substance is considered to be fully mixed over the river cross-section, can be estimated by employing the advection–dispersion equation and its associated Gaussian distribution solution. The order of magnitude of the distance from a single point source to the zone of complete transverse mixing can be approximated by equating the lateral standard deviation σ of the substance's transverse concentration distribution to the width of the river:

$$\sigma = \sqrt{2D_y \Delta t} \approx B$$ (11.40)

where D_y = the longitudinal dispersion coefficient [L^2 T^{-1}] and B = the width of the river channel [L], and Δt is the travel time between the start of the release of the substance and complete mixing in the river [T] given by L_m/u_x, where L_m = mixing length [L] and u_x = average flow velocity [L T^{-1}]. So, the mixing length can be estimated by:

$$L_m \approx \frac{B^2 u_x}{2D_y} \qquad (11.41)$$

Example 11.6 Transverse dispersion

A stream has a depth of 0.6 m, a width of 5.0 m, and a flow velocity of 12 cm s^{-1}. Estimate the mixing length for a single source introduced to the stream from the stream bank.

Solution
First, estimate the shear velocity using Equation (11.35):

$$u_* \approx 0.10 \, u_x = 0.10 \times 0.12 = 0.012 \text{ m s}^{-1}$$

Second, estimate the transverse dispersion coefficient for a natural stream, using Equation (11.39):

$$D_y = 0.6 \, H \, u_* = 0.6 \times 0.6 \times 0.012 = 4.32 \cdot 10^{-3} \text{ m}^2 \text{ s}^{-1}$$

Finally, estimate the mixing length using Equation (11.41):

$$L_m \approx \frac{B^2 u_x}{2D_y} = \frac{5^2 \times 0.12}{2 \times 4.32 \cdot 10^{-3}} = 347 \text{ m}$$

The stream parameters in this example are the same as in Example 11.3. Although in that example we assumed instantaneous mixing, in reality it takes at least 347 m for the effluent to be mixed across the stream width. In order to describe the transport of contaminants between the point source input and L_m, where the contamination cloud is three-dimensional, the transport equation cannot be simplified. For distances beyond L_m, a one-dimensional longitudinal dispersion model is sufficient.

11.3.5 Numerical dispersion

If we solve the advection equation (Equation 11.14) in two dimensions using a numerical technique, we usually discretise space by defining fixed points in a so-called Cartesian coordinate system. Such methods using a fixed grid are known as Eulerian methods (like the numerical integration method named after the mathematician Leonhard Euler (1707–1783); see Box 11.II). In this case, apparent dispersion may also occur as an artefact of the numerical solution technique. Such dispersion is called numerical dispersion and does not occur in reality. This mixing is a result of calculating average concentrations in discrete spatial units (grid cells or model elements) and time-steps. For example, if we use a two-dimensional, dynamic raster GIS we may calculate the average concentration in a square grid cell (x,y) at a given time-step t as function of the concentration and water flow velocity in the grid cell itself and its adjacent cells at time-step $t = t$ -1. If we consider a simple situation of advective movement of a solute with a steady state water flow in x direction which is constant over space and time, the concentration in cell (x,y) is given by:

$$C(x,y,t) = \frac{1}{V(x,y,t)} \cdot (C(x,y,t-1) \cdot V(x,y,t-1)$$

$$+ C(x-1,y,t-1) \cdot u_x(x-1,y,t-1) \cdot H(x-1,y,t-1) \cdot \Delta y \cdot \Delta t$$

$$- C(x,y,t-1) \cdot u_x(x,y,t-1) \cdot H(x,y,t-1) \cdot \Delta y \cdot \Delta t) \tag{11.42}$$

where C = concentration [M L^{-3}], V = volume [L^3], u_x = flow velocity [L T^{-1}], H = water depth [L], Δy = flow width, i.e. grid cell size [L], and Δt = time-step size [T]. The first term between brackets represents the mass present in the grid cell in the previous time step and the second and third terms represent respectively the mass input from the upstream cell and the mass outflow to the downstream cell. Because the flow field (i.e. V, H, and u_x) is constant over space and time and $V = H \cdot \Delta x \cdot \Delta y$, we may simplify Equation (11.42) to:

$$C(x,y,t) = C(x,y,t-1) + \frac{u_x \cdot \Delta t}{\Delta x} \cdot (C(x-1,y,t-1) - C(x,y,t-1)) \tag{11.43}$$

If we assume a block front of a concentration of 1 mg l^{-1} entering the system with an initial concentration of 0 mg l^{-1} on the left-hand side ($x = 0$) at $t = 0$ with a velocity of 0.1 m s^{-1} in direction x, it should take 10 seconds before the front of the concentration reaches 1 m. (Note that we are considering advective transport only.) However, if we discretise the system with a grid cell size of 1 m and a time step of 1 s, Equation (11.43) shows that after 1 s (one time-step), the concentration at x = 1 m has already increased to 0.1 mg l^{-1}. After 150 s, the front should have moved 15 m, but the results of the simple GIS-based model shows that the block front is considerably blurred (Figures 11.11a and 11.11b). In two dimensions, a similar effect can be observed if we consider an instantaneous release of a substance from a local point source in the upper left corner of the system, which is transported in east/south-eastern direction. In this case, the concentration spreads in two directions (Figures 11.11c and 11.11d).

To minimise the undesired effects of numerical dispersion, it is often better to define a volume of space containing a fixed mass of fluid and to let the boundaries of these cells move in response to the dynamics of the fluid. The differential equation is transformed into a form in which the variables are the positions of the boundaries of the cells rather than the quantity or concentration of a substance in each cell. These methods are known as Lagrangian methods or particle tracking methods. A well-known example of such a method is the method of characteristics (MOC) for solute transport in groundwater (Konikow and Bredehoeft, 1978). Thonon *et al.* (2007) have implemented this method in a GIS-based model that simulates sediment transport and deposition over a floodplain.

The first step in the method of characteristics involves placing a swarm of traceable particles or points in each grid cell, resulting in a uniformly distributed set of points over the area of interest. Usually, four or nine particles are used per grid cell. An initial mass is assigned to each particle; it depends on the initial concentration and the water volume in the grid cell. For each time-step, each individual particle is moved over a distance that is determined by the product of the time-step size and the flow velocity at the location of the point, which is derived by interpolating the flow velocities in the grid cell under consideration and its adjacent cells (Figure 11.12). After all particles have been moved, the concentration in each grid cell is computed as the total mass of all particles located within the grid cell, divided by the water volume. This concentration represents the average concentration as a result of advective transport. Subsequently, using a Eulerian method, the method of characteristics solves the transport due to dispersion. Alternatively, dispersion can be simulated using a Lagrangian method. The random walk method, for example, approximates advective transport in the same manner as the method of characteristics, after which the particles are moved both parallel and perpendicularly to the flow direction

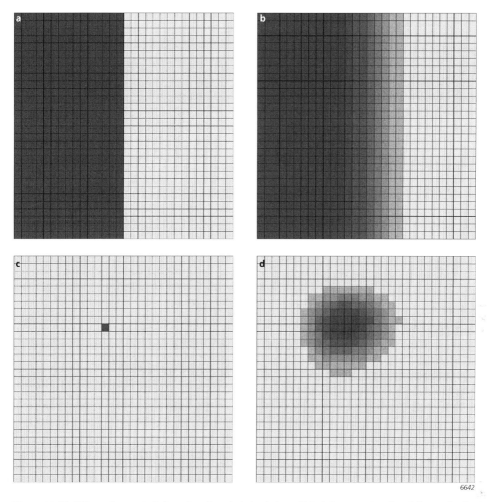

Figure 11.11 Effects of numerical dispersion a) analytical solution of block front moving from left to right due to advective transport; b) numerical solution; c) analytical solution of a pulse injection moving from the upper left corner in south-east direction; d) numerical solution.

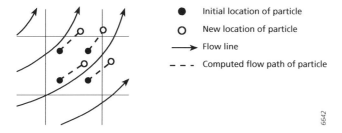

Figure 11.12 Method of characteristics (MOC) (Konikow and Bredehoeft, 1978).

according to a bivariate Gaussian distribution with a mean of zero and standard deviations of $\sqrt{2D_x\Delta t}$ and $\sqrt{2D_y\Delta t}$ for longitudinal and transverse dispersion, respectively (see above).

11.4 MULTI-FLUID FLOW

In some contamination problems two or more fluids are involved. Examples are groundwater and organic liquids (dense non-aqueous phase liquids (DNAPLs) and light non-aqueous phase liquids (LNAPLs)) in the saturated zone, or air, water, and organic liquids in the unsaturated zone. Problems involving transport of organic liquids are more complex than transport of dissolved chemicals. The most important features of multi-fluid flow include the differences in densities of the fluids, their limited miscibility, and the interference with one another during flow. If the organic liquid is less dense than water (LNAPLs), the liquid tends to float on the water; if it is denser than water (DNAPLs), the liquid tends to sink. Figure 11.13 illustrates the behaviour of DNAPLs and LNAPLs in the unsaturated and saturated zones.

When two immiscible fluids are in contact with a solid surface, one fluid will preferentially spread over the solid surface at the expense of the other fluid. The fluid that is most spread over the solid surface is called the 'wetting fluid', whereas the other fluid the 'non-wetting fluid'. The relative degree to which a fluid will spread on or coat the solid surface is called the 'wettability' of the fluid. The wettability depends on the properties of the fluids and the solid surface and reveals itself in the contact angle, i.e. the angle subtended by the liquid–liquid interface and the solid surface (Figure 11.13). A fluid that produces a contact angle of less than 90° is wetting relative to water and a fluid that produces a contact angle of more than 90° is non-wetting relative to water. Fluids with contact angles ranging between 70° and 110° are considered neutrally wetting (Anderson, 1986). In general, NAPLs

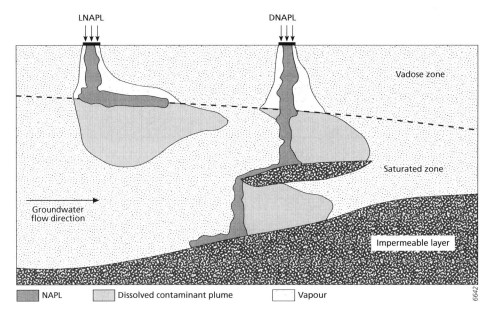

Figure 11.13 Behaviour of LNAPLs and DNAPLs in the subsurface environment.

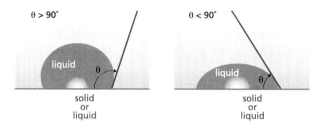

Figure 11.14 Contact angle between a liquid and a solid or another immiscible liquid. If the contact angle is > 110°, the liquid has a low wettability (left); if the contact angle is < 70°, the liquid has a high wettability (right).

produce a contact angle of more than 110° and are thus non-wetting relative to water. This brings about that in a soil or sediment that contains both water and a NAPL, water spreads preferentially across solid surfaces and occupies the smaller pore spaces, while the NAPLs are restricted to the larger pores. A few rules of the thumb to determine whether a fluid is wetting or non-wetting are given in Table 11.3

The interference of the fluids during flow results in a reduction of the hydraulic conductivity K_s (see Box 11.III). In general, the hydraulic conductivity is dependent on both the properties of the fluid (water) and the properties of the media (aquifer materials):

$$K_s = \frac{k\rho g}{\mu} \tag{11.44}$$

where K_s = the saturated hydraulic conductivity [L T^{-1}], k = the intrinsic permeability [L^2], ρ = fluid density [M L^{-3}], g = gravitational acceleration constant [L T^{-2}], and μ = dynamic viscosity [M L^{-1} T^{-1}]. The properties of the medium are embedded in the intrinsic permeability k, and the properties of the fluid are contained in the density and viscosity parameters. The flow rate of a fluid depends not only on the hydraulic conductivity, but also – according to Darcy's law (see Box 11.III, Equation 11.IIIh) – on the gradient in hydraulic head. The hydraulic head consists of three components: the pressure head, the elevation head, and the velocity head. Because flow velocity in groundwater is slow, the velocity head is ignored, so the hydraulic head can be expressed as:

$$h = \frac{P}{\rho g} + z \tag{11.45}$$

where h = hydraulic head [L], P = pressure [M L^{-1} T^{-2}], and z = the elevation [L]. Combining Equation (11.43), Equation (11.44), and Darcy's law (Equation 11.IIIh) for one-dimensional flow of fluid i in a homogeneous medium gives:

$$q_i = \frac{k_i}{\mu_i}\left(\nabla P_i + \rho_i g \nabla z\right) \tag{11.46}$$

where q_i = the flow of the ith fluid per unit area [L T^{-1}], k_i = the effective permeability of the medium to the ith fluid [L^2], μ_i = dynamic viscosity of the ith fluid [M L^{-1} T^{-1}],

Table 11.3 Wetting and non-wetting fluids in different mixtures of air, water, and NAPLs.

Mixture	Wetting fluid	Non-wetting fluid
Air–water	water	air
Air–NAPL	NAPL	air
Water–NAPL	water	NAPL

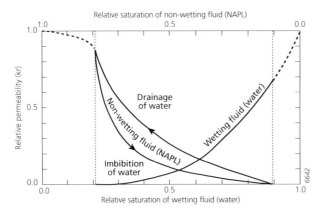

Figure 11.15 Typical relative permeability curves.

∇P_i = gradient in pressure head [M L^{-2} T^{-2}], ρ = density of the *i*th fluid [M L^{-3}], and ∇z = gradient of the elevation [-]. The relative permeability to the *i*th fluid is:

$$k_{r,i} \quad = \quad \frac{k_i}{k} \tag{11.47}$$

where $k_{r,i}$ = the relative permeability of the *i*th fluid [-], which ranges between zero and one, and k = the intrinsic permeability [L^2]. The relative permeability k_{ri} depends on the relative saturation, the wettability of the fluid, and whether the fluid is imbibed into or drains from the aquifer material. Figure 11.15 shows the effect of the relative saturation of the fluids – in this case an NAPL and water. The relative permeability is equal to one if the porous medium is fully saturated with one fluid. Note that the relative permeabilities rarely sum to one. This implies that in the case of multi-fluid flow, the permeability for a fluid depends not only on the properties of the porous aquifer material, but also on the properties of the other one or more fluids present in the aquifer. In other words, the medium through which the fluid flows is comprised of both the aquifer material and the other fluids. If both an NAPL and water are present in an aquifer, some quantity of either the NAPL or water is not capable of flowing below a given saturation threshold, since the fluid is not connected across the pore network below this saturation threshold. This threshold is called a residual saturation, and the residual saturation of water is typically greater than for the NAPL. At residual saturation of water, the water is held in the narrowest parts of the pores (pendular saturation) and at residual saturation of the NAPL, the NAPL is held as an isolated blob in the centre of the pore (insular saturation) (see Figure 11.16). Another feature of the relative permeability curves depicted in Figure 11.14 is the hysteretic character of the curve for the NAPL. For the same saturation, the relative permeability for the NAPL when water is displaced by the NAPL (drainage) is greater than when the NAPL is displaced by water (imbibition). This is because during drainage the NAPL flows mainly through the larger pores that contribute most to the permeability. In contrast, during imbibition, water flows through the larger pores, whilst the NAPL occupies the smaller pores.

Example 11.7 Effective hydraulic conductivity

An underground oil storage tank leak has contaminated the shallow, unconfined groundwater and the oil is floating on top of the water table. After the leak has been stopped, the oil is contained and recovered by pumping the groundwater. A deeper pump

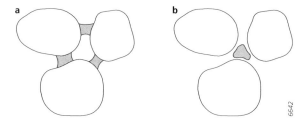

Figure 11.16 Examples of fluid saturation: a) pendular saturation of water; b) insular saturation of the NAPL. Adapted from Domenico and Schwarz (1998).

lowers the water table and produces a hydraulic gradient. A second pump collects the floating oil. After some time, the relative saturation of water in the oil-contaminated zone has increased to 50 percent. At this relative saturation, the relative permeability is 0.10 for water and 0.12 for oil. Calculate the effective hydraulic conductivities of the aquifer material for water and oil, given that the hydraulic conductivity for water is 3 m d^{-1}. The dynamic viscosity and density for oil and water are given in the table below.

	Dynamic viscosity (Pa s)	Specific density (kg m^{-3})
Water	$1 \cdot 10^{-3}$	1000
Oil	0.07	737

Solution
First, use Equation (11.44) to calculate the intrinsic permeability of the aquifer material that is independent of the fluid:

$$K_{s,water} = \frac{k\rho g}{\mu} = 3 \text{ m d}^{-1} \implies \frac{k \times 1000 \times 9.8}{1 \cdot 10^{-3}} = 3$$

$$k = \frac{0.003}{9800} = 3.06 \cdot 10^{-7} \text{ m}^2$$

Then, calculate the hydraulic conductivity for oil:

$$K_{s,oil} = \frac{3.06 \cdot 10^{-7} \times 737 \times 9.8}{0.07} = 0.032 \text{ m d}^{-1}$$

The effective hydraulic conductivities for water and oil (corrected for the presence of the other liquid) are

$$K_{eff,water} = 0.1 \times 3 = 0.3 \text{ m d}^{-1}$$

$$K_{eff,oil} = 0.12 \times 0.032 = 0.0038 \text{ m d}^{-1}$$

Note that by comparison with water, the effective conductivity of oil is considerably lower, primarily because oil is more viscous. Oil is less dense than water, which lowers the conductivity, but not to the degree that the viscosity decreases it. The effect also prevails in the flow rates of both fluids (see Equation 11.46). Given the hydraulic gradient induced by the pumping, the water will flow at a much faster rate than oil, because it is less viscous.

Example 11.8 Residual concentration

A spill of one barrel (= 159 l) of a DNAPL spreads slowly in a shallow aquifer with a porosity of 28 percent. The residual concentration of the DNAPL in the aquifer material is 18 percent. Calculate the maximum aquifer volume in m^3 that will be contaminated by the DNAPL.

Solution
The proportion of the maximum contaminated aquifer volume that will be occupied by the DNAPL equals $0.18 \times 0.28 = 0.0505$. Thus, the maximum contaminated aquifer volume is $0.159 \text{ m}^3/0.0505 = 3.15 \text{ m}^3$.

EXERCISES

1. In a small river with an average width of 3 m and an average depth of 0.75 m the stream flow velocity is 6.5 cm s^{-1} and the phosphate (PO_4^{3-}) concentration is 0.23 mg l^{-1}. Calculate the inorganic phosphorus (P) load of this river in tonnes y^{-1}.

2. Explain why numerical environmental models are often preferable to analytical solutions.

3. The discharge of a river with a cross-sectional area of 10 m^2 is 2 m^3 s^{-1}. At time t = 0 a dye tracer is released into the river. Calculate the time needed for the tracer to reach a measurement location 500 m downstream from the release point, assuming that the tracer is subject only to advection.

4. A river has a discharge of 16 m^3 s^{-1} and a sodium concentration of 44.5 mg l^{-1}. A tributary discharges into this river at a rate of 3 m^3 s^{-1}. The sodium concentration in the tributary is 16 mg l^{-1}. Calculate the sodium concentration in the river downstream from the inflow of the tributary.

5. A flow reactor with a cross-sectional area of 0.8 m^2 is filled with sandy sediment with a dispersivity α of 0.01 m. The sediment is fully saturated with water with a Cl concentration of 0 mg l^{-1}. At time t = 0 a continuous inflow of water with a Cl concentration of 250 mg l^{-1} is started with a flow velocity of 1 m d^{-1}.
   a. Calculate the dispersion coefficient.
   b. Does this coefficient refer to longitudinal or to transverse dispersion? Give reasons for your answer.
   c. Calculate the Cl concentration at x = 1.0 m after 18 hours, using Equation (11.32).
   The experiment is repeated with a water flow velocity of 0.5 m d^{-1}.
   d. Calculate the Cl concentration at x = 1.0 m after 36 hours.
   e. Compare the answers of questions c) and d) and explain the outcomes.
   f. Explain why dispersion is linearly related to the groundwater flow velocity (see also Figure 11.4).

6. In the same flow reactor as in question 5, in which the flow velocity was set at 0.75 m d^{-1}, a mass of 1 g of an inert tracer is released instantaneously.
   a. Sketch the concentration field in the flow reactor as function of distance after 1 day.
   b. Calculate the maximum concentration after 1 day.
   c. Sketch the concentration evolution as function of time at x = 1.0 m
   d. Calculate the maximum concentration at x = 1.0 m.

   e. Calculate the Peclet number for the situation in question d. What conclusion can you draw about the contribution of dispersion to overall transport?

Molecular diffusion was not considered in this calculation.

   f. Describe qualitatively the effect of molecular diffusion on the above calculations (hint: see Table 11.1)

7. Estimate the mixing length in a river with the following characteristics: width = 4 m, water depth = 1 m, bed slope = 0.0001, and average flow velocity = 31 cm s^{-1}.

8. Describe the principles of the method of characteristics to minimise numerical dispersion.

9. After a spill of 50 litres of a NAPL into a water-saturated soil with a porosity of 30 percent, the maximum relative saturation of the NAPL in the soil reaches 60 percent. From Figure 11.14 determine:

   a. The relative permeability of the soil with respect to water.

   b. The relative permeability of the soil with respect to the NAPL.

The NAPL is slowly replaced by water (see the imbibition curve in Figure 11.14).

   c. From Figure 11.14 determine the residual saturation of the NAPL.

   d. Explain why the NAPL will not be completely replaced by water in the contaminated soil and a NAPL residue will remain.

   e. Calculate the volume of soil contaminated by the NAPL at residual saturation.

10. A landfill is situated just above the water table on a permeable, sandy subsoil. An aquitard is present at about 40 m depth and dips to the west at 3 percent. Groundwater in the unconfined aquifer flows east at a velocity of 10 m y^{-1}.

   a. Sketch the chloride concentrations in the leachate in a west–east cross-section of the unconfined aquifer 10 years after construction of the landfill.

Leaking containers filled with tetrachloromethane (specific density = 1.59 g cm^{-3}) and toluene (specific density = 0.87 g cm^{-3}) are present in the landfill.

   b. In the same figure, sketch the plumes of tetrachloromethane and toluene.

   c. Based on your knowledge of organic pollutants (see Chapter 9) and the behaviour of NAPLs in the subsurface, discuss the fate of the tetrachloromethane and toluene in the long term.

Sediment transport and deposition

12.1 INTRODUCTION

Suspended matter plays an important role in surface water quality. The amount of suspended matter determines the turbidity of the water, and so the underwater light climate. Moreover, suspended fine sediments are an important vector for contaminant transport in surface waters, since many chemicals adsorb to the negatively charged surfaces of clay minerals and organic matter (see Chapter 4). The sediment may already be contaminated as a result of accidental spills, atmospheric deposition, or the application of fertiliser or pesticides on soils susceptible to erosion. Soil erosion and deposition (sedimentation) may accordingly result in an ongoing transfer of contaminants from terrestrial ecosystems to surface water, bed sediments, and floodplains. Moreover, during transport the sediment may be contaminated by discharges into surface water and the subsequent exchange between the soluble and particulate phase. To be able to predict the fate of sediment-associated chemicals, it is therefore not only necessary to understand the adsorption reactions between sediment particles and the soluble phase, but also to understand the fate of the sediment itself.

The processes of erosion and sedimentation come about particularly in two different environments, namely on hill slopes and in surface water bodies. The basic principles of sediment transport are essentially the same in both environments, but the processes of sediment detachment or erosion may differ. On slopes, a considerable part of the sediment detachment is due to the impact of raindrops falling on the soil surface. As soon as overland flow occurs, the detached sediment can be transported downhill; this water flowing over the ground surface cushions the impact of the raindrops. The runoff water itself can bring about sediment detachment, due to the flow shear stresses occurring at the soil–water interface. In river channels, this is the main process of sediment detachment, though the collapse of river banks and other forms of mass movement (e.g. debris flows) can supply significant inputs of sediment into a river channel. In lakes, the sediment detachment occurs mainly under the influence of bottom shear stresses induced by waves. Sediment detachment of bed sediments in rivers and lakes is also often referred to as resuspension. If the shear stresses are sufficiently small, the sediment settles, i.e. is deposited. The various processes involved are described in detail in the following sections, considering single rainfall or flood events (time scales from minutes to days). Subsequently, models for long-term erosion and deposition (time scales of years to decades) are discussed.

12.2 TRANSPORT EQUATION

For modelling sediment transport in the short term (single hydrological events), the general one-dimensional advection–dispersion equation for surface water, including the erosion and sedimentation of suspended solids, reads:

$$\frac{\partial C}{\partial t} = -\bar{u}_x \frac{\partial C}{\partial x} + D_x \frac{\partial^2 C}{\partial x^2} - \frac{J_s}{H}$$ (12.1)

where J_s = net sediment flux from the water to the bottom [M L^{-2} T^{-1}], and H = water depth [L]. As mentioned above, the erosion and sedimentation rate are governed by the shear stress at the soil–water interface. If this bottom shear stress τ_b exceeds a certain critical value $\tau_{b,e}$, then net erosion occurs. If the shear stress drops below another critical value $\tau_{b,d}$, then net sedimentation occurs. In general, there are four situations:
1. erosion ($\tau_b > \tau_{b,e}$; $J_s < 0$);
2. neither erosion nor sedimentation ($\tau_{b,d} \leq \tau_b \leq \tau_{b,e}$; $J_s = 0$);
3. hindered sedimentation ($\tau_b < \tau_{b,d}$; $J_s > 0$);
4. free sedimentation ($\tau_b = 0$; $J_s > 0$).

For cohesive sediments, i.e. fine sediments or a mixture of fine and coarse sediments in which the particles adhere to each other, the critical bottom shear stress for erosion $\tau_{b,e}$ is larger than the critical shear stress for deposition $\tau_{b,d}$. This implies that deposition and erosion of cohesive sediments do not occur simultaneously (Parchure and Mehta, 1985). For non-cohesive sediments, the critical bottom shear stress for erosion $\tau_{b,e}$ may be equal to or less than the critical shear stress for deposition $\tau_{b,d}$. In this case, deposition and erosion may occur simultaneously and situation 2 (neither erosion nor sedimentation) does not occur.

12.3 BOTTOM SHEAR STRESS

In turbulent flowing waters the bottom shear stress is given by:

$$\tau_b = \rho_w \, g \, H \, S = \rho_w \, u_*^2$$ (12.2)

where τ_b = the bottom shear stress [M L^{-1} T^{-2}], ρ_w = the density of water [M L^{-3}], g = the gravitational acceleration constant (= 9.8 m s^{-2}) [L T^{-2}], S = the river slope [-], H = the depth of the river channel [L], and u_* = the shear velocity [L T^{-1}] (see also Equation 11.34).

Example 12.1 Critical shear stresses for erosion and deposition in rivers

A river has a channel 10 m wide that is uniformly rectangular in cross-section. The river has the following stage–discharge rating curve that gives the relationship between channel depth and river discharge:
$Q = 3.0 \times H^{1.6}$

where Q = the discharge (m^3 s^{-1}) and H = the water depth (m). The critical shear stress for deposition = 0.45 N m^{-2} and the critical shear stress for erosion = 0.95 N m^{-2}. Calculate the discharge below which deposition occurs and the discharge above which erosion occurs.

Solution
The bottom shear stress τ_b is given by Equation (12.2):
$$\tau_b = \rho_w \, u_*^2 = 1000 \, u_*^2$$

The shear velocity u_* can be estimated using Equation (11.35):
$$u_* = 0.10 \, u_x$$

Hence,

$$\tau_b = 10\, u_x$$

$$u_x = \left(\frac{\tau_b}{10}\right)^{0.5}$$

The flow velocity is also given by

$$u_x = \frac{Q}{A} = \frac{Q}{10\,H}$$

The stage–discharge relationship can be rewritten in

$$H = \left(\frac{Q}{3.0}\right)^{\frac{1}{1.6}} = 0.503\,Q^{0.625}$$

Combining the three latter equations gives

$$u_x = \frac{Q}{5.03\,Q^{0.625}} = \left(\frac{\tau_b}{10}\right)^{0.5}$$

$$Q^{0.375} = 5.03 \cdot \left(\frac{\tau_b}{10}\right)^{0.5}$$

$$Q = 74.28 \cdot \left(\frac{\tau_b}{10}\right)^{1.333}$$

Thus, the discharge at which the bottom shear stress equals the critical shear stress for deposition is

$$Q = 74.28 \cdot \left(\frac{0.45}{10}\right)^{1.333} = 1.2 \text{ m}^3\text{ s}^{-1}$$

and the discharge at which the bottom shear stress equals the critical shear stress for erosion is

$$Q = 74.28 \cdot \left(\frac{0.95}{10}\right)^{1.333} = 3.2 \text{ m}^3\text{ s}^{-1}$$

In lakes it is more difficult to determine the shear stress due to wind-induced waves at the interface between bed sediment and water, because of the complex three-dimensional water movement. Nevertheless, there are relatively simple relationships for deriving the water flow velocity at the lake bottom. These relationships can then be used to derive the shear stress. Water moves in an orbital motion (circular path) as a wave passes by; if the water is not too deep, this orbital motion diminishes to a to-and-fro motion over the lake bottom. The maximum water velocity during this motion is thus a function of the water depth and the wave height, length and period:

$$u_{b,\max} = \frac{\pi H_w}{T_w} \frac{1}{\sinh\left(2\pi H / L_w\right)} \tag{12.3}$$

where $u_{b,\max}$ = the maximum velocity at the bottom [L T^{-1}], H_w = the wave height [L], T_w = the wave period [T], sinh = the hyperbolic sine function, which is defined as

$\sinh(x) = \dfrac{e^x - e^{-x}}{2}$, H = water depth [L], and L_w = wave length [L].

The wave characteristics can be estimated using the following empirical equations (Lijklema, 1991):

$$T_w = 7.54 \, \frac{W}{g} \, k_1 \, \tanh\left(\frac{\gamma_1}{k_1}\right) \tag{12.4}$$

where tanh = the hyperbolic tangent function, defined as $\tanh(x) = \dfrac{e^x - e^{-x}}{e^x + e^{-x}}$,

$k_1 = \tanh\left(0.833\left(\dfrac{gH}{W^2}\right)^{0.375}\right)$ and $\gamma_1 = 0.077\left(\dfrac{gF}{W^2}\right)^{0.25}$, H = water depth (m),

W = wind speed (m s^{-1}), and F = wind fetch length (m).

$$H_w = 0.283 \, \frac{W^2}{g} \, k_2 \, \tanh\left(\frac{\gamma_2}{k_2}\right) \tag{12.5}$$

where $k_2 = \tanh\left(0.53\left(\dfrac{gH}{W^2}\right)^{0.75}\right)$, and $\gamma_2 = 0.0125\left(\dfrac{gF}{W^2}\right)^{0.42}$.

$$L_w = \frac{gT_w^2}{2\pi} \qquad\qquad \text{for deep water } (H > 0.5 \, L_w) \tag{12.6}$$

$$L_w = \frac{gT_w^2}{2\pi} \, \tanh\left(\frac{2\pi H}{L_w}\right) \qquad \text{for shallow water} \tag{12.7}$$

Note that Equation (12.7) needs to be solved iteratively, since the term L_w is found on both sides of the equation. The maximum bottom shear stress can subsequently be found using:

$$\tau_b = \rho_w \, C_f \, u_{b,\max}^2 \tag{12.8}$$

where C_f = a friction coefficient [-], that can be estimated using the following empirical relation:

$$C_f = 0.4\left(\frac{K_n}{A_b}\right)^{0.75} \tag{12.9}$$

where K_n = a measure of the bottom roughness (m), and A_b = the amplitude of the wave motion at the bottom (m):

$$A_b = \frac{H_w}{2 \sinh(2\pi H / L_w)} \tag{12.10}$$

The measure of the bottom roughness K_n is defined as the so-called D90 value of the bed sediment particles, i.e. the 90th percentile of the particle size distribution. If the bottom surface roughness is determined by irregularities, such as ripples or aquatic vegetation, the value of K_n is adjusted to an equivalent particle size diameter.

Example 12.2 Bottom shear stress in lakes

Estimate the bottom shear stress τ_b for a shallow lake with the following characteristics: water depth = 90 cm, wind fetch length = 300 m, wind speed = 17 m s^{-1} (Beaufort scale wind force 7), D90 of the bed sediment = 70 μm. Given the critical shear stress for both erosion and deposition = 0.10 N m^{-2}, say whether the sediment is settling or eroding.

Solution
First, calculate the parameters k_1, k_2, γ_1, and γ_2:

$$k_1 = \tanh\left(0.833\left(\frac{9.8\times0.90}{17^2}\right)^{0.375}\right) = \tanh(0.833\times0.0305^{0.375}) = \tanh(0.225) = 0.221$$

$$k_2 = \tanh\left(0.53\left(\frac{9.8\times0.90}{17^2}\right)^{0.75}\right) = \tanh(0.53\times0.0305^{0.75}) = \tanh(0.039) = 0.039$$

$$\gamma_1 = 0.077\left(\frac{9.8\times300}{17^2}\right)^{0.25} = 0.077\times10.17^{0.25} = 0.138$$

$$\gamma_2 = 0.0125\left(\frac{9.8\times300}{17^2}\right)^{0.42} = 0.0125\times10.17^{0.42} = 0.033$$

Second, calculate the wave height H_w and wave period T_w:

$$H_w = 0.283\times\frac{17^2}{9.8}\times0.039\tanh\left(\frac{0.033}{0.039}\right) = 0.325\tanh(0.846) = 0.325\times0.689 = 0.224 \text{ m}$$

$$T_w = 7.54\times\frac{17}{9.8}\times0.221\tanh\left(\frac{0.138}{0.221}\right) = 2.891\tanh(0.624) = 2.891\times0.554 = 1.60 \text{ s}$$

To estimate the wave length L_w, first make an initial estimation by using Equation (12.6):

$$L_w = \frac{9.8\times1.60^2}{2\pi} = 3.99 \text{ m}$$

In this case, the water depth H is less than the half of the wave length L_w, so Equation (12.7) should be used to estimate the wave length. Because L_w appears on both sides of the equation, it should be solved iteratively. For the first estimation, the above initial estimation is used. The iteration is continued until L_w no longer changes:

1st iteration: $L_w = 3.99\tanh\left(\frac{2\pi\times0.90}{3.99}\right) = 3.99\tanh(1.417) = 3.99\times0.889 = 3.55 \text{ m}$

2nd iteration: $L_w = 3.99\tanh\left(\frac{2\pi\times0.90}{3.55}\right) = 3.99\tanh(1.593) = 3.99\times0.921 = 3.67 \text{ m}$

3rd iteration: $L_w = 3.99\tanh\left(\frac{2\pi\times0.90}{3.67}\right) = 3.99\tanh(1.541) = 3.99\times0.912 = 3.64 \text{ m}$

4th iteration: $L_w = 3.99 \tanh\left(\dfrac{2\pi \times 0.90}{3.64}\right) = 3.99 \tanh(1.554) = 3.99 \times 0.914 = 3.65$

5th iteration: $L_w = 3.99 \tanh\left(\dfrac{2\pi \times 0.90}{3.65}\right) = 3.99 \tanh(1.549) = 3.99 \times 0.914 = 3.65$ m

After the 5th iteration with L_w = 3.65 m, we have achieved convergence. Now, the amplitude of the wave motion at the lake bottom A_b is calculated using Equation (12.10):

$$A_b = \frac{0.224}{2 \sinh(2\pi \times 0.90 / 3.65)} = \frac{0.224}{2 \sinh(1.549)} = \frac{0.224}{2 \times 2.247} = 0.050 \text{ m}$$

The maximum water velocity at the lake bottom due to wave motion is then estimated using Equation (12.3):

$$u_{b,\max} = \frac{\pi \times 0.224}{1.60} \frac{1}{\sinh(2\pi \times 0.90 / 3.65)} = 0.440 \times \frac{1}{2.247} = 0.196 \text{ m s}^{-1}$$

The friction coefficient is calculated using Equation (12.9) (note that K_n = D90 = 70 µm = 70·10^{-6} m):

$$C_f = 0.4 \left(\frac{70 \cdot 10^{-6}}{0.050}\right)^{0.75} = 0.4 \times 0.0014^{0.75} = 0.4 \times 7.2 \cdot 10^{-3} = 2.9 \cdot 10^{-3}$$

Finally, calculate the bottom shear stress τ_b using Equation (12.2):

$$\tau_b = 1000 \times 2.9 \cdot 10^{-3} \times 0.196^2 = 0.111 \text{ N m}^{-2}$$

As the bottom shear stress is slightly larger than the critical shear stress for erosion and deposition, which is 0.10 N m^{-2}, some of the bottom sediment is resuspended into the water column.

12.4 SEDIMENT DEPOSITION

If the shear stress at the soil–water interface is less then the critical value $\tau_{b,d}$, the sediment in the water column settles and the net sediment flux J_s (see Equation 12.1) becomes a deposition flux, which can be formulated as:

$$J_d = \alpha \, w_s \, C_b \tag{12.11}$$

where J_d = sedimentation flux [M L^{-2} T^{-1}], α = a delay factor [-], w_s = the maximum settling velocity of the suspended sediment when τ_b = 0 [L T^{-1}], and C_b = the suspended sediment concentration just above the soil–water interface [M L^{-3}]. If the variation in suspended sediment concentration over the vertical profile is small (this holds for turbulent streams), the vertically averaged concentration (C) can be adopted for C_b. Note that in this case the removal of sediment from the water column follows first-order kinetics (combine Equation 12.11 (C_b = C) with the transport Equation 12.1). If only sedimentation is considered, then the differential equation becomes:

$$\frac{\partial C}{\partial t} = -\frac{J_s}{H} = -\frac{\alpha \, w_s}{H} C \tag{12.12}$$

The analytical solution for Equation (12.12) reads:

$$C(t) = C_0\, e^{-\frac{\alpha\, w_s}{H} t} \qquad (12.13)$$

where C_0 = initial suspended sediment concentration at t = 0.

If no sedimentation occurs ($\tau_b > \tau_{b,e}$), α equals zero. In stagnant water ($\tau_b = 0$), α equals 1. In the case of hindered deposition ($0 \leq \tau_b \leq \tau_{b,d}$), α can be described as function of the bottom shear stress, for example a linear relation:

$$\alpha = 1 - \frac{\tau_b}{\tau_{b,d}} \qquad (12.14)$$

The critical bottom shear stress for sedimentation $\tau_{b,d}$ is in the order of $0.06 - 1.0$ N m^{-2} (Van Rijn, 1989). In some models, α is described as a linear relation with the vertically averaged flow velocity (which is, in turn, proportional to the square root of the bottom shear stress):

$$\alpha = 1 - \frac{u_x}{u_{x,d}} \qquad (12.15)$$

where $u_{x,d}$ = the critical flow velocity for sedimentation.

Example 12.3

A stream 120 cm deep has an average flow velocity of 0.2 m s^{-1}. The average settling velocity of the suspended sediment is 5.8 m d^{-1} and the critical shear stress for deposition is 1.0 N m^{-2}. Calculate the distance needed to reduce the sediment concentration in the stream water by 75 percent if dispersion can be neglected.

Solution
First, estimate the shear stress in the stream, using Equations (11.35) and (12.2):

$$u_* = 0.1 \times 0.2 = 0.02 \ \text{m s}^{-1}$$

$$\tau_b = 1000 \times 0.02^2 = 0.4 \ \text{N m}^{-2}$$

Second, calculate the delay factor α, using Equation (12.14):

$$\alpha = 1 - \frac{0.4}{1.0} = 0.6$$

If the sediment concentration has been reduced by 75 percent of the initial value, the $C(t)$: $C(0)$ ratio equals $1 - 0.75 = 0.25$. Thus, according to Equation (12.13):

$$\frac{C(t)}{C(0)} = e^{-\frac{\alpha\, w_s}{H} t} = 0.25$$

The constant in the exponent is

$$\frac{\alpha\, w_s}{H} = \frac{0.6 \times 5.8}{1.2} = 2.9 \ \text{d}^{-1}$$

Therefore
$e^{-2.9t} = 0.25$
$-2.9t = \ln(0.25) = -1.386$
$t = 0.478\text{d} = 0.478 \ \text{d} \times 86400 \ \text{s d}^{-1} = 41299 \ \text{s}$

During this time, the water has travelled a distance of 5702 s × 0.2 m s^{-1} = 8260 m. Thus, the suspended sediment concentration has diminished by 75 percent over a distance of 8260 m.

If the water flow around the particles is laminar, the settling velocity w_s can be described by Stokes's law. Stokes's law assumes that the upward force according to Archimedes' principle and the shear stresses during settling are in equilibrium with the gravity force. For spherical particles, Stokes's law is:

$$w_s = \frac{1}{18} \frac{(\rho_s - \rho_w) g d^2}{\mu} \tag{12.16}$$

where ρ_s = the density of the solid particles [M L^{-3}], ρ_w = the density of water [M L^{-3}], μ = the dynamic viscosity of water (= 1.14 10^{-3} kg m^{-1} s^{-1}) [M L^{-1} T^{-1}], g = the gravitational acceleration constant (= 9.8 m s^{-2}) [L T^{-2}], and d = the diameter of the particles. As mentioned above, Equation (12.16) is only valid for laminar flow around the particle. The flow regime (laminar or turbulent flow) is described by the Reynolds number. For flow around the particles the Reynolds number is defined as:

$$Re = \frac{w_s \, \rho_w \, d}{\mu} \tag{12.17}$$

where Re = Reynolds number [-]. If the Reynolds number is sufficiently small (< 1) then the flow around the particles is laminar. If the Reynolds number is greater than 1 (in general, this is the case for particles larger than about 100–150 μm) then the settling velocity can be approximated by (Fair *et al.*, 1968):

$$w_s = \sqrt{\frac{4}{3 \, C_d} \frac{(\rho_s - \rho_w) g \, d}{\rho_w}} \tag{12.18}$$

where C_d = the Newton's drag coefficient [-], which is estimated by:

$$C_d = \frac{24}{Re} + \frac{3}{\sqrt{Re}} + 0.34 \tag{12.19}$$

If the Reynolds number exceeds 2000 (in general, this is the case for solid particles larger than about 1.4 mm), then C_d becomes independent of Re. In this case, the settling velocity can be approximated by:

$$w_s = \sqrt{3.3 \, (\rho_s - \rho_w) g \, d} \tag{12.20}$$

Particles that are not spherical have a somewhat slower settling velocity (up to a factor of about 1.4 less).

As can be seen from Equations (12.16), (12.18), and (12.20), the settling velocity depends on both the diameter and the specific density of the particles. Considering the variety in the nature and size of the particles, it is practically impossible to formulate one mass balance for all suspended matter. Usually, suspended matter is subdivided on the basis of the nature and size of the particles, or a mass balance is formulated for the most abundant class of suspended matter. Apart from that, the particle size distribution and densities of the particles in natural waters are difficult to measure directly. An important feature of fine particles is that they tend to coagulate to form flocs. The flocculation process occurs if the thickness of the 'diffuse layer' surrounding the particle, which is governed by the chemical composition of the water, decreases, enabling the particles to approach each other more closely (see Section 4.2.3). The flocculation process is also favoured by the presence of sticky biofilms. Flocs may be formed when particles collide; the frequency of collisions between particles increases with increasing turbulence, but the shear stresses induced by turbulence around the flocs themselves can break up large flocs. Given a constant shear stress, the particle size distribution in the water column will achieve a steady state, in which the creation and destruction of flocs are in equilibrium. Increasing the shear stress increases the rate of flocculation and shortens the time to steady state, but decreases the median floc size. As flocs are very porous, they are considerably less dense than the individual particles. Nevertheless, the effect of the larger particle size prevails and the settling velocity of the flocs is generally faster than the settling velocity of the individual particles. Hence, flocculation results in an enhanced sedimentation.

12.5 SEDIMENT EROSION

If the bottom shear stress at the soil–water interface exceeds the critical shear stress for erosion, a net sediment flux from the soil or bed sediment to the water column occurs. Like the deposition flux, the erosion flux can be formulated as a function of the shear stress:

$$J_e = M \left(\frac{\tau_b}{\tau_{b,e}} - 1 \right) \tag{12.21}$$

where J_e = the erosion flux [M L^{-2} T^{-1}], and M = the erosion flux when $\tau_b = 2\tau_{b,e}$. Just as in the case of sedimentation, the erosion flux can be modelled as a linear function of the flow velocity instead of the bottom shear stress. The critical bottom shear stress for erosion $\tau_{b,e}$ increases with the degree of consolidation of the sediment (which is inversely related to the water content of the sediment), the organic matter content, the clay content, and the activity of soil organisms. The value for $\tau_{b,e}$ is in the order of 0.05–0.5 N m^{-2} (Van Rijn, 1989).

Equation (12.21) is only valid for consolidated sediments in which the critical shear stress for erosion $\tau_{b,e}$ is constant over the sediment profile (Mehta *et al.*, 1989). However, in surface waters an unconsolidated sediment layer with a high water content (> 50 %) is usually present between the consolidated sediment and the water column. Parchure and Mehta (1985) described the erosion or resuspension rate for these soft cohesive sediment layers using the following empirical expression:

$$J_e = \varepsilon_f \, e^{\beta \sqrt{\tau_b - \tau_{b,e}}} \tag{12.22}$$

where ε_f = resuspension flux constant [M L^{-2} T^{-1}], and β = empirical constant [L$^{0.5}$ T M$^{-0.5}$]. The critical shear stress for erosion usually increases with depth and if $\tau_{b,e}$ becomes larger, the resuspension process can be described using Equation (12.21). Mehta *et al.* (1989) reported values of 8.3 m N$^{-0.5}$ and 13.6 m N$^{-0.5}$ for β for lake sediments. The accompanying values for ε_f range from 0.07·10^{-5} kg m^{-2} s^{-1} to 0.53·10^{-5} kg m^{-2} s^{-1}, respectively.

On slopes, erosion also occurs due to the impact of falling raindrops; this is also referred to as ***splash detachment***. The erosion flux as a result of splash detachment is a function of the kinetic energy of the rainfall, the depth of the surface water layer, and the stability of soil aggregates present at the soil surface. The kinetic energy can arise from both direct rainfall and drainage from leaves (throughfall). The LISEM model (LImburg Soil Erosion Model) (De Roo *et al.*, 1996; Jetten, 2013) uses the following equation to estimate splash detachment (Jetten, 2013):

$$E_s = \left(\frac{2.82}{A_s} K_e \, e^{-1.48H} + 2.96 \right) P \tag{12.23}$$

where E_s = the splash erosion (g s^{-1} m^{-2}), A_s = the aggregate stability (median number of drops required to decrease the aggregate by 50 percent), K_e = rainfall or throughfall kinetic energy (J m^{-2}), H = the depth of the water layer on the soil surface (mm), P = the amount of rainfall (mm). The kinetic energy of free rainfall and leaf drainage from the plant canopy can be estimated from:

$$K_{e,r} \quad = \quad 8.95 + 8.44 \cdot \log(I)$$

$$K_{e,l} \quad = \quad 15.8 \cdot \sqrt{h} - 5.87$$

(12.24)

where $K_{e,r}$ = rainfall kinetic energy (J m^{-2}), $K_{e,l}$ = kinetic energy of leaf drainage (J m^{-2}), I = the rainfall intensity (mm h^{-1}), and h = the height of the plants (m).

12.6 LONG-TERM SOIL EROSION AND DEPOSITION

The physically-based soil erosion models mentioned above predict the sediment losses and gains at the time scale of a single runoff event. These models are less adequate at predicting sediment transport in the long term because upscaling from event-based predictions to long-term, e.g. annual, predictions is time-consuming and prone to uncertainties. A more effective way to predict long-term soil erosion on slopes is to use empirical models that predict soil erosion and deposition on the basis of topography and soil properties.

One of the most popular simple, empirical soil erosion models is the Universal Soil Loss Equation (USLE), which is based on statistical analysis of soil erosion data collected from small soil erosion plots in the USA (Wischmeier and Smith, 1978). The USLE predicts soil loss on a slope by multiplying a series of numbers, each representing a key factor contributing to soil erosion. The USLE in formula form is as follows:

$$E = R \cdot K \cdot L \cdot S \cdot C \cdot P$$

(12.25)

where E = the mean annual soil loss [M L^{-2} T^{-1}], R = the rainfall erosivity factor, K = the soil erodibility factor, L = the slope length factor, S = the slope steepness factor, C = the crop management, and P = the erosion control practice factor. The slope length L and slope steepness factor S are combined to produce a single index LS that represents the ratio of soil loss on a given slope to the soil loss from a standard erosion plot 22 m long and with a slope of 5°, for which LS = 1.0. The value of LS can be estimated from:

$$LS \quad = \quad \left(\frac{l}{22.13}\right)^n \left(0.065 + 0.045s + 0.0065s^2\right)$$

(12.26)

where l = the slope length (i.e. the horizontal distance from the divide or field boundary) (m) and s = the slope gradient (%). The value of n is varied according to the slope gradient. Morgan (1995) provides the details for the other factors in the USLE. A well-known disadvantage of the USLE is that the model is less appropriate for sites in Europe because of the differences in climate and soil between the USA and Europe. In particular, the model should not be used to determine the soil erodibility factor K of European loess soils without first carrying out some fundamental modifications.

Another model to predict annual soil loss from field-sized areas on slopes is the Revised Morgan, Morgan and Finney method (Morgan *et al.*, 2001). This model separates the soil erosion process into a water phase and a sediment phase. The water phase is considered in order to calculate the kinetic energy of the rainfall and the volume of overland flow; these are required in order to be able to predict the detachment of soil particles by raindrop impact and transport capacity, respectively. The processes of splash transport and detachment by runoff are ignored. An overview of the operating functions of this model and typical values for the input parameters is given by Morgan (2001).

A common shortcoming of the above described long-term soil erosion models is that they ignore sediment deposition. Govers *et al.* (1993) have proposed an alternative one-dimensional model for erosion on a slope, which also includes sediment deposition. In addition, the model accounts for soil redistribution due to splash erosion, soil creep, and tillage. Because overland flow concentrates in rills, i.e. small channels where the water flows faster and is deeper, the model makes a distinction between rill erosion and interrill erosion, i.e. erosion on the land between the rills. The erosion rate is modelled as a function of slope gradient and length:

$$E_r = a\,\rho_b\,s^b\,l^c \qquad (12.27)$$

where E_r = the rill erosion rate per unit area per unit time (kg m^{-2} y^{-1}), ρ_b = the dry bulk density of the soil (kg m^{-3}), s = the sine of the slope, l = slope length (m), a, b, c are empirical constants. From a field study of rill erosion in the loam belt in central Belgium, the mean values of a, b, and c were found to be $3{\cdot}10^{-4}$, 1.45 and 0.75 respectively (Govers *et al.*, 1993). For the dry bulk density a mean value of 1350 kg m^{-3} can be assumed. The interrill erosion rate is assumed to depend only on the local slope:

$$E_{ir} = d\,\rho_b\,s^e \qquad (12.28)$$

where d and e are empirical constants, for which values of $1.1{\cdot}10^{-3}$ and 0.8 can be assumed (Govers *et al.*, 1993). The transport capacity is considered to be directly proportional to the potential for rill erosion:

$$T_c = f\,E_r \qquad (12.29)$$

where T_c = transport capacity (kg m^{-1} y^{-1}), and f = an empirical constant (m). In the model, the eroded sediment is routed downslope until the transport capacity is reached. If the accumulated erosion exceeds the transport capacity, the excess sediment is deposited. Accordingly, a sediment mass balance for each location along the slope is formulated, taking into account the supply of sediment from upslope areas, the local soil erosion and deposition, and the losses to the downslope areas.

The diffusion process as a consequence of splash erosion, soil creep, and tillage operations is modelled by assuming that the resultant soil movement is proportional to the sine of the slope angle:

$$E_d = g\,\frac{\partial s}{\partial x} \qquad (12.30)$$

where E_d = the erosion rate per unit area attributable to diffusion processes (kg m^{-2}), x = the distance from the divide (m), and g = a coefficient (kg m^{-1}). For small slope angles, the sine of the slope angle s is approximately equal to the tangent of the slope angle (the difference for slopes up to 14 percent ($\approx$ 8°) is less than 1 percent), so $\partial s/\partial x$ is approximately equal to the profile curvature (i.e. the concavity/convexity in the direction of the slope).

The governing equations of the above long-term soil erosion model can be implemented in a raster GIS relatively easily in order to calculate the spatial distribution of soil erosion and deposition. The advantage of a GIS implementation of the model is that the GIS can also be used to derive the model input parameters related to the topography from a gridded digital elevation model (DEM) (see Burrough and McDonnell, 1998) (e.g. slope gradient, slope length, profile curvature, and local drainage direction network, i.e. a converging network in which the grid cells are connected in the direction of steepest downhill slope). The SEDEM model presented by Van Rompaey *et al.* (2001) and the WEPP model (USDA, 1995, 2012) are examples of such a spatially distributed soil erosion and deposition model.

Example 12.4 Soil erosion and deposition

Consider a long, straight slope of 6 percent. Estimate the distance from the divide (slope length) at which the rill erosion rate equals the interrill erosion rate and the distance at which the transport capacity is exceeded. Assume a value of $f = 170$ m and for the other parameters use the values given in the text above for the loam belt in central Belgium.

Solution
First, calculate the sine of the slope:
$s = \sin(\arctan(0.06)) = 0.0599$

Note that for gentle slopes, the sine of the slope approximates the tangent of the slope (in this example 6 percent = 0.06). Second, use equation (12.28) to calculate the interrill erosion, which is independent from the slope length ($d = 1.1 \cdot 10^{-3}$ and $e = 0.8$):

$$E_{ir} = 1.1 \cdot 10^{-3} \times 1350 \times 0.0599^{0.8} = 0.156 \text{ kg m}^{-2} \text{ y}^{-1}$$

For the slope length at which the rill erosion rate equals the interrill erosion rate, the following applies:

$$E_r = a\, \rho_b\, s^b\, l^c = 3 \cdot 10^{-4} \times 1350 \times 0.0599^{1.45} \times l^{0.75} = E_{ir} = 0.156 \text{ kg m}^{-2} \text{ y}^{-1}$$

$$0.0068\, l^{0.75} = 0.156$$

$$l^{0.75} = 22.9$$

$$l = 65 \text{ m}$$

Thus, at 65 m from the divide, the rill erosion rate has increased to the same value as the interrill erosion rate. At smaller slope lengths, the interrill erosion dominates the erosion process, whereas further downslope, rill erosion prevails.

To calculate the slope length at which the transport capacity is exceeded, we must first calculate the total cumulative erosion as a function of slope length. The total erosion is simply the sum of rill erosion and interrill erosion:

$$E_{tot} = E_r + E_{ir} = 0.0068\, l^{0.75} + 0.156 \text{ kg m}^{-2} \text{ y}^{-1}$$

The total cumulative erosion is the integral of the total erosion with respect to the slope length:

$$\int E_{tot}\, dl = \frac{0.0068}{1.75} l^{1.75} + 0.156\, l + p = 0.0039\, l^{1.75} + 0.156\, l + p$$

where p = an integration constant. In order to satisfy the boundary condition $E_{tot} = 0$ at $l = 0$, we must set $p = 0$. Hence,

$$\int E_{tot}\, dl = 0.0039\, l^{1.75} + 0.156\, l$$

The transport capacity as function of slope length is calculated using Equation (12.29) with $f = 170$ m:

$$T_c = 170 \, E_r = 170 \times 0.0068 \, l^{0.75} = 1.156 l^{0.75}$$

To find the slope length at which the total cumulative erosion equals the transport capacity, the last two equations should be equated. The slope length can then be found by iteration or 'trial and error'. The equations can also be solved graphically by plotting both the total cumulative erosion and transport capacity as function of slope length (see Figure 12.1): the slope length corresponding to the point of intersection is the slope length at which the total cumulative erosion equals the transport capacity. From Figure 12.1 it can be seen that this slope length is approximately 156 m. Thus, in this case, sediment deposition commences at 156 m from the divide and the erosion is transport-limited at distances greater than this.

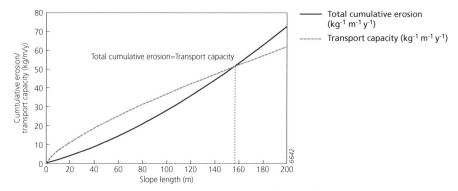

Figure 12.1 Total cumulative erosion and transport capacity as function of slope length given a straight slope of 6 percent.

EXERCISES

1. a. Calculate the bottom shear stress τ_b for a river with the following characteristics: width = 4 m, water depth = 1 m, bed slope = 0.0001, and average flow velocity = 31 cm s-1 (see question 6 in Chapter 11).
   b. Given the critical shear stress for erosion $\tau_{b,e}$ = 1.0 N m^{-2} and the critical shear stress for deposition $\tau_{b,d}$ = 0.5 N m^{-2}, indicate whether or not there is erosion or deposition of sediment.

2. Name three factors that control the critical shear stress for erosion or deposition.

3. Calculate the bottom shear stress τ_b for a shallow lake with the following characteristics: water depth = 1 m, wind fetch length = 300 m, wind speed = 15 m s^{-1}, D90 of the bed sediment = 70 μm.

4. The water column above 1 ha of floodplain contains 125 mg l^{-1} suspended solids. The effective settling velocity of the sediment is 1.75 10^{-6} m s^{-1} (= $\alpha \cdot w_s$). Assume that the water above the floodplain is continuously being refreshed and the suspended solids concentration in the river water remains the same.

    a. Calculate the total amount of sediment that is deposited in the 1 ha of floodplain during one day.

    b. What would happen if water stopped flowing over the floodplain?

5. Describe the limitations of Stokes's law for calculating the settling velocity of suspended solids in natural surface waters.

6. Name two main effects of flocculation on sediment deposition.

7. A hill slope in the loam belt of Belgium is 150 m long (measured from the local drainage divide) and can be subdivided into three sections of 50 m, each with a different gradient. The slope angles of the sections are:

Section	Slope angle (%)
1	5
2	10
3	3

    a. Estimate the long-term interrill erosion rate, rill erosion rate, and transport capacity at the end of each section (i.e. 50 m, 100 m and 150 m from the drainage divide), given $f = 120$ m. Note that the slope angles are given in percentages.

    b. Give the cumulative total erosion rate and transport capacity as function of slope length for each section.

    c. Evaluate the fate of the eroded sediment on this hill slope.

13

Chemical transformation

13.1 INTRODUCTION

As well as being transported via advection and dispersion, chemical substances may undergo a wide variety of chemical, physical, and biological transformation processes (see Chapter 2), which must be accounted for in the transport equations. To deal with this, the advection–dispersion equation is extended with a reaction term r:

$$\frac{\partial C}{\partial t} = -\bar{u}_x \frac{\partial C}{\partial x} + D_x \frac{\partial^2 C}{\partial x^2} \pm r \qquad (13.1)$$

where r = the rate of change in dissolved or particulate concentration due to physical, chemical or biological reactions [M L^{-3} T^{-1}]. This one-dimensional Equation (13.1) can, for example, be used to calculate the evolution of the chemical concentration of a degradable pollutant in a river downstream of an industrial wastewater discharge.

If the transformation processes proceed at a faster rate than the transport process, we may assume that the reactions subsystem is in equilibrium. In this case the concentration of the substance at a given location is largely governed by the reaction equilibrium. If the reaction rate is slow, we also must consider the process kinetics that describe the change in concentration resulting from the reaction as a function of time, i.e. the reaction rate. The rates of physical transformation processes such as volatilisation and radioactive decay vary considerably and depend on the physical properties of the chemical. Unlike volatilisation, radioactive decay rates are independent of physico-chemical environmental factors such as temperature, pH, or redox conditions. By comparison with transport processes, chemical acid–base and complexation reactions are usually fast, but redox reactions (including most biochemical transformations) are slow. The rates of chemical dissolution and precipitation processes are very variable and some may be quite slow. Even if the reaction rate is slow and the system is not in equilibrium, it is always useful to compute the equilibrium state of the system, to know where the system is heading.

Extending the advection–dispersion equation (see Section 11.3.2) with a first-order reaction, in which the chemical is removed from solution, gives:

$$\frac{\partial C}{\partial t} = -\bar{u}_x \frac{\partial C}{\partial x} + D_x \frac{\partial^2 C}{\partial x^2} - kC \qquad (13.2)$$

Figure 13.1 shows the evolution concentration profile after a pulse release into a river according to Equation (13.2). Just as in Figure 11.6, the centre of the mass travels at a velocity u_x and as a result of longitudinal dispersion the Gaussian curve becomes broader while travelling downstream. In addition, the area of the Gaussian curve, which is proportional to the total mass transported, decreases because of the chemical removal. Figure 13.2 shows the shape of the plume resulting from the continuous input into groundwater of a contaminant subject to decay (compare Figure 11.10).

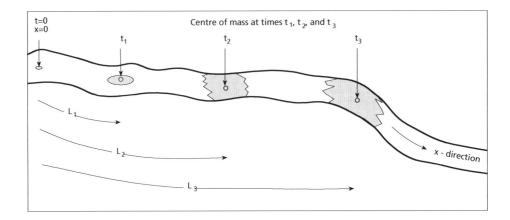

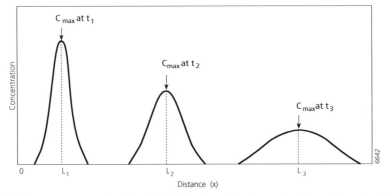

Figure 13.1 Downstream propagation of a pulse injection into a river due to advection and dispersion. In addition, the injected chemical is subject to first-order removal. Note that the chemographs become broader while travelling downstream (dispersion) and their size decreases.

13.2 SORPTION EQUILIBRIUM AND KINETICS

Adsorption and desorption reactions between solute and the surfaces of solids play a very important role in the retention or even immobilisation of chemicals by solids. The sorption mechanisms of dissolved constituents onto solid particles include cation exchange at negatively charged surfaces of clay minerals, and hydrophobic sorption of organic compounds to organic coatings or organic matter. These mechanisms are controlled by the physico-chemical properties of both the solute and the sorbent; many solids can preferentially adsorb some types of dissolved constituents.

In Section 2.5.4 we saw that when water containing a dissolved chemical is mixed with a solid medium, the total mass of the chemical partitions between the solution and the solid. The following equation represents a mass balance of this process in a given volume of water:

$$C_{tot} = C_w + C_s \cdot \frac{M_s}{V} \quad \Rightarrow \quad C_s = (C_{tot} - C_w) \cdot \frac{V}{M_s} \tag{13.3}$$

where C_{tot} = the total concentration of the chemical in the water [M L^{-3}], C_w = the concentration of the chemical in solution [M L^{-3}], C_s = the concentration of the chemical

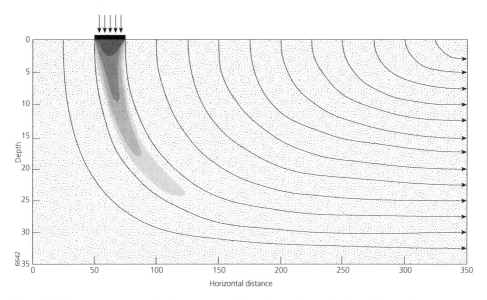

Figure 13.2 Downstream propagation in groundwater of a continuous input of a solute subject to decay.

adsorbed to the solid [M M^{-1}], M_s = the mass of the solids [M], V = the volume of the solution [L^3]. Box 13.1 gives an overview of how to determine the ratio between mass of solids and the solution volume (M_s/V; see Equation 13.3) for groundwater, soil, and surface water.

In the unsaturated zone and in groundwater, where the water is flowing sufficiently slowly, local equilibrium between the solution and the sediment can be assumed. This means that the partitioning between the dissolved phase and the adsorbed phase can be modelled using an isotherm model, for example the Freundlich or Langmuir isotherm models or a simple distribution coefficient (linear Freundlich isotherm) (see Section 2.5.3). For hydrophobic organic chemical pollutants that preferentially sorb to organic matter, the ***organic carbon–water partition coefficient*** K_{oc} can be used. This partition coefficient expresses the ratio between the chemical concentration sorbed to organic carbon [M M^{-1}] and the chemical concentration in water [M L^{-3}]. The organic carbon–water partition coefficients K_{oc} are commonly reported in chemical factsheets (e.g. EPA, 2013; ATSDR, 2013) or can be estimated from the octanol–water partition coefficient K_{ow} (see Section 2.5.4). Table 13.1 shows some relationships between K_{oc} and K_{ow} for various types of organic compounds. The partition or distribution coefficient of a hydrophobic organic constituent between the bulk sediment and water can be estimated by multiplying K_{oc} by the weight fraction of organic carbon in the sediment f_{oc} [M M^{-1}]:

$$K_d = f_{oc} K_{oc} \tag{13.4}$$

This Equation (13.4) can be used for sediments in which the organic carbon fraction f_{oc} is larger than about 0.001 (= 0.1 percent), because in these sediments sorption of organic compounds to organic matter prevails. In sediments with smaller organic carbon fractions, direct sorption to mineral surfaces may become important, so K_{oc} becomes less adequate for predicting sorption in sediment.

Box 13.I Determination of the ratio between solid mass and solution volume

In the saturated zone (groundwater), the mass of the solids divided by the solution volume can be calculated by:

$$\frac{M_s}{V} = \frac{(1-n)\,\rho_s}{n} = \frac{\rho_b}{n} \tag{13.Ia}$$

where M_s = the mass of the solids [M], V = the solution volume [L³], n = the water-filled porosity of the sediment [-], ρ_s = the sediment particle density (approximately 2.65 g cm⁻³ for mineral sediments, much smaller for organic sediments) [M L⁻³], and ρ_b = the dry bulk density of the bed sediment (defined as weight of dry solids divided by the volume of the bulk soil or sediment, so $\rho_b = (1-n)\cdot\rho_s$ [M L⁻³].

In the unsaturated zone in soil, the mass of the solids divided by the solution volume is given by:

$$\frac{M_s}{V} = \frac{\rho_b}{\theta} = \frac{\rho_w}{w} \tag{13.Ib}$$

where θ = the soil volumetric moisture content [-], ρ_w = density of water ($\approx$ 1.00 g cm⁻³ for fresh water), and w = the soil gravimetric moisture content.

If the solids are in suspension in surface water, the mass of the solids divided by the solution volume represents the concentration of suspended matter:

$$\frac{M_s}{V} = SS \tag{13.Ic}$$

where SS = the suspended solids concentration [M L⁻³]. For exchange between surface water and bed sediments it is usual to assume an active top layer in which the sediment mass is fully able to interact with the overlying water:

$$\frac{M_s}{V} = \frac{A\,d\,(1-n)\rho_s}{A(H+n\,d)} = \frac{d\,\rho_b}{H+n\,d} \tag{13.Id}$$

Where A = a standard surface area over which the exchange takes place, d = the depth of the active top sediment layer (usually in the order of between 1 and 5 mm) [L], and H = the water depth [L].

Table 13.1 Relationships between K_{oc} and K_{ow} (K_{oc} and K_{ow} in l kg⁻¹) (source: Lymann *et al.*, 1990).

Class of organic compounds	Regression equation
Aromatics, polynuclear aromatics, triazines, and dinitroaniline herbicides	log K_{oc} = 0.937 log K_{ow} − 0.006
Mostly aromatic or polynuclear aromatics; two chlorinated	log K_{oc} = 1.00 log K_{ow} − 0.21
s-Triazines and dinitroaniline herbicides	log K_{oc} = 0.94 log K_{ow} + 0.02
Variety of pesticides	log K_{oc} = 1.029 log K_{ow} − 0.18
Wide variety of organic compounds, mainly pesticides	log K_{oc} = 0.544 log K_{ow} +1.377

Note: units of K_{oc} and K_{ow} are in l kg⁻¹

We may write the rate law for mass transfer from solution to the adsorbed phase in groundwater as (see also Box 13.I) as:

$$r = \frac{M_s}{V}\frac{\partial C_s}{\partial t} = \frac{\rho_b}{n}\frac{\partial C_s}{\partial t} \tag{13.5}$$

where r = the sorption rate [M L^{-3} T^{-1}], ρ_b = the dry bulk density of the bed sediment, and n = the water-filled porosity of the sediment [-]. If we assume local equilibrium and a linear Freundlich isotherm and differentiate it with respect to time, we obtain:

$$\frac{\partial C_s}{\partial t} = K_d \frac{\partial C_w}{\partial t} \tag{13.6}$$

Combining Equations (13.5) and (13.6) gives:

$$r = \frac{\rho_b}{n} K_d \frac{\partial C_w}{\partial t} \tag{13.7}$$

When we enter this reaction rate into the general transport equation (Equation 13.1), where $C = C_w$, we obtain:

$$\frac{\partial C}{\partial t} = -\bar{u}_x \frac{\partial C}{\partial x} + D_x \frac{\partial^2 C}{\partial x^2} - \frac{\rho_b}{n} K_d \frac{\partial C}{\partial t} \Rightarrow \left[1 + \frac{\rho_b}{n} K_d\right]\frac{\partial C}{\partial t} = -\bar{u}_x \frac{\partial C}{\partial x} + D_x \frac{\partial^2 C}{\partial x^2} \tag{13.8}$$

The bracketed term on the left-hand side is a constant called the ***retardation factor*** R_f:

$$R_f = \left[1 + \frac{\rho_b}{n} K_d\right] \tag{13.9}$$

So, the transport equation with sorption reaction becomes:

$$\frac{\partial C}{\partial t} = -\frac{\bar{u}_x}{R_f}\frac{\partial C}{\partial x} + \frac{D_x}{R_f}\frac{\partial^2 C}{\partial x^2} \tag{13.10}$$

The retardation factor as defined above is a number equal to 1 in absence of adsorption (K_d = 0) or greater than 1 if there is adsorption. It has the effect of slowing down the chemical transport (retardation). The retardation factor can also be computed as the ratio of the mean velocity of the water to the mean velocity of the chemical. For example, if a column experiment is carried out using a chemical that is readily sorbed by the mineral fraction in the column (e.g. potassium), the breakthrough curve shows a delayed response (Figure 13.3). The first curve in Figure 13.3 is the breakthrough curve of a conservative substance that does not sorb. The slight S-shape is due to dispersion. The time taken for the half of the input concentration to pass is equal to the mean residence time of the solution in the column. The second curve shows the breakthrough curve of a substance subject to sorption. So, the travel time of the substance is increased by a factor that equals the retardation factor R_f relative to the travel time of the conservative substance. It is also more S-shaped because more time has elapsed for dispersion.

Note that the simple retardation factor should only be applied when a single distribution coefficient K_d is adequate to describe sorption. It is conceivable that this will not always be the case, because the sorption isotherm may not be linear or the sorption kinetics may be slow. In the case of a convex isotherm, the distribution coefficient and the retardation increase with decreasing concentration. This means that low concentrations are transported at a slower rate than high concentrations. Consequently, the front of the breakthrough curve remains sharp (see Figure 13.3). On the other hand, it also results in a delayed release of the residual substance after the bulk of the chemical substance has passed. In such cases, the breakthrough

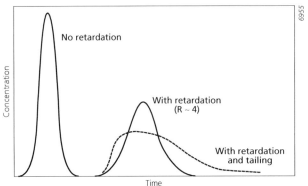

Figure 13.3 Breakthrough curves of a conservative substance and a substance subject to linear sorption or retardation (solid line). The dotted line represents the breakthrough curve of a substance subject to non-linear sorption according to a convex isotherm. Note the self-sharpening tendency of the front and the tailing at the end of the breakthrough curve.

curve derived from a column experiment exhibits 'tailing' and the concentration $C = C_0$ is reached much later than predicted by Equation (13.10) (see Figure 13.3). In the case of an accidental release of a chemical into groundwater, this may lead to much slower removal of the chemical from the aquifer than was predicted using a linear adsorption isotherm. A concave isotherm causes low concentrations to be transported faster and has the reverse effects: the breakthrough front is spread out and the declining limb of the breakthrough curve is relatively steep. Nevertheless, sorption is most commonly modelled with a single parameter (K_d or R_f), because this is easy to implement in chemical transport models. Figure 13.4 shows the plume in groundwater of a substance subject to sorption. Comparison with Figure 11.10 shows that the contaminant has been transported downstream for only about half the distance compared to an inert substance that undergoes advection and dispersion only. This implies that in this case the retardation factor R_f is about 2.

Example 13.1 Retardation

An aquifer underneath a waste disposal site is contaminated by 1,4-dichlorobenzene, a chlorinated hydrocarbon commonly used as an air freshener for toilets and refuse containers, and as a fumigant for control of moths, moulds, and mildews. The log octanol–water partition coefficient for 1,4-dichlorobenzene is log K_{ow} = 3.43. The aquifer has a bulk density of 1675 kg m^{-3}, a porosity of 0.3, and an organic carbon content of 0.1 percent. The groundwater flows at an average horizontal velocity of 50 m y^{-1}. Estimate the horizontal velocity of the 1,4-dichlorobenzene plume.

Solution
First, estimate the organic carbon–water partition coefficient K_{oc} using the empirical relationship between K_{oc} and K_{ow} listed in Table 13.1:

$$\log K_{oc} = 0.937 \log K_{ow} - 0.006 = 0.937 \times 3.43 - 0.006 = 3.21$$

$$K_{oc} = 1614 \ \text{l kg}^{-1}$$

Second, use Equation (13.4) to estimate the distribution coefficient K_d:
$$K_d = f_{oc} \ K_{oc} = 0.001 \times 1614 = 1.614 \ \text{l kg}^{-1} = 1.614 \cdot 10^{-3} \ \text{m}^3 \ \text{kg}^{-1}$$

Third, use Equation (13.9) to estimate the retardation factor:

$$R_f = 1 + \frac{1675}{0.3} \times 1.614 \cdot 10^{-3} = 10$$

Thus, the migration velocity of the 1,4-dichlorobenzene plume is 9.9 times less than the average groundwater flow velocity:

$$\frac{\bar{u}_x}{R_f} = \frac{50}{10} = 5 \text{ m y}^{-1}$$

So far in this section, we have assumed equilibrium between the dissolved and adsorbed phases. However, if the reaction rates are slow compared to the transport rate, reaction kinetics has to be taken into account. For example, in surface waters, it takes usually several hours to days before equilibrium is reached between the solute and the adsorbed phase, because the mass of solids is usually small compared to the water volume. During this time the solution is moved over a considerable distance due to water flow, so it is therefore necessary to account for the sorption kinetics.

In adsorption–desorption reactions, the rate of change in the solute concentration is the sum of the rate of removal by adsorption and the rate of production by desorption (Stumm and Morgan, 1996). If both the adsorption and desorption reactions are simulated using first-order kinetics, the differential equation is:

$$\frac{dC_w}{dt} = -k_{ads}C_w + k_{des}C_s \qquad (13.11)$$

where k_{ads} = adsorption rate constant [T^{-1}] and k_{des} = desorption rate constant [M L^{-3} T^{-1}]. At equilibrium, the net change of C_w equals zero, so:

$$\frac{dC_w}{dt} = 0 \quad \Rightarrow -k_{ads}C_w + k_{des}C_s = 0 \qquad (13.12)$$

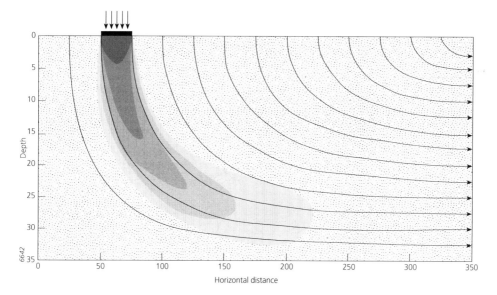

Figure 13.4 Downstream propagation in groundwater of a continuous input of a solute subject to sorption/retardation.

Assuming a linear Freundlich isotherm, Equation (13.12) can be rewritten as:

$$k_{des} = \frac{k_{ads}}{K_d} \tag{13.13}$$

Subsequently, substitution of Equation (13.13) into Equation (13.11) yields:

$$\frac{dC_w}{dt} = -k_{ads}\left(C_w - \frac{C_s}{K_d}\right) \tag{13.14}$$

The term on the right between brackets represents the concentration of a chemical in solution at equilibrium with C_s. If the system is not too far from equilibrium or C_s is barely affected by the adsorption–desorption reaction (which is true for large K_d values), this term may be approximated by the final equilibrium concentration. The differential equation then reads:

$$\frac{dC_w}{dt} = -k_{ads}\left(C_w - C_{eq}\right) \tag{13.15}$$

where C_{eq} = the equilibrium concentration [M L^{-3}]. The analytical solution of Equation (13.15) reads (see also Box 11.I):

$$C_w(t) = C_{eq} + \left(C_0 - C_{eq}\right)e^{-k_{ads}\,t} \tag{13.16}$$

Example 13.2 Sorption kinetics

Stormy winds agitate the water in a lake, and the turbulent conditions resuspend the PCB-contaminated lake bed sediments into the water column. As a result, the suspended sediment concentration in the water increases to 50 mg l^{-1} and the PCB concentration in the suspended sediment increases to 22 mg kg^{-1}. The initial dissolved PCB concentration before the resuspension of the bed material was 18 ng l^{-1}. Calculate the dissolved PCB concentration in the lake water 4 hours after the resuspension event, given the first-order adsorption rate constant k_{ads} = 0.2 h^{-1} and the distribution coefficient K_d = 1.1·10^5 l kg^{-1}. Assume that the suspended sediment concentration and the PCB concentration in the suspended sediment have increased instantaneously.

Solution
First, calculate the equilibrium concentration of the PCB in the dissolved phase. To do so, first calculate the total PCB concentration (dissolved and adsorbed) in the water (see Equation 13.3):

$$C_{tot} = C_w + C_s \cdot SS = 18 + 22 \times 50 = 1118 \text{ ng l}^{-1}$$

At equilibrium, the following applies:

$$C_{tot} = C_w + C_s \cdot SS = C_w + C_w \cdot K_d \cdot SS = C_w(1 + K_d \cdot SS)$$

In the above equation, C_w represents the equilibrium concentration, so we may therefore write:

$$C_{eq} = \frac{C_{tot}}{1 + K_d \cdot SS} = \frac{1118}{1 + 1.1 \cdot 10^5 \times 50 \times 10^{-6}} = 172 \text{ ng l}^{-1}$$

Use Equation (13.16) to calculate the dissolved PCB concentration after 4 hours:

$$C_w = 172 + (18 - 172)e^{-0.2 \times 4} = 172 - 154e^{-0.8} = 172 - 169 = 103 \text{ ng } l^{-1}$$

13.3 BIOLOGICAL PRODUCTION AND DEGRADATION

Organic chemicals and some inorganic compounds (e.g. NH_4, NO_3, H_2S) are susceptible to biological transformation processes. Microorganisms, especially bacteria and fungi, obtain their energy by oxidation (primarily organic matter, but also organic pollutant chemicals and sulphide species such as FeS_2, H_2S). Under oxic conditions, the biological degradation processes involves oxygen, but under anoxic conditions, nitrate, sulphate, and reduced organic carbon act as respective oxidants as the conditions become increasingly reducing. Some organic chemicals in which the carbon is in a fairly oxidised state (e.g. chlorinated hydrocarbons) are reduced under reducing conditions. The microorganisms responsible for the biological transformation processes form biofilms on the solid surfaces (e.g. the river bed and aquatic vegetation in the case of surface waters, and the aquifer matrix in the case of groundwater. A biofilm consists of the microorganisms attached by an extracellular 'glue' made of polysacharides. By forming a biofilm, the microorganisms remain at one place and take advantage of the supply of nutrients by advective flow, instead of relying solely on diffusive transport, as happens when they are suspended in the water. Because most biochemical transformation processes take place in biofilms, they play a key role in the environmental fate of biodegradable substances in both surface water and groundwater.

The changes in chemical concentrations that result from biochemical reactions are usually modelled using first-order kinetics. The first-order rate constant is commonly assumed to be proportional to the active cell density in the biofilm [cells L^{-3}]. Although, as mentioned above, biofilms are very important for biotransformation, they are rarely explicitly accounted for in distributed transport models, but the influence of the cell density or number of microorganisms in the biofilm is commonly incorporated in the first-order rate constants. The advantage of this approach is that the model becomes simpler and requires fewer data. The disadvantage is that the rate constants become dependent on the system. In addition to the cell density, the total microbial activity and, accordingly, the first-order rate constants depend on environmental factors such as the redox regime, temperature, and concentration of the substrate. The substrate consists of nutrients (reductants, including organic carbon) needed for energy supply and the growth of the organisms. In addition, an oxidant (mostly oxygen, but some microorganisms use nitrate or sulphate) is needed to oxidise the nutrients. If the oxidant or one or more of the substrate components is in limited supply, the organism's growth and the biodegradation process are hampered. A pollutant may be the principal substrate, a co-substrate, or oxidant. The dependence on temperature is particularly relevant in surface waters where the water temperature may vary considerably throughout the day and the year. In groundwater, the effects of the water temperature are of minor importance, since groundwater temperatures are relatively constant throughout the year. The mathematical expressions used to account for the effects of substrate limitation and temperature in the first-order rate constants are elaborated further below.

Table 13.2 demonstrates that the first-order rate constants for the degradation of organic chemicals in soil and groundwater may vary by up to two orders of magnitude. Likewise, the denitrification rates in groundwater are subject to substantial variation. In sandy aquifers in the Netherlands, where nitrate is the main oxidant for the degradation of sediment organic matter, the first-order denitrification rate constant ranges from less than $3.5 \cdot 10^{-5}$ d^{-1} to about $4 \cdot 10^{-4}$ d^{-1} (Uffink, 2003). The organic matter content of these aquifers is very low (usually less than 0.5 percent). The value of the denitrification rate may increase significantly – up to

Table 13.2 Minimum and maximum biodegradation rate constants in soil and groundwater (source: EPA, 1999b; Mathess, 1994).

Chemical group	First-order degradation rate constant k	
	Minimum	Maximum
	d^{-1}	d^{-1}
Benzene*	0	0.071
Toluene*	0	0.186
Trichloroethylene*	0.00082	0.04
Vinyl chloride*	0	0.0582
Phenol*	0	0.2
Halogenated hydrocarbons	0.00016	0.0036
Carbamate	0.0011	0.35
Aniline	0.0022	0.073
Urea	0.00098	0.031
Organic phosphorus compounds	0.0012	0.12
Triazine	0.0018	0.032

* anaerobic groundwater (EPA, 1999b)

about 0.01 d^{-1} – if the concentrations of sediment organic matter or dissolved organic carbon increase and the redox potential decreases: for example, in polluted plumes from septic tanks. The denitrification rate constants may be even larger in organic sediments. Because the aquifer properties and their accompanying rate constants vary so widely, and information on these parameters is often scarce, the use of first-order kinetics in regional-scale groundwater quality modelling is limited (see Vissers 2006).

In surface waters, the first-order constants for nitrification and denitrification may also change considerably in space and time. Table 13.3 shows that these constants may vary by over several orders of magnitude for different river systems. As mentioned above, limitation of one of the substrate components or oxidant may hamper biodegradation. A common expression for the effect of substrate limitation is the ***Monod equation***, also referred to as ***Michaelis–Menten kinetics***:

$$k = \frac{S}{M_n + S} k_{\max} \tag{13.17}$$

where S = substrate concentration [M L^{-3}], $k_{\max}$ = maximum rate constant when the substrate is not limiting [T^{-1}], M_n = Monod half-saturation concentration or Michaelis constant [M L^{-3}], which is the substrate concentration for which $k = 0.5 \cdot k_{\max}$. Figure 13.5 shows the rate constant as function of the substrate concentration. This Equation (13.17) was originally derived by Michaelis–Menten for modelling uptake kinetics of organisms growing on a substrate and subsequently applied by Monod to quantify microbial population dynamics in a system. The equation can also be used to describe substrate or oxidant limitation in biodegradation reactions. For example, the effect of oxygen limitation on nitrification is described by:

$$\frac{dNH_4}{dt} = -\frac{DO}{M_n + DO} k_{n,\max} \, NH_4 \tag{13.18}$$

Table 13.3 Rate constants for nitrification (k_n) and denitrification (k_d) in rivers reported in the literature.

k_n	k_d	River	Source
d^{-1}	d^{-1}		
0.5–0.8		'Normal range'	Veldkamp and Van Mazijk (1989)
0.1–0.6		'Normal range'	Thomann (1972)
> 1		'Smaller streams'	Thomann and Mueller (1987)
0.5–3.0		River Trent, UK	Garland (1978)
0.7		Willamette River, Oregon, USA	Rickert (1982)
0.2–4.4		Speed River, Ontario, Canada	Gowda (1983)
0.0–1.9		River Rhine, Netherlands	Admiraal and Botermans (1989)
0.1	0.0–0.1	'Normal range'	EDS (1995)
0.04–0.2	0.0–1.0	'Model documentation values (maximum range)'	EPA (1985)
3.0–4.0	1.75	Biebrza River, Poland (summer)	Van der Perk (1996)
0.0	1.25	Langbroekerwetering, Netherlands (winter)	Van der Perk (1996)
0.5	0.5	Regge River, Netherlands	Van den Boomen *et al.* (1995)
3.0	1.75	Grindsted River, Denmark	Bach *et al.* (1989)
0.66–5.20	0.57–5.28	Upper reach South Platte River (annual range normalised to 20 °C)	Sjodin *et al.* (1997)
	0.1	Bedford Ouse River, UK	Whitehead *et al.* (1981)
	0.05	River Thames, UK	Whitehead and Williams (1982)
	0.455	Mississippi River, USA ($Q < 28.3$ m^3 s^{-1}) (nitrogen loss rate rate constant)	Alexander *et al.* (2000)
	0.118	Mississippi River, USA (28.3 m^3 s^{-1} $< Q < 283$ m^3 s^{-1}) (nitrogen loss rate rate constant)	Alexander *et al.* (2000)
	0.051	Mississippi River, USA (283 m^3 s^{-1} $< Q < 850$ m^3 s^{-1}) (nitrogen loss rate rate constant)	Alexander *et al.* (2000)
	0.005	Mississippi River, USA ($Q > 850$ m^3 s^{-1}) (nitrogen loss rate rate constant)	Alexander *et al.* (2000)

where NH_4 = ammonium concentration [M L^{-3}], DO = dissolved oxygen concentration [M L^{-3}], $k_{n,max}$ = maximum nitrification rate constant [T^{-1}]. The Monod half-saturation concentration is approximately 0.5 mg O$_2$ l^{-1} for most biochemical oxidation reactions with oxygen as limiting oxidant. Moreover, in biodegradation reactions of some organic pollutants, the pollutant itself is one of the substrate components or can even constitute the major substrate component. In this case, the rate constant depends on the concentration of the pollutant itself, a co-substrate (reductant), and the oxidant:

$$k = \underset{pollutant}{\frac{C}{M_n + C}} \cdot \underset{\substack{co\text{-}substrate \\ (reductant)}}{\frac{S}{M_n + S}} \cdot \underset{oxidant}{\frac{O_2}{M_n + O_2}} k_{max} \qquad (13.19)$$

In the environmental temperature range of 10 to 30 °C, an increase of temperature generally results in an enhancement of microbial activity. The rate of biochemical reactions increases exponentially with temperature, so the percentage increase per degree is approximately constant (Thomann and Mueller, 1987):

$$k_{T_1} = k_{T_2} \, \theta^{(T_1 - T_2)} \qquad (13.20)$$

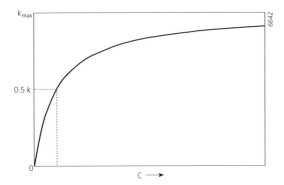

Figure 13.5 First-order rate constant as function of substrate concentrations (Michaelis–Menten kinetics).

Table 13.4 Temperature coefficients for some biological and biochemical processes (source: EPA, 1985; Thomann and Mueller, 1987; Schnoor, 1996).

Process	θ
Nitrification	1.04
Denitrification	1.08
Biochemical oxygen demand	1.047
Sediment oxygen demand	1.04
Algae photosynthesis	1.065
Bacterial respiration	1.03
Zooplankton respiration	1.06

where k_T = rate constant at temperature T [T^{-1}], θ = constant temperature coefficient [-] greater than 1.0 and usually within the range of 1.0–1.10. Usually, a reference temperature of 20 °C is used to report the values of k:

$$k_T = k_{20}\, \theta^{(T-20)} \tag{13.21}$$

Table 13.4 lists some temperature coefficients for different biological and biochemical processes. Equation (13.21) can be used for a temperature range of approximately from 10 °C to 30 °C.

Example 13.3 Nitrification rate

A wastewater treatment plant discharges its effluent into a stream at a constant rate of 40 l s^{-1}. The stream has a depth of 0.6 m, a width of 5.0 m, and a flow velocity of 12 cm s^{-1} (see Example 11.3). The ammonium (NH$_4^+$) concentration in the stream just upstream from the effluent outfall is 0.22 mg l^{-1} and the NH$_4^+$concentration of the effluent water is 18 mg l^{-1}. The water temperature in the stream is 16 °C and the dissolved oxygen concentration in the stream water is 6 mg l^{-1}. The maximum nitrification rate constant at 20 °C $k_{n,max,20°}$ = 1.8 d^{-1}, the temperature coefficient for nitrification θ_n = 1.04, the Monod half-saturation concentration M_n = 2.0 mg l^{-1}.

Calculate the NH$_4^+$ concentration at 2.5 km downstream from the effluent outfall and the distance at which the NH$_4^+$ concentration has dwindled to the level upstream from the outfall (0.22 mg l^{-1}). Assume instantaneous mixing and neglect dispersion.

Solution

First, calculate the discharge of the stream just upstream from the effluent outfall (see Example 11.4):

$$Q_{stream} = 0.6 \text{ m} \times 5 \text{ m} \times 0.12 \text{ m s}^{-1} = 0.36 \text{ m}^3 \text{ s}^{-1}$$

The ammonium concentration directly downstream from the outfall is (see Equation 11.17):

$$C_{downstream} = \frac{0.22 \times 0.36 + 18 \times 0.04}{0.36 + 0.04} = \frac{0.80}{0.4} = 2.0 \text{ mg l}^{-1}$$

Second, calculate the actual nitrification rate constant corrected for the dissolved oxygen concentration (see Equation 13.18) and water temperature (see Equation 13.21):

$$k_n = \frac{DO}{M_n + DO} \cdot \theta^{T-20} \cdot k_{n,max,20°C} = \frac{6}{2+6} \times 1.04^{-4} \times 1.8 = 0.75 \times 0.85 \times 1.8 = 1.15 \text{ d}^{-1}$$

The time it takes for the water to travel 2.5 km is

$$t = \frac{2500}{0.12} \times \frac{1}{86400} = 0.24 \text{ d}$$

where 86 400 = the number of seconds per day. The ammonium concentration at 2.5 downstream of the outfall is

$$C(t) = C_0 e^{-k_n t} = 2.0 \times e^{-1.15 \times 0.24} = 2.0 \times 0.76 = 1.5 \text{ mg l}^{-1}$$

The distance downstream from the outfall at which the NH_4^+ concentration is back to the upstream concentration of 0.22 mg l^{-1} can be derived as follows:

$$C(t) = = 2.0 \times e^{-1.15 t} = 0.22$$

$$-1.15 t = \ln(0.22 / 2.0) = -2.21$$

$$t = 1.92 \text{ d}$$

The distance that the water travels in 1.92 days is

$$x = 1.92 \times 86400 \times 0.12 = 19907 \text{ m} = 19.9 \text{ km}.$$

Thus, the influence of the wastewater outfall is discernable up to nearly 20 km downstream from the outfall.

In rivers and lakes, manifold complex and irreversible biogeochemical reactions that do not depend on the concentration of the reactant in the overlying surface water occur in the bed sediments. Therefore, the production and subsequent release of substances to the overlying water is often modelled using **zero-order kinetics**, because the release is limited by physical constraints such as the size of the area over which the release occurs. Examples of such zero-order reactions are methane production and the release of mineralisation products (e.g. NH_3, PO_4^{3-}) from anaerobic sediments. The general differential equation for zero-order production or decay is:

$$\frac{dC_w}{dt} = k_0 \tag{13.22}$$

where k_0 = zero-order rate constant [M L^{-3} T^{-1}] which has a positive sign in the case of production and a negative sign in the case of decay. Integration of this rate expression results in a straight rising line for which the tangent of the slope equals the zero-order rate constant. Note that if the production rate is expressed as a flux density, i.e. mass production per unit area, the zero-order rate constant can be calculated as follows:

$$k_0 = J \cdot \frac{A}{V} \tag{13.23}$$

where J = flux density [M L^{-2} T^{-1}], A = surface area over which production occurs [L^2], and V = volume of water body [L^3]. The surface area divided by the water volume can be approximated by $1/H$ where H = the water depth [L]. Then the differential equation becomes:

$$\frac{dC_w}{dt} = \frac{J}{H} \tag{13.24}$$

Combination of a zero-order production process (e.g. release of nutrients from bed sediments) and a first-order removal process yields the following differential equation:

$$\frac{dC_w}{dt} = -kC_w + \frac{J}{H} = -k\left(C_w - \frac{J}{H\,k}\right) = -k\left(C_w - C_{eq}\right) \tag{13.25}$$

where $k \cdot C_w$ represents the zero-order production term. So, the combination of zero-order production and first-order removal results in a differential equation similar to the equation for adsorption–desorption reactions (see Equations 13.15 and 13.16). Furthermore, the combination of zero-order production and first-order adsorption–desorption results in an analogue expression with an equilibrium concentration. This concept of equilibrium concentration (equilibrium phosphate concentration; *EPC*) is widely applied in the literature, particularly for modelling phosphate exchange with sediments (e.g. Froelich, 1988; House *et al.*, 1995). The same concept can also be used for modelling the ammonium exchange with bed sediment using an equilibrium ammonium concentration (*EAC*)

Table 13.5 Rate constants for ammonium adsorption (k_f), and phosphate fixation (k_p), the equilibrium ammonium concentration (*EAC*), and equilibrium phosphate concentration (*EPC*) in rivers reported in the literature.

k_f	*EAC*	k_p	*EPC*	River	Source
d^{-1}	mg l^{-1}	d^{-1}	mg l^{-1}		
		0.2–0.7		'normal range'	Thomann (1972)
0.5–1.0	0.15–0.65	1.25	0.20–0.25	Biebrza River, Poland	Van der Perk (1996)
1.0	1.25	0.4	0.6	Langbroekerwetering, Netherlands	Van der Perk (1996)
		0.25–0.37		Nepean River, Australia	Simmons and Cheng (1985)
			0.3–3	-	Nichols (1983)
			0.02–0.14	-	Froelich (1988)
			0.33	Mississippi River, USA	Wauchope and McDowell (1984)
			0.006	Bear Brook, USA	Meyer (1979)
			0.12	Colorado River, USA	Mayer and Gloss (1984)
			0.01–0.0025	Duffin Creek and Nottawasaga River, Ontario, Canada	Hill (1982)
			0.01–0.55	Avon catchment, UK	Jarvie *et al.* (2005)
			0.001-0.60	Wye catchment, UK	Jarvie *et al.* (2005)

(e.g. Van der Perk, 1996). The value of the equilibrium concentration depends on the sediment composition with respect to grain size distribution and organic matter content, and phosphorus/nitrogen loading of the river or lake. Table 13.5 lists some literature values of the first-order constant k and the equilibrium concentration for ammonium (*EAC*) and phosphate (*EPC*) in selected rivers (see Equation 13.25).

Example 13.4 Phosphate release

In a 75 cm deep channel draining a Dutch polder, the equilibrium phosphate (PO_4^{3-}) concentration *EPC* and the PO_4^{3-} fixation rate constant were measured to be 0.6 mg l^{-1} and 0.4 d^{-1}, respectively. Assuming that the PO_4^{3-} fixation process can be entirely attributed to exchange between the bed sediments and the overlying water, calculate the corresponding zero-order PO_4^{3-} release rate from the bed sediments.

Solution
From Equation (13.25), it can be seen that the *EPC* equals

$$EPC = \frac{J}{H\,k}$$

Thus,

$$\frac{J}{0.75 \times 0.4} = 0.6 \text{ mg l}^{-1} \ (= \text{g m}^{-3})$$

$J = 0.18$ g PO_4^{3-} m^{-2} d^{-1} = 180 mg PO_4^{3-} m^{-2} d^{-1}

EXERCISES

1. Derive the analytical solution of the following differential equations (hint: compare Box 11.I)

    a. $\dfrac{dC}{dt} = -kC$

    b. $\dfrac{dC}{dt} = -k\left(C - C_{eq}\right)$

2. The mass balance of a biodegradable substance in a small well-mixed lake is:

    $$\frac{dM}{dt} = QC_{in} - QC - kVC$$

    where M = mass of biodegradable substance [M], Q = discharge through lake [L^3 T^{-1}], C_{in} = concentration of biodegradable substance in inflowing stream [M L^{-3}], C = concentration of biodegradable substance in lake [M L^{-3}], k = biodegradation rate constant [T^{-1}], V= lake volume. The steady state solution of the concentration is

    $$C = \frac{C_{in}}{1 + kT}$$

    where T = residence time of water in lake [T].
    Derive this steady state solution from the differential equation given above.

3. In an 80 cm deep river, ammonium release from the bed sediment occurs at a rate of 16 mg m^{-2} d^{-1}. The nitrification rate constant is 0.4 d^{-1} and the denitrification rate constant is 0.8 d^{-1}. If internal and external sources and sinks other than release from bed sediments, nitrification, and denitrification can be ignored, and the above rate parameters remain constant, the system tends to equilibrium.
   a. Calculate the ammonium concentration in mg l^{-1} at equilibrium.
   b. Calculate the nitrate concentration in mg l^{-1} at equilibrium.
   c. Name three sources or sinks apart from direct anthropogenic sources (e.g. effluent discharges, agriculture) which may have been overlooked.

4. In a laboratory experiment the following rate equation for the decomposition of *o*-xylene was determined (in μg l^{-1} d^{-1}):

$$\frac{dS}{dt} = \frac{-45S}{1000 + S}$$

   a. What does *S* mean
      - in this particular equation?
      - in general?
   b. In a graph plot the concentration against time (in days) given an initial *o*-xylene concentration of 1500 μg l^{-1}?
   c. How long does it take for the *o*-xylene concentration to fall to 70 μg l^{-1}?

5. A column experiment is carried out to determine the potassium adsorption characteristics of a sediment. For this purpose, a cylinder of inert material is filled with sediment, through which a KCl solution is percolated. The cylinder is 1 m long and 0.45 m in diameter. The sediment has a porosity of 0.32 and the bulk density is 1600 kg m^{-3}.

Time (day)	Cl$^-$ (mg l^{-1})	K$^+$ (mg l^{-1})	Time (day)	Cl$^-$ (mg l^{-1})	K$^+$ (mg l^{-1})	Time (day)	Cl$^-$ (mg l^{-1})	K$^+$ (mg l^{-1})
1	15	1	11	24	5	21	135	5
2	15	1	12	33	5	22	143	12
3	15	1	13	45	6	23	142	19
4	15	1	14	50	5	24	148	21
5	15	1	15	60	6	25	146	23
6	17	1	16	65	5	26	149	22
7	16	2	17	70	6	27	147	24
8	18	3	18	90	5	28	148	26
9	20	3	19	110	5	29	150	26
10	23	4	20	125	6	30	149	25
						31	150	25

   a. Draw the breakthrough curves of chloride and potassium (i.e. a graph in which the relative concentrations (= (C(t) – C(0))/(C$_{max}$ – C(0)) are displayed as function of time).
   b. Determine the water flow velocity in m d^{-1}.
   c. Assuming a linear Freundlich isotherm for K sorption, calculate the retardation factor R_f and the distribution coefficient K_d.
   d. Calculate the amount of potassium that is adsorbed within the column.
   The experiment is repeated to study the behaviour of nitrate.

    e. In the above figure sketch the breakthrough curve of nitrate. Briefly explain the processes responsible for the position and shape of the curve.

6. A road accident involving a tanker causes a spill of 10 000 l of benzene into the soil. Quick cleanup recovers most of the substance, but some benzene reaches the shallow water table. Some of this benzene dissolves in the groundwater. The log octanol–water partition coefficient log K_{ow} of benzene is 2.13. The aquifer material consists of sandy sediment with a bulk density of 1.8 g cm^{-3} and a porosity of 0.3. The weight fraction of organic carbon in the aquifer sediment is 0.8 percent. The local horizontal groundwater flow velocity is 12 m y^{-1}.
    a. Estimate the maximum horizontal distance over which the benzene is transported in 10 years. Assume that the benzene transport is only affected by retardation. (Hint: choose an appropriate relation between K_{ow} and K_{oc} from Table 13.1).
    b. Will the aquifer be contaminated over the entire aquifer depth? Motivate your answer.
    c. Discuss other processes that affect the evolution of the benzene plume in the subsurface environment.

7. Explain why a convex isotherm results in tailing of the breakthrough curve.

8. A canal discharges effluent from a wastewater treatment plant. At two locations in the canal, water samples were collected and analysed for ammonium, nitrate, and phosphate. The first sampling location was near the effluent outflow and the second sampling location was 1 km further downstream. Flow velocity and water temperature were also measured. The data are given in the table below.

Location	NH_4^+ (mg l^{-1})	NO_3^- (mg l^{-1})	PO_4^{3-} (mg l^{-1})	T (°C)	U_x (m s^{-1})
Near effluent discharge	<< d.l.*	15.0	2.6	16.5	0.12
1 km downstream from effluent discharge	<< d.l.*	13.5	2.1	16.5	0.12

* d.l. = detection limit

    a. Using the above data, calculate the denitrification rate constant at 20 °C.
    b. Calculate the distance needed to lower the nitrate concentration below 10 mg l^{-1}, assuming that the other parameters remain constant.
    By means of a field experiment the equilibrium phosphate concentration in the canal was determined to be 0.5 mg l^{-1}.
    c. Calculate the phosphate fixation rate constant.

14

Gas exchange

14.1 HENRY'S LAW

So far, we have considered the mass exchange within or between the liquid or dissolved phase and the solid phase. However, gas solution, exsolution, and volatilisation can cause significant mass exchange of atmospheric gases (O_2, N_2), gases produced by the decomposition of organic matter (CO_2, CH_4, NH_3, H_2S), and organic chemicals between the liquid phase and the atmosphere. The concentration of dissolved gas in solution is commonly modelled using **Henry's law**, which relates the concentration of dissolved gas in solution to the partial pressure of that gas in an atmosphere in contact with the solution and is an example of a chemical equilibrium constant (see Chapter 2):

$$K_H \quad = \quad \frac{P}{C_{aq}} \tag{14.1}$$

where K_H = the Henry's law constant usually expressed in units like atm l mol^{-1} or atm m^3 mol^{-1} (note that some textbooks define the Henry's law constant as the reciprocal of the Henry's law constant given here; compare Equation 2.9), P = partial pressure of the chemical in the gas phase (atm), and C_{aq} = the equilibrium molar concentration of the chemical in solution (mol l^{-1} or mol m^3). The Henry's law constant can also be expressed in dimensionless form, by dividing the gas concentration in air by the aqueous concentration in the same units (mol l^{-1}, mol m^3, mg l^{-1}, or mg m^{-3}):

$$K'_H \quad = \quad \frac{C_{air}}{C_{aq}} \tag{14.2}$$

where K'_H = the dimensionless form of the Henry's law constant (-) and C_{air} = the molar concentration of the gas in air (mol l^{-1} or mol m^3). The relation between partial pressure P and the concentration in air is given by the ideal gas law:

$$C_{air} \quad = \quad \frac{n}{V} \quad = \quad \frac{P}{RT} \quad \Rightarrow$$

$$K'_H \quad = \quad \frac{K_H}{RT} \tag{14.3}$$

where n = the number of moles of the chemical in air (mol), V = the air volume (l or m^3), P = the partial vapour pressure of the gas (atm), R = the gas constant (= 0.082058 (l atm mol^{-1} K^{-1})), and T = the absolute temperature (K). The Henry's law constant is tabulated in many handbooks. It can also be estimated by dividing the vapour pressure of a chemical at a given temperature by its aqueous solubility at that temperature. Table 14.1 lists the dimensionless Henry's law constant and vapour pressure for some selected chemicals.

14.2 THIN FILM MODEL

If the system is not in equilibrium, a gas flux across the water–air interface occurs to counteract the non-equilibrium state. For surface water bodies, this process of gas exchange is often modelled according to the ***thin film model*** (or stagnant layer model). This model is based on the assumption that a dissolved chemical has a uniform concentration throughout both the surface water body and the overlying atmosphere, except in two very thin layers – one in the water and one in the air – at the water's surface (Figure 14.1). Within these very thin layers, it is assumed that the eddies responsible for turbulent diffusion are suppressed and that the chemical transport is solely through molecular diffusion, governed by Fick's first law (Equation 11.16):

$$J = -D\frac{dC}{dx} = -\frac{D}{\delta}\Delta C \tag{14.4}$$

where J = the flux density [M L^{-2} T^{-1}], C = the chemical concentration [M L^{-3}], D = the molecular diffusion coefficient [L^2 T^{-1}], and δ = the film thickness [L]. At steady state, the flux through the water film is equal to the flux through the air film, so:

$$J = -k_a\left(C_{sa} - C_a\right) = -k_w\left(C_w - C_{sw}\right) \tag{14.5}$$

$$k_a = \frac{D_a}{\delta_a} \tag{14.6}$$

$$k_w = \frac{D_w}{\delta_w} \tag{14.7}$$

where k_a = the gas exchange coefficient for air [L T^{-1}], k_w = the gas exchange coefficient for water [L T^{-1}], D_a = the molecular diffusion coefficient in air [L^2 T^{-1}], D_w = the molecular diffusion coefficient in water [L^2 T^{-1}], δ_a = the air film thickness [L], δ_w = the water film thickness [L], C_a = the chemical concentration in air [M L^{-3}] (Note the difference in dimensions compared to C_{air} in Equation (14.2)!), C_{sa} = the chemical concentration in air at

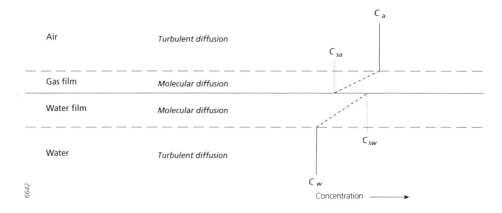

Figure 14.1 Concept of the thin film model: The resistance of gas exchange across the air–water interface is assumed to lie in two thin stagnant layers – one in the air and the other in the water. Within the thin layer, molecular diffusion occurs, driven by the concentration gradients. Outside the layers, turbulent diffusion occurs, which causes the concentration gradients to be negligible. Adapted from Hemond and Fechner-Levy (2000).

the water–air interface [M L^{-3}], C_w = the chemical concentration in water [M L^{-3}], and C_{sw} = the chemical concentration in water at the water–air interface [M L^{-3}]. The thickness of the gas and water films cannot be measured directly, so the gas exchange coefficients should be determined empirically. Subsequently, the thickness of the films can be estimated from the gas exchange coefficient and independent knowledge of the molecular diffusion coefficients. Typical vales for δ_w range between 20 and 200 μm and typical values for δ_a are in the order of 1 cm (Hemond and Fechner-Levy, 2000).

If Henry's law applies exactly at the water–air interface, then:

$$J \;=\; -k_a \left(K'_H \, C_{sw} - C_a \right) \tag{14.8}$$

From Equation (14.5), it appears that:

$$C_{sw} \;=\; \frac{J}{k_w} + C_w \tag{14.9}$$

By combining the Equations (14.8) and (14.9), we can solve for the mass flux J across the water–air interface in terms of the bulk phase concentrations, Henry's law constant, and the gas exchange coefficients for water and air:

$$J = \left(\frac{-1}{1/k_w + 1/(K'_H \cdot k_a)} \right) \left(C_w - \frac{C_a}{K'_H} \right) = -k_L \left(C_w - \frac{C_a}{K'_H} \right) = -k_L \left(C_w - C_{eq} \right) \tag{14.10}$$

where k_L = the overall gas exchange coefficient [L T^{-1}], C_{eq} = the equilibrium concentration of the chemical in solution [M L^{-3}]. The equilibrium concentration C_{eq} is an important term if the chemical under consideration is a constituent of the atmosphere, like oxygen, nitrogen, or carbon dioxide. In the case of anthropogenic contaminants, this term often approximates zero.

For the vertically averaged concentration in a surface water body, which is influenced by gas exchange between the surface water and the atmosphere, we may formulate the following mass balance:

$$\frac{dC_w}{dt} \;=\; \frac{J}{H} \;=\; -\frac{k_L}{H}\left(C_w - \frac{C_a}{K'_H} \right) \tag{14.11}$$

where H = the water depth [L]. Note the similarities between Equation (14.11) and Equations (13.15) and (13.25). Analogous to electrical resistance, the reciprocal of the overall gas exchange coefficient k_L represents the overall resistance against gas exchange. This overall resistance consists of two components (see Equation 14.10):

$$\frac{1}{k_L} \;=\; \frac{1}{k_w} + \frac{1}{K'_H \, k_a} \tag{14.12}$$

The first term of the right-hand side of the equation represents the liquid phase resistance of the water film, the second term represents the gas phase resistance of the air film. If the gas is soluble in water, then the dimensionless Henry's law constant K'_H is small and the gas exchange is controlled by the gas phase resistance. This is the case if K'_H is much smaller than 0.01. In this case, the resistance of the water film may be ignored. Examples are H_2O evaporation, SO_2, NH_3, H_2O, most pesticides, PAHs, and long chain organic molecules. If the gas is hardly soluble in water (e.g. O_2, N_2, most volatile organic solvents), then K'_H is large and the water film resistance controls the transfer (Schnoor, 1996). This is the case if K'_H is much greater than 0.01. If the value of K'_H is in the order of 0.01, both resistances

Table 14.1 Molecular weight, vapour pressure, aqueous solubility, and the Henry's constant of selected chemical compounds at 20 °C (after Schnoor, 1996).

Chemical	Molecular weight	Vapour pressure	Solubility	K'_H
	g mol^{-1}	atm	mg l^{-1}	-
Water	18	0.0231		
Oxygen	32	-	9.2	25.6
Ammonia	17			0.00073
trans-1,2-Dichloroethene	97	0.3432	600	2.29
Carbon tetrachloride	154	0.1267	800	1.01
Toluene	92	0.0294	478	0.231
Benzene	78	0.125	1680	0.183
Chloroform	120	0.32	7800	0.124
Chlorobenzene	113	0.0115	475	0.113
m-Dichlorobenzene	147	0.00172	111	0.0935
p-Dichlorobenzene	147	0.000792	69	0.0694
o-Dichlorobenzene	147	0.00132	134	0.0605
Bromoform	253	0.00713	2780	0.0268
Naphthalene	128	0.0003	30	0.021
Nitrobenzene	123	0.0002	1900	0.0025
PCB (Aroclor 1254)	327	1.00 10^{-7}		0.000446
Dieldrin	385	3.564 10^{-9}	2	0.000035
Alachlor	269.8	2.904 10^{-8}	242	1.30 10^{-6}
Atrazine	215.7	1.32 10^{-9}	33	5.60 10^{-7}

contribute to limiting the gas exchange and the complete expression must be used (Equation 14.12). Table 14.1 lists the molecular weights, vapour pressures, aqueous solubilities, and dimensionless Henry's law constants of certain chemical compounds.

Example 14.1 Volatilisation

The dissolved concentration of benzene (C_6H_6) in a 1.5 m deep pond is 0.2 mg l^{-1}. The pond water temperature is 20 °C. Given the gas exchange coefficient of $1.2 \cdot 10^{-5}$ m s^{-1}, calculate the flux density of benzene from the surface of the pond, and the volatilisation half-life of benzene in the pond.

Solution
Table 14.1 gives a gas exchange coefficient for benzene of 61.1 m d^{-1}, which is equal to $7.1 \cdot 10^{-4}$ m s^{-1}. If we assume that the benzene concentration in the air above the pond is virtually zero, Equation (14.10) reduces to:

$$J = -k_L \cdot C_w = -1.2 \cdot 10^{5} \times 0.2 = -2.4 \cdot 10^{6} \text{ g m}^{-2} \text{ s}$$

The volatilisation half-life is found by integrating Equation (14.11). The analytical solution of Equation (14.11) with $C_a = 0$ is

$$C_w = C_0 e^{k_L/Ht}$$

After one half-life

$$\frac{C_w}{C_0} = e^{-k_L / H\, t_{1/2}} = 0$$

Thus,

$$t_{1/2} = \frac{\ln(2)}{k_L / H} = \frac{\ln(2)}{2.4 \cdot 10^{-6} / 1.5} = 4.33 \cdot 10^5 \text{ s} = 5.01 \text{ d}$$

This implies that half of the benzene has been volatilised after approximately 5 days.

14.3 REAERATION

In the case of oxygen exchange between surface water and the atmosphere, the term k_L/H in Equation (14.11) is referred to as the ***reaeration constant*** k_r. The mass balance for exchange of oxygen between surface water and the atmosphere is then:

$$\frac{dDO}{dt} = -k_r \left(DO - DO_{sat} \right) \tag{14.13}$$

where DO = dissolved oxygen concentration in surface water [M L^{-3}], DO_{sat} = saturated (equilibrium) dissolved oxygen concentration in surface water [M L^{-3}], and k_r = reaeration constant [T^{-1}]. The saturated dissolved oxygen concentration DO_{sat} depends on the water temperature. The following empirical expression is commonly used to determine DO_{sat} in mg l^{-1} as function of temperature (Lijklema, 1991):

$$DO_{sat} = 14.652 - 0.41022\,T + 0.007991\,T^2 - 0.0000777774\,T^3 \tag{14.14}$$

where T = the water temperature in °C.

Oxygen transfer across the air–water interface is controlled on the water side, because the value of K'_H is much larger than 0.01 (see Table 14.1). This means that the overall gas exchange coefficient depends on the hydrodynamic characteristics of the air–water interface and the flow regime. The ***surface renewal model*** – an alternative model for gas exchange across the air–water interface – assumes that turbulent eddies bring small parcels of water to the water surface. As long as these parcels are at the surface they can equilibrate with the atmosphere. After equilibrium is reached, the chemical flux across the air–water interface stops. If fresh parcels replace the depleted parcels more quickly, the flux becomes larger. In deep flowing waters, the renewal rate in the mixing depth near the water surface is approximately equal to the average flow velocity divided by the water depth. This assumption leads to the O'Connor and Dobbins (1958) formula for estimation of the gas exchange coefficient:

$$k_L = \sqrt{\frac{D_w\, u}{H}} = 3.90 \sqrt{\frac{u}{H}} \tag{14.15}$$

where u = the average flow velocity (m s^{-1}) and H = the water depth (m). The surface renewal model thus assumes that the gas exchange coefficient is proportional to the square root of the molecular diffusion coefficient in water D_w, in contrast to the thin film model, which assumes that the gas exchange coefficient is proportional to D_w (see Equation 14.4). Besides this theoretically derived Equation (14.15), a number of empirical relationships have been derived based on data from natural streams (see Jha *et al.*, 2001). Since oxygen concentration is one of the most important ecological parameters in surface water, most relationships focus on oxygen transfer (reaeration). A commonly used relationship for estimation of the gas exchange coefficient for oxygen is the Churchill *et al.* (1962) relationship:

$$k_L = 5.05\,u^{0.97}\,H^{-0.67} \tag{14.16}$$

with the gas exchange coefficient in m d^{-1}, the flow velocity u in m s^{-1}, and the water depth H in m. This Equation (14.16) is valid in the range $0.6 < H < 3.35$ m and $0.6 < u < 1.52$ m s^{-1}. In lakes, the turbulence is mainly induced by wind. The mechanisms that drive the reaeration are the shear stress at the water surface, wave formation, drop formation, and wave breaking. The latter two mechanisms only occur at high wind speeds. Banks (1975) derived the following relationship for the oxygen exchange coefficient from field studies:

$$k_{L,W} = 0.0864\left(8.43\,W_{10}^{0.5} - 3.67\,W_{10} + 0.43\,W_{10}^2\right) \tag{14.17}$$

where $k_{L,W}$ = the oxygen exchange coefficient due to wind (m d^{-1}), and W_{10} is the wind speed at 10 m above the water surface (m s^{-1}). According to this equation, the reaeration of the lake water stops if the wind speed is zero. Ruys (1981) disputed this and replaced Banks's equation in the range from $W_{10} = 0$ m s^{-1} to $W_{10} = 1.82$ m s^{-1} by:

$$k_{L,W} = 0.37 + 0.09\,W_{10} \tag{14.18}$$

which represents the tangent of Equation (14.17) at $W_{10} = 1.82$ m s^{-1}. So, this equation predicts a gas exchange coefficient for oxygen of $K_L = 0.37$ m d^{-1} at zero wind speed. Banks and Herrera (1977) demonstrated that rain can also contribute considerably to oxygen exchange in a lake through an increase of turbulence and a large dissolved oxygen concentration in the raindrops. They related the gas exchange coefficient to the energy that a rain shower adds to the lake water:

$$B = 10^{-6}\left(2390 + 0.103\,Z\right)i^{1.26} \tag{14.19}$$

$$k_{L,P} = 24.45\,B \tag{14.20}$$

where $k_{L,P}$ = the oxygen exchange coefficient due to precipitation (m d^{-1}), B = the energy added to the water body per unit time per unit area in W m^{-2}, Z = the altitude of the lake surface above sea level (m) and i = the rainfall intensity in mm h^{-1}. The gas exchange coefficient for oxygen at 20 °C in a lake in the case of both wind and rain can be estimated by:

$$k_L = k_{L,W} + k_{L,P} - 0.047\left(k_{L,W} \cdot k_{L,P}\right) \tag{14.21}$$

Besides flow velocity, wind speed, and rainfall intensity, the gas exchange coefficient is also affected by water temperature and the presence of contaminants. Contaminants in the water generally cause a decrease of the gas exchange coefficient, particularly if surface-active substances (e.g. detergents), or floating film-forming substances (e.g. oil) are involved. However, at present we lack the knowledge to be able to quantify these effects for a general situation. A rise in the water temperature results in an increase of the molecular diffusion coefficient and a decrease of the stagnant film thickness at the water surface. This in turn results in an increase in the gas exchange coefficient of between 0.5 percent and 3.0 percent per °C. Usually a value of 2.4 percent per °C is adopted (Thomann and Mueller, 1987) (compare Equation 13.20):

$$k_L(T) = k_L(20°) \cdot (1.024)^{T-20} \tag{14.22}$$

where T = the water temperature (°C).

Example 14.2 Reaeration

In a small lake at an altitude of 100 m above sea level, the dissolved oxygen concentration is 4 mg l^{-1}, the water temperature is 13 °C, and the wind speed 10 m above the water surface is 7.5 m s^{-1}. Estimate the oxygen flux density during a rain shower with an intensity of 10 mm h^{-1}.

Solution
First, calculate the oxygen exchange coefficient at 20 °C due to wind. As the wind speed at 10 m is greater than 1.82 m s^{-1}, use Equation (14.17):

$$k_{L,W} = 0.0864 \left(8.43 \times 7.5^{0.5} - 3.67 \times 7.5 + 0.43 \times 7.5^2\right) = 1.706 \text{ m d}^{-1}$$

Second, calculate the oxygen exchange coefficient at 20 °C due to precipitation, using Equations (14.19) and (14.20):

$$B = 10^{-6} \left(2390 + 0.103 \times 100\right) \times 10^{1.26} = 0.0437$$
$$k_{L,P} = 24.45 \times 0.0437 = 1.068 \text{ m d}^{-1}$$

Thus, wind contributes more to the reaeration than precipitation. The resulting oxygen exchange coefficient at 20 °C is estimated using Equation (14.21):

$$k_{L,20°C} = 1.706 + 1.068 - 0.047 \times 1.706 \times 1.068 = 2.689 \text{ m d}^{-1}$$

The oxygen exchange coefficient at 13 °C is calculated using Equation (14.22):

$$k_{L,15°C} = 2.689 \times 1.024^{-7} = 2.277 \text{ m d}^{-1}$$

Then estimate the saturated dissolved oxygen concentration at 13 °C, using Equation (14.14):

$$DO_{sat} = 14.652 - 0.41022 \times 13 + 0.007991 \times 13^2 - 0.0000777774 \times 13^3 = 10.5 \text{ mg l}^{-1} (= \text{g m}^{-3})$$

Finally, determine the oxygen flux density across the air–water interface (see Equation 14.10):

$$J = -k_L(DO - DO_{sat}) = -2.277 \times (7 - 10.5) = 7.97 \text{ g m}^{-2} \text{ d}^{-1} = 0.092 \text{ mg m}^{-2} \text{ s}^{-1}.$$

14.4 GAS EXCHANGE CONTROLLED ON THE WATER SIDE

The empirical relationships presented above can, in principle, also be used for gases other than oxygen, provided that the gas exchange is controlled on the water side ($K'_H \gg 0.01$) and the gas is non-reactive. The gas exchange coefficient k_L must be corrected for the appropriate molecular diffusion coefficient. As mentioned above, the gas exchange coefficient is proportional to the molecular diffusion coefficient or the square root of the molecular diffusion coefficient, depending on the gas exchange model adopted (thin film model versus surface renewal model, respectively). The molecular diffusion coefficient in turn is approximately proportional to the square root of the molecular weight (Hemond and Fechner-Levy, 2000). This implies that:

$$k_L(A) \ = \ k_L(O_2) \cdot \left(\frac{D_{w,A}}{D_{w,O_2}} \right)^{\beta} \ \approx \ k_L(O_2) \cdot \left(\frac{\sqrt{MW_{O_2}}}{\sqrt{MW_A}} \right)^{\beta} \tag{14.23}$$

where $k_L(A)$ = the gas exchange coefficient for gas A [L T^{-1}], $k_L(O_2)$ = the oxygen exchange coefficient [L T^{-1}], $D_{w,A}$ = the molecular diffusion coefficient of gas A in water [L^2 T^{-1}], $D_{w,O2}$ = the molecular diffusion coefficient of oxygen in water [L^2 T^{-1}], MW_{O2} = molecular weight of oxygen (= 32 g mol^{-1}), MW_A = molecular weight of gas A (g mol^{-1}), and β = an exponent [-] between 0.5 (surface renewal model) and 1.0 (thin film model). Usually a value of 0.7 is used for β.

> **Example 14.3 Gas exchange controlled on the water side**
>
> Estimate the gas exchange coefficient at 20 °C for carbon tetrachloride (CCl$_4$) for a 3.5 m deep river with an average flow velocity of 73 cm s^{-1}.
>
> *Solution*
> First, estimate the gas exchange coefficient for oxygen using the Churchill *et al.* relationship (Equation 14.16):
>
> $$k_L = 5.05 \times 0.73^{0.97} \times 3.5^{-0.67} = 1.61 \ \text{m d}^{-1}$$
>
> Then, the gas exchange coefficient for carbon tetrachloride can be estimated using Equation (14.23). The molecular weight of carbon tetrachloride is 154 and that of oxygen is 32 (see Table 14.1). Thus,
>
> $$k_L = 1.61 \left(\frac{\sqrt{32}}{\sqrt{154}} \right)^{0.7} = 1.61 \times 0.577 = 0.93 \ \text{m d}^{-1}$$

14.5 GAS EXCHANGE CONTROLLED ON THE AIR SIDE

If the gas exchange is controlled on the air side ($K'_H \ll 0.01$), the gas exchange coefficient can be estimated using (Lijklema, 1991):

$$k_L(A) \ = \ 1.607 \, W_{10} \left(\frac{D_{a,A}}{D_{a,H_2O}} \right)^{0.5} \tag{14.24}$$

where $k_L(A)$ = the gas exchange coefficient for gas A in m d^{-1}, W_{10} = the wind speed at 10 m above the water surface (m s^{-1}), $D_{a,A}$ = the molecular diffusion coefficient of gas A in air [L^2 T^{-1}], and $D_{a,H2O}$ = the molecular diffusion coefficient of water vapour in air [L^2 T^{-1}] (= 2.4·10^{-5} m^2 s^{-1} at 20 °C). Other empirical expressions for gas exchange controlled on the air side are given by Schwarzenbach *et al.* (1993), Schnoor (1996), and Thibodeaux (1996).

Note that because of our incomplete understanding of the air–water gas exchange processes and the fact that so many factors influence the exchange rate, the empirical relationships presented above for estimating the gas exchange coefficient are subject to considerable uncertainty. So, the relationships should not be applied blindly. The most accurate determination of the gas exchange coefficient requires careful field experiments in the water body of interest, using a tracer gas (e.g. propane (C_3H_8)).

Example 14.4 Gas exchange controlled on the air side

The wind speed at 10 m above the surface of a river is 3.5 m s^{-1}. Estimate the gas exchange coefficient at 20 °C for nitrobenzene ($C_6H_5NO_2$), given that its molecular diffusion coefficient in air is 7.6 10^{-2} cm^2 s^{-1}.

Solution
The molecular diffusion coefficient of water vapour at 20 °C is 2.4·10^{-5} m^2 s^{-1} = 0.24 cm^2 s^{-1}. The gas exchange coefficient for nitrobenzene is calculated using Equation (14.24):

$$k_L = 1.607 \times 3 \times \left(\frac{0.076}{0.24}\right)^{0.5} = 2.7 \text{ m d}^{-1}$$

14.6 GAS EXCHANGE IN THE SUBSURFACE ENVIRONMENT

Obviously, the abovementioned relationships between the gas exchange coefficient and wind speed and rainfall intensity apply to surface water bodies in direct contact with the atmosphere. Nevertheless, Henry's law (Equation 14.1) and the thin film model (Equation 14.5) are also applicable to soil water and groundwater. Because the movement of soil water and groundwater is slow and laminar, the surface renewal model is inappropriate to describe gas exchange in the subsurface. Moreover, because the water movement is slow, equilibrium may be assumed between the soil air and the aqueous phase (soil water and groundwater near the water table) and in most cases Henry's law is sufficient for calculating the distribution of chemicals between the aqueous and gas phases. However, the soil air and free atmosphere differ in their chemical composition. The most notable difference is the smaller oxygen concentration and larger carbon dioxide concentration in the soil air due to decomposition of soil organic matter. The average oxygen content of the atmosphere is about 21 volume percent and the oxygen content in soil generally varies between 9 and 21 volume percent. The average CO_2 content of the free atmosphere is 0.03 volume percent and generally ranges between 0.021 and 0.044 volume percent above land, although the concentrations may be larger around industries, cities, thermal springs, and volcanoes (Mathess, 1994). In general, the CO_2 content of soil air is 10 to 100 times larger, with frequent values between 0.2 and 5 volume percent. In areas with natural CO_2 releases or intensive anthropogenic soil contamination by organic pollutants (e.g. waste disposal sites), values of up to 25 volume percent or larger are found. In general, natural gases occurring in soil air with elevated concentrations compared to the free atmosphere are – like CO_2 – products of the oxic or anoxic decomposition of organic matter, or radioactive decay products. Besides CO_2, these gases include nitrous oxide (N_2O)), hydrogen sulphide (H_2S), methane (CH_4), and radon (Rn).

Gas exchange between the upper soil layers (typically up to about 2 m depth) and the free atmosphere is primarily due to Fickian diffusion as a result of differences in gas concentration, although advective gas transport may also occur due to fluctuations in groundwater level, infiltrating rainwater, or biogas generation. If the gas flow velocity is zero and the gas concentrations are small (less than 5 volume percent), the gas flux due to diffusion can be calculated using:

$$J = D* \cdot \frac{\left(C(z) - C(0)\right)}{z} \tag{14.25}$$

where J is the gas flux across the soil–air interface [M L^{-2} T^{-1}], $D*$ = the effective diffusion coefficient [L^2 T^{-1}], z = soil depth [L], $C(z)$ = the gas concentration at depth z [M L^{-3}], and $C(0)$ = the gas concentration at the soil–air interface (z = 0) [M L^{-3}]. The flux J is positive for the chemical moving from soil to air. If the gas concentration becomes larger than 5 volume percent, the diffusion process itself leads to a significant apparent velocity in the gas-filled pores. In this case the gas flux can be calculated using (Thibodeaux, 1996):

$$J = D* \cdot \frac{MW}{z} \frac{P_T}{RT} \ln\left(\frac{P_T - P(z)}{P_T - P(0)}\right) \tag{14.26}$$

where MW = the molecular weight of the chemical (g mol^{-1}), P_T = the total pressure (atm), R = the gas constant (= 0.082058 (l · atm)/(mol · K), and T = the absolute temperature (K), $P(z)$ = the partial vapour pressure of the gas at depth z (atm), and $P(0)$ = the partial vapour pressure of the gas at the depth z (atm). The effective diffusion coefficient $D*$ is derived by correcting the chemical specific molecular diffusion coefficient D_a for effects of the temperature, effective porosity of the soil, and soil water content. The porosity and water content of the soil reduce the flow area and increase the flow path length (tortuosity). A useful model for describing these effects on the effective gas diffusion coefficient is (see Thibodeaux, 1996):

$$D* = \frac{(n-\theta)^{10/3}}{n^2} D_a \tag{14.27}$$

where n = the total soil porosity [-] and θ = the volumetric soil water content [-]. Note that $n - \theta$ represents the volumetric air content of the soil. This empirical Equation (14.27) does not have calibration constants and gives only a rough estimate (accurate within a factor of about 5). One use of the vapour transport models for gas exchange across the soil–air interface (Equations 14.25 and 14.26) is for estimating volatile gas emissions from landfills. As mentioned before, the concentration of anthropogenic chemicals in the free atmosphere is approximately zero, so the flux can be estimated using the chemical concentration in the soil air at a given depth, the chemical specific diffusion coefficient (and molecular weight in some cases), and the porosity and volumetric water content of the soil.

Example 14.3 Gas exchange across the soil–air interface.

The soil vapour concentration of tetrachloroethene (perchloroethylene or PCE) at 5 m below the soil surface is 150 mg m^3. Given a soil porosity of 0.30 and a volumetric soil water content of 0.12, estimate the PCE flux across the soil–air interface due to diffusion. The molecular weight of PCE is 165.83 and its diffusion coefficient in air is 7.2·10^{-2} cm^2 s^{-1}.

Solution
The effective gas diffusion coefficient in soil can be estimated using Equation (14.27):

$$D* = \frac{(0.30 - 0.12)^{10/3}}{0.30^2} \times 7.2 \cdot 10^{-2} = 0.037 \times 7.2 \cdot 10^{-2} = 2.7 \cdot 10^{-3} \text{ cm}^2 \text{ s}^{-1} = 2.7 \cdot 10^{-7} \text{ m}^2 \text{ s}^{-1}$$

he PCE concentration of 150 mg m³ corresponds to $150/165.83 = 0.90$ mmol m³ $= 0.90 \cdot 10^{-6}$ mol l⁻¹. The partial pressure is calculated using the ideal gas law (Equation 2.15):

$$\frac{n_i}{V} = \frac{P_i}{RT} \quad \Rightarrow \quad P_i = RT \frac{n_i}{V}$$

where R = the gas constant ($= 0.0821$ l atm mol⁻¹ K⁻¹). Thus, the partial pressure is

$$P_i = 0.0821 \times 293.15 \times 0.90 \cdot 10^{-6} = 2.17 \cdot 10^{-5} \text{ atm}$$

The PCE concentration is thus $2.17 \cdot 10^{-3}$ volume percent, which is considerably lower than 5 volume percent. Therefore, Equation (14.25) can be used to estimate the gas flux across the soil–air interface:

$$J = 2.7 \cdot 10^{-7} \times \frac{150}{5} = 8.1 \cdot 10^{-6} \text{ mg m}^{-2} \text{ s}^{-1} = 0.70 \text{ mg m}^{-2} \text{ d}^{-1}$$

EXERCISES

1. Explain why if the water film controls the gas exchange, an increase in the water turbulence increases the gas exchange between water and atmosphere (think of stirring), and why this does not affect the gas exchange when the air film controls the gas exchange.

2. What is the distance needed for a stream 60 cm deep to reduce the oxygen deficit by 50 percent of the initial value at flow velocities of respectively 10 cm s⁻¹ and 20 cm s⁻¹?

3. A lake 2 m deep is situated at 1500 m above sea level. The water temperature is 12 °C and the dissolved oxygen saturation is 80 percent. The wind speed at 10 m above the water surface is 9 km h⁻¹ and it is raining at an intensity of 20 mm h⁻¹.
   a. What contributes more to reaeration: wind or rain?
   b. Calculate the time needed to increase the dissolved oxygen saturation to 90 percent.

4. Estimate the gas exchange coefficient at 20 °C for toluene ($C_6H_5CH_3$) for a 2 m deep river with an average flow velocity of 1.5 m s⁻¹.

5. The total concentration of NH_3 and NH_4^+ in a small pond is 0.5 mg N l⁻¹. The temperature of the air and water is 20 °C and the wind speed at 10 m above the water surface is 4 m s⁻¹. The following parameters are given:
   Molecular diffusion coefficient of NH_3 in water $D_{NH3} = 1.76 \cdot 10^{-9}$ m² s⁻¹;
   Henry's law constant $K_H = 0.02$ atm·l mol⁻¹;
   Base constant K_b for the reaction $NH_3 + H_2O \leftrightarrow NH_4^+ + OH^- = 1.75 \cdot 10^{-5}$.
   a. Calculate the NH_3 concentration in the pond at pH = 7 and pH = 8.
   b. Calculate the volatilisation rate of NH_3 from the pond at both pH values (hints: evaluate both air-side controlled and water-side controlled NH_3 exchange).

Model calibration and validation

15.1 INTRODUCTION

In the previous chapters, a wide range of both physically based and empirical mathematical equations has been presented that may be used to model the transport and fate of substances in soil and water. The equations presented can be used alone, to tackle simple problems, or can be combined to construct complex spatio-temporal environmental models. Figure 15.1 shows an example of a water quality model including many of the transport processes presented in Chapters 12, 13, and 14. The mathematical equations listed in Figure 15.1 were coupled to the general one-dimensional advection–dispersion equation and implemented numerically in computer code. Subsequently, the computer model was used to simulate the short-term dynamics of suspended sediment and dissolved nutrients in the surface water of a rural catchment (Van der Perk, 1998). Note that although models may look impressively

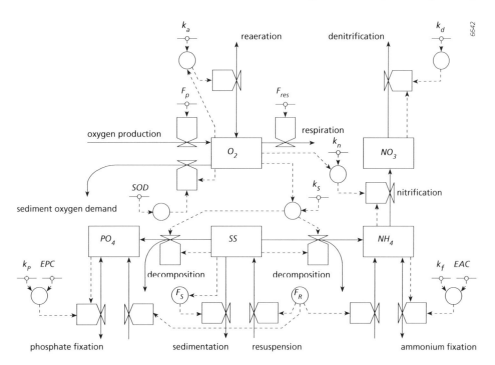

Figure 15.1 Schematic overview of a water quality model (Van der Perk, 1998). Rectangles represent state variables (see Figure 15.2), arrows and schematic valves represent first-order differential equations (see Figure 15.2a), and circles represent model parameters.

Suspended solids

$$\frac{dSS}{dt} = F_r - F_s - \frac{O_2}{M_{SS} + O_2}\theta_s^{T-20} k_s\, SS$$

$$F_r = \frac{m}{z}\frac{U_x - U_{crit1}}{U_{crit1}} \qquad\qquad if \quad U_x > U_{crit1}$$

$$F_r = -\frac{V_s}{z}\frac{U_{crit2} - U_x}{U_{crit2}}\, SS \qquad\qquad if \quad U_x < U_{crit2}$$

Phosphate

$$\frac{dPO_4}{dt} = -k_p(PO_4 - EPC) + \frac{p}{1-p}\frac{F_r}{\rho_w}PO_4 s + \frac{95}{31}\frac{O_2}{M_{SS} + O_2}\theta_s^{T-20}k_s\, SS\, P_{SS}$$

Ammonium

$$\frac{dNH_4}{dt} = -\frac{O_2}{M_n + O_2}\theta_n^{T-20}k_n\, NH_4 - k_f(NH_4 - EAC) + \frac{p}{1-p}\frac{F_r}{\rho_w}NH_4 s +$$

$$+ \frac{18}{14}\frac{O_2}{M_{SS} + O_2}\theta_s^{T-20}k_s\, SS\, N_{SS}$$

Nitrate

$$\frac{dNO_3}{dt} = -\theta_d^{T-20}k_d\, NO_3 + \frac{62}{18}\frac{O_2}{M_n + O_2}\theta_s^{T-20}k_n\, NH_4$$

Oxygen

$$\frac{dO_2}{dt} = \theta_r^{T-20}k_a(O_2 sat - O_2) - \frac{O_2}{M_{SS} + O_2}\theta_s^{T-20}k_s\, SS\, s_{O2/SS} - \frac{64}{18}\frac{O_2}{M_n + O_2}\theta_n^{T-20}k_n\, NH_4$$

$$- \frac{O_2}{M_{SS} + O_2}\theta_{SOD}^{T-20}\frac{SOD}{z} + F_p - F_{resp}$$

$$k_a = 3.94\frac{U_x^{0.5}}{z^{0.5}} \qquad\qquad if \quad 3.94\frac{U_x^{0.5}}{z^{0.5}} > K_{r\,min}$$

$$k_a = \frac{K_{r\,min}}{z} \qquad\qquad if \quad 3.94\frac{U_x^{0.5}}{z^{0.5}} \leq K_{r\,min}$$

$$O_2 sat = \gamma_1 + y_2 T + \gamma_3 T^2$$

Figure 15.2a Differential equations for a water quality model (Van der Perk 1998).

State variables

NH_4	= ammonium concentration	$[M.L^{-3}]$
NO_3	= nitrate concentration	$[M.L^{-3}]$
O_2	= dissolved oxygen concentration	$[M.L^{-3}]$
PO_4	= phosphate concentration	$[M.L^{-3}]$
SS	= suspended solids concentration	$[M.L^{-3}]$

Input variables

$NH_4 s$	= ammonium concentration for interstitial water	$[M.L^{-3}]$
$PO_4 s$	= phosphate concentration for interstitial water	$[M.L^{-3}]$
SOD	= maximum sediment oxygen demand at 20 °C	$[M.L^{-2}.T^{-1}]$
T	= temperature (in °C)	$[\theta]$

Parameters

EAC	= equilibrium ammonium concentration	$[M.L^{-3}]$
EPC	= equilibrium phosphate concentration	$[M.L^{-3}]$
F_p	= oxygen production rate due to photosynthesis	$[M.L^{-3}.T^{-1}]$
$F_{res\ p}$	= oxygen removal rate due to respiration	$[M.L^{-3}.T^{-1}]$
k_a	= reaeration rate constant at 20 °C	$[T^{-1}]$
k_d	= denitrification rate constant at 20 °C	$[T^{-1}]$
k_f	= ammonium fixation rate constant	$[T^{-1}]$
k_n	= maximum nitrification rate constant at 20 °C	$[T^{-1}]$
k_p	= phosphate fixation rate constant	$[T^{-1}]$
k_s	= maximum suspended solids decomposition rate constant at 20 °C	$[T^{-1}]$
K_{rmin}	= minimum oxygen mass transfer constant	$[L.T^{-1}]$
M	= erosion rate parameter	$[M.L^{-2}.T^{-1}]$
M_n	= Monod half saturation concentration nitrification	$[M.L^{-3}]$
M_{SS}	= Monod half saturation concentration decomposition SS	$[M.L^{-1}]$
N_{ss}	= nitrogen content suspended solids	[-]
p	= moisture content stream bed sediment	[-]
P_{ss}	= phosphorus content suspended solids	[-]
$s_{O2/SS}$	= stoichiometric coefficient oxidation SS	[-]
$U_{crit\ 1}$	= critical flow velocity for resuspension	$[L.T^{-1}]$
$U_{crit\ 2}$	= critical flow velocity for settling	$[L.T^{-1}]$
V_s	= settling velocity suspended solids	$[L.T^{-1}]$
γ_1	= empirical constant	$[M.L^{-3}]$
γ_2	= empirical constant	$[M.L^{-3}.\Theta^{-1}]$
γ_3	= empirical constant	$[M.L^{-3}.\Theta^{-2}]$
θ_d	= temperature coefficient denitrification	[-]
θ_n	= temperature coefficient nitrification	[-]
θ_r	= temperature coefficient reaeration	[-]
θ_s	= temperature coefficient decomposition SS	[-]
θ_{sod}	= temperature coefficient SOD	[-]
ρ_w	= density of water	$[M.L^{-3}]$
18/14	= stoichiometric coefficient NH4:N	[-]
62/18	= stoichiometric coefficient NO_3:NH_4	[-]
64/18	= stoichiometric coefficient oxidation NH_4 ($2O_2$:NH_4)	[-]
95/31	= stoichiometric coefficient PO_4:P	[-]

Figure 15.2b Nomenclature for Figure 15.2a (Van der Perk, 1998)

Internal functions

F_r	= resuspension rate	$[M.L^{-3}.T^{-1}]$
F_s	= settling rate	$[M.L^{-3}.T^{-1}]$
O_2sat	= dissolved oxygen saturation concentration	$[M.L^{-3}]$
U_x	= mean flow velocity	$[L.T^{-1}]$
z	= water depth	$[L]$

Figure 15.2b (continued) Nomenclature for Figure 15.2a (Van der Perk, 1998)

complex at first glance, they always remain simplified mathematical representations of the real world.

The construction of complex environmental models usually proceeds in stages. The first step is to propose a mathematical model, based on a combination of theoretical and empirical considerations and heuristics. The next step is to determine and check whether the mathematical model is a faithful representation of the conceptual model. If applicable, this step also involves a test to check if the numerical implementation of the mathematical model is a faithful representation of the mathematical model. This procedure is called model **verification**, which denotes the establishment of truth. Model verification is only possible for the mathematical steps in the model construction.

In Section 10.2 we saw that besides the set of mathematical equations, it is also necessary to input variables and parameters in the model, to perform mathematical modelling. The values of the model input variables, for instance time series of mass discharge inputs and water temperature, or digital maps of aquifer properties, are usually based on field data but may also be hypothetical. The values of the model parameters, such as for example rate constant or equilibrium coefficients, are usually chosen initially from laboratory studies or the literature. Often, these values of the model parameters must be adjusted to improve the model results to a satisfactory level. To evaluate whether the model results are satisfactory, a data set of experimental or field observations of the model's state variables is needed. In addition, some criteria are needed against which the model's performance can be evaluated. If the model results correspond favourably with the observations, then the model is accepted; if they do not, the model parameters should be further adjusted to fit the observations better. This procedure of tuning the model parameters is called model **calibration**.

If the model has been calibrated, i.e. the errors are within an acceptable range, the model is often tested using a second, independent, data set of experimental or field observations: for example, observations from another year or a different site. At this stage, the model parameters are not adjusted any more. This procedure, called model **validation**, provides reassurance that the model is performing adequately and fulfils the purpose for which it was constructed. In this case too, prescribed criteria are needed so it can be decided whether to accept or reject the model. If the model fails to reproduce the observed values within acceptable limits, the model must be redesigned and the procedure of model calibration described above should be repeated.

A verified, calibrated, and validated model does not necessarily imply that the model includes all major processes and that that the processes are formulated correctly. Real-world systems are intractable and the processes governing the transport and fate of chemicals in the environment usually depend on more factors than the models account for, but the degree of influence of those factors is often obscure. For example, most models used in river water quality management that account for the sinks and sources of dissolved oxygen and the nitrogen and phosphorus cycles, do not take account of the adverse effects of accidental toxin discharges on biochemical transformation rates. These models may perform well under regular conditions but are not intended to predict water quality after accidents. In other words, these models have not been validated for circumstances in which accidental toxin spills occur. So, in these circumstances, the model predictions will deviate from reality. Such deviations are generally acceptable if their cause is evident and the spills reoccur infrequently. On the other hand, the effects of increased toxin concentrations on the biochemical transformation rates are of scientific interest and may be incorporated and verified in some ecotoxicological models. This example demonstrates that it is unfeasible for a model to be validated for all possible and less likely conditions. This stresses the need for the range of environmental conditions for which the model has proven to be adequate to be explicitly described. The only way to gain confidence in the model's results and to understand its limitations is to test the model repeatedly.

In this chapter, the procedure of calibration and validation of environmental models will be further elaborated upon, with special reference to the criteria for an adequate model. These criteria are, in principle, the same for both model calibration and validation. Subsequently, some aspects of the model choice are discussed from the viewpoints of the purpose of the model and the interdependence between model structure and uncertainty.

15.2 MODEL PERFORMANCE CRITERIA

In order to evaluate whether the model's performance is satisfactory, some criteria for model calibration and validation should be established *a priori*. How well the model should fit the observed data depends on the nature of the observations and the desired use of the model. The simplest evaluation method is to visually compare the model's predictions and the observed values. Both the predicted and the observed values are then plotted against time or one or two spatial dimensions and the similarity of the lines is assessed (Figure 15.3a). It is also possible to plot the predicted values against the observed values and evaluate whether the points are close to the 1:1 line (Figure 15.3b). The 'soft' criteria include that the predicted values should be close to the observed values and that the predicted values should not systematically deviate above or below the observed values (in statistical terms: the residuals, i.e. the difference between predicted and observed values, should be randomly distributed, with zero mean). Visual comparison is often used in manual '***trial and error***' calibration, which entails adjusting the model parameters by hand on the basis of logic and heuristics until the model's predictions satisfactorily resemble the observations. This method is useful, especially for finding out more about the model's behaviour and the sensitivity of the model outcomes to variations in the model parameters. The main disadvantage of the trial and error calibration procedure is, however, that it remains uncertain whether the calibrated model parameter values are the statistically best (i.e. optimal) values.

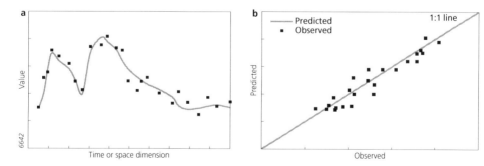

Figure 15.3 Visual comparison between model predictions and observed values: a) predicted and observed values plotted against time; b) predicted values plotted against observed values.

To obtain optimal and reproducible calibrated parameter values, a more quantitative
approach is preferable. Such an approach requires the model performance criteria to be
quantified, formalised in an ***objective function***. In mathematical terms, this objective
function describes the difference between observed values and model predictions, given
a set of model parameter values; these differences could be the absolute differences or the
squared differences, for example. The outcome of the objective function thus varies with
varying values for the model parameter values. The calibration procedure aims at finding
the minimum of this function; once this has been done, the associated parameter values are
considered to be the best estimates. Often, the sum of the square of the differences between
predicted and observed values is used as an objective function:

$$f(\alpha) = \sum_{i=1}^{n} \left(\tilde{x}_i - \hat{x}_i(q) \right)^2 \qquad (15.1)$$

where $f(\alpha)$ = the objective function, α = the combination of parameter values, n = number
of observations, $\tilde{x}_i$ = ith observed value, and $\hat{x}_i(q)$ = ith predicted value with parameter
combination q. Minimising this objective function (Equation 15.1) is also referred to
as ***least squares fitting***; it yields statistically unbiased estimates of the parameter values,
provided that the residuals are normally distributed. However, environmental concentrations
often have positively skewed (lognormal) distributions. Therefore, both the predicted and
observed values are usually logtransformed by taking their logarithm before calculating the
residuals. If the model predicts two or more variables for which observed data are available
for calibration, the least squares criterion (Equation 15.1) can also be used, but the residuals
of the different variables should be weighted proportional to the reciprocal of the means or
variances of the respective variables, otherwise the variable with the largest absolute values
dominates the calibration result.

   Several methods are available for finding the minimum of the objective function.
In all of them, the ranges within which the parameter values are allowed to vary must
be defined *a priori*, based on the literature and hard physical limits. For example, if the
denitrification rate constant in a surface water quality model has been calibrated, we know
that this constant has a value larger or equal than zero and based on the literature values
(see Table 13.3) we know that these values rarely exceed the value of 2.0 d^{-1}. An *a priori*
range for this parameter could thus be 0 – 2.0 d^{-1}. The most straightforward method for
finding the minimum of the objective function is the so-called 'brute force' method: this
method divides the range of each parameter to be calibrated into a number of discrete
steps. The model is then run for each possible parameter combination, and the objective
function is evaluated. Figure 15.4 shows, for example, the surface of the objective function
for the different parameter combinations of a simple one-dimensional, two-parameter
model of phosphate concentrations in a river, which includes first-order phosphate removal
and dilution by inflowing groundwater (Van der Perk, 1997). The minimum value of the
objective function can easily be found, but the discrete steps may cause the estimate of the
best parameter combination to become inaccurate. Another disadvantage of this method
may be the large computational effort, especially when calibrating a complex model.
Allowing only a few parameters to vary in a few discrete steps already leads to considerable
number of parameter combinations and thus model runs. This may cause the method to
become very time-consuming. Other, more sophisticated, methods search for the minimum
of the objective function more efficiently. In general, these methods start with a user-
defined initial parameter estimate after which the minimum is searched for iteratively. This
automated calibration is also referred to as ***inverse modelling***. There are several computer
tools available for automated calibration, e.g. PEST (WHI, 1999) and UCODE (Poeter and
Hill, 1998).

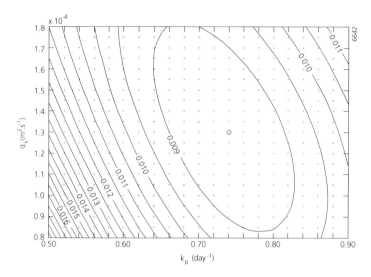

Figure 15.4 Objective function plotted against parameter values of a two-parameter model of riverine phosphate concentrations (Van der Perk, 1997). The best parameter combination (minimum value for the objective function) is indicated by.

Figure 15.4 also reveals two other features of the objective function surface, which tell us more about the uncertainty of the parameter estimate. Firstly, it is notable that the minimum value of the objective function is located in a broad oval-shaped depression in the objective function surface. This implies that many other parameter combinations perform comparably well as the best fitting parameter combination. If only a few percent additional residual error are allowed compared to the residual error of the best fit, then a considerable number of other parameter combinations appear to be reasonable and to result in an equally good model fit. Furthermore, it appears to be difficult to identify parameter q_s, because the reasonable values range over almost the entire predefined range. Thus, the objective function is not very sensitive for perturbations of the parameter q_s. The second feature of the objective function surface is that the oval depression is slightly tilted, which means the parameters are correlated. An increase in the value for q_s and a simultaneous decrease in the value for k_p results in the same value for the objective function, i.e. model performance. In general, the model parameters become more difficult to identify uniquely if the correlation between the parameters increases. Such a ***sensitivity analysis*** of the objective function, which includes both an assessment of the residual error in the parameter values and the correlations between parameters, thus provides useful information on the identifiability of the parameters.

To validate a model in a quantitative manner, various statistical criteria for model acceptance or rejection can be established. A criterion for acceptance of a catchment-scale model of nitrogen leaching into groundwater could be: 'In at least 95 percent of the observations, the predictions of nitrate concentrations (in mg l^{-1}, rounded off to one decimal place) in the upper groundwater should not exceed a factor of two'. A widely used indicator for the model performance is the squared Pearson's correlation coefficient (R^2) between predictions and observations. The Pearson's correlation coefficient is a general measure of the interrelation between two attributes; it varies between −1 and +1. A criterion for acceptance of the model might be that the R^2 must be greater than, for example, 0.7. The disadvantage of R-squared as criterion is that it is merely a measure of the strength of a relationship, and does not give any information about the absolute deviation of predicted from observed values.

In theory, the correlation coefficient could be +1, but the model predictions could be a factor of 2 too large and/or could have a constant offset. To overcome this shortcoming, the Nash efficiency coefficient has been introduced (Nash and Sutcliffe, 1970). The Nash efficiency coefficient is defined as 1 minus the ratio of the residual sum of squares $\sum_{i=1}^{n} (\tilde{x}_i - \hat{x}_i(q))^2$ to the original sum of squares of the observations $\sum_{i=1}^{n} (\tilde{x}_i - \bar{x})^2$ where $\bar{x}$ = the mean of the observations, and may therefore vary between -∞ and +1. A value of 1 implies a perfect model; a value less than zero implies that the mean of the observations is, on average, a better estimate than the model prediction.

A paired Student's *t* test can be used to test if the average difference between observed and predicted values differs significantly from zero. This test is particularly convenient to use in combination with the Pearson's squared correlation coefficient (R^2). The test statistic *t* is:

$$t = \frac{\bar{x} - \overline{x(q)}}{s.d./\sqrt{n}} \tag{15.2}$$

where $\overline{x(q)}$ = the mean of the model predictions, *s.d.* = the sample standard deviation of the mean differences between the observed and predicted values. The model is accepted if:

$$P(t \geq t_0) \geq \alpha \tag{15.3}$$

where t_0 = the critical *t* value for $n - 1$ degrees of freedom. The critical *t* values are also tabulated in standard statistical textbooks.

Example 15.1 Model validation

A water quality model is used to predict the nitrate concentrations in six different lakes. The model is validated against observed nitrate concentrations. The observed and predicted concentrations are given in the table below and Figure 15.5. Validate the model by evaluating the squared Pearson's correlation coefficient and the Nash efficiency coefficient, and by performing a paired Student's *t* test.

Lake	Observed nitrate concentration (mg l⁻¹)	Predicted nitrate concentration (mg l⁻¹)
i	$\tilde{x}_i$	$\hat{x}_i(q)$
1	1.1	1.4
2	0.63	0.7
3	0.67	0.56
4	0.87	0.99
5	0.5	0.88
6	0.32	0.34
Average	0.682	0.812

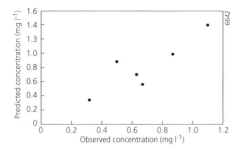

Figure 15.5 Predicted versus observed nitrate concentration

Solution
The squared Pearson's correlation coefficient (R^2) can be obtained from a statistical package or a spreadsheet. In this case, the R^2 is 0.777, which can be considered reasonable. The Nash efficiency coefficient is calculated by

$$Nash\ e.c. = 1 - \frac{\sum_{i=1}^{n} (\tilde{x}_i - \hat{x}_i(q))^2}{\sum_{i=1}^{n} (\tilde{x}_i - \bar{x})^2}$$

The calculation of the nominator and denominator in the above equation is given in the table below.

Lake i	$\tilde{x}_i - \hat{x}_i(q)$	$(\tilde{x}_i - \hat{x}_i(q))^2$	$\tilde{x}_i - \bar{x}$	$(\tilde{x}_i - \bar{x})^2$
1	-0.30	0.0900	0.418	0.1750
2	-0.07	0.0049	-0.052	0.0027
3	0.11	0.0121	-0.012	0.0001
4	-0.12	0.0144	0.188	0.0355
5	-0.38	0.1444	-0.182	0.0330
6	-0.02	0.0004	-0.362	0.1308
Sum		0.266		0.377

Thus,

$$Nash\ e.c. = 1 - \frac{0.266}{0.377} = 0.294$$

The Nash efficiency coefficient is considerably smaller than the squared Pearson's correlation coefficient. This is because the predicted values deviate slightly from the 1:1 line. On average, the predicted values are larger than the observed values (see also the average observed and predicted nitrate concentrations in the table above).

To test whether the average difference between the observed and predicted nitrate concentrations differs significantly, we first calculate the t statistic using Equation (15.2) with $\bar{x}$ = 0.682 mg l^{-1}, $\overline{x(q)}$ = 0.812 mg l^{-1}, and $s.d.$ = 0.182:

$$t = \frac{0.682 - 0.812}{0.182 / \sqrt{6}} = -1.75$$

The critical t for 5 degrees of freedom is 2.571 at a significance level of 0.05. This means that to be able to reject the null hypothesis that the mean difference between the observed and predicted values differs significantly from zero, t must be 2.571 or less (or −2.571 or less). In this case, $t = -1.75$, so we cannot reject the null hypothesis. However, note that a decision not to reject the null hypothesis does not necessarily mean that the null hypothesis is true, only that there is insufficient evidence against the hypothesis. Given the reasonably large R^2 and the fact that we cannot reject the null hypothesis that the mean difference between the observed and predicted values differs significantly from zero, we may decide to accept the model. It is possible that the deviation from the 1:1 line is a random effect given the small sample size of the example ($n = 6$). As in all statistical tests, more confidence can be gained by examining a larger sample of lakes.

15.3 CONSIDERATIONS AFFECTING MODEL CHOICE

The outcomes from model verification, calibration, and validation may justify reconsidering the model's structure, i.e. the set of mathematical equations, including the mathematical solution technique. If the model parameters have to be calibrated, the calibration data set needs to contain sufficient information to identify these parameters. It has been demonstrated (Van der Perk, 1997) that given a calibrated data set, the model parameters become less identifiable if the model becomes more complex; this is because both the errors in the parameter estimates and their mutual correlation increase. This, in turn, enhances the uncertainty of the model outcomes unless the correlations between the errors in the model parameters are taken into account. On the other hand, the accuracy of the model outcome generally increases with increasing model complexity, up to a maximum. Beyond this maximum, the effect of increasing uncertainty due to a poorer parameter identifiability becomes manifest. This means that more complex models do not necessarily yield better model outcomes and that a single 'most adequate model' can be found for a given set of calibration data. This supports the general plea for the development and use of simple and straightforward models that describe the most relevant processes and contain no redundant parameters (process descriptions).

Note that the model input variables are also often modelled: for example, through interpolation of measurements – either a simple interpolation using external data such as soil maps, or a data-driven interpolation or simulation (see Burrough and McDonnell, 1998). In some cases the input variables must be simulated using another physically-based model, and so this simulation becomes part of the model structure. This is, for example, the case when time series of rainfall simulated by an atmospheric circulation model are input into a regional groundwater quality model or a catchment-based phosphorus transport model. Whether a spatially distributed model or a 'lumped' model is chosen depends on the degree of spatial variability relative to the degree of uncertainty in the end result. If the spatial variability is much larger than the uncertainty in the model result, then spatially distributed modelling makes sense from a predictive perspective.

In conclusion, a model structure is mostly chosen based on an assessment of the purpose of the model, prior knowledge from the literature or experience, a heuristic evaluation of the important processes, and the availability of data. However, it is not always possible to choose the desired model structure: for example, when a commercial software package with fixed process descriptions is used. Even when an appropriate model has been chosen, the quality of the model output always depends on the quality of the input. Therefore, one should beware of relying too heavily on sophisticated model output and be open to a critical assessment of both model structure and model input.

EXERCISES

1. Describe in brief the similarities and differences between model calibration, verification, and validation.

2. Explain in your own words the need for model verification, calibration, and validation.

3. What are the disadvantages of 'trial and error' model calibration?

4. A model has four model parameters to be calibrated. The brute force method is used and the predefined ranges of model parameter values are subdivided into six discrete steps. How many model runs are needed to calibrate the model? How many model runs would be needed if the model had six parameters to calibrate?

5. The following table presents the observed and simulated concentration data from a biodegradation experiment.

Time (d)	Observed (mg/l)	Simulated (mg/l)	Time (d)	Observed (mg/l)	Simulated (mg/l)
1	2.61	2.28	13	0.76	0.48
2	2.54	1.98	14	0.76	0.43
3	1.48	1.72	15	0.52	0.39
4	1.64	1.50	16	0.35	0.35
5	1.48	1.31	17	0.24	0.32
6	1.48	1.14	18	0.47	0.29
7	0.85	1.00	19	0.56	0.27
8	1.29	0.88	20	0.19	0.25
9	0.90	0.77	21	0.52	0.23
10	0.95	0.68	22	0.07	0.22
11	0.74	0.61	23	0.04	0.20
12	0.82	0.54	24	0.15	0.19

  a. Calculate the squared Pearson's correlation coefficient (R^2).
  b. Calculate the Nash efficiency coefficient.
  c. Perform a Student's t test.
  d. Discuss the performance of the model, using the outcomes of questions a–c.

6. Explain why a complex model does not necessarily perform better than a simple model.

Part IV
Patterns of substances in soil and water

The genesis and evolution of spatio-temporal patterns of contaminants in soil and water are determined by a complex interaction between the spatial and temporal variation of various factors. The factors are: inputs of pollutants in soil and water, soil or sediment composition, physico-chemical behaviour of the contaminants, and transport rates. Parts I–III of this book provided an introduction to these individual factors. These governing factors and the resulting spatio-temporal pollution patterns must be known and understood if effective sampling and monitoring networks are to be designed for identifying, assessing, and delineating areas affected by environmental pollution, and if contaminated sites are to be managed successfully. Moreover, the analysis and interpretation of patterns of soil and water pollution may reveal the most important pollution sources. This, the final part of the book, deals with the various phenomena that lead to the spatial and temporal differentiation of contaminant concentrations in soil, groundwater, and surface water at a range of spatial and temporal scales. The phenomena are elucidated by means of various selected examples of observed patterns of contamination in the different environmental compartments.

Patterns in the soil and in the vadose zone

16.1 INTRODUCTION

In the previous chapters, various causes that give rise to differences in environmental concentrations of contaminants and their partitioning over the different phases (e.g. solid, liquid, gas, adsorbed, and dissolved phases) have been discussed. These causes include spatial and temporal variation in 1) the natural background concentration of contaminants, 2) the amounts and rates of contaminant inputs, and 3) the transport and chemical transformation processes the contaminants are subject to. The natural background concentration of chemicals in soil and the vadose zone depends on factors such as the natural composition of the parent material and the nature and intensity of soil-forming processes (see Section 1.3.2).

What most affects contaminant levels apart from the variation due to natural composition of the soil are the past and present contaminant inputs to the environment. The transport and fate of contaminants in soil and in the vadose zone are controlled by many factors, including climatic factors and the physico-chemical properties of soil and of the contaminant itself. After arriving in the soil, contaminants may be transported over land by runoff water, be leached into the soil profile, volatilise into the atmosphere, be taken up by plants or other soil organisms, or be broken down. Soil properties that affect the rate of contaminant transport and fate have been discussed in detail in the previous chapters and include soil depth, slope gradient, infiltration characteristics, soil texture, total porosity, pore size distribution, permeability, microbial population density and diversity, organic matter content, cation exchange capacity, soil pH, redox potential, and temperature. All these factors and properties vary in space and time to various extents and the superposition of these spatially and temporally varying phenomena results in characteristic, but often complex, spatio-temporal patterns of contaminants in soil and in the vadose zone. Such patterns may be smooth, with concentrations that vary continuously in time and space: for example, a contamination pattern that arises from atmospheric deposition. Alternatively, they may be discontinuous, with concentrations that exhibit crisp boundaries (see Burrough, 1993): for example, a contamination pattern that arises from different pesticide application rates on different fields in a patchy agricultural landscape. Analysis of the spatial and temporal patterns in soil and the vadose zone may yield site-specific information about the most important pollution sources, but also fundamental scientific knowledge and understanding about the complex interactions between the various factors and processes affecting the transport and fate of pollutants in soil. This knowledge is essential to identify, assess, and delineate areas affected by soil pollution, and for the effective management of contaminated sites and areas.

The concepts of scale and dimension discussed in Section 1.5 also apply to the aspects of spatio-temporal patterns in soil and in groundwater and surface water. Although the spatial patterns of contaminants in soil and the vadose zone extend over three dimensions, many studies have only considered one or two dimensions. In general, studies on spatial variation of soil contamination focus either on the lateral variation due to, for example, spatial differences in contaminant immissions at the soil surface, or vertical variation over the soil profile due to, for example, temporal variation of contaminant immissions. The level of contamination is mostly expressed in concentrations per unit mass, but in some studies of the lateral variation, the contaminant concentrations are integrated over the mass of soil present in a soil profile to obtain amounts per unit area, often referred to as deposition density. One application of this unit of measurement is to express radioactive contamination of soil due to atmospheric deposition (Bq m^{-2}). To convert concentration to deposition density, the following expression can be used:

$$C_d = \int_d C\, \rho_b \qquad\qquad (16.1)$$

where C_d = deposition density [M L^{-2}] or [T^{-1} L^{-2}], C = concentration [M L^{-3}] or activity concentration [T^{-1} L^{-3}], ρ_b = dry bulk density of the soil, d = depth of soil profile. If the dry bulk density is constant over the soil profile, the solution to Equation (16.1) is:

$$C_d = \overline{C} \times \rho_b \times d \qquad\qquad (16.2)$$

where $\overline{C}$ = depth-averaged concentration [M L^{-3}] or activity concentration [T^{-1} L^{-3}].

In the following sections the causes and effects of various kinds of variation of soil contamination will be illustrated and further discussed on the basis of a number of selected case studies.

16.2 NATURAL VARIATION IN BACKGROUND CONCENTRATIONS

Elements and substances that occur naturally in the environment exhibit concentration patterns in soil that are closely linked to bedrock lithology, geological mineralisation (i.e. the hydrothermal deposition of metals in ore bodies), and soil type (e.g. Bini *et al.*, 1988; Salminen and Gregorauskien, 2000; Rawlins *et al.*, 2003; Myers and Thorbjornsen, 2004). Concentrations of many substances, including metals and nutrients, are generally higher in fine-textured soils than in coarse-textured soils. Metal concentrations also tend to be higher in geologically mineralised areas. In recent years, many geological surveys have mapped the spatial patterns of elements in various environmental media, including the topsoil and the subsoil, The maps have been published in geochemical atlases, which provide data at national scale (e.g. Gustavsson *et al.*, 2001; Caritat and Cooper, 2011; Rawlins *et al.*, 2012; Mol *et al.*, 2012) or at the international, continental scale (e.g. Salminen *et al.*, 2005; Reimann *et al.*, 2012).

Because contaminants mostly enter the soil from above through atmospheric deposition, the application of fertilisers, manure, and pesticides on agricultural land, or the deposition of contaminated sediments on floodplains, it is in the topsoil that contaminant concentrations are generally highest. There are obvious exceptions, for example when contaminants are buried underground or supplied from below through upward seepage of contaminated groundwater. In areas where these exceptions are not present and if the contaminants have not migrated into deeper soil layers, the concentration of an element or substance in the uncontaminated subsoil is generally considered to be an appropriate proxy for the natural background concentration. However, it is important to recognise that natural background concentrations may vary within a soil profile as a result of natural soil-forming processes, including deposition of rainwater components, aerosols, sediments, and organic matter and the translocation of substances within the soil.

Figure 16.1 shows two example maps from the FOREGS geochemical atlas of Europe (Salminen *et al.*, 2005), which depict the spatial distribution of lead and mercury in the subsoil (samples were taken from the C horizon – a 25 cm thick section within a depth range of 50 to 200 cm). This figure shows that high Pb content is found in subsoil in northern Portugal and Galicia, southern Portugal and the Spanish Sierra Morena, the Massif Central and the Massif Armoricain in France, the Alps, the Black Forest in southwest Germany, the Ore mountains in the border area between Germany and the Czech Republic, the Tyrrhenian fringe of Italy, the karstic coastal areas of Croatia and Slovenia, Slovakia, the Attica region in Greece, and the Dalarna-Jämtland region (Sweden) in the Central Scandinavia Baltic Shield. In the majority of these areas the mercury values are also elevated, but the mercury content in the subsoil is also high in southern Spain, western Austria, southern and western Germany, and eastern Slovakia. The lead and mercury values are notably low in the metamorphic basement rocks of Fennoscandia and, consequently, in the Pleistocene glacial drift derived from these rocks in the northern Germany, Poland, Denmark and the Baltic states. Low values are also found throughout parts of central and eastern Spain and northeastern Greece where there are granite, gneiss, schist, and sedimentary rocks. These maps clearly demonstrate that lithology and geological mineralisation are important factors influencing the occurrence of high concentrations of lead and mercury – and other metals – in the soil (De Vos *et al.*, 2006).

16.3 VARIATION DUE TO CLAY AND ORGANIC MATTER CONTENT

The contaminant concentration in a soil or sediment sample depends not only on the natural background concentration and the amount of contaminant immitted, but also on the adsorption properties of the soil material. For organic contaminants, the adsorption properties are largely determined by the organic matter content, whereas for metals they depend on both the organic matter content and clay content. Because the clay and organic matter content in soils can vary over several orders of magnitude, their spatial variation laterally and vertically may be the dominant cause of spatial variation in environmental concentrations of contaminants in soils and sediments (see Van der Perk and Van Gaans, 1997). For example, Figure 16.2 shows the resemblance of spatial patterns of clay content and the concentrations of As and Zn within the topsoil of a field near Pozzuolo del Friuli, Italy (De Zorzi *et al.*, 2002).

The relation between organic matter and clay content on the one hand and contaminant concentrations on the other represents the natural preferential adsorption to the surfaces of clay minerals and organic matter. It holds for naturally occurring background concentrations as well as for elevated concentrations in polluted soils. However, in polluted soils, the quantitative relationships may be very different from those in unpolluted soils (Salomons and Förstner, 1980; Middelkoop, 1997). Middelkoop (1997) demonstrated that the effect of clay and organic matter content becomes stronger with increasing degree of pollution of river sediments from the Waal and Meuse rivers in the Netherlands. Using multiple linear regression he evaluated the relationships between clay (particle size < 2 μm) and organic matter content and heavy metal concentrations (Cu, Cd, Pb, and Zn) for three subsets of sediments. The first two subsets comprised recent overbank deposits from washlands (embanked floodplain sections) along the river Meuse and river Waal (a distributary of the river Rhine), respectively, with the Meuse sediments being more polluted than the Waal sediments. A third subset comprised more or less unpolluted pre-industrial sediment that had been deposited between about 1500 AD and 1750 AD. Although these sediment samples were not contaminated by historical industrial activities, due to pre-industrial human activities they might have slightly higher metal concentrations than natural background values. For instance, in Meuse sediments dating from Roman times (Subatlanticum) Tebbens *et al.* (2000) found Pb concentrations that were higher than natural background values; these

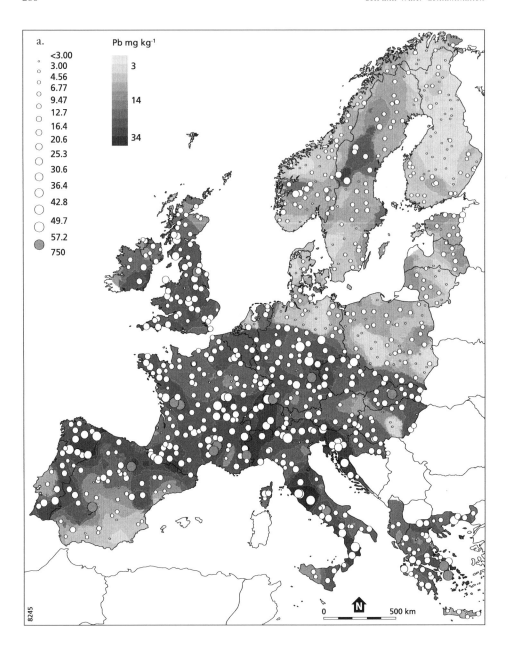

elevated values are probably attributable to atmospheric pollution due to ore mining and smelting in Roman times. In his study, Middelkoop (1997) did a regression analysis on each subset, using the following regression equation:

$$Y = a + bL + cOM + \varepsilon \tag{16.3}$$

where Y = concentration of the constituent in question (mg kg^{-1}), L = clay content (%), OM = organic matter content (%), a, b, c = regression coefficients (-), and ε = error with mean of 0.

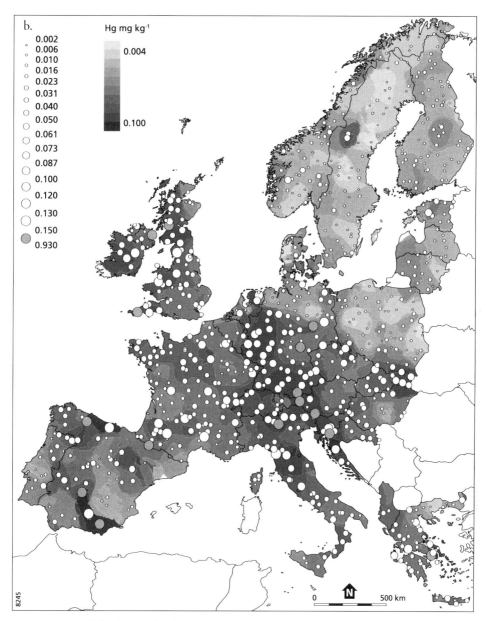

Figure 16.1 Spatial distribution of a) Pb and b) Hg in the European subsoil (source: FOREGS (Salminen *et al.*, 2005)).

The regression coefficients for the four heavy metals and three subsets considered are listed in Table 16.1; the regression surfaces are depicted in Figure 16.3.

Table 16.1 and Figure 16.3 demonstrate that the regression surfaces for the relatively uncontaminated fossil sediment are relatively flat for all four metals. This indicates that at background concentrations, the metals are only partly associated with clay and organic matter. The increments *b* and *c* increase strongly with increasing degree of pollution,

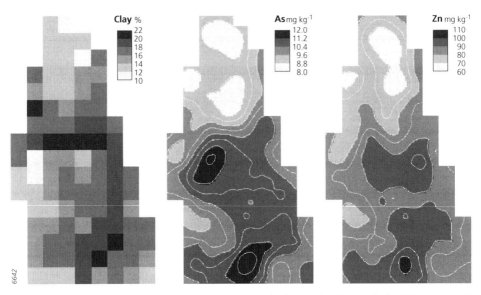

Figure 16.2 Patterns of clay content and the concentrations of As and Zn within the topsoil of a field near Pozzuolo del Friuli, Italy (source: Paolo De Zorzi, personal communication).

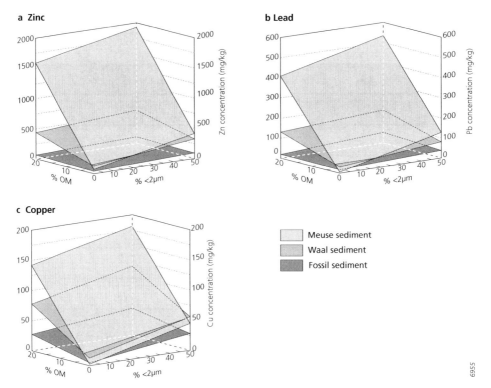

Figure 16.3 Regression surfaces of metal concentrations as function of clay and organic matter content (Middelkoop, 1997).

implying a strong sorption of the enriched metals by clay and organic matter. The coefficients related to organic matter increase more than the coefficients for clay, particularly for Pb, implying that in unpolluted sediments, metals are more associated with the mineral fraction from which they are released by weathering. In polluted sediments, metal ions adsorb to organic matter more than in unpolluted sediments. However, this does not necessarily mean that metals bond preferentially to organic matter. The way metals are released into the fluvial environment may also have determined the relationships in the sediments (Förstner and Wittman, 1983). In the case of these polluted overbank deposits, the metals have mainly been derived from discharges from wastewater treatment plants and industrial outfalls in upstream parts of the catchments (Vink and Behrendt, 2002). During transport by the river water, the metals can readily partition between the river water solution and the suspended matter that consists both of mineral components and organic matter. The reason that the metals seem to prefer the organic matter in polluted sediments rather than in unpolluted sediments (in which the metal ions originate principally from the mineral fraction) could be that the polluted mineral and organic suspended matter have been deposited concurrently on the floodplain.

As noted above, the spatial variation in particle size distribution and organic matter content of soils may dominate the patterns of contaminants in soil. This source of variation may largely obscure regional trends of soil contamination from other sources. The effect of variations of clay and organic matter content can be filtered out by standardising the contaminant concentrations to a standardised soil or sediment with a given clay and organic matter content. Equation (16.3) can be used to standardise the concentrations to a standard soil with given percentages of clay and organic matter. The concentration in the standard sediment can be obtained by calculating (see Van der Perk and Van Gaans, 1997):

Table 16.1 Regression coefficients between metal concentrations and percentages of clay and organic matter content and R^2 for three subsets of overbank deposits of the rivers Meuse and Waal, the Netherlands (see regression Equation 16.3) (Middelkoop, 1997).

	a	b	c	R^2
Cd				
Fossil (< 1750 AD)*	-	-	-	-
Waal	0.64	0.02	0.11	0.69
Meuse	-0.94	0.05	0.49	0.65
Cu				
Fossil (< 1750 AD)	5.5	0.46	0.99	0.66
Waal	15	0.82	3.05	0.88
Meuse	0.4	0.78	7.12	0.94
Pb				
Fossil (< 1750 AD)	22.0	0.50	0.25	0.48
Waal	38.2	0.75	4.71	0.73
Meuse	4.18	2.45	20.50	0.95
Zn				
Fossil (< 1750 AD)	46.1	1.00	1.00	0.66
Waal	163.0	3.35	13.87	0.71
Meuse	67.6	6.75	76.20	0.95

* Regression could not be performed, because Cd concentrations in fossil sediment were below the estimation limit (0.20 mg kg^{-1})

$$Y_{st} = \frac{Y_m}{Y_r} \left(a + L_{st}\, b + OM_{st}\, c \right) \tag{16.4}$$

where Y_{st} = standardised concentration (mg kg^{-1}), Y_m = measured concentration (mg kg^{-1}), Y_r = regression prediction of concentration (mg kg^{-1}), L_{st} = clay content of standard soil (%), and OM_{st} = organic matter content of standard soil (%). In the Netherlands, the standard soil is usually defined as a soil with 25 percent clay and 10 percent organic matter. However, different percentages of clay and organic matter content may also be assumed for the standard soil.

16.4 LATERAL VARIATION

16.4.1 Introduction

The principal cause of lateral spatial variation of contaminants in the soil and vadose zone is spatial differences in inputs and losses of contaminants at the soil surface or from groundwater. As noted, the transport processes in the soil and vadose zone are generally vertically oriented, so lateral dispersal of contaminants from a point source contamination is of minor importance. Therefore, point source pollution in soil and vadose zone can usually be considered to be horizontally confined. Conversely, diffuse inputs of contaminants extend over much larger areas. Examples of important diffuse sources that contribute to soil contamination are applications of agricultural fertiliser and pesticide, atmospheric deposition, and the deposition of contaminated sediment. In contrast to the direct anthropogenic immissions of fertiliser and pesticide application, contaminant immissions into soil due to atmospheric deposition and deposition of contaminated sediment are controlled by natural transport processes driven by wind or water flow.

16.4.2 Effects of fertiliser and pesticide application

As a consequence of anthropogenic inputs of fertilisers and pesticides, the topsoil of agricultural soils is enriched with a wide range of contaminants such as nitrogen, phosphorus, heavy metals, and chlorinated hydrocarbons. These contaminants comprise the persistent organic and inorganic residues of the agrochemicals. Ferguson *et al.* (2003) demonstrated that the application of agrochemicals (including livestock manure, inorganic fertilisers and lime, pesticides, sewage sludge, irrigation water, industrial by-product 'wastes' and composts) is the second most important source of heavy metals in agricultural soils in England and Wales. In these soils, these inputs contribute to an estimated 51% of the total annual Zn input, 61% of the total Cu input, 22% of the total Pb input, and 47% of the total Cd input. However, in individual fields where livestock manure and sewage sludge are extensively applied, they could be the major source of heavy metals (Ferguson *et al.*, 2003). Consequently, differences in present and past land use and management between different fields or agricultural areas are likely to be reflected in the spatial pattern of contaminants in the topsoil.

   The contaminant concentrations in soil resulting from application of agrochemicals may have a strong seasonal variation. This seasonal variation will be influenced by the times of application and crop harvest, the crop uptake rate, and the rates of transport and chemical transformation processes (e.g. leaching, decomposition, and retention). In the long term, persistent impurities in agrochemicals (e.g. heavy metals) or other residues may build up in the topsoil of farmland. Many residues (for example, phosphorus and heavy metals) exhibit a great adsorptive affinity for soil solids. Others (for example, nitrate and some pesticide compounds) are very soluble in water and are primarily found in dissolved form in the pore water solution of the vadose zone. These substances are particularly susceptible to leaching to groundwater. The field variability of soil and pore water composition and related biogeochemical processes is known to be quite high and to occur across very short spatial (e.g. several metres)) and temporal scales (e.g. days) (e.g. Parkin, 1993; Daniels *et al.*, 2001; Eghball *et al.*, 2003; Cox *et al.*, 2003). There are many reasons for this; they include the complex, interrelated processes associated with nutrient cycles, short-range spatial variability of the soil properties such as soil texture, organic matter content, and pore size distribution, and spatial and temporal variability of soil management practices (e.g. ploughing, fertiliser application, sowing, and harvest), and climatic factors such as temperature, precipitation, and evapotranspiration. Short-range spatial variability and short-term temporal variability may interfere with the ability to detect regional patterns of variation: for example, differences in nitrate concentrations in pore water between fields. To reveal regional patterns of concentrations that are subject to high variability, the sample support (see section 1.5) can be increased by bulking samples or averaging the results of sample analysis.

Field-based studies on patterns of organic pollutants in soil at the landscape or regional scale as a consequence of agricultural activities are very rare, probably because such studies require a large number of field samples and the costs of laboratory analysis are high. Studies on spatial variability of inorganic pollutants in soil are more common (e.g. Sauer and Meek, 2003; Spijker, 2005).

Spijker (2005) presented an analysis of the natural and man-induced geochemical patterns in soils of the rural part of the province of Zeeland, the Netherlands. Samples of young Holocene marine clayey sediments were collected at two depths (the plough zone at 5–30 cm and the C-horizon at about 40–80 cm) at 270 locations on agricultural land and analysed for the elemental composition. This study revealed that for the major elements (Al, Si, Ca, Mg, S) the largest significant source of small-scale variation occurred at the field scale (about 300 m). The major source of variation for heavy metals generally appeared to be at a smaller scale than for the major elements, probably because of local human interference. Before the regional patterns of soil contamination could be analysed, the natural variation due to differences in soil properties, notably clay matter content, had to be removed first. To do so, a linear regression between the different elements and the aluminium (Al_2O_3) content was performed for the deeper samples from the C-horizon. This regression is similar to the regression described in the previous section (see Equation 16.3), but in this case the Al_2O_3 content was used as a proxy for clay content. Organic matter was omitted from the regression, because the organic matter contents in the samples were quite low (about 3 percent maximum) and did not contribute much to the variation in element concentrations. The enrichment in the topsoil for each location was determined by calculating the difference between the actual measured concentration and the regression prediction based on the Al_2O_3 content of the topsoil. In this way, it was shown that the topsoil was enriched in Cd, Sb, As, Cu, Pb, Zn, and P_2O_5. The strong correlation between Cd and phosphorus (see Figure 16.4) suggests that the Cd enrichment was primarily the consequence of input as impurities of phosphate fertilisers. Figure 16.5 shows the regional variation in Cd enrichment: compared with the rest of the province, the western part of the central peninsula (Walcheren and Noord-Beveland) is more enriched in Cd whereas the south (Zeeuws-Vlaanderen) is less enriched in Cd. These regional differences might reflect differences in fertilisation rates, or fertilisers containing different amounts of Cd.

16.4.3 Effects of atmospheric deposition

Deposition of atmospheric pollutants (e.g. acidifying component, POPs, heavy metals) is often a major source of diffuse soil contamination, especially in non-agricultural areas,

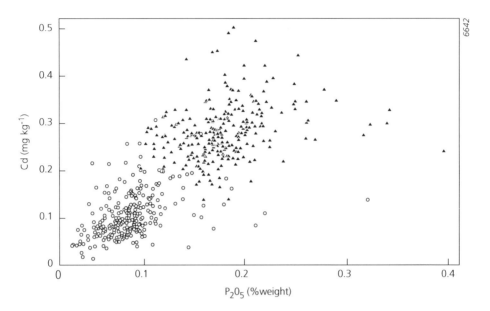

Figure 16.4 Relation between cadmium and phosphorus concentrations in the topsoil (Δ) and the C-horizon (o) in the rural areas of the Province of Zeeland, the Netherlands (Spijker, 2005).

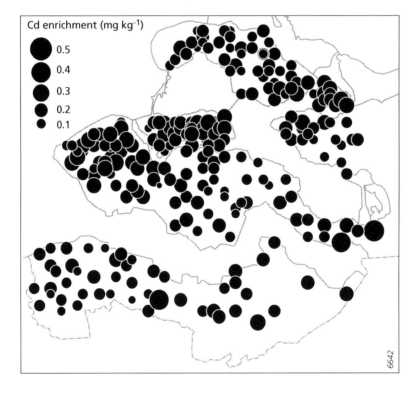

Figure 16.5 Spatial pattern of cadmium enrichment in the topsoil in the rural areas of the Province of Zeeland, the Netherlands (Spijker, 2005).

but also on agricultural land. For example, in England and Wales, atmospheric inputs to agricultural soils account for 49% of the total annual Zn input, 39% of the total Cu input, 78% of the total Pb input, and 53% of the total Cd input (Ferguson et al., 2003). Atmospheric deposition onto soils varies over a broad range of spatial scales and depends on the proximity to the source, precipitation, and landscape structure. The relation to distance to the pollution source can be exemplified by the spatial distribution of ^{137}Cs deposition onto soil following the Chernobyl accident in Ukraine, April 1986 and the Fukushima nuclear accident in Japan, March 2011 (see Figures 8.3 and 8.4, respectively). The areas most affected by the Chernobyl accident were Ukraine, Belarus, and the European part of Russia (see Figure 8.3). Two factors influenced the patterns of continental deposition of ^{137}Cs farther away from Chernobyl. The first was the atmospheric circulation patterns during and immediately after the accident. The second was the incidence of local rainstorms which flushed the radioactive dust from the atmosphere to settle on vegetation and soil as the radioactive cloud passed overhead. The Chernobyl accident released radioactivity over a relatively short period of time: about ten days. The first day accounted for about 25 percent of the total radioactive emission. But it has also been shown that supra-regional deposition patterns of previous bomb-derived ^{137}Cs associated with the testing of nuclear weapons, which occurred during a much longer time span in the late 1950s and 1960s, are determined by annual rainfall (Basher and Mathews, 1993; Owens and Walling, 1996; Bernard *et al.*, 1998).

Prolonged releases from point sources of atmospheric pollutants also result in atmospheric deposition tending to decrease with increasing distance from the pollution source. However, the resulting pattern is usually smoother compared with momentary releases, because it is comprised of a superposition of patterns developed under various atmospheric conditions of wind direction, wind speed, and precipitation rates. Nevertheless, the resulting pattern often clearly reflects the prevailing wind directions. For example, in the USA, many power plants are located in the Midwest, and here and throughout the east coast, acid deposition derived from SO_2 emissions from these power plants is common. Likewise, large areas of Scandinavia, Central and Eastern Europe, and parts of China have been adversely affected by deposition of acidifying components originating from upwind emissions.

Figure 16.6 shows an example of soil contamination due to past atmospheric emissions from an incinerator in Sint Niklaas, Belgium (De Fré et al., 1992). This figure depicts the

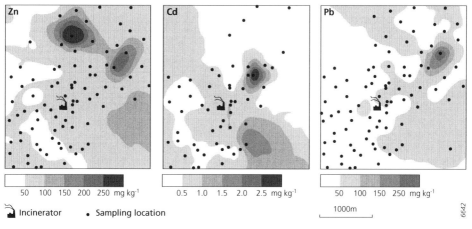

Figure 16.6 Interpolated zinc, cadmium, and lead concentrations in the topsoil around an incinerator near Sint Niklaas, Belgium (De Fré *et al.*, 1992).

resulting concentration patterns of the heavy metals Zn, Cd, and Pb in the topsoil in the vicinity of the incinerator. The heavy metals have dispersed predominantly north-eastwards, due to the prevailing south-western winds. The maximum concentrations of heavy metals occur at a distance from 1.5 to 2.0 kilometres away from the incinerator, where the plume from the chimney most often reaches the soil surface. It is conspicuous that the maximum concentrations do not coincide for the different metals. The maximum levels of Cd are found nearer the source than Zn and Pb. This is probably due to different transport ranges of the different particle size fractions to which the metals bind. Zinc and lead vaporise during incineration and condense mainly onto the smallest airborne particles that have the largest specific surface. Cadmium does not vaporise and, as a consequence, remains also associated to the somewhat larger particle fractions. These larger particles are transported over shorter distances than the smaller particles (De Fré *et al.*, 1992).

At the local scale, deposition is affected by aerodynamic surface roughness controlled by local surface topography and land cover (Bachhuber *et al.*, 1987; Draaijers, 1993). The presence of hills and transitions in the height and structure of vegetation cover causes airflow perturbations which alter the rates of dry and wet atmospheric deposition. Draaijers (1993) studied the effect of canopy and forest edge structure on deposition amounts in the Netherlands, using the so-called throughfall method. Throughfall refers to the precipitation water dripping from the leaves, needles or branches and falling through the canopy gaps. By measuring the amount and composition of the throughfall water using gutters of 5 m long and a total collection area of 0.054 m^2, the total deposition (i.e. wet + dry + occult deposition) flux could be estimated. Strong correlations were found between net throughfall fluxes of SO_4^{2-}, NO_3^-, NH_4^+, Na^+, and Cl^- and the roughness and leaf area of forest canopies. Draaijers (1993) also observed an exponential increase of the net through fluxes towards the forest edge, although there was a considerable scatter around this general trend (Figure 16.7). Because the sample support (i.e. gutter length) was of the same order of magnitude of one tree crown, this scatter was largely attributed to short-range variability associated with the leaf area and the occurrence of branches and canopy gaps above the gutters. The width of the zone with enhanced net throughfall fluxes approximated 5 edge heights. The net throughfall increase of SO_4^{2-}, NO_3^-, NH_4^+ near the forest edges (on average, a factor of 2) was smaller than that of Na^+ and Cl^- (on average, a factor of 5). These differences were attributed to the sources for these ions in throughfall. In the Netherlands, net throughfall of SO_4^{2-}, NO_3^-, NH_4^+ occurs predominantly through dry deposition of gases and particles smaller than 1 μm, whereas net throughfall of Na^+ and Cl^- mostly occurs by dry deposition of sea salt particles larger than 1 μm. In the Draaijers study, differences between forests in net throughfall increase in forest edges were attributed to differences in forest density and edge aspect. Net throughfall fluxed was found to be positively correlated to forest density; forest edges exposed to prevailing wind directions (south and south-west) exhibited enhanced dry deposition.

Draaijers *et al.* (1994) and Weathers *et al.* (2001) have discussed the relevance of these forest edge effects for atmospheric deposition in complex and patchy forested landscapes. As forest edges can function as significant traps both for airborne nutrients and for pollutants from adjacent agricultural or urban landscapes, an important factor in quantifying contaminant inputs to the soil surface through atmospheric deposition is the landscape structure and fragmentation. In the Netherlands, almost 80 percent of all forest complexes are smaller than 5 ha, and more than 70 percent of all individual forest stands are smaller than 1.5 ha. Assuming a forest edge of five edge heights, at least 50 percent of the total forest area in the Netherlands is affected by edge effects (Draaijers, 1993).

The atmospheric deposition of acidifying components (NO_x, NH_3, SO_2) has not only caused soil pH to fall, but has also caused secondary pollution through the loss of Ca from the soil profile and the mobilisation of toxic aluminium. These effects are especially noticeable in poorly buffered soils that are low in bicarbonate, soil organic matter, cation

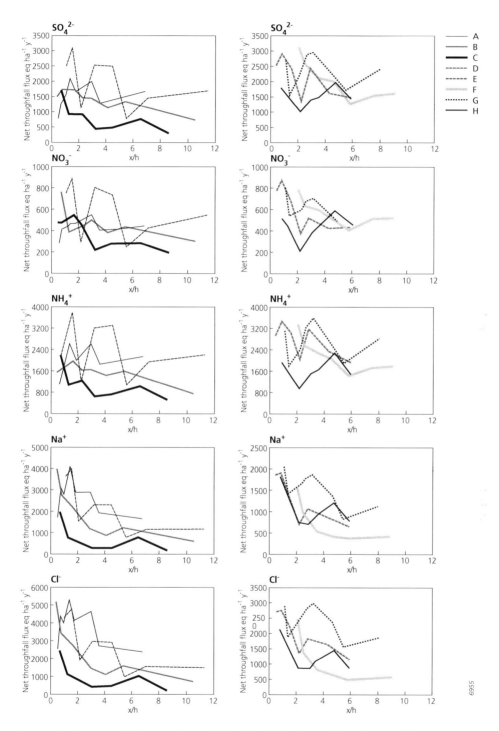

Figure 16.7 Net throughfall flux gradients of SO_4^{2-}, NO_3^-, NH_4^+, Na^+, and Cl^- as function of distance from the forest edge for eight forest edges located in the Netherlands. The distance to the forest edge is expressed in edge heights. Measurements were performed during one year in the period 1989–1991 (Draaijers, 1993).

exchange capacity and base saturation. In the USA and Europe, the emission and atmospheric deposition rates of acidifying components peaked in the 1970s and 1980s. Since then, the rates of atmospheric deposition of these and other compounds have declined as a result of environmental policies. For example, total annual emissions of SO_2 in the USA dropped from 35.0 million tonnes in the 1970s to 15.1 million tonnes in 2005 and in Western Europe they decreased from 29.3 million tonnes in the 1970s to 6.2 million tonnes in 2005 (Smith *et al.*, 2011). A recent study in a number of spruce forests in the northeastern USA (Lawrence *et al.*, 2013) showed only a modest recovery of the soil in response to the declining acid deposition, but not at all sampling sites. The study showed that, in general, exchangeable aluminium in the topsoil began to disappear when the mobilisation of aluminium decreased and the topsoil was replenished by decaying plant litter, which has low levels of aluminium. However, calcium levels in the soil remained low, because the soil material is not rich in this element and weathers very slowly. This study demonstrates that soil recovery is slow process. It is also important to note that globally, the total emissions and deposition of acidifying components have not decreased as dramatically as they have in North America and Europe. For example, in China, the fastest growing economy in the 2000s, the total annual emission of SO_2 rose from 7.3 million tonnes in the 1970s to 32.7 million tonnes in 2005 (Smith *et al.*, 2011). The average annual bulk nitrogen deposition also rose from 13.2 kg N ha^{-1} y^{-1} in the 1980s to 21.1 kg N ha^{-1} y^{-1} in the 2000s. In the industrialised and agriculturally intensified regions of China, current nitrogen deposition rates match the peak levels of deposition in Western Europe in the 1980s (Liu *et al.*, 2013).

16.4.4 Effects of soil erosion and deposition

Erosion and deposition of soil particles is an important vector for contaminant redistribution in catchments (Stone, 2000; Walling and Owens, 2003). Obviously, this effect only applies to sediment-associated contaminants. The various processes involved in the erosion and deposition depend on landscape position and characteristics such as slope gradient, slope length, soil infiltration capacity, soil erodibility, and vegetation (see Chapter 12). If the topsoil has been enriched with contaminants as a consequence of agricultural inputs or atmospheric deposition, the contaminants may be translocated from hill slopes to sedimentation areas downstream, such as valley bottoms or floodplains. This leads to contaminant losses in erosion areas and gains in sedimentation areas and, consequently, to patterns of contaminants and nutrients that are closely related to landscape topography (e.g. ; Van der Perk *et al.*, 2004; 2007; Xiaojun, 2010). In addition to soil erosion induced by rainfall events, soil particles are also moved laterally by tillage. The net translocation is downslope and its rate depends on the slope gradient. As in the case of rainfall-induced soil erosion, this results in a net loss of soil material and associated contaminants on convex parts of the landscape and a net gain on concave parts (Lobb *et al.*, 1995; Quine *et al.*, 1999; Heckrath *et al.*, 2005).

Van der Perk *et al.* (2002) investigated the effect of soil erosion and deposition on the spatial distribution of Chernobyl-derived ^{137}Cs deposition values in the arable part of the small Mochovce catchment (1.4 km^2) situated in a hilly part of the Danube Lowlands in western Slovakia. This study combined a straightforward long-term sediment redistribution model presented by Govers et al. (1993) (see Section 12.6) and geostatistical interpolation of point samples of ^{137}Cs activity in soil to distinguish the effects of sediment erosion and deposition from other sources of variation in ^{137}Cs. Figure 16.8 shows the interpolated pattern of ^{137}Cs deposition density. Enhanced ^{137}Cs activity in soil was found in the bottoms of the side-valleys and in a small floodplain area in the central part of the study area, whereas diminished ^{137}Cs activities were found on the steep slopes in the eastern part of the catchment. At the scale of the entire catchment, soil erosion and deposition accounted for only 25 percent of the total variation in ^{137}Cs activity in soil. This rather low value could be

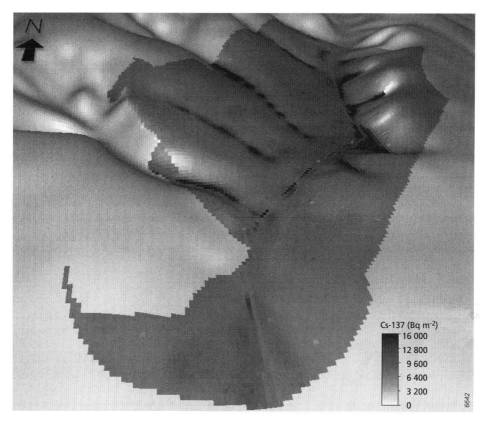

Cs-137 (Bq m⁻²)

16 000
12 800
9 600
6 400
3 200
0

6642

Figure 16.8 Interpolated ^{137}Cs deposition density in the Mochovce catchment, Slovakia (Bq m^{-2}) (Van der Perk *et al.*, 2002).

attributed to floodplain sedimentation (which was not included in the erosion–deposition model), to the initial pattern of ^{137}Cs deposition, and to short-range spatial variation.

As was demonstrated by the above ^{137}Cs study in Slovakia, floodplain deposition can be a substantial source of variation of contaminants in catchments. The amounts and spatial variability of deposition of sediment-associated contaminants on floodplains during overbank flooding depend on several factors, including frequency and duration of the flooding, suspended sediment concentration in the main river channel, and the flow patterns and stream velocities in the floodplain. The spatial variability of floodplain deposition is related to floodplain topography. In general, high deposition rates are found on low-lying floodplains that are frequently inundated for relatively long periods. The fine sediments that carry the greatest contaminant load tend to accumulate in the lower parts of the floodplain (Leenaers, 1989; Leenaers *et al.*, 1990). At the same time, the deposition rates generally decrease with distance from the source, i.e. the river channel (Burrough *et al.*, 1996). Figure 16.9 illustrates these general trends for a floodplain along the river Meuse, the Netherlands.

However, the relationships between floodplain contamination on the one hand and floodplain elevation and distance to the main channel on the other are essentially empirical and often only apply to small stretches of floodplain. They may become inapplicable if the floodplain area is larger and has an irregular topography with depressions, backwaters, minor dikes, and other objects that influence sedimentary patterns. In these areas, a physically-based approach is required to explain the patterns of floodplain sedimentation and contamination.

Zn (mg kg⁻¹)
- < 125
- 125- 250
- 250- 500
- 500-1 000
- >1 000

Figure 16.9 Interpolated zinc concentrations for the Meuse floodplain near Meerssen, South Limburg, the Netherlands (Burrough *et al.*, 1996).

Middelkoop and Van der Perk (1998) combined a hydrodynamic model that calculated two-dimensional water flow patterns over a floodplain with a straightforward floodplain sedimentation model. This enabled them to relate hydrodynamic flow patterns to observed amounts of deposited sediment in several floodplain sections along the river Waal and to model both sediment accumulation during single flood events and average annual sediment accumulation rates. Figure 16.10 shows the typical calculated flow patterns of the flood water over a stretch of washland with a minor dike at various discharge stages; Figure 16.11 shows the interpolated sediment accumulation observed by Middelkoop and Asselman (1998) on the same stretch of washland after the flood of December 1993.

Figures 16.10 and 16.11 demonstrate that in washlands with minor dikes the amount of sediment deposited is principally controlled by sediment supply from the main channel by advective transport, rather than by floodplain elevation or duration of inundation. The inundation frequency and duration is almost the same in the entire area and is controlled by the crest height of the minor dike. As soon as the river water rises above the lowest crest height, the entire area is inundated. The lower part of the minor dike in the western part of the washland allows water and sediment to enter this part of the floodplain. Although the eastern part of the washland is submerged, the sediment supply from the river channel is zero until the river water rises above the dike crest in the eastern part of the area. Only then does the river water flow over the entire floodplain section, supplying the eastern part of the washland with sediment. Because the duration of sediment supply is much shorter here than in the western part of the washland, the eastern part receives less sediment (see Figure 16.11).

It is clear that the input of sediment-bound contaminants on floodplain soils is directly related to the sediment input to soil. Middelkoop *et al.* (2003) demonstrated that the heavy metal concentrations in fresh sediment standardised for clay and organic matter content (see Section 16.2) do not vary significantly within and among stretches of washlands. Nevertheless, the absolute concentrations vary with particle size distribution. In general, the clay and organic matter content of freshly deposited sediment increases with distance from the sediment source (Thonon, 2006), i.e. along the flow lines as depicted in Figure 16.10. As

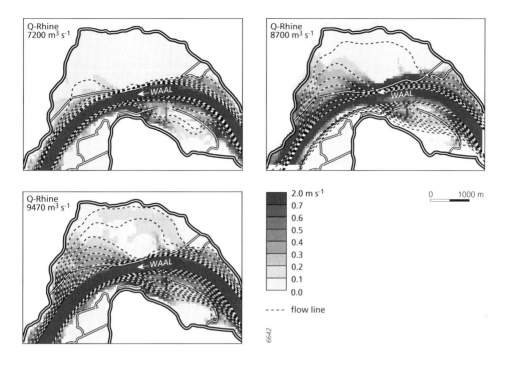

Figure 16.10 Typical flow patterns of the flood water in the Bemmelsche Waard washland near Nijmegen, the Netherlands for different discharge stages of the river Rhine. Shading indicates flow velocities (Middelkoop, 1997).

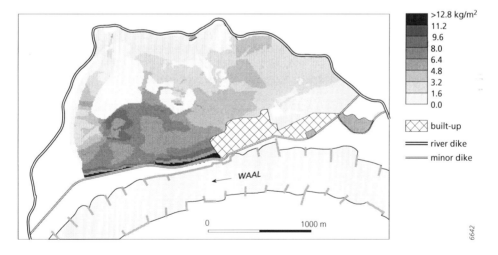

Figure 16.11 Observed sediment accumulation on the Bemmelsche Waard washland after the flood of December 1993 (Middelkoop, 1997).

is the case with the above study of ^{137}Cs, the heavy metal content in the topsoil increases with contaminant input (Van der Perk *et al.*, 1992).

16.5 VERTICAL VARIATION

16.5.1 Introduction

Vertical variation of contaminants over the soil profile occurs as a result of temporal variation of contaminant inputs and losses at the soil surface and subsequent vertical transport processes that redistribute the contaminants over the soil depth. When considering the development of soil contaminant profiles, it is useful to distinguish between inputs of contaminants only and inputs of contaminated sediment. Although any contaminant input implies an input of mass to the soil, the input of contaminants by atmospheric deposition or application of agrochemicals does not usually increase the mass or volume of a soil body significantly. However, erosion or deposition of sediment does lead to a change in soil profile structure. Long-term application of mulch or sewage sludge may alter the soil profile structure. Obviously, the soil profile structure is altered when land is excavated or when land is raised with imported soil material or waste materials.

The concentration profile is continuously being altered by transport processes in soil, particularly by leaching and bioturbation. In permeable soils in temperate climates, leaching of dissolved substances due to gravity-driven vertical flow is typical in the unsaturated zone. In semi-arid and arid climates, vertical upward migration of solutes by capillary action predominates. Bioturbation, i.e. the reworking of soil material by burrowing animals such as earthworms, arthropods, moles, and voles, causes dispersion of contaminants in the soil profile. The effect of the dispersive action of bioturbation on the vertical distribution of contaminants in soil was discussed in Section 11.3.3 (see Figure 11.9). In addition, in agricultural soils vertical mixing also occurs as a consequence of tillage operations, which – in the long term – result in a more or less uniform redistribution of substances over the plough depth (usually 20–30 cm).

16.5.2 Effects of leaching

The leaching of solutes through the soil profile is affected by a set of physical, chemical, and biological processes, including infiltration, evapotranspiration, root uptake, advection, dispersion, sorption, decay, and volatilisation. The soil properties that govern these processes are often heterogeneous in nature, for example due to the presence of soil horizons, cracks, and root channels. One particular consequence of the heterogeneities in physical soil parameters is the occurrence of preferential flow. Preferential flow gives rise to a rapid leaching of contaminants through the vadose zone, because the percolating flow is concentrated in macropores or fingers (Keesstra *et al.*, 2012; see also Section 3.2.3). Bowman *et al.* (1994) found, for example, that as much as 80 percent of the water percolating through a soil profile actually passes through only about 20 percent of the cross-sectional area of the profile. Furthermore, the rapid preferential flow and the exchange between the preferential flow paths and the soil matrix cause solute breakthrough curves in the vadose zone to exhibit earlier arrival and longer tailing than predicted by percolation models that assume flow through a homogeneous, unsaturated medium.

In addition to using laboratory experiments and model simulations, scientists often study field-scale solute transport in the vadose zone with tracer experiments. For this purpose, chemical tracers are put in or on the soil and then the spread of the tracer concentrations in soil and groundwater is monitored. Hendriks *et al.* (1999) carried out an experiment using a conservative bromide tracer on the Bouwing experimental farm in the centre of the Netherlands from November 1994 to May 1995. The soil was alluvial with a light to medium-heavy clay texture (clay content varied between 31–44 percent in the top 120 cm). The soil swells and shrinks, which causes shrinkage cracks to form under dry conditions. An

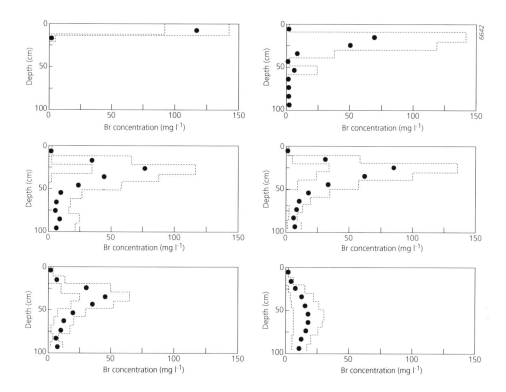

Figure 16.12 Concentration distribution of bromide tracer in an alluvial clay soil in the Netherlands on six sampling dates. Measured values are means of 16 measurements ± one standard deviation. Adapted from Hendriks *et al.* (1999).

amount of 12.0 g m^{-2} Br$^-$ was applied under wet conditions when no shrinkage cracks were visible at the soil surface. After the bromide application, the vertical distribution of bromide concentrations in the top 1 m of the soil was monitored at 16 randomly distributed locations within the experimental plot (16 m × 90 m). The plot was ploughed two days after bromide application. During these two days there was very little precipitation (0.2 mm), so most of the bromide was still present in the topsoil (see Figure 16.12a). The ploughing inverted the topsoil, such that bromide was displaced to a depth just above the plough depth of 25 cm. Therefore, five days after bromide application, the largest amounts of bromide in the soil were observed at depths of 0.15 and 0.25 m, but with great spatial variability (Figure 16.12b). After ploughing, the bromide peak remained basically in the same position, because of the poor permeability of the plough pan. Despite the very poor permeability of the clayey soil matrix, the bromide peak at 25 cm depth decreased and the bromide concentrations below the plough depth increased slowly (Figures 16.12c-f). The spatial variability in bromide contents also decreased considerably, probably due to the lateral redistribution of water and bromide in the soil matrix. The total amount of bromide integrated over the soil profile decreased with increasing rainfall and within three weeks of application and after 29 mm of precipitation, enhanced bromide concentrations were observed in the shallow groundwater (1.2–1.4 m depth) (Figure 16.13). This suggests that a substantial proportion of the bromide was rapidly transported from the plough layer to deeper in the soil profile and groundwater via shrinkage cracks and permanent macropores. Scott *et al.* (2000) have reported similar results for alluvial soils in the Mississippi valley, Louisiana, USA.

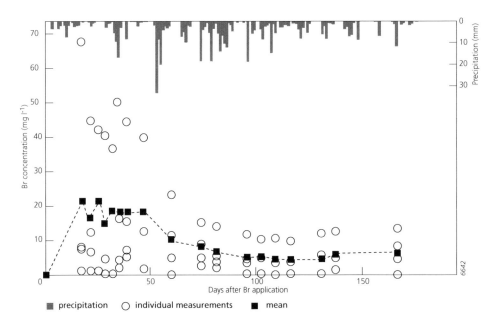

Figure 16.13 Precipitation and observed bromide concentrations in the shallow groundwater (four observations per date) below the experimental tracer plot in the Netherlands. Adapted from Hendriks *et al.* (1999).

Despite a broad consensus about the importance of preferential flow paths for solute transport, techniques for assessing its effect at spatial scales exceeding the local scale (> 1 km) are still poorly developed. The main reason is the lack of basic soil information relevant for ascertaining preferential flow paths, (such as quantitative data on soil structure) at the regional scale and beyond. This is why the effect of preferential flow on the transport and fate of solutes in regional scale assessments has often been ignored, even in recent studies. Nevertheless, mapping or modelling solute leaching at the regional scale may provide valuable information on factors controlling the rate and extent of leaching. For example, Tiktak *et al.* (2006) performed a model-based assessment of the vulnerability to pesticide leaching at the European scale. In the model they developed, pesticide leaching in soil is described with the advection-dispersion equation. Adsorption onto the soil particles is described using a linear Freundlich isotherm with a distribution coefficient proportional to soil organic matter content. Degradation of pesticides is described as a first-order rate process for which the rate constant depends on temperature, soil moisture, and depth in soil. Figure 16.14 shows the leaching concentrations for two model pesticides (substances A and B) with different physico-chemical properties at 1 m depth and at a 10 km × 10 km spatial resolution and assuming a spatially uniform annual application rate of 1 kg ha^{-1} one day after crop emergence. The two pesticide substances A and B differ in their affinity to organic matter (substance A: K_{om} = 60 l kg^{-1}; substance B: K_{om} = 10 l kg^{-1}; note that this organic matter–water partition coefficient K_{om} is analogous to the organic carbon–water partition coefficient K_{oc}; see Section 13.2, Equation 13.4) and the substance's degradation rate (substance A: degradation half-life at 20 °C DT_{50} = 60 d; substance A: DT_{50} = 20 d). In general, the modelled leaching concentrations increase concomitantly with increasing precipitation and decrease with increasing organic matter content. The leaching maps also show that the short-range variability of the leaching concentration is considerable. This is mainly because pesticide leaching depends on the soil organic matter content, which varies greatly over short distances. Furthermore, despite the relatively high temperatures that promote pesticide degradation, and the low precipitation surpluses, the modelled leaching

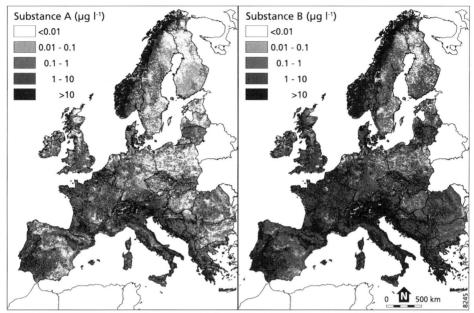

Figure 16.14 Modelled leaching concentrations of two pesticide substances A and B at 1 m soil depth in Euope. (Tiktak *et al.*, 2006).

concentrations are relatively high in many areas of southern Europe. This can be explained by the often extremely low organic matter contents (< 1%) in these areas (Tiktak *et al.*, 2006).

16.5.3 Effects of contamination history

Changes in contaminant input over time may become manifested in the vertical distribution of contaminant concentrations in the soil profile. In Section 16.4.3, the evolution of the concentration profile following a momentary input of a conservative tracer was discussed (Figure 16.12). It is to be expected that periodic input of soluble chemicals will similarly be reflected in the soil profile, provided that sufficient water percolates downward through the soil and the solutes undergo little sorption to solid surfaces. In the case that the chemicals are strongly sorbed to the soil solids, translocation of soil material by bioturbation or tillage operations is the predominant manner of vertical transport. However, these mixing processes homogenise the topsoil and blur the vertical variations in concentrations. The contamination history will then only be reflected in an increased deposition density over time, while the evolution of contamination inputs over time will not be expressed in vertical concentration gradients in the soil profile. In the case of contaminants with high affinity for soil solids, the contaminant input into soil is associated with sediment input. Historical variations in inputs of contaminants with high affinity for soil solids may in this way result in vertical concentration profiles in the sediment deposit: for example, in floodplain deposits or bed sediments (e.g. Ayres and Rod, 1986; Winkels *et al.*, 1998; Middelkoop, 1997, 2000; Tebbens *et al.*, 2000). In this case, the vertical concentration profile largely reflects the evolution of sediment contamination over time.

Middelkoop (1997) reconstructed the history of heavy metal contamination of Rhine sediments using undisturbed sediment cores from three dike breach ponds located in the floodplains along the river Waal. These ponds had undergone more or less continuous deposition of sediment when the washlands had flooded sufficiently rapidly, so he was able to achieve a temporal resolution of less than five years. The sediments were dated by

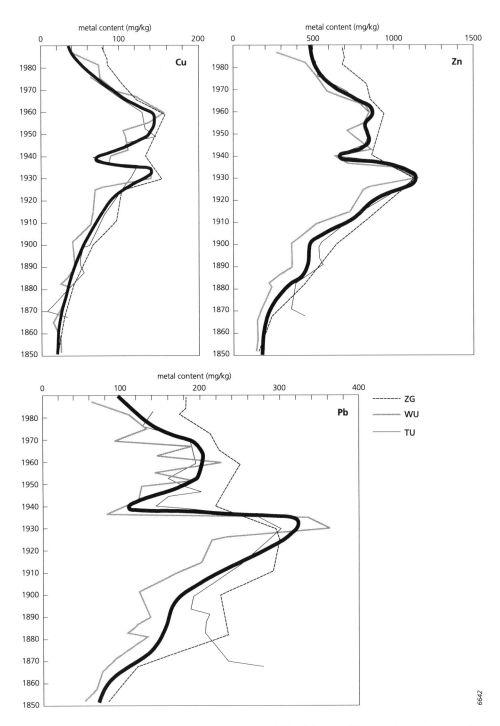

Figure 16.15 Major trends in contamination by Cu, Pb, and Zn of the river Rhine floodplain sediments since 1850. Metal concentrations have been standardised to a standard sediment containing 40 percent clay and 8 percent organic matter (Middelkoop, 1997).

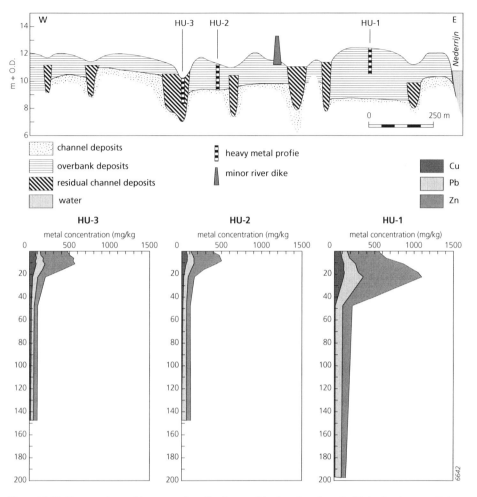

Figure 16.16 Cross-section and heavy metal profiles for a washland section along the Nederrijn river near Huissen, the Netherlands (Middelkoop, 1997).

means of radioactive dating using ^{210}Pb. The dating results were supplemented by historical information about the ages of the ponds. Figure 16.15 shows the reconstructed trends for the metals Zn, Pb, Cd, and Cu since 1850, standardised for clay and organic matter content. The concentrations in 1850 represent the pre-industrial levels, which were almost as low as the natural background concentrations. Figure 16.15 shows that the maximum contamination by heavy metals occurred around 1930, after which a dip occurred during the Second World War (1940–1945). A second peak occurred in the 1960s followed by a consistent decrease of metal pollution in the 1970s and 1980s.

Figure 16.16 shows three heavy metal profiles along a transect perpendicular to the Nederrijn river, which is, after the river Waal, the second largest distributary of the river Rhine in the Netherlands. Further inspection of this figure, and comparison with Figure 16.15, reveals the following:

- The contaminated layer is thickest near the main river channel, where the washland is not protected by a minor dike. The inundation frequency and duration, and the supply of sediment during flooding is greater than in the areas further away from the river, where a minor dike protects the washland from low floods.

- The maximum concentrations in the soil profile are also greatest near the main river channel. The reasons for this are twofold. First, the part of the washland section near the river receives more input of contaminated sediment. After deposition, the fresh sediment is mixed with the topsoil through the dispersive action of bioturbation and, possibly, tillage (although most of the washlands are used as grasslands with little tillage). This dispersion effect causes the maximum concentration to decrease (compare Figure 11.9). Second, heavy metal contamination of suspended sediments tends to decrease with increasing discharge: for example, due to dilution with less contaminated sediments from slopes (see also Chapter 18). As a consequence low-lying washlands that are not protected by a minor dike are already inundated during relatively low discharges and thus receive sediments that are more contaminated than parts of the washlands that are only inundated during high discharges (Middelkoop *et al.*, 2003).
- The concentration in the top 10 cm of the soil is lower near the main river channel than further away from the river. Again, this can be attributed to the larger sediment inputs on the washland near the river. The decreased contamination levels of the Rhine sediments since the early 1970s have caused the heavy metal contamination in the topsoil to diminish gradually. This effect is less pronounced in the areas that receive less sediment, because the thinner layer of less contaminated sediment has been mixed with the more contaminated sediments underneath.
- The decreased concentrations during the 1940s are not visible in either of the profiles. This is probably because the decrease in concentrations occurred only during a relatively short period; the mixing processes in the soil profile have blurred the two distinct concentration peaks of the 1930s and the 1960s into one peak.

It should also be mentioned that sediment deposition on floodplains is a discontinuous process that only occurs during high discharge events. If such events do not happen for a prolonged period of years, a significant part of the contamination history will not be reflected in the soil profile. Therefore, the peak concentration in the profile of a floodplain soil does not necessarily represent the year when the concentrations in suspended sediment were at their maximum (Delsman, 2002).

16.6 FURTHER READING ON CONTAMINANTS IN SOIL

- Mirsal, I.A., 2008, *Soil Pollution: Origin, Monitoring & Remediation*, 2nd edition, (Berlin: Springer).
- Ginn, J.S., and Russel Boulding, J., 2003, *Practical Handbook of Soil, Vadose Zone, and Ground-Water Contamination: Assessment, Prevention, and Remediation*, 2nd edition, (Boca Raton FL: Lewis/CRC).
- Merrington, G., Winder, L., Parkinson, R., and Redman, M., 2002, *Agricultural Pollution: Environmental Problems and Practical Solutions*, (New York: Spon Press).

EXERCISES

1. Explain briefly:
   a. which factors control the natural background concentrations of substances in soils;
   b. why heavy metals are usually more mobile in sandy soils than in loamy soils;

    c. why the pH in soil pore water is generally lower near a forest edge than in the centre of a forest;

    d. why heavy metals are usually more mobile in sandy soils than in loamy soils;

    e. why heavy metal concentrations correlate positively with the Al_2O_3 concentration in soil;

    f. why preferential flow paths promotes the leaching of contaminants from the soil root zone.;

    g. why tillage causes a net loss of soil material and associated contaminants on convex parts of the landscape and a net gain on concave parts;

    h. why the topsoil of floodplains along the river Rhine is less contaminated at locations with high sediment deposition rates than at locations with low deposition rates;

    i. why the ^{137}Cs deposition density in soil is greater on concave parts than on convex parts of a slope.

2. Spijker (2005) studied the anthropogenic metal enrichment in the topsoil in the Province of Zeeland, the Netherlands, by collecting soil samples from two depths (0–5 cm and 40–80 cm) at each sampling location. These samples were analysed for several parameters, including Cd and Al_2O_3. The figure below shows the Cd concentration (in mg kg^{-1}) plotted against the Al_2O_3 concentration (in %) for both the topsoil samples (Δ) and the deeper samples ($\circ$).

    a. What is probably the reason for the topsoil samples containing more Cd than the deeper samples?

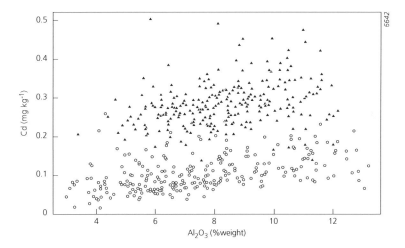

    b. In a topsoil sample the Cd concentration is 0.43 mg/kg and the Al_2O_3 concentration is 11 percent. Estimate the anthropogenic enrichment of Cd in this topsoil sample using the information from the graph below.

    c. What are the two critical assumptions in this method for calculating the anthropogenic Cd enrichment?

3. A vineyard soil was sampled at intervals of 10 cm depth. The soil sample for each interval was bulked, homogenised, and analysed for Cu concentrations (Cu is a constituent of Bordeaux mixture, a widely used fungicide in viniculture). The table below gives the resulting data. Calculate the Cu deposition density in g m^{-2} for the top 40 cm of the soil, given that the bulk density of the soil is 1350 kg m^{-3}.

Depth	Cu concentration (mg kg^{-1})
0–10 cm	152
10–20 cm	103
20–30 cm	17
30–40 cm	8

4. Regression analysis of Zn concentrations against clay and organic matter content in a contaminated region yielded the following regression equation:
   Zn = 25.6 + 5.5 clay (%) + 16.5 O.M. (%)
   A new soil sample from the same region contains 12 percent clay, 3.5 percent organic matter, and 174 mg kg^{-1} zinc. Calculate the standardised concentration for a standardised soil with 25 percent clay and 10 percent organic matter (hint: see equation 16.4). Is the zinc contamination of this sample more or less than the area average?

5. Name two drawbacks of using contaminant depth profiles in floodplain soils to reconstruct the contamination history of a river.

Patterns in groundwater

17.1 INTRODUCTION

Contaminant patterns in groundwater, as in soil, arise from spatial and temporal variations in the contaminant inputs, groundwater flow, and chemical interactions between the sediment (or bedrock) and groundwater. Groundwater and its associated dissolved and colloidal substances are transported along groundwater flow lines. Each groundwater flow line is a line whose tangent at any point is parallel to the groundwater flow velocity. Groundwater flow lines intersect perpendicularly with equipotential lines, i.e. lines along which the hydraulic potential remains constant. The network of groundwater flow lines and equipotential lines is called a flow net (Hubbert, 1940) and the volume bounded by arbitrarily selected flow lines is called a stream tube. This implies that the amount of water flowing through a stream tube is constant along the stream tube. Exchange of dissolved substances across the boundaries of a stream tube is limited and, by definition, occurs only due to molecular diffusion and transverse mechanical dispersion.

Figure 17.1 depicts the concept of a stream tube. Note that stream tubes do not necessarily have to be regular in shape and smooth. Complex large-scale whirl patterns may occur in groundwater flow lines, as a result of the occurrence of anisotropy and macroscale permeability heterogeneities in the aquifer material: for example, due to sedimentary structures or the presence of palaeosols (Hemker *et al.*, 2004).

17.2 HYDROLOGICAL SYSTEMS ANALYSIS

An analysis of the spatial arrangement of flow lines or stream tubes in the subsurface gives valuable insight into the movement of groundwater and related processes. This approach is called hydrological systems analysis and was developed by Tóth (1963) and later refined by Engelen and Kloosterman (1996). The definition of a groundwater system used by various

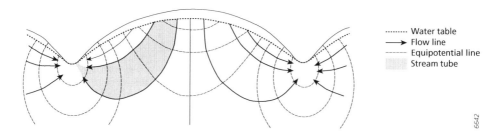

········ Water table
——→ Flow line
– – – Equipotential line
▧ Stream tube

6642

Figure 17.1 Concept of flow nets and stream tubes (after Hubbert, 1940).

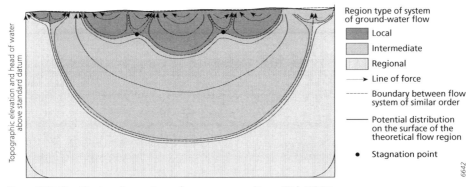

Figure 17.2 Classification of groundwater flow systems according to Tóth (1963).

researchers, including Tóth (1963) and Engelen and Kloosterman (1996), has varied. In this book, we define a groundwater flow system as a three-dimensional closed system through which groundwater flows from recharge areas where groundwater percolates downward to discharge areas where upward seepage occurs. This definition is similar to the definition of a catchment of a spatially continuous discharge area. Thus, hydrological systems analysis links groundwater discharge areas, usually rivers, lakes, or wetlands located in topographic lows, to their respective groundwater recharge areas, usually located in topographic highs. Groundwater catchments and surface water catchments coincide often, but not always. Groundwater catchment boundaries may deviate from those of surface water catchments in the case of tilted aquifers (geologically controlled), large differences in the base level of groundwater discharge over short distances (geomorphologically controlled), or artificial groundwater abstraction (human controlled).

Figure 17.2 shows that groundwater flow systems can occur in hierarchically nested configurations. Based on the hierarchical level of the groundwater flow systems, they can be divided into three types: local, intermediate or subregional, and regional flow systems (Tóth, 1963). Local flow systems link discharge areas to nearby, adjacent recharge areas, whereas intermediate and regional flow systems link discharge areas with discrete recharge areas farther away from the discharge area. Local flow systems are particularly likely in areas with pronounced local relief. They are influenced by seasonal recharge, and the flow velocities in these systems are often much greater than those in intermediate and regional flow systems. Because of these greater flow velocities and the shorter travel distances, the residence time in local systems is often many times shorter than in regional systems. Figure 17.2 also illustrates that, except in groundwater discharge areas, the groundwater travel distance (and, accordingly, the groundwater age) increase with depth, at least in homogeneous aquifers.

Because of the shorter residence times, groundwater discharge areas receive most of their groundwater inflows from local flow systems. In areas where local relief is negligible, local or intermediate flow systems are generally absent, but the groundwater gradient creates a regional flow system. However, if areas of low relief are intensively drained by a dense network of streams, or ditches and canals, such as in the agricultural lowlands of north-west Europe, local systems predominate. For example, Figure 17.3 shows a map of groundwater flow systems in Salland area in the east of the Netherlands (Vissers, 2006). The area consists of a shallow sandy unconfined aquifer of 40 to 80 m thickness stretching westwards from an ice-pushed ridge (i.e. a ridge composed of local, essentially non-glacial material pushed up in an ice age by an advancing glacier) with no surface drainage to a low-lying, intensively drained area. Figure 17.3a shows the flow systems and the transit distance of groundwater, i.e. the distance that the groundwater will travel at the time of infiltration. The dark areas represent the infiltration of recharge areas (long travel distance), and the white areas the

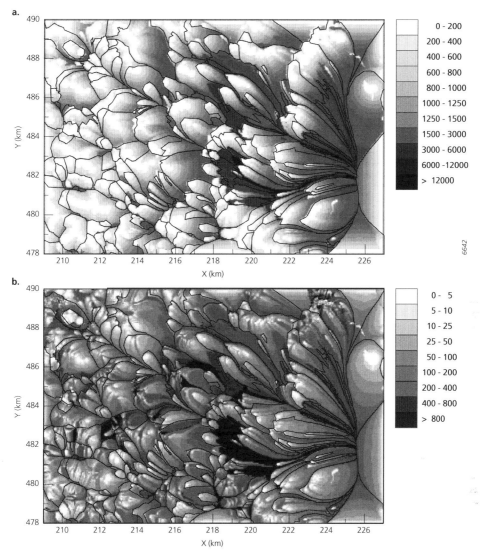

Figure 17.3 Groundwater flow systems in the Salland area with a) groundwater travel distance (m) and b) groundwater travel time (y) Adapted from Vissers (2006).

groundwater discharge areas (short travel distance). Figure 17.3b shows the age that the groundwater will reach from the point of infiltration. These figures demonstrate that in this area most of the groundwater does not travel large distances and is rarely more than 100 years old. Note the relation between groundwater travel time and travel distance. Most groundwater is discharged within several hundreds of metres from the point of infiltration within the same delineated areas. This implies that most of the groundwater flow systems are local. The only exceptions are the isolated dark-shaded areas in the centre of the study area, which do not contain discharge areas within the same polygon. These areas represent recharge areas of a regional groundwater flow system.

17.3 HYDROCHEMICAL SYSTEMS ANALYSIS

From Figure 17.2 it can be seen that a groundwater system can be described as a set of stream tubes. For groundwater quality studies it is useful to define the stream tubes on the basis of a geochemically homogeneous groundwater recharge area. Among the factors that determine chemical homogeneity of an area are, for example, soil type and land use (Vissers *et al.*, 1999; Broers, 2002; Vissers, 2006). Boundaries between soil types and fields with different land use can thus be followed along the groundwater flow lines that form the boundaries of the stream tubes underground (see Figure 10.4). In addition to spatial variation, the chemical inputs at the soil surface may also vary with time: for example, seasonal fluctuations in the mineralisation of organic matter, and seasonal and annual variations in net precipitation causing differences in leaching rates; changes in land cover or crop type causing variations in fertiliser rates; or long-term changes in atmospheric deposition rates. This temporal variation of chemical inputs is subsequently propagated along the flow lines (see also Figure 10.4), but the magnitude of the fluctuations may fade due to longitudinal mixing. Furthermore, the position of stream tubes may shift in time as a result of changes in the groundwater flow pattern induced by climate change or by human disturbances of groundwater flow: for example, due to groundwater abstraction or artificial drainage. Such shifts may thus also cause considerable temporal changes in groundwater composition near the displaced stream tube boundaries.

During transport along the groundwater flow lines, the interactions between water and sediment cause groundwater to be exposed to a sequence of chemical conditions that vary in pH, redox potential, temperature, etc. (see also Section 3.3.3). These chemical conditions are governed by the geochemical composition of the sediment or bedrock and are thus geologically defined. The minerals (clay minerals, carbonate minerals, reactive iron, pyrite in particular; see Van Gaans *et al.*, 2011) and the organic matter in the sediment act as a buffering agent. If the buffering agent is abundant (for example, organic matter in organic sediments, or calcite in limestone rocks) the chemical conditions can be considered as fixed in space and time, at least at the time scale of centuries. This results in steady transitions in groundwater composition at the boundary between geological layers. If the reaction rate is fast compared to the groundwater flow rate, for example in the case of most acid–base reactions, the transition in groundwater is abrupt and coincides with the geological boundary. In contrast, if the reaction rate is slow, as for most redox reactions, the transition in groundwater composition is more gradual. On the other hand, if the buffering agent is limited, for example calcite in sandy sediments or base cations adsorbed to exchange sites, the transition in the chemical composition of groundwater migrates gradually along the groundwater flow lines.

Stuyfzand (1999) has summarised the typical hydrochemical development of groundwater in the direction of groundwater flow and has derived a sequence of hydrochemical facies in stream tubes. He distinguished the following processes as illustrated in Figure 17.4:

1. From strong fluctuations in groundwater composition, mainly caused by seasonal atmospheric and biological variations, to a stable water composition, due to dispersion.
2. From polluted groundwater containing, for instance, SO_4^{2-}, NO_3^-, K^+, heavy metals, tritium, and organic pollutants, to unpolluted, due to a) elimination processes, such as filtration, acid buffering, sorption, and decay; and b) the older age of the water exposed to a smaller pollution load.
3. From acidic to basic water due to weathering and buffering reactions with the porous medium, which consume acids like H_2CO_3, H_2SO_4, and HNO_3 and produce HCO_3^- (alkalinity).

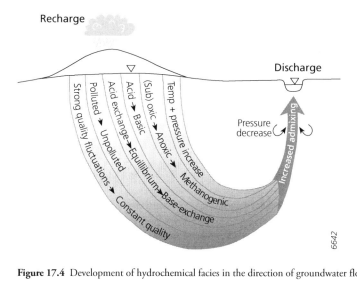

Figure 17.4 Development of hydrochemical facies in the direction of groundwater flow (after Stuyfzand, 1999).

4. From oxic to anoxic–methanogenic conditions due to the oxidation of organic matter in a system closed from the atmosphere. This process brings about a typical order of consumption of oxidants (see Section 4.3.4; Figure 4.10) in water and an increase in alkalinity.
5. From low to elevated pressure and water temperature due to downward migration of groundwater in groundwater recharge areas. In groundwater discharge areas the reverse occurs. An increase of temperature and pressure may lead to K^+ and Mg^{2+} depletion due to recrystallisation of clay minerals and to an increase of overall salinity due to a decrease of the cation exchange capacity.

Along the groundwater flow lines, substance concentrations can vary by up to several orders of magnitude, due to differences in chemical inputs or the chemical processes. Towards the groundwater discharge areas, the groundwater flow lines converge and the stream tubes become narrower and increasingly mixed.

Figure 17.5 illustrates in more detail the concept of the hydrochemical development of groundwater, using as an example the largely unconfined chalk aquifer of Dorset, southern England (Edmunds and Shand, 2008a). Here, the background concentrations of Cl^- are controlled by rainwater chemistry after allowing for evapotranspiration and decline gradually inland from the coast. The deepest part of the aquifer is separated from present-day groundwater circulation, and in some parts minor residual salinity derived from older saline water may be present. The major control on the composition of groundwater, however, is the geochemistry of the chalk sediment. Interactions between groundwater and the chalk produce relatively alkaline, hard water, rich in Ca^{2+} and HCO_3^- ions. Variations in the natural groundwater composition also take place with increasing residence time, a major cause being redox changes occurring as groundwater moves beneath the Tertiary confining cover. In this deeper part of the chalk aquifer, high natural concentrations of Fe^{2+} and Mn^{2+} may be found (Edmunds and Shand, 2008a).

The above description of the development of groundwater composition demonstrates that, as with soils, bedrock geology is a major factor that controls the variation in the natural composition of groundwater. The potassium concentrations in European bottled water as shown in Figure 17.6 (Reimann and Birke, 2010) provide an illustrative example of the effect

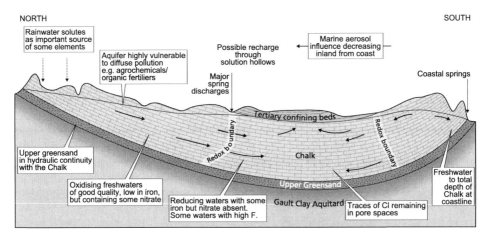

NORTH   SOUTH

Figure 17.5 Conceptual diagram of the chalk aquifer of Dorset, highlighting the main geochemical processes controlling groundwater composition. Adapted from Edmunds and Shand (2008a).

of bedrock geology on groundwater composition. The 'hot spots' of high K concentrations in northern Portugal, the eastern Pyrenees and the Massif Central in France are related to Hercynian granite intrusions. Enhanced K concentrations are also found in the young alkaline volcanic rocks in Italy. Furthermore, many wells in the Carpathian Mountains and Dinaric Alps also exhibit elevated K concentrations. The high K concentrations in northeastern Europe are probably related to the deep sedimentary basins from which the mineral water is extracted. However, the higher concentrations in bottled mineral water could also be explained by a cultural bias in taste, with a tendency for Eastern Europeans to prefer stronger-tasting, more mineralised water (Reimann and Birke, 2010).

In addition to bedrock geology, the origin of groundwater (rainwater or infiltrated surface water or relict sea water) may also an important factor that determines the variation in groundwater composition, especially in coastal areas. Van den Brink *et al.* (2007) showed that bedrock geology and groundwater provenance are the two major factors accounting for the vast majority of the variation in chemical composition of shallow groundwater at the supra-regional scale (> 100 km) in the Netherlands. Land use appeared less important and emerged as a significant factor only for the chemical composition of recently (post-1950; see Section 17.5) infiltrated groundwater at regional and local scales. This insight led to the development of a European aquifer typology intended to facilitate the assessment of the major groundwater composition at the continental scale (Wendland *et al.*, 2008; Pauwels *et al.*, 2009). This typology subdivides the European aquifers into eight major types, based on aquifer rock. Each of these aquifer types has its typical range with respect to porosity, permeability, and hydrochemistry. Some primary aquifer types are further subdivided according to secondary criteria that address geochemical and hydrological factors in particular, which are partly controlled by bedrock geology. For example, the sands and gravels aquifer type is further subdivided to account for variation in salinity, redox conditions, and carbonate content. The latter two factors are mainly controlled by the groundwater residence time, which appears to be related to the depositional environment. In general, due to their long groundwater residence times and sufficient reduction capacity, glacial sands and gravel deposits exhibit the typical behaviour of anoxic, reduced aquifers. By contrast, fluvial sand and gravel deposits generally exhibit oxidised conditions, although exceptions occur, such as the fine and organic-rich deposits in the Rhine-Meuse delta, which favour reduced conditions.

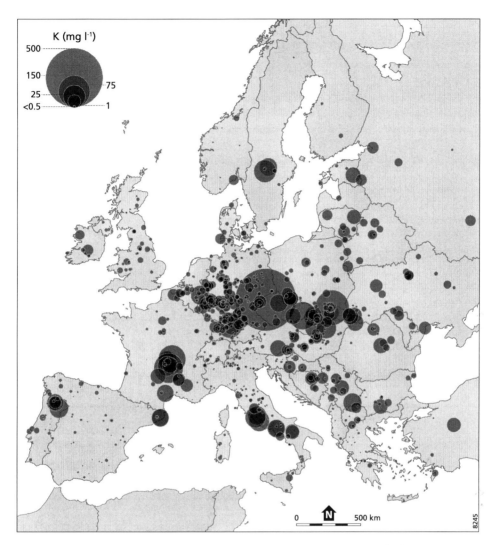

Figure 17.6 Potassium (K^+) concentrations in European bottled water (source: Reimann and Birke, 2010).

In the case of point source pollution of groundwater, the contaminants come from a discernible, spatially confined, and discrete source: usually spills from storage tanks or containers, underground injection of waste, or effluents from factories or landfills. Groundwater flow brings about the development of a contaminant plume in downgradient direction. Several schematic examples of such plumes were given in Chapters 11 and 13 (e.g. Figures 11.10, 13.2, and 13.4). If the contaminant is an NAPL, the plume of pure liquid develops virtually independently from the groundwater flow. LNAPLs float on the water table, whereas DNAPLs sink through the aquifer until they reach an impermeable layer (see Figure 11.13). As NAPLs are sparingly soluble in water, the dissolved phase does migrate with groundwater flow, also resulting in the formation of a plume. As noted above, the dissolved compounds are subjected to a number of physico-chemical processes while being transported along the groundwater flow lines in the plume.

Table 17.1 Criteria for designation of aquifer typologies (Wendland *et al.*, 2008).

Primary aquifer types	Additional secondary criteria
Sands and gravels	Saline influence or hydrodynamic conditions (residence time)
Marls and clays	
Sandstones	Geological age
Chalk	
Limestones	Hydrodynamic conditions (degree of karstification, residence time)
Volcanic rocks	
Schists and shales	
Crystalline rocks	

To summarise from the above: spatially homogeneous bodies of groundwater composition are delimited either by land use boundaries or by reaction boundaries that may be fixed or shift in time. The spatial displacemeniet of these homogeneous groundwater bodies may cause substantial temporal changes in groundwater composition at a given location. The unravelling of the position of stream tubes as function of space and time is a crucial step in understanding regional patterns and trends in groundwater composition. In the next sections, observed contaminant patterns and trends in groundwater from a selection of case studies will be discussed in the light of the patterns and trends in contaminant inputs and the physical and chemical processes mentioned above.

17.4 EFFECTS OF LATERAL VARIATION IN CONTAMINANT INPUTS

As the transport through the vadose zone is primarily vertically downward, lateral differences in contaminant inputs into groundwater are reflected in the contaminant concentrations in the shallow groundwater. This is obviously true for point source pollution at the local scale, but also for diffuse pollution at the regional (> 10 km) and supra-regional scales (> 100 km). Pebesma and De Kwaadsteniet (1994, 1997) mapped the quality of the shallow groundwater at the supra-regional, national scale in the Netherlands in 1991, giving a good insight into the influence of diffuse source pollution on groundwater quality. They interpolated the dissolved concentrations of a variety of substances that had been observed in the Dutch national and provincial groundwater quality monitoring networks. As an example from this study, Figure 17.7a shows the median zinc concentration in groundwater at a resolution of 4 km × 4 km. The zinc concentrations in groundwater are highest in the vicinity of a zinc smelter in the sandy area in the south of the Netherlands (compare Figure 17.7b). The smelting process in the zinc factory caused considerable zinc emissions to the atmosphere up to the early 1970s. Subsequent atmospheric deposition and the use of zinc ash to pave roads and farmyards caused substantial inputs of heavy metals into the soil in the wide vicinity of the zinc smelter. In general, most of the zinc inputs at the soil surface are retained in the vadose zone, particularly in clayey soils. However, in sandy sediments, some of the zinc is leached to groundwater, which causes the elevated zinc concentrations in this region.

At the supra-regional scale, as in the above example of median concentrations at a spatial resolution of 4 km, groundwater flow is less important for the spatial distribution of contaminants because it usually takes decades to centuries for groundwater to travel such a long distance. This is generally longer than the age of most pollution. Furthermore, as noted in the previous section, in densely drained areas like the Netherlands, most of the groundwater recharge is discharged within such distance. Therefore the contaminant

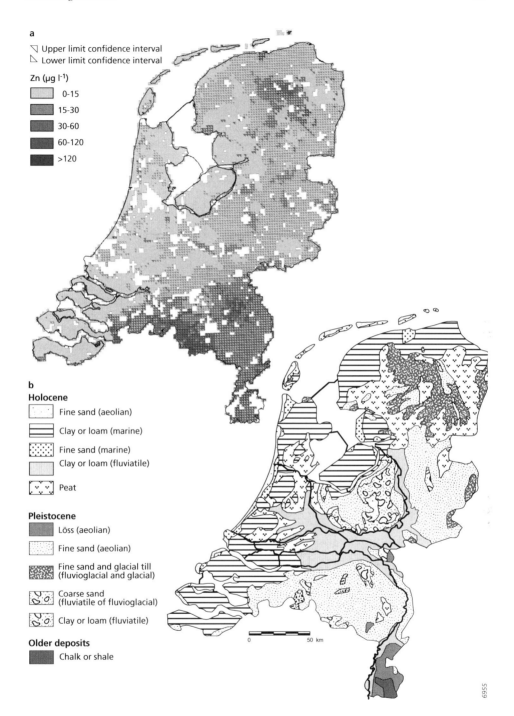

Figure 17.7 a) Zinc concentration in shallow groundwater (5–17 m depth) in the Netherlands in 1991. The concentrations are visualised as the 95 percent confidence interval for the median value in grid cells of 4 km × 4 km (Pebesma and De Kwaadsteniet, 1994); b) Simplified surficial deposits map of the Netherlands (after Steur *et al.*, 1987).

concentrations in groundwater within an area or grid cell can usually be directly related to the land use class or soil type within that same area.

At the regional scale or below, groundwater flow does become important. Due to the bending and the convergence of flow lines and stream tubes, the spatial differences in contaminant inputs at the soil surface that occur over several hundreds of metres are mirrored in downgradient vertical variation in groundwater composition to a depth of several metres (see Figures 17.1 and 10.3). This effect can be illustrated by Cl^- concentration–depth profiles in groundwater. Since Cl^- is conservative, the concentration–depth profile is not interfered with by chemical reactions, but only by differences in Cl^- input.

Figure 17.8 shows the Cl^- concentration–depth profile in a single multilevel well (A1) located in a sandy, unconfined aquifer at the bottom of an ice-pushed ridge in the Salland area in the Netherlands (see Figure 17.3) (Vissers *et al.*, 1999). The well is situated in a patch of forest and the land cover in the upgradient groundwater recharge area consists of alternating patches of forest and heather. The evapotranspiration rate for forest is greater than for heather. Consequently, the Cl^- concentration in groundwater underneath the forest is also greater, due to the concentration effect of evapotranspiration (see box 5.II). Compared with the average Cl^- concentration in precipitation, the Cl^- concentration of infiltrating water is about four to five times higher underneath forest and about two times higher underneath heather. In the Salland region, the average Cl^- concentration in rainwater is about 4 mg l^{-1}, so the Cl^- concentrations are 16–20 mg l^{-1} in groundwater originating from forests and about 8 mg l^{-1} in groundwater originating from heather. Considering these different concentrations, five different stream tubes originating from different patches of land can be distinguished in the concentration–depth profile. In the top stream tube the Cl^- concentration is around 5 mg l^{-1}, which suggests a recent intensive recharge by rainwater that was not affected by the concentration effect of evapotranspiration. The deeper stream tubes refer to recharge areas farther away and, as the age of groundwater increases with depth, to land cover occurring in the recharge areas further back in time. It should thus be emphasised that the Cl^- concentration reflects the land cover at the time of infiltration, which does not necessarily correspond with the present land cover. An important point about the Salland area is that in the course of the 20th century, much of the heather was planted to trees. So, it is possible that the recharge areas of the stream tubes linked to heather are now under forest or other land cover types.

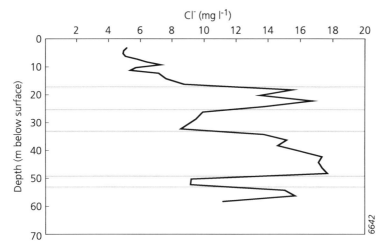

Figure 17.8 Cl^- concentration depth profile in a well (A1) in Salland, the Netherlands, with subdivision into stream tubes originating from different patches of forest and heather Adapted from Vissers *et al.* (1999).

17.5 EFFECTS OF TEMPORAL VARIATION IN CONTAMINANT INPUTS

Temporal changes in contaminant inputs in groundwater recharge areas lead to changes in concentrations along groundwater flow lines. In the long term at the regional scale or above, the trends follow the general trends in diffuse anthropogenic emissions, particularly emissions from agriculture. Since 1950, the influence of agriculture on groundwater quality has been manifested as increased levels of nitrate in shallow groundwater in many countries in Europe (e.g. Wendland, 1992; Roux, 1995; Dufour, 2000) and in North America (e.g. Burkart and Stoner, 2002). The relation between nitrate concentrations in groundwater and fertiliser application rates can be clearly revealed by determining the age (i.e. groundwater travel time since recharge) of each groundwater sample through dating techniques rather than by plotting the average nitrate concentration in groundwater against time (Bronswijk and Prins, 2001). Figure 17.9 shows the trend in nitrate concentration in recharging groundwater in Denmark (Figure 17.9a) and the Netherlands (Figure 17.9b) between 1940 and 2000.

In Denmark, the median nitrate concentration rose considerably from about 13 mg l^{-1} in the mid-1950s to about 60 mg l^{-1} in 1980 (Hansen *et al.*, 2011). In the sandy areas of the province of Noord-Brabant, the Netherlands, the increase in nitrate levels up to the 1980s was even more pronounced than in Denmark (see Figure 17.9b) and the median nitrate concentration reached a level of about 160 mg l^{-1} by the mid-1980s (Visser *et al.*, 2008). The groundwater in the sandy areas in the south and centre of the Netherlands is considered to be amongst the most nitrate-polluted in Europe. For comparison, the arithmetic mean of the nitrate concentrations in Western Europe was 25 mg l^{-1} in 1993 (Lindinger and Scheidleder, 2004).

The rise in nitrate concentrations is largely attributable to increased fertiliser and manure application on agricultural land and increased atmospheric deposition of nitrogen compounds. In addition, increased drainage of agricultural land may promote nitrate leaching to groundwater. Although artificial drainage usually decreases the local rate of

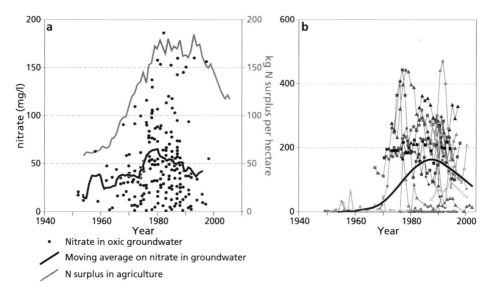

Figure 17.9 a) Trend in nitrate concentration in recharging concentration in groundwater in a) Denmark (Hansen *et al.*, 2011) and b) the province of Noord-Brabant, the Netherlands (Visser *et al.*, 2008) in the period 1950–2001. The different symbols in Figure b) represent different groundwater wells.

groundwater recharge, it has a widespread effect on the groundwater level. When the groundwater level falls, the upper soil horizons become more aerated, which, in turn, results in enhanced nitrate production through ammonification and subsequent nitrification. It may also cause water deficiencies in the root zone, so that nitrate is taken up by the vegetation less efficiently. In sandy recharge areas in particular, where the shallow groundwater is usually well aerated, nitrate is very mobile and is barely affected by denitrification. The large scatter around the general trend line is caused by differences in manure and fertiliser application rates or differences in nitrate leaching rates between the measurement locations or the years. Nitrate leaching increases with increasing net precipitation and this effect is generally stronger than the dilution effect brought about by increased net precipitation. As a result, wet years are reflected as zones of higher nitrate concentrations in groundwater, whereas dry years appear as zones of lower nitrate concentrations.

Since the mid-1980s, the nitrate concentrations in recharging groundwater have been declining, both in Denmark and in the Netherlands. This downward trend is mainly due to decreasing inputs of fertilisers as a result of EU policy to protect water against pollution by nitrate from agricultural sources. This decreasing trend in recharging groundwater has not yet become apparent in a decreasing trend in nitrate concentrations at all monitoring locations, because the discharge year of the sampled groundwater was before the mid-1980s. In about one-third of the national monitoring locations in Europe the trend in nitrate concentrations is upwards, in another third the trend is stable, and in the last third the trend is downwards (EC, 2011). In some regions, a decreasing trend has also become apparent in the mean nitrate concentrations over larger areas. For example, the mean nitrate concentrations in shallow groundwater below agricultural land in the central part of the Netherlands (province of Gelderland) have been declining since 1993 (Province of Gelderland, 2013). The delay in response of the mean nitrate concentration to the decline in nitrate concentrations in the recharging groundwater can be mainly attributed to the travel time between the points of infiltration and the locations of the groundwater wells. Nevertheless, The nitrate concentrations in the deep groundwater have continued to rise, since the deeper groundwater is older than the shallow groundwater and includes more water that infiltrated during the 1970s and 1980s when the fertiliser application rates and the corresponding nitrate leaching rates were at their maximum.

At the local scale, temporal changes in contaminant input may be the result of, for example, a leaking storage tank, remediation of point sources of pollution, recurrent fertiliser application on agricultural land, or landuse changes. Since most of these changes usually take place rather abruptly, they lead to sharp concentration gradients in the direction of groundwater flow. Ryan and Stokman (2001) studied the dynamics of nitrate in several

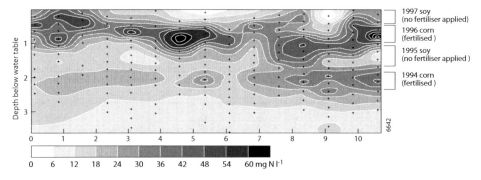

Figure 17.10 Nitrate concentrations in a cross-section on Strathmere Farm, Ontario, Canada, as measured in May 1997. The sampling locations are indicated by black dots (source: Cathy Ryan, personal communication).

vertical cross-sections in a shallow sandy aquifer in southern Ontario, Canada. They installed fifteen multi-level wells down to about 3.5 m below the water table across an 11 m long transect on an arable field perpendicular to the direction of groundwater flow. Figure 17.10 shows the observed pattern of nitrate in one of the cross-sections. Successive years of fertiliser application can be seen as nitrate-rich zones in the cross-sections. The differences in depth of the nitrate-rich zones originating from fertiliser application one and three years before groundwater sampling suggest a vertical groundwater flow velocity of about 0.5 m y^{-1}. The horizontal groundwater flow velocity in the shallow aquifer amounts to about 50 m y^{-1}. This means that the lowest nitrate-rich band originates from fertiliser applied at about 150 m upgradient from the cross-section.

17.6 EFFECTS OF DISPERSION

The process of dispersion tends to level out spatial differences in concentrations. In Section 11.3 we noted that given a certain concentration gradient, the magnitude of dispersion expressed in terms of the dispersion coefficient or dispersivity depends on the scale at which the process is studied. At the scale of contaminant plumes (spatial resolution ~ 0.1 m), regional scale (spatial resolution ~ 10–100 m), or larger scales, dispersion in groundwater is dominated by macroscopic dispersion due to heterogeneities in the hydraulic conductivity of the sediment (Domenico and Schwarz, 1996; Zheng and Gorelick, 2003). As dispersion is driven by the concentration gradient, it particularly brings about concentration changes in time at locations where the concentration gradient is large: for example, at the edges of a contaminant plume.

Contaminant concentrations are generally highest in the leachate just below the contaminant point source. All compounds in the leachate entering the aquifer will be diluted as the leachate mixes with the uncontaminated groundwater due to longitudinal and transverse dispersion. The dispersion process can be made visible using a dye tracer in a two-dimensional, analogous aquifer model, as shown in Figure 17.11. The figure shows a simplified cross-section of a shallow sandy aquifer about 60 m thick, from a groundwater recharge area on an ice-pushed ridge on the right to a discharge area in an alluvial area on the

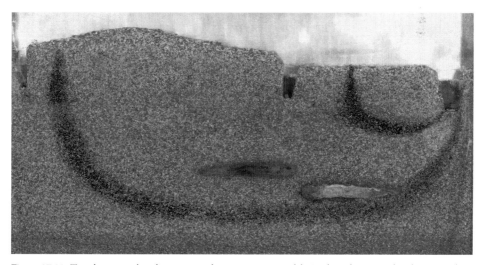

Figure 17.11 Two-dimensional analogous groundwater transport model. A coloured tracer makes dispersion along groundwater flow paths visible.

left. The tracer has dispersed both longitudinally and vertically. It should be emphasised that in reality, dispersion occurs in three dimensions; the lateral dispersion cannot be shown in the two-dimensional model in Figure 17.11. Leachate migration should thus be considered as a three-dimensional plume developing in a three-dimensional aquifer.

Figure 17.12 shows a vertical cross-section of observed Cl⁻ concentrations in a leachate plume in a relatively homogeneous, sandy aquifer from the Vejen landfill in Denmark (Lyngkilde and Christensen, 1992). Obviously, the elevated chloride concentrations originate from the landfill. Since Cl⁻ is a conservative substance, the plume is only affected by advection and dispersion. Figure 17.12 shows that the Cl⁻ plume is diluted in downgradient direction and that the dilution is much greater longitudinally than vertically. This phenomenon of a greater longitudinal than transverse dispersion, also in the horizontal plane, is typical for many landfill leachate plumes (Christensen *et al.*, 2001), so most leachate plumes are not much wider than the landfill.

Rivett *et al.* (2001) performed a controlled field experiment to study the development of the dissolved plume from a continuous, block-shaped multi-component DNAPL source. This experiment was carried out in a 10 m thick unconfined sandy aquifer located in Ontario, Canada. The DNAPL source contained trichloromethane (TCM), trichloroethene (TCE), and perchloroethene (PCE). TCM behaved essentially conservatively and was barely degraded or adsorbed. Figure 17.13 shows the development of the TCM plume in time. In this experiment the longitudinal dispersion was also much greater than the transversal dispersion, just as in the landfill leachate plume described above. This caused the TCM plume to be long and narrow. Apart from the effect of dispersion, the TCM plume exhibits some density-induced sinking. Figure 17.13 shows that the zone of maximum concentrations is below the groundwater flow line through the contaminant source. Because the TCM solution was denser than the surrounding, uncontaminated groundwater, the plume was diverted slightly downwards

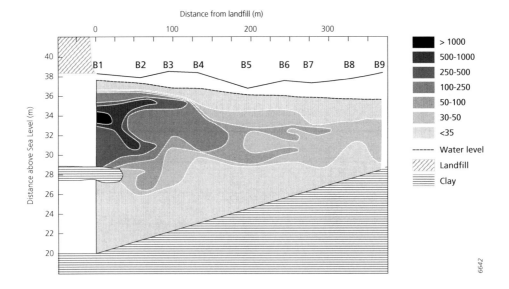

Figure 17.12 Cl⁻ concentrations in a longitudinal, vertical cross-section in the Vejen landfill leachate plume. Adapted from Lyngkilde and Christensen (1992).

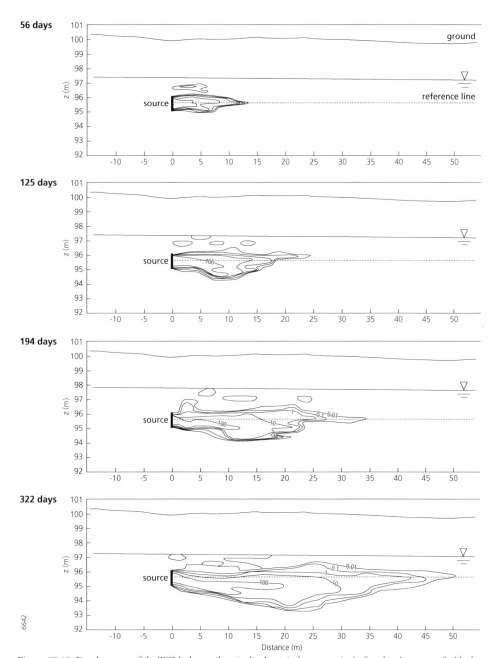

Figure 17.13 Development of the TCM plumes (longitudinal, vertical cross-section) after the placement of a block-shaped source. Concentrations are given in mg l⁻¹. Adapted from Rivett *et al.* (2001).

The two above examples show that dispersion mainly occurs longitudinally and that transverse dispersion is often small. It is for this reason that seasonal variation (as for example shown in Figure 17.10) fades out in downgradient direction, whereas spatial variation in chemical inputs at the soil surface can be traced over long distances along the groundwater flow lines.

17.7 EFFECTS OF RETARDATION

Retardation occurs when substances are subject to sorption onto mineral or organic aquifer materials and are therefore transported more slowly than the average groundwater flow velocity. Examples of such substances are cations or many organic pollutants. Net adsorption or desorption with the aquifer materials occurs when the dissolved phase is not in equilibrium with the solid phase of the aquifer. This is the case if the geochemical properties of the aquifer material vary along the groundwater flow line or when the composition of the percolating water changes in time: for example, near the infiltration front of agriculturally polluted water or at the edge of contaminant plumes. Since sorption processes are generally fast compared to groundwater flow, local equilibrium will usually be attained between the solid and dissolved phases.

Figure 17.14 shows an example of the effects of retardation on the transport of agriculturally derived K^+. The figure depicts the depth profiles of K^+ and Cl^- concentrations measured in a multi-level well (A2) located in a sandy aquifer in the Salland area underneath agricultural land (Vissers *et al.*, 1999). This well is approximately 1 km from well A1 whose Cl^- profile was presented in Figure 17.8. Manure spreading has clearly influenced the K^+ and Cl^- profiles: the upper groundwater is polluted by both K^+ and Cl^-. Some of the K^+ could be potentially derived from desorption from the exchange sites. The K^+ at the adsorption complex would then be exchanged when the TDS concentration of the infiltrating agriculturally polluted water increases. However, Griffioen (2001) has shown that by far the most K^+ in the Salland area is directly derived from agricultural inputs of manure. Comparison of the K^+ and Cl^- profiles reveals that the infiltration depth of anthropogenic Cl^- is more than 20 m (compare the Cl^- concentrations in Figure 17.8), whereas the K^+ has not been transported deeper than 15 m below the surface. This illustrates that K^+ has been retarded compared to Cl^-, which behaves almost conservatively.

As noted above, retardation also affects the advective and dispersive transport in solute plumes from point sources. DeSimone *et al.* (1997) studied the early development of a plume of wastewater effluent in a phreatic aquifer on Cape Cod, Massachusetts, USA. The aquifer was composed of alternating layers of fine and coarse-grained sands. This plume originated from a surface discharge of septage (i.e. waste from septic tanks) from infiltration beds, which started in February 1990 and consisted mainly of major dissolved constituents (Cl^-, Ca^{2+}, Mg^{2+}, Na^+, K^+, HCO_3^-) and nitrate. It spread mainly in the 1.5–18 m thick

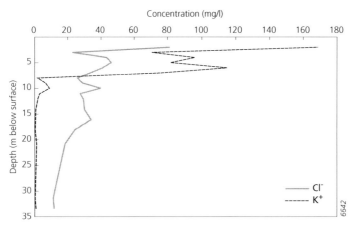

Figure 17.14 K^+ and Cl^- concentration depth profiles in a well (A2) in Salland, the Netherlands, in 1996. Data from Vissers (2006).

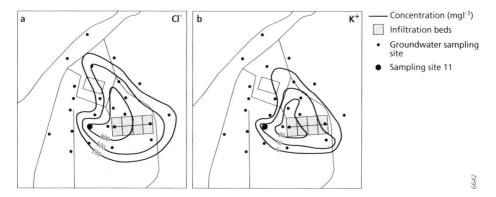

Figure 17.15 Distribution of a) Cl⁻ and b) K⁺ in groundwater after 34 months of wastewater effluent discharge (December 1992) on Cape Cod, Massachusetts, USA. Lines show equal concentrations in mg l⁻¹ and dots show the groundwater sampling sites. The thick dot represents sampling site 11 (see Figure 17.16). Adapted from DeSimone *et al.* (1997).

coarse-grained unit of the aquifer. After 34 months of septage discharge, the Cl⁻ plume in groundwater extended approximately 60–140 m in downgradient direction (Figure 17.15a). The K⁺ plume extended only 40–100 m due to retardation (Figure 17.15b). The irregular shape of the plume in plan view probably resulted from the spatial variation of the thickness of the various sediment layers.

The development of the concentrations of Cl⁻, Na⁺, Ca²⁺, and K⁺ in time at site 11 (see Figure 17.15) is depicted in Figure 17.16. Site 11 is located 56 m downgradient from the infiltration beds. The figure shows that Na⁺ is only slightly retarded and that K⁺ is most retarded. Retardation factors could be estimated from these breakthrough curves by taking the time required for the concentration of a constituent to reach one-half the median effluent concentration and dividing it by the corresponding time for Cl⁻. The retardation factors derived in this manner were 0.93 for Na⁺, 1.38 for Ca²⁺, and 1.76 for K⁺ (K⁺ concentrations extrapolated over time) (DeSimone *et al.*, 1997). The retardation factor of less than 1 for Na⁺ is physically impossible (see Section 13.2) and is probably due to uncertainties in the estimation of the median concentration in the effluent or initial groundwater. Nevertheless, Figure 17.16 suggests a small net increase of Na⁺ concentrations during transport of the

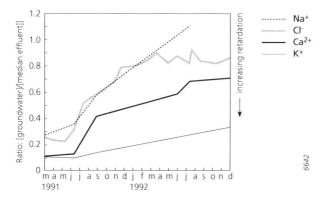

Figure 17.16 Breakthrough curves of Cl⁻, Na⁺, Ca²⁺, and K⁺ in groundwater at site 11 (see Figure 17.13; about 11 m below water table). Adapted from DeSimone *et al.* (1997).

effluent-contaminated groundwater. This may have been gained from sediments through exchange with H^+, K^+, Ca^{2+}, and from other cations in the plume, although exchangeable sodium made up only less than 2.5 percent of the pre-discharge exchangeable cations (DeSimone *et al.*, 1997).

17.8 EFFECTS OF ACID–BASE REACTIONS

By nature, rainwater has a slightly acidic pH, ranging from 5.6 to 5.8 due to the presence of carbonic acid. Normally, the acid rainwater is neutralised as it passes through the vadose zone. In areas with heavy rainfall and acidic soils with little buffering capacity, the pH of infiltrating water largely reflects the rainwater pH values. Anthropogenic factors including acid rain, waste disposal, and afforestation may, however, lower the pH of the infiltrating water. While groundwater flows through the aquifer, its acidity becomes partly or wholly neutralised by interaction with the aquifer material. This acid buffering is accomplished in two important ways: a) carbonate mineral dissolution and b) silicate mineral dissolution (Bricker and Rice, 1989). Both dissolution processes consume H^+ ions and produce dissolved base cations (Ca^{2+}, Mg^{2+}, K^+, and Na^+) and HCO_3^- ions, and thus an increase of the alkalinity. Furthermore, acidity can be neutralised through exchange with base cations on the cation exchange complex.

Carbonate mineral dissolution is one of the key processes controlling groundwater composition, particularly in aquifers made up exclusively of Ca and Mg carbonate rocks such as limestone and dolomite. Not only carbonate rocks, but also sandy and sandstone aquifers may contain sufficient carbonate minerals, especially calcite, to neutralise acidity. The carbonates react readily and quickly with groundwater; this is directly manifested in a sharp increase of dissolved Ca^{2+} and HCO_3^- concentrations at the calcite dissolution front. For example, Figure 17.17 shows the Ca^{2+} and HCO_3^- depth profiles measured in a multi-level well (A10) located in a sandy aquifer in the Salland area underneath a forest, not far from the Cl^- profile presented in Figure 17.8 (Vissers *et al.*, 1999). Both profiles show a sharp increase in the Ca^{2+} and HCO_3^- concentrations at about 6.5 metres below the surface. In the deeper part of the sandy aquifer, small amounts (up to 1 percent weight) of calcite are present (Frapporti *et al.*, 1995), whereas in the upper 6 metres of the aquifer all carbonates have been removed by the infiltrating groundwater since the deposition of the sandy sediments (in this case, late Pleistocene). In this decalcified zone above the calcite dissolution front, the base saturation has declined over time as an increasing proportion of H^+ and Al^{3+}, rather than base cations, has occupied the exchange sites, and the dissolved base cations have been removed by the infiltrating water. It is especially this zone that is susceptible to accelerated acidification by acid precipitation. In 1996, the pH of the groundwater in the decalcified zone ranged between 4.0 and 4.2, whereas in the deeper part beneath the calcite dissolution front, the pH was above 7. At 14 m below the soil surface a stream tube boundary is present, below which the Ca^{2+} and HCO_3^- concentrations are approximately 150 mg l^{-1} and 360 mg l^{-1}, respectively. This water originates from agricultural land and is polluted due to manure application. The ammonium in the manure is nitrified in the vadose zone, which brings about an additional input of acid (HNO_3) to groundwater (see Equation 6.2). This additional acid dissolves more calcite compared to the 'pristine' groundwater in the upper stream tube.

In contrast to carbonate dissolution, silicate mineral dissolution is a very slow process. This causes the changes in groundwater chemistry as a result of silicate weathering to be less apparent. Nevertheless, silicate weathering is the major mechanism of acid buffering in aquifers free from carbonate minerals (Appelo and Postma, 1996). In addition to buffering acidity and producing silicic acid (H_4SiO_4) and HCO_3^-, silicate mineral weathering releases

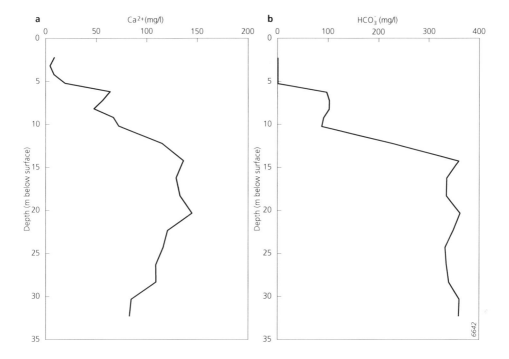

Figure 17.17 a) Ca^{2+} and b) HCO_3^- concentration depth profile in a well (A10) in Salland, the Netherlands, in 1996. Data from Vissers (2006).

Al, Fe and base cations. Silicic acid and base cations are relatively soluble and mobile, whereas Al and Fe are relatively insoluble and immobile. Figure 17.18 shows the depth profiles of HCO_3^-, the molar Na/Cl ratio, and dissolved silica in a borehole of up to 40 m depth, drilled into a fractured aquifer composed of Lower Palaeozoic mudstone, shale, and greywacke bedrock in the headwater catchment areas of the River Severn in central Wales (Neal *et al.*, 1997). Normally, these bedrock types, which are free of carbonates except for some rare calcite veins, are assumed to be relatively unreactive, but the increase of HCO_3^-, Na^+, and dissolved silica with depth, i.e. with groundwater age, clearly demonstrates the effect of the weathering of silicate compounds on the groundwater composition. Contrary to the HCO_3^- profile shown in Figure 17.17, the increase in HCO_3^-, Na^+, and dissolved silica with depth shown in Figure 17.18 is gradual, which indicates that the process of silicate weathering is slow.

17.9 EFFECTS OF REDOX REACTIONS

Reduction and oxidation processes control the distribution of species such as O_2, Fe^{2+}, SO_4^{2-}, NO_3^-, NH_4^+, and CH_4 in groundwater, both under natural and anthropogenically influenced conditions. Under natural conditions, the redox potential of the infiltrating water is high, as near the soil surface the water is more or less in equilibrium with atmospheric oxygen. While groundwater is travelling through the vadose and saturated zone, oxidation of organic matter causes the dissolved oxygen concentration to diminish. When oxygen is depleted, the conditions become anoxic and other oxidants are successively used for the oxidation of organic matter (see Section 4.3). In order of decreasing redox potential, these oxidants include NO_3^-, $Mn(IV)$, $Fe(III)$, SO_4^{2-}, and CO_2 (see Figure 4.10).

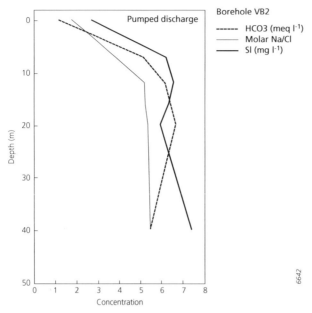

Figure 17.18 Depth profile of HCO_3^-, Na/Cl ratio and dissolved silica in a borehole in central Wales. Adapted from Neal *et al.* (1997).

The occurrence and distribution of redox-sensitive species in the shallow groundwater is thus largely controlled by the amount and reactivity of organic matter of the aquifer or the confining layer (Hartog *et al.*, 2004). To illustrate this, Figure 17.19 shows the spatial distribution of nitrate and ammonium in the shallow groundwater in the Netherlands (Pebesma and De Kwaadsteniet, 1994). Groundwater in unconfined, sandy aquifers poor in organic matter is generally oxic. In general, the ammonium concentration is low, because it is nitrified to nitrate in the well-aerated soil before it reaches the groundwater. In the oxic groundwater, nitrate is not denitrified and remains in the groundwater. Therefore, in the sandy areas (compare Figure 17.7b) dissolved inorganic nitrogen derived from fertiliser and manure application and atmospheric deposition is mainly present as nitrate. In the oxic groundwater, it is not denitrified and remains in the groundwater. The ammonium concentration is low, because it is nitrified to nitrate in the well-aerated soil before it reaches the groundwater. In contrast, in the west of the Netherlands, the aquifer is confined by a confining layer consisting of clay and peat (see Figure 17.7b). Here, the redox potential of the groundwater is much lower than in the sandy areas, so nitrate is removed by denitrification. The low redox potential also prevents the nitrification of ammonium derived from agriculture or the decomposition of organic matter. Consequently, the ammonium concentrations are relatively high in the west of the Netherlands.

Redox reactions are generally slow. If the aquifer contains only small quantities of organic matter, the stepwise decrease of the redox potential can be observed as a spatially distinct sequence of redox zones, each with its own associated groundwater composition. As the groundwater age and travel distance increase with depth, in a single well the consecutive redox zones can be observed, with redox potential decreasing with increasing depth. The redox zones can be identified by their characteristic oxidant (e.g. O_2 or NO_3^-) or reductant (e.g. Mn^{2+}, Fe^{2+}). This is illustrated in Figure 17.20, which depicts the concentration–depth profiles of some redox-sensitive species as observed in a well underneath agricultural land in the Salland area (Vissers *et al.*, 1999). The upper 12 m of the groundwater is polluted by agricultural nitrate. With increasing depth, NO_3^- is the first to disappear (at approximately 14 m below the surface), rapidly followed by an increase of Fe^{2+}, and disappearance of SO_4^{2-} at greater depth. Note the lower concentrations of Fe^{2+} in the sulphate reduction zone, caused by the insolubility of iron sulphides.

If nitrate-polluted water enters an aquifer that contains sufficient amounts of pyrite, the pyrite can act as an electron donor in the denitrification of nitrate. The disappearance of nitrate is then accompanied by an increase in sulphate concentrations. Figure 17.21 shows an example of this process in the Oostrum aquifer in the north of Limburg Province, the Netherlands. This aquifer contains 0.1–0.9 percent weight of pyrite and is nearly calcite free (< 0.1 percent weight) (Broers and Buijs, 1997). At approximately 10 m below the water table, nitrate is reduced by oxidation of pyrite. Below the denitrification front the sulphate concentration increases. In addition, the concentrations of the trace elements As, Co, and Ni also increase below the denitrification front. These trace elements can be derived from the pyrite, in which they are present as substitutions for Fe or S in the crystal lattice (e.g. Co and Ni for Fe, As for S). However, pyrite oxidation alone is not sufficient to explain the increase in trace element concentrations. Broers and Buijs (1997) found that desorption of these elements from the adsorption complex is an additional mechanism for the increase of trace element concentrations in the groundwater. This is because the oxidation of pyrite by nitrate produces not only sulphate, but also acid:

$$FeS_2 \;+\; 3\,NO_3^- \;+\; 2\,H_2O \;\rightarrow\; Fe(OH)_3 \;+\; 1.5\,N_2 \;+\; H^+ \;+\; 2SO_4^{2-} \qquad (17.1)$$

In this case of calcite-poor sediments, the H^+ ions cause desorption of the trace elements from the sediment. In the case of carbonate-bearing aquifer sediments, the acid released would have been buffered by the dissolution of the carbonate minerals, thereby increasing the hardness of the groundwater. This illustrates that after denitrification by pyrite oxidation, agriculturally polluted groundwater can still be identified by the enhanced SO_4^{2-} concentration, possibly accompanied by increased hardness or trace metal concentrations.

Redox zoning also develops downgradient of landfills, due to the decomposition of organic-rich materials in the leachate. Compared with pristine groundwater, leachate contains high concentrations of the following redox species: DOC, Fe(II), CH_4, and NH_4^+. In this case, the redox zones appear in the reverse order, as the decomposition of dissolved organic material creates anoxic conditions in groundwater just below the landfill. The anoxic plume enters the oxic aquifer, where the redox potential increases progressively due to dispersion and depletion of the dissolved organic matter. Figure 17.22 illustrates the redox zoning downgradient of a landfill site near Grindsted, Denmark: methanogenic (methane-producing) conditions close to the landfill, through SO_4^{2-}, Fe(III), Mn(IV), and NO_3^- reducing conditions, to oxic conditions farthest away from the landfill. Recent investigations have shown that redox zones may overlap and the redox processes sometimes occur simultaneously, although one process usually dominates in terms of actual rates (Christensen *et al.*, 2001).

From Figure 17.22 it is clear that in the leachate plume downgradient from the landfill, Fe^{2+} and Mn^{2+} are mobilised through reduction of Fe(III) and Mn(IV), respectively. In the sulphate reduction zone close to the landfill, however, the Fe(II) concentrations are lower due to the precipitation of pyrite. The mobilised Fe and Mn species will migrate downgradient until the redox potential has risen to a critical level. At this point, the Fe^{2+} and Mn^{2+} will be re-oxidised and will precipitate as oxyhydroxides.

Thus, redox reactions occur in particular at the fringe of a landfill leachate plume, which is typically some decimetres to a few metres thick. These reactions involve the oxidation of dissolved organic carbon (DOC), CH_4, Fe(II), Mn(II), and NH_4^+ from leachate and reduction

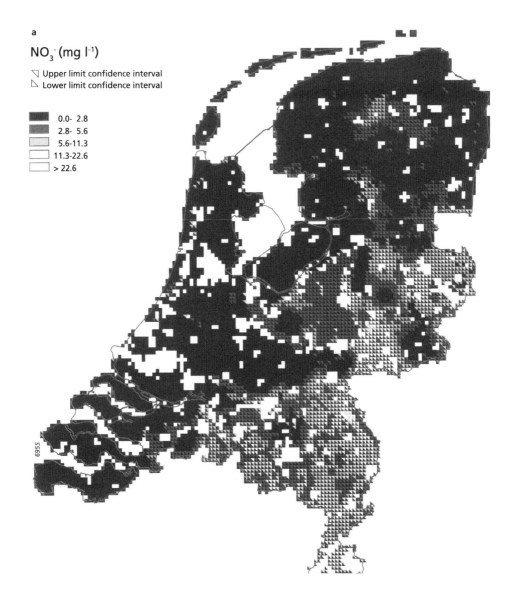

a

NO$_3^-$ (mg l^{-1})

▽ Upper limit confidence interval
△ Lower limit confidence interval

■	0.0- 2.8
▨	2.8- 5.6
▨	5.6-11.3
□	11.3-22.6
□	> 22.6

of O$_2$, NO$_3^-$, and SO$_4^{2-}$ from pristine groundwater (Hunter *et al.*, 1998). However, redox-sensitive cations (Fe(II), Mn(II), and NH$_4^+$) and organic micro-pollutants in the leachate plume are also subject to retardation, which hampers their advective and dispersive transport across the plume fringe (Griffioen, 1999). Van Breukelen and Griffioen (2004) performed a detailed study of the hydrochemical gradients at the top fringe of the Banisveld landfill leachate plume, the Netherlands. In the period 1998–2001, the top of the plume rose by approximately 0.6 m in response to intensified drainage by a nearby brook. Figure 17.23 shows the depth profiles of some solutes including redox-sensitive species sampled in 2001 (Van Breukelen and Griffioen, 2004). Due to upward advective transport induced by the increased drainage and differential retardation, the fronts of cations are clearly separated in space. The following cation fronts follow the conservative tracer front: 1) Na$^+$, 2) NH$_4^+$, and K$^+$, and 3) Ca^{2+}, Mg^{2+}, Fe^{2+}, and Mn^{2+}. Thus, cation exchange results in the spatial separation of NH$_4^+$ and in particular

b

NH$_4^+$ (mg l^{-1})

▽ Upper limit confidence interval
△ Lower limit confidence interval

▉ 0.0-0.2
▓ 0.2-0.4
▒ 0.4-1.0
░ 1.0-4.0
☐ >4.0

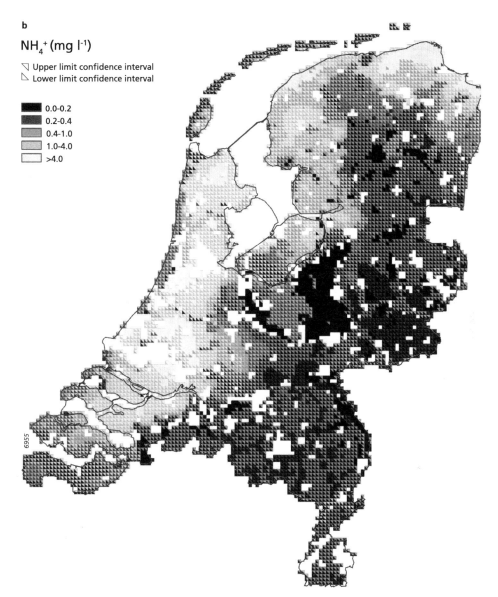

6955

Figure 17.19 a) NO$_3^-$ and b) NH$_4^+$ concentration in shallow groundwater (5–17 m depth) in the Netherlands in 1991. The concentrations are visualised as the 95 percent confidence interval for the median value in grid cells of 4 km × 4 km (Pebesma and De Kwaadsteniet,1994); see Figure 17.5b for surficial sediments.

of Fe(II) and Mn(II) from potential oxidants outside the plume. Note that the type of cation exchanger present in the aquifer determines the extent of retardation of NH$_4^+$ relative to Fe(II) and Mn(II). Illitic clay minerals bring about high retardation factors for NH$_4^+$, whereas organic matter brings about high retardation factors for divalent and trivalent metals (Van Breukelen and Griffioen, 2004). Non-sorbing organic solutes (e.g. DOC and methane) are transported across the plume fringe without retardation, so they may be oxidised before the cations arrive. Van Breukelen *et al.* (2004) showed that the flow of pH-neutral leachate through the slightly

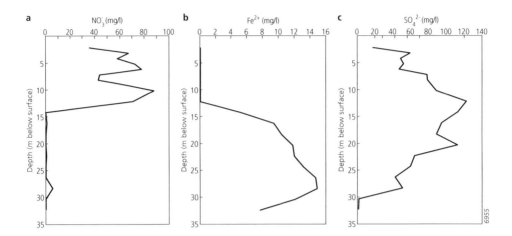

Figure 17.20 Sequence of redox zones as reflected by groundwater composition: NO_3^-, Fe^{2+}, and SO_4^{2-} concentration depth profiles in a well(A10) in Salland, the Netherlands . Data from Vissers (2006).

acidic aquifer triggers release of H^+ ions due to cation exchange (including proton buffering). The consequent pH decrease causes a decrease of bicarbonate and an associated increase in dissolved carbon dioxide and partial CO_2 pressure. The increase in total gas pressure probably triggers degassing of methane, which is rapidly transported into the pristine groundwater, where it redissolves. Consequently, methane is the principal reductant consuming soluble oxidants in pristine groundwater, thereby limiting the oxidation of other solutes including

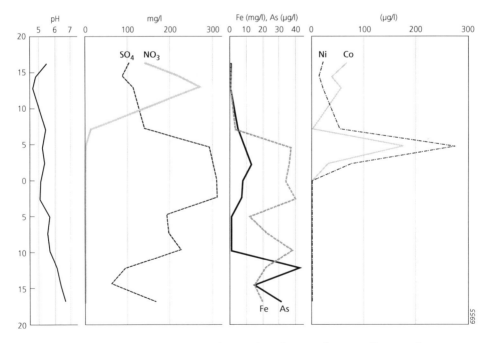

Figure 17.21 Depth profiles of pH, NO_3^-, SO_4^{2-}, As, Fe^{2+}, Co^{2+}, and Ni^{2+} in a profile in a well (WP40) near Oostrum, the Netherlands. Adapted from Broers and Buijs (1997).

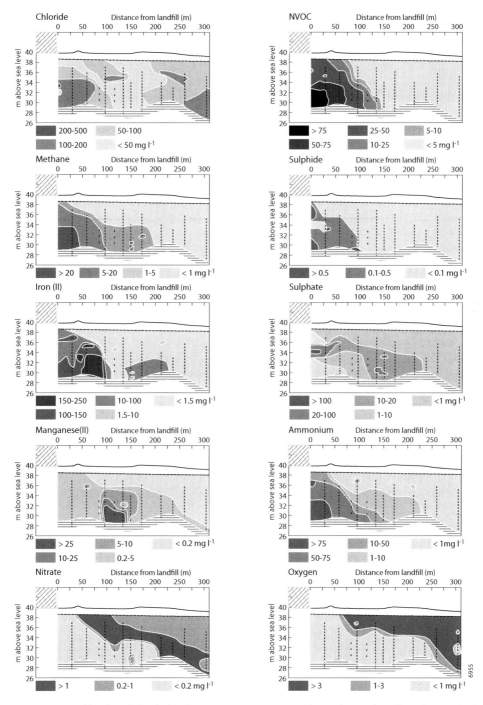

Figure 17.22 Chloride and dissolved redox-sensitive species in groundwater downgradient from the Grindsted landfill site, Denmark. Non-volatile organic carbon (NVOC) is given in mg C l^{-1}, NO$_3^-$ and NH$_4^+$ in mg N l^{-1}, and sulphide and SO$_4^{2-}$ in mg S l^{-1}. Adapted from Bjerg *et al.* (1995).

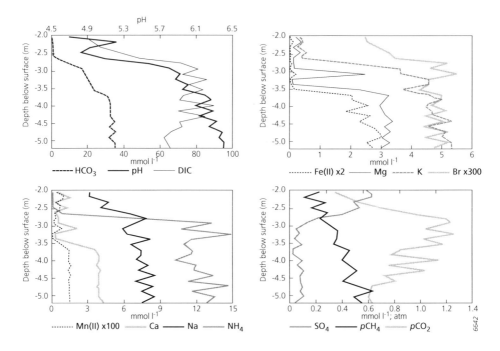

Figure 17.23 Hydrochemical depth profiles for a well (M2) in the top of the Banisveld landfill leachate plume, the Netherlands, in 2001. Adapted from Van Breukelen and Griffioen (2004).

organic micro-pollutants at the fringe of this landfill leachate plume (Van Breukelen and Griffioen, 2004).

After the source of a leachate plume has been depleted or eradicated, the original redox potential of the aquifer will gradually recover. This natural attenuation will obviously also be reflected in the concentration of redox-sensitive species. Repert *et al.* (2006) conducted a study on the long-term natural attenuation of carbon and nitrogen in a groundwater plume in the Cape Cod phreatic aquifer (the same site as depicted in Figure 17.15) after removal of the pollutant source. This source had consisted of treated wastewater that was nevertheless rich in nitrate and dissolved organic carbon (DOC) and had infiltrated into infiltration beds, but emissions of the wastewater were stopped in December. Figure 17.24 shows the downgradient development contaminant plume for DOC, nitrate and ammonium during 8.5 years after removal of the source of contamination. Along the transect, the DOC, NO_3^-, and NH_4^+ concentrations decreased with time but remained higher than the natural background concentrations. The highest DOC concentrations were typically in the uppermost zone directly below the infiltration beds, although pulses of DOC between 200 and 300 µM C travelled through the plume transect. Mobilisation of DOC from sediment organic matter in the infiltration beds remained a DOC source even after the cessation of wastewater emissions. The NO_3^- concentrations in groundwater decreased considerably during the study period: from values of more than 1000 µM in the oxic/suboxic zone prior to December 1995 and shortly thereafter to about 230 µM in 2004. The anoxic, nitrate-free zone at the centre of the plume persisted for the entire period of sample collection and even grew bigger directly after the oxygenated wastewater source had been removed. In 2002 the nitrate-free zone started to shrink, in response to uncontaminated groundwater entering the plume area. However, there were still relatively high concentrations of NO_3^- (150-200 µM) in the shallow groundwater,

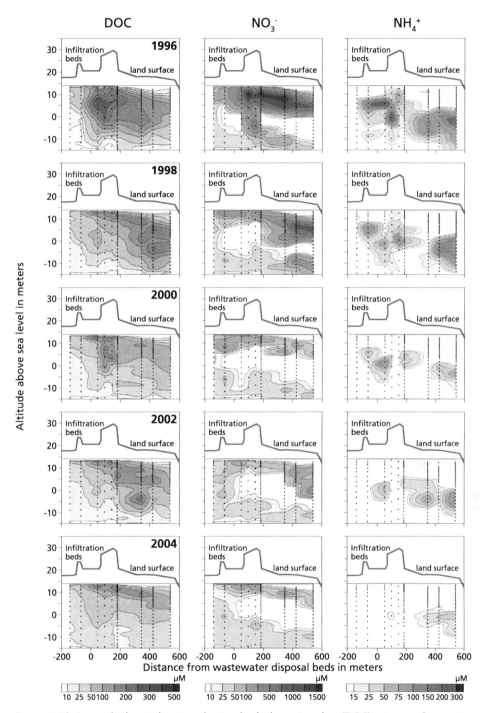

Figure 17.24 Longitudinal vertical sections showing dissolved organic carbon (DOC), nitrate, and ammonium concentrations in groundwater downgradient from infiltration beds on Cape Cod, Massachusetts, USA in 1996 (0.5 yrs), 1998 (2.5 yrs), 2000 (4.5 yrs), 2002 (6.5 yrs), and 2004 (8.5 yrs post-cessation). Adapted from Repert *et al.* (2006).

which were probably derived from the oxidation of sediment organic matter. NH_4^+ concentrations in the groundwater also decreased throughout the plume transect after cessation of wastewater emissions, from peak concentrations ranging from 100 to >400 μM in the anoxic zone in 1996 to less than 90 μM at the distal end of the plume transect in 2004 (Repert *et al.*, 2006). It is likely that after the source has been fully depleted and aerobic conditions have taken over in the aquifer, the aquifer will have permanently changed features. For example, Fe and Mn oxyhydroxides may have been depleted in the central part of the plume and enriched further downgradient due to the migration of reduced Fe and Mn (Christensen *et al.*, 2000).

17.10 FURTHER READING ON CONTAMINANTS IN GROUNDWATER

- Appelo, C.A.J. and Postma, D., 2005, *Geochemistry, Groundwater and Pollution*, 2nd edition, (Leiden NL: Balkema/Taylor and Francis).
- Ginn, J.S. and Russel Boulding, J., 2003, *Practical Handbook of Soil, Vadose Zone, and Ground-Water Contamination: Assessment, Prevention, and Remediation*, 2nd edition, (Boca Raton FL: Lewis/CRC).
- Mayer, A.S. and Hassanizadeh S.M. (Eds.), 2005, *Soil and Groundwater Contamination: Nonaqueous Phase Liquids*, (Washington DC: American Geophysical Union).

EXERCISES

1. Explain briefly:
   a. why chloride concentrations in groundwater are generally lower below heather than below forests;
   b. why dissolved iron disappears in the sulphate reduction zone in groundwater;
   c. why the Na^+: K^+ ratio increases from the centre to the edge of a landfill leachate plume;
   d. why the alkalinity of groundwater increases sharply at a certain depth in a sandy aquifer;
   e. why groundwater in peatlands contains no nitrate.

2. Name three fundamental causes of the temporal changes in groundwater composition observed in a groundwater monitoring well.

3. Explain why in some regions, the mean nitrate concentartions in groundwater do not show a decreasing trend in repsonse to the reduction in fertiliser application rates

4. The concentrations of Ca^{2+} measured in groundwater abstracted from a well rise over time. Give two possible causes of this.

5. From Figure 17.8 estimate the evapotranspiration rates for forest and heather, given an average annual precipitation of 780 mm y^{-1} and a chloride concentration in rainwater of 4 mg l^{-1} (hint: see Box 5.I).

6. Sketch the isolines for the redox potential (in mV) for the same cross-section downgradient from the Grindsted landfill site as depicted in Figure 17.22 (hint: see Figure 4.10).

Patterns in surface water

18.1 INTRODUCTION

The patterns of surface water composition are controlled by a wide range of factors and processes, including catchment geology, diffuse and point source inputs from anthropogenic sources, catchment topography and hydrology, and in-stream processes (see e.g. Jarvie *et al.*, 2002). As we have seen in the previous chapters, catchment geology determines the background concentrations in soils and groundwater, and thus also in surface water. Catchment topography and hydrology play important roles in the transfer of contaminants from the catchment surface to the drainage network. After contaminants have entered the surface water via overland or underground pathways or a combination of both, in-stream biogeochemical processes modify their concentrations. Thus, to understand, interpret, or predict patterns in surface water composition, it is important to understand the processes and patterns in soil and groundwater, and the physical and biogeochemical conditions affecting the surface water (see e.g. Bouwman *et al.*, 2013).

Surface water differs from groundwater by having a stronger interaction with the atmosphere, a weaker interaction with sediments, and by being penetrated by sunlight. As a result, the nature and rate of the various biogeochemical processes that act in surface water differ from those in groundwater. Generally, the redox potential is higher in surface water than in groundwater, because of the presence of relatively high concentrations of dissolved oxygen. The oxygen originates from diffusion of atmospheric oxygen or photosynthesis by primary producers (e.g. algae and submerged aquatic macrophytes) during daylight. Photosynthesis simultaneously extracts dissolved CO_2 (carbonic acid) from the water, which raises the pH. Therefore, surface waters are mostly neutral to basic (compare Figure 2.3), except for water bodies that are susceptible to acidification in poorly buffered catchments.

Compared to groundwater, surface water flows fast. Whereas groundwater travels at the rate of centimetres per day or even less, surface water travels at a rate in the order of metres to tens of metres per minute. Thanks to the fast transport rate of surface water, the reaction kinetics usually become apparent in the spatial variation in surface water composition. Accordingly, water pollutants that are discharged instantaneously or continuously into surface waters are rapidly carried downstream and can assert their influence across distances up to hundreds of kilometres, as unfortunately demonstrated by large accidental spills into rivers: for example, the Sandoz accident in the river Rhine in 1986 (see Figure 10.4) or the Ajka alumina sludge spill near Kolontár, Hungary, in October 2010.

Whereas rivers have a pronounced downstream flow of water, lakes do not. Instead of gravity, wind is the main force generating and driving currents in lakes. The downwind current near the lake surface typically moves at a rate of 2 to 3 percent of the average wind speed (Hemond and Fechner-Levy, 2000). Since the water flowing downwind cannot accumulate indefinitely at the downwind lake shore, a return current arises, usually at a greater depth. In large lakes the flow pattern may become very complex, as the water flow is

100 km

6642

Figure 18.1 General water circulation pattern in Lake Michigan, USA.

affected by variations in wind speed and direction, the shape of the lake, inflowing streams, variations in water density and, particularly in very large lakes, by the Coriolis force caused by the rotation of the Earth (Hemond and Fechner-Levy, 2000). In the northern hemisphere, the Coriolis force deflects surface currents to the right; in the southern hemisphere it deflects them to the left. Figure 18.1 shows the general water circulation in Lake Michigan, USA, as an example of a complex pattern of water movement in large lakes produced by variability of winds, complex lake geometry, and the Coriolis force.

Because rivers are long, shallow and narrow water bodies, it is commonly assumed that dissolved and particulate matter is uniformly mixed across the cross-sectional area of the channel. Therefore, rivers are usually considered as one-dimensional objects. Only in river studies at the local scale, for example in the case of transverse mixing of point discharges within the river channel or dispersal of sediment and associated contaminants over floodplains, is the second or even third dimension taken into consideration. Obviously, lakes are much wider than rivers, so in most lake studies, the second lateral dimension is also taken into account. Furthermore, in deep lakes, where vertical variation of the physico-chemical properties of lake becomes important, the third (vertical) dimension is also considered.

18.2 SPATIAL VARIATION IN RIVER WATER COMPOSITION

18.2.1 Effects of diffuse sources

It has long been recognised that besides direct point source discharges, contaminant transport in and from catchments is controlled by climate, geology, topography, and diffuse anthropogenic inputs (see e.g. Dillon and Kirchner, 1975; Grobler and Silberbauer, 1985; Qu *et al.*, 1993; De Caritat, 1996; McKee *et al.*, 2001). In pristine rivers, the transport of substances is largely determined by the interrelated geochemical composition and weathering rate of the bedrock, which can be traced back to geology and climate factors,

Table 18.1 Heavy metal loads and the share from diffuse sources in the river Rhine at the Lobith monitoring station, the Netherlands (river basin size 159 715 km^2; mean discharge 2201 m^3 s^{-1}) and the river Elbe at the Snackenburg monitoring station, Germany (river basin size 125 160 km^2; mean discharge 712 m^3 s^{-1}) in the period 1993-1997 (source: Vink and Behrendt, 2002).

	Cd	Cu	Hg	Pb	Zn
Rhine					
Total load (t y^{-1})	8.4	483	3.9	269	2256
Diffuse share (%)	76	70	70	71	75
Elbe					
Total load (t y^{-1})	9.6	268	4.7	198	1638
Diffuse share (%)	62	70	51	75	64

such temperature and precipitation (Meybeck, 2003). In human-influenced rivers, most of the total concentration or load of many substances can be apportioned to diffuse sources from agricultural land, urban areas, or atmospheric deposition (Meybeck, 2002; Novotny, 2003). For example, between 1985 and 1990, 58 percent of the N load and 36 percent of the P load in the river Rhine came from diffuse sources. In the Elbe basin, diffuse emissions were estimated to contribute to 66 percent of the N load and 37 percent of the P load in the same period (De Wit, 2001). Vink and Behrendt (2002) apportioned the sources for heavy metals transported by the same rivers. Table 18.1 lists the total load and share from diffuse sources for the metals Cd, Cu, Hg, Pb, and Zn in the period 1993-1997. In the Rhine basin, diffuse source inputs dominate the total transport and contribute to more than 70 percent of the total load. In the Elbe basin, between 51 percent (for Hg) and 74 percent (for Pb) of the total heavy metal load is supplied by inputs from diffuse sources. The diffuse hydrological pathways that contribute most include erosion and runoff from urban areas.

Diffuse inputs enter surface water over large areas. Accordingly, the spatial patterns of concentrations in surface waters that arise from diffuse inputs are also diffuse and barely observable in catchments smaller than a few square kilometres. They may only emerge at the regional scale and higher, where distinct spatial differences in diffuse emissions occur between regions or catchments. Figure 18.2 shows the contribution of diffuse (agricultural and background) and point sources to the total riverine N export rate per unit area in three large European catchments around the year 2000 (EEA, 2005). The largest N loads from both point and diffuse sources occur in the North Sea catchment, because here the population density is high and agriculture is intensive (and thus there is an N surplus resulting from fertilisation and manuring). Although the proportion of agricultural land in the Danube catchment is similar to that in the North Sea catchment, here the agricultural land use is much less intensive and so the diffuse loads from agriculture are also less than those in the North Sea catchment. The N load in the Baltic Sea catchment is relatively small. The majority of the N load in this catchment area is derived from Germany, Denmark and southern Sweden. The N losses from Poland, Russia and the Baltic States are much less, because of less intensive agriculture in these parts of the catchment. The N inputs from Finland and central and northern Sweden is also small, because in these areas the proportion of agricultural land is relatively small (EEA, 2005). The same spatial distribution of N in Europe can be noticed in Figure 18.3, which depicts observed annual average NO$_3^-$ concentrations in large European rivers in 1994–1996 (EEA, 1999). The NO$_3^-$ concentration in unaffected rivers (these occur mainly in Scandinavia and Scotland) was 0.1–0.5 mg l^{-1}. In the Nordic countries, 70 percent of the sites had NO$_3^-$ levels below 0.3 mg l^{-1}. Excluding the Nordic rivers, 68 percent of the river stations had mean NO$_3^-$ concentrations exceeding

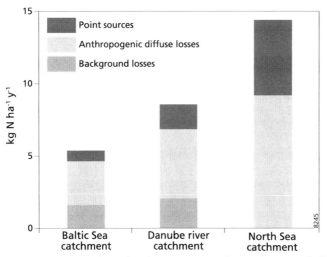

Figure 18.2 Sources of N in three large European catchments (> 300 000 km²): Source apportionment for annual nitrogen load per hectare (EEA, 2005).

1 mg l⁻¹. The highest NO_3^- concentrations were found in rivers in the intensive agricultural regions in the northern part of Western Europe.

It is not only the overall magnitude of diffuse sources that determines the quality of water; also important is the extent to which the diffuse contaminants are able to reach the surface water. This is largely controlled by the availability of water, the directness of the pathways linking the terrestrial part of the catchment to the river network, and the transport rates along these pathways. So-called *critical source areas* of diffuse pollution occur where a pollutant source in the landscape coincides with active hydrological transport mechanisms (Pionke *et al.*, 2000; Gburek and Sharpley, 1998). In addition, critical source areas should be hydrologically connected to the stream network (for definitions of hydrological connectivity, see Michaelides and Chappell (2009)). The spatial delineation of the critical source areas of dissolved and sediment-associated contaminants is difficult, as it relies on the assessment of factors such as the connectivity and sediment delivery from soils to the stream channel, which are highly variable in space and time (e.g. Beven *et al.*, 2005). The effect of hydrological events on the temporal variation of river water quality will be discussed further in Section 18.3.

In general, the riverine export of contaminants from catchments with similar diffuse loadings is larger for catchments with higher rainfall (e.g. De Wit, 1999), since here the soils are more intensely hydrologically connected to the streams and the transport rates are faster. Between-year variation in contaminant export from catchments can also often be largely attributed to differences in rainfall or river discharge between the years (e.g. Laznik *et al.*, 1999). Contaminant transfer is also generally greater in steep catchments with shallow, impermeable soils than in permeable lowland catchments. Moreover, contaminant transfer is also quicker in impermeable catchments with short water transit times, so here the water quality responds rapidly to hydrological events triggered by intensive rainstorms or snowmelt. In addition, contaminant transfer to surface water is also determined by retention along the pathways. Substances susceptible to degradation, such as pesticides, or substances susceptible to sorption to soil and aquifer materials, such as heavy metals and phosphorus, reach the surface water network less easily than more mobile substances such as Cl⁻ and, to a lesser extent, NO_3^-. Both the transport time and retention along the surface and subsurface pathways cause areas close to the surface water network to contribute more to diffuse

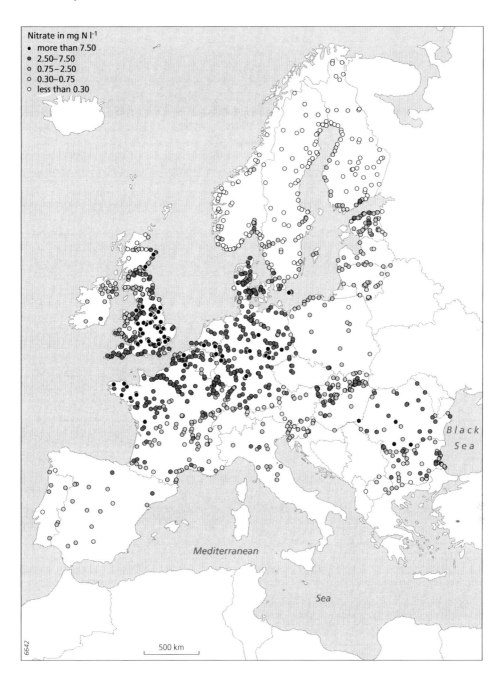

Figure 18.3 Annual average NO$_3^-$ concentrations in large European rivers in 1994–1996 (EEA, 1999).

contamination of river water than areas near the drainage divide. Catchment disturbance by agricultural or urban developments also affect the distance the water travels and the retention of contaminants along the flow paths. For example, Sudduth *et al.* (2013) demonstrated that undisturbed forested catchments possess a significant capacity to remove NO$_3^-$ after it passes below the soil root zone and that land use changes tend to reduce both the length of the flow paths and the nitrate removal capacity along the flow paths.

During transport through the river network, contaminant loads are attenuated by physical and biogeochemical processes. Seitzinger *et al.* (2002) related N removal in sixteen eastern US river networks to the Strahler stream order (Figure 18.4). In the Strahler stream classification system, a first-order stream has no tributaries; where two 1st order channels join, a stream of 2nd order is formed and where a 2nd order stream confluences with another 2nd order stream, a 3rd order stream is formed, and so on (Strahler, 1952). In general, the contribution of each stream order to total stream length in a river network decreases with increasing stream order. Seitzinger *et al.* (2002) found that reach-specific N removal (and thus the first-order denitrification constant: see Section 13.3) is greatest in the smaller, lower-order streams (Figure 18.4a). This can be explained by the increase of N removal with decreasing water depth (Alexander *et al.* 2000; see Table 13.3) and by the fact that lower-order streams are shallower than higher-order streams. In contrast to the reach-specific N removal, the total mass of nitrogen removed per unit stream length increases with increasing stream order, since the discharge and N load increase with increasing stream order (Figure 18.4b). Overall, 37% to 76% of the total N input to the rivers studied by Seitzinger *et al.* (2002) is removed in the river network, and approximately half of this is removed in 1st through 4th order streams, which account for 90% of the total stream length (Figure 18.4c). The other half is removed in higher-order rivers, which account for only about 10% of the total stream length (Seitzinger *et al.*, 2002). This means that the proportion of the N inputs that is removed to the catchment outlet is smaller for the higher-order streams, because in

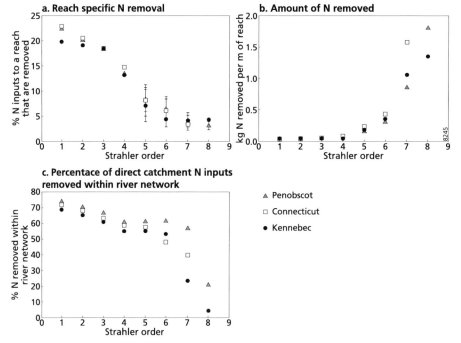

Figure 18.4 Relationship between Strahler stream order for the Penobscot, Connecticut and Kennebec river networks and a) the proportion of N input to a reach that is removed within that reach (average ± s.d.), b) the average amount of N (kg) removed per meter of stream/river with uniform direct catchment N loading to river (100 kg N km^{-2} y^{-1}), and c) the proportion of N input into a particular Strahler order that is removed within the river network. Adapted from Seitzinger *et al.* (2002).

such streams there is less reach-specific removal, and the flow paths and associated transit times to the catchment outlet are shorter. This is the main reason that downstream areas close to the catchment outlet generally contribute much more to the diffuse catchment N export than the headwaters far away from the catchment outlet.

18.2.2 Effects of point sources

Point source discharges of pollutants into rivers can have major adverse effects on downstream water quality. The increase in concentrations of the substances discharged is most apparent in the reach directly downstream of the discharge, since most substances are affected by physical and chemical processes during downstream transport, such as dilution, volatilisation, sorption to suspended and bed sediments, biochemical decay, uptake by organisms, or exchange between the stream or river channel and the hyporheic zone. These processes are generally referred to as the self-cleaning capacity of a river and generally lead to a decrease in pollutant concentrations in downstream direction. The distance at which the concentration of a substance reaches a certain fraction of the peak concentration depends on the nature of the processes acting upon the substance and the rates at which these processes occur.

The downstream effects of point source discharges on the concentrations of Cl^-, PO_4^{3-}, NH_4^+, and NO_3^- are illustrated in Figure 18.5, which depicts the longitudinal concentration profiles of these four substances in the upper reach of Biebrza river in north-east Poland in June 1994 (Van der Perk, 1996). Two steady inflows of untreated domestic wastewater were present at 0.6 km and 7.5 km downstream from the local origin of the longitudinal profile. The discharge and flow velocity of the Biebrza river and the concentrations of Cl^-, PO_4^{3-}, NH_4^+, and NO_3^- in the wastewater remained approximately constant in time during the sampling period, and therefore steady state conditions could be assumed. The discharge

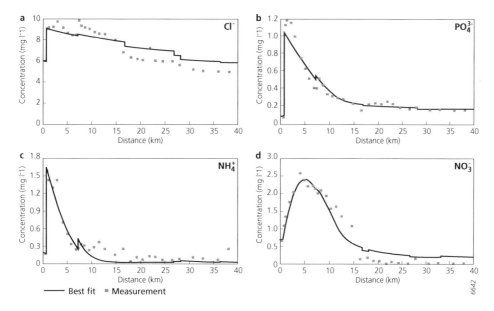

Figure 18.5 Longitudinal concentration profiles of a) Cl^-, b) PO_4^{3-}, c) NH_4^+, and d) NO_3^- in the Biebrza river, Poland, in June 1994. The dots represent observed concentrations and the line represents modelled concentrations (Van der Perk, 1996).

at 0.6 km has a particularly clear impact on the concentrations of the four substances. The Biebrza river flows through a relatively undisturbed peatland without agricultural inputs, so that diffuse pollution sources are almost absent. Since Cl^- is a conservative substance, it is not removed by adsorption or biochemical processes, nor is it produced. Only dilution by groundwater discharge into the river and inflowing tributaries affects the Cl^- concentration (Figure 18.4a). The local Cl^- concentration in groundwater amounted to about 1 mg l^{-1}. As the influx of groundwater was only moderate (approximately 0.035 m^3 s^{-1} per kilometre of river), the Cl^- concentration decreased rather slowly in downstream direction. The PO_4^{3-} concentration (Figure 18.4b) decreased much faster, since PO_4^{3-} is removed through adsorption onto stream bed sediments and particulate matter, and through biological uptake by algae and aquatic macrophytes, with adsorption onto bed sediments being the major mechanism for PO_4^{3-} retention (Van der Perk, 1996). Similarly to PO_4^{3-}, the NH_4^+ concentration (Figure 18.4c) also decreased to pre-discharge concentrations within a relatively short distance. Although adsorption to bed sediments played a relevant role in the decline in NH_4^+ concentrations, nitrification was identified as being the main process of NH_4^+ removal. The nitrification of NH_4^+ is also manifested in the gradual increase of NO_3^- concentrations (Figure 18.4d) in the first few kilometres downstream of the waste water discharge at 0.6 km. In this stretch of the river, the production of NO_3^- by nitrification exceeded the NO_3^- removal by denitrification. At about 6 km downstream of the wastewater discharge, the NH_4^- concentration was depleted to such an extent that nitrification could not compensate for the NO_3^- lost by denitrification. In the stretch of river further downstream from the peak NO_3^- concentration, the denitrification process prevails, so the NO_3^- concentration falls to pre-discharge concentration within 15 to 20 km downstream from the wastewater discharge.

In steady state conditions, as in the above example in which a continuous inflow of wastewater occurs in a river with constant discharge, the effect of longitudinal dispersion on the spatial distribution of concentrations is insignificant (Thomann, 1973; Fisher *et al.*, 1979). The Peclet number, which expresses the ratio of mass transport by advection to that by dispersion (see Section 11.3.3), is usually much larger than 1 for most rivers. Moreover, the concentration gradients in longitudinal direction that result from dilution and biochemical processes are so small that the net dispersive flux is negligible by comparison with the advective flux.

As with diffuse sources, point sources located in the downstream parts of river catchments contribute more to the total contaminant export from catchments than those in the headwaters, because the shorter flow path to the catchment outlet and deeper water result in less contaminant removal. Moreover, in many parts of the world, the population density tends to increase in downstream direction with the largest population centres be in coastal areas, i.e. near the catchment outlets (Vörösmarty *et al.*, 2010) . These areas are characterised by large contaminant inputs into surface waters from urban runoff and industrial and domestic wastewater treatment facilities. Destouni *et al.* (2008) demonstrated that small, near-coastal catchment areas may generate large nutrient and mass loads entering the sea that are as large or larger than river loads from large river basins.

18.3 TEMPORAL VARIATION IN RIVER WATER COMPOSITION

18.3.1 Short-term dynamics

Rivers and streams are very dynamic environments. Concentrations of the substances they carry may exhibit a typical and complex response to increased runoff events. The response is the result of many processes that occur simultaneously during such hydrological events, including dilution by rainwater or meltwater, increased inputs from overland flow, increased supply of groundwater discharge, mobilisation of particulate and dissolved material from the river bed and terrestrial part of the catchment, and increased inputs from urban runoff and, possibly, sewage overflows. This means that river water is a dynamic mixture of water entering the streams through different hydrological pathways, each having their characteristic transit time and chemical and isotopic signature (Kendall and McDonnell, 1998). The statistical distribution of catchment water transit times (of both surface and subsurface pathways) follows a power-law distribution characterised by a long tail of long transit times (Kirchner *et al.*, 2000; McGuire and McDonnell, 2006; Wörmann *et al.*, 2007). Such power-law distributions indicate self-similar fractal behaviour in the temporal scaling of surface–subsurface interactions. This means that the majority of dissolved contaminants will initially be flushed out rapidly, but the delivery of low-level contamination to streams continues for a long time.

The contribution of the quick runoff pathways, such as overland flow, is larger in relatively impermeable catchments with thin soils, or in catchments with wet soils and shallow water tables. Overland flow originates mostly from the same small parts of the catchment, and these comprise less than 10% (usually 1–3%) of the catchment area; on such areas, only 10–30% of the rainfall produces overland flow (Freeze, 1974). Overland flow has a short transit time and therefore is usually poor in major dissolved phase constituents, has a low pH, but is rich in dissolved organic matter and may carry sediments and associated substances. On the other hand, groundwater has a much longer transit time and is often characterised by higher concentrations of weathering products, such as base cations and silica, Cl⁻ concentrations that are higher than in rainwater (due to evapotranspiration: see Section 17.3), and high alkalinity (Jenkins *et al.*, 1994). However, not all the water reaching the surface water via subsurface pathways has a long transit time. For example, rainwater that infiltrates into soils that are artificially drained by tile drainage, or groundwater in shallow ephemeral perched aquifers that arise during the rainy season can be rapidly transferred to the stream (e.g. Ocampo *et al.*, 2003).

Figure 18.6 gives an example of the typical response of the Ca^{2+} concentration to hydrological events as measured at 7-hourly intervals in the small Upper Hafren stream near Plynlimon in Wales. This stream drains an area of relatively undisturbed moorland overlying poorly permeable bedrock (Neal *et al.*, 2012, 2013). As a result, the stream response to rainfall is 'flashy', with the rising limb to peak flow typically lasting less than an hour, and the stream water is acidic with low solute concentrations. During baseflow conditions, the Ca^{2+} concentrations are normally between 0.70 and 0.85 mg l⁻¹, but decrease quickly in response to the sudden increases in streamflow during runoff events, due to dilution by low-calcium rainwater. During the recession limb of the hydrograph, the stream water Ca^{2+} concentration slowly recovers to pre-event values.

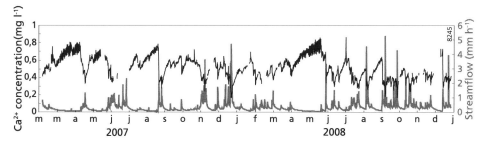

Figure 18.6 Response of the Ca^{2+} concentration to hydrological events in the Uper Hafren stream near Plynlimon, Wales (source: Neal *et al.*, 2012, 2013).

The areas that contribute most to increased runoff events are those that are hydrologically active and connected to the streams via quick runoff pathways (for example, through overland flow or tile drainage). They are generally located in relatively narrow zones along the stream channel, but their surface areas vary between and during rainstorm events. Therefore, these areas are usually referred to as variable source areas (Hewlitt and Hibbert, 1967; Beven and Kirkby, 1979; McDonnell, 2003). If the topsoil in these areas is contaminated, these variable source areas also act as ***critical source areas*** of diffuse pollution (see Section 18.2.1), as the pollutants may be quickly transferred to the river network and thereby contribute to river water contamination (e.g. Doppler *et al.*, 2012). Note that critical source areas do not necessarily coincide with variable source areas, as critical source areas may also refer to areas that act as pollutant source via slow subsurface pathways.

Clearly, to be able to understand the response of river water composition to runoff events it is essential to identify the hydrological pathways within the catchment and their associated transit times. Conversely, the chemical or isotopic composition of the stream water can be used to subdivide hydrographs into components of flow: a method called hydrograph separation (Robson and Neal, 1990; Wels *et al.*, 1991).

18.3.2 Hysteresis response of dissolved concentrations to changes in discharge

The timing of the arrival of water from various hydrological pathways, which brings about the typical temporal patterns in chemical transport from catchments, is controlled by topography, groundwater gradients and permeability of the aquifer, and surface resistance to flow. Furthermore, spatial variations in the occurrence of the hydrological pathways due to variations in rainfall, soil type, and land use across the catchment, can also significantly affect the temporal patterns in streamwater chemistry (Walling and Webb, 1986). These temporal variations become visible in characteristic cyclical shapes in the relationship between instantaneous water discharge and instantaneous concentration at a measurement site during a single storm event (Q–C relationship) (Evans and Davis, 1998). Such Q–C relationships can be characterised by 1) the direction of the relationship (negative: the concentration decreases in response to increased discharge; positive: the concentration increases with discharge; zero: no response), 2) the timing of the peak or trough concentration relative to peak discharge (before, at the same time, or after peak discharge), and 3) the difference in concentrations occurring during the same discharge during the rising limb of the hydrograph compared to the falling limb. These characteristics determine the nature of the hysteresis of the Q–C relationship (clockwise or anticlockwise). Figure 18.7 shows some hypothetical concentration responses to increased discharge. For example, if the concentration increases with discharge, the peak concentration lags behind the peak discharge, and the concentrations are higher during the falling limb of the hydrograph than during the rising limb (Figure 18.7c), then an anti-clockwise hysteresis loop is obtained (Figure 18.7d).

Both the direction of the Q–C relationship and the direction (clockwise/anticlockwise) of the hysteresis pattern can be used to infer the flow pathways dominating during individual storm events (Evans and Davies, 1998). They reflect changes in water composition resulting from changes in the contribution from groundwater, soil water, and surface event water (overland flow) during storms. For example, if the concentrations of dissolved substances in overland flow are higher than in subsurface pathways, they will increase with increasing volume of overland flow, so the concentration will peak early during storms with a positive, clockwise Q–C response. Conversely, if it is the concentrations of chemicals in groundwater discharge that are higher, then the overland flow during storms dilutes the in-stream concentrations and produces negative, often anticlockwise Q–C responses. Figure 18.8 shows an example of the response of different phosphorus forms to storm events in the Den Brook catchment, Devon, UK (Haygarth *et al.*, 2004). The temporal trend of reactive P (<0.45 μm)

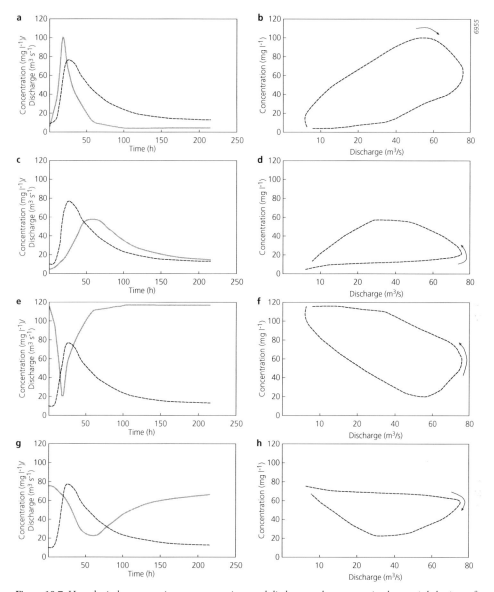

Figure 18.7 Hypothetical concentration responses to increased discharge and accompanying hysteresis behaviour of the Q–C relationship.

(i.e. dissolved ortho-PO$_4^{3-}$) shows only a slight negative response to the main storm event on 23 January 2002. In contrast, total P increased strongly with discharge and exhibits a clockwise hysteresis pattern. The different responses imply there are different processes of mobilisation of different forms of P. Reactive P (< 0.45 μm) reflects the soluble species of P that can be more easily mixed in the matrix flow paths of the soil. The total P includes P forms that are not soluble and thus reflects the particles with P attached that have been physically detached by the kinetic energy of rainfall (Haygarth *et al.*, 2004). The response of particulate matter (suspended sediment) to hydrological events is discussed in further detail in Section 18.3.4.

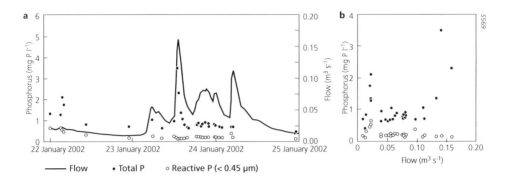

Figure 18.8 Response of different forms of P to storm events in the Den Brook catchment, Devon, UK, in January 2002. Adapted from Haygarth *et al.* (2004).

The hysteresis pattern can change from storm to storm and across seasons (Walling and Webb, 1986). For example, when the store of readily soluble substances in soil has become exhausted over time due to progressive leaching or uptake by vegetation during the growing season, the concentration response may decrease during the successive storm events. On the other hand, mineralisation of litter at the end of the vegetative period may have the opposite effect. Rising groundwater levels during winter and spring can also influence hysteresis patterns, as the contribution of groundwater to flow increases. Furthermore, variations in hysteresis between storm events can also be caused by different patterns of rainfall across the catchment. Consequently, although hysteresis patterns can be generalised, the Q–C relationship in a river may not follow the same pattern between storms.

18.3.3 Release of old water

Storm events not only cause increases in inputs from overland flow, but may also mobilise substantial amounts of 'old' or pre-event water, which may even dominate the increase in discharge. Eshleman *et al.* (1992) performed a chemical hydrograph separation based on chloride concentrations as a means of quantifying the absolute and relative contributions of 'new' rainwater and 'old' groundwater to the storm hydrograph. Figure 18.9a shows a hydrograph and chemograph of Cl^- during a 5.2 cm rainfall event in the Reedy Creek catchment in Virginia, USA, during spring 1991. The rainstorm caused substantial Cl^- dilution in the creek, since Cl^- concentrations are much lower in precipitation than in baseflow. Provided that the Cl^- concentrations in rainwater and groundwater do not vary during the storm or vary spatially across the catchment, measurements of the Cl^- concentration in the two components and in the mixture over time can be used to back-calculate the contributions from 'old' and 'new' water. Figure 18.9b shows that 'old' water is almost totally responsible for the increase in discharge in this catchment during the storm event. Though negative, the hysteresis loop is clockwise (Figure 18.7c), which implies that the main contribution of event water is during the falling limb of the hydrograph. This contradicts the general assumption that the rising limb of the hydrograph is dominated by channel precipitation and quick overland flow, as explained in the previous paragraph. This contradicts the general assumption that the rising limb of the hydrograph is dominated by channel precipitation and quick overland flow, as explained in Section 18.3.1. In contrast to the Cl^- concentration, the concentrations of reactive chemical species (such as Ca^{2+} or H^+) in old water may be highly variable. The combination of rapid mobilisation of old water

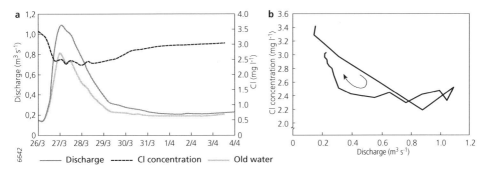

Figure 18.9 Hydrograph separation of a storm event in the Reedy Creek catchment, Virginia, USA, during spring 1991: a) hydrograph and chloride concentrations, b) contribution of 'old' water to storm runoff, c) hysteresis loop Data from Eshleman *et al.* (1992).

with variable runoff chemistry is what Kirchner (2003) referred to as 'a double paradox' in catchment hydrology and geochemistry.

A number of physical mechanisms can explain the rapid groundwater response of systems such as Reedy Creek. In some systems, infiltrating rain triggers the conversion of the tension-saturated capillary fringe into phreatic water and, accordingly, a rapid and disproportionately large rise in the water table (Sklash and Farvolden, 1979). A rapid rise of the water table may also be caused by rapid drainage of soil water through soil cracks and macropores (Bleuten, 1988).In turn, the rise of the water table increases the groundwater gradient and, accordingly, the discharge towards the stream. In many soils, this process may be amplified by a progressive increase in the lateral saturated hydraulic conductivity towards the soil surface (Bishop et al., 2004). Another mechanism that contributes to the delivery of 'old' water to a stream under storm conditions is the entrainment of 'old' soil water by overland flow or hyporheic exchange. Overland flow in riparian areas occurs in patches, whereby water may infiltrate locally into the soil, mix with shallow soil water, and subsequently re-emerge at the surface. Consequently, the overland flow may contain varying portions of 'old' soil water (Hornberger *et al.*, 1998).

18.3.4 Sediment dynamics

Sediment concentrations also display hysteresis behaviour in response to hydrological events. The sediment concentration response is the result of increased sediment detachment rates, increased transport capacity of both overland and channel flow, and the increased surface areas across which overland flow and sediment detachment take place during hydrological events. As a consequence, the Q–C relationship for sediment is always positive. This implies that Q–C relationships for sediment-associated substances, like the example of total P given above (see Figure 18.8), are also usually positive. Sediment transport rates are a function not only of the transport capacity of a river but also of sediment availability. During the course of a hydrological event, the sediment supply usually dwindles, so that at equal discharge the sediment concentrations are lower during the falling limb of the hydrograph than during the rising limb. Therefore, the Q–C relationships for sediment often exhibit a clockwise hysteresis loop. As well as being affected by sediment depletion, the timing of maximum sediment transport depends on the upstream spatial patterns of sediment supply and also on the subsequent mixing and routing of water and sediment from the different source areas. For the same reasons, the sediment concentration response to increased discharge may vary between different hydrological events. The first hydrological event after a long relatively dry period (which often happens during winter) produces a pronounced response of the sediment concentration. However, due to sediment depletion, the concentration response is often much less during the second or third hydrological event later during the year.

Asselman (1999) studied sediment concentration dynamics during 112 periods of increased discharge in the river Rhine near Rees, Germany. She distinguished the following four types of hysteresis loops (see Figure 18.10) and related them to time of the year and sediment source areas:

1. Counter-clockwise hysteresis (type 1). This type occurs primarily during minor hydrological events in summer. It was found 20 times (18 percent of all floods studied, but only during 8 percent of the high discharge periods exceeding 4000 m^3 s^{-1}). The upstream tributaries Neckar and Main deliver much more sediment than tributaries downstream.
2. Absence of hysteresis (type 2) occurs mainly when the rivers Neckar, Main and Mosel supply similar amounts of sediment. This type of hysteresis occurred 26 times (17 percent of the high discharge periods exceeding 4000 m^3 s^{-1}). This type of hysteresis occurs more often during summer.
3. Moderate clockwise hysteresis (type 3) occurs when sediment concentrations gradually increase with increasing discharge. Most sediment is delivered by the river Mosel. This type of hysteresis was found 62 times (55 percent of all peaks studied, but in 69 percent of the peaks exceeding 4000 m^3 s^{-1}) and occurs more often during winter.
4. Pronounced clockwise hysteresis (type 4) is found when sediment concentrations increase very rapidly, even before the sharpest increase in discharge occurs. This type of hysteresis occurred only four times during high discharge periods with peak discharge exceeding 4000 m^3 s^{-1} and seems to be related to early sediment supply from the tributaries just upstream from the measurement location.

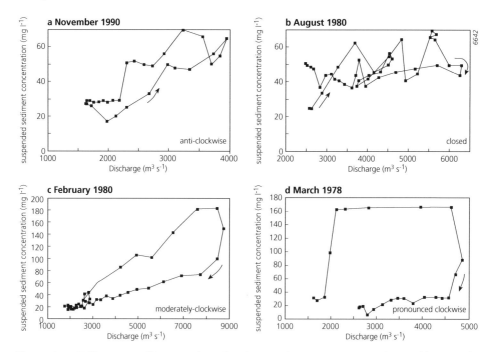

Figure 18.10 Different types of hysteresis observed in the river Rhine near Rees. For each type of hysteresis the percentage of the floods with maximum discharge exceeding 4000 m^3 s^{-1} and that were characterised by this type of hysteresis are given. Adapted from Asselman (1999).

In small urbanised catchments, the response of suspended sediment to runoff events may be even more complex, as demonstrated by Lawler *et al.* (2006). In their study in the urban headwaters of the river Tame, UK, counter-clockwise hysteresis responses of stream water turbidity were observed during the majority of the runoff events, with the result that for a given flow level, turbidity increased during the falling limb of the hydrograph (see Figure 18.7c-d). This is contrary to the response found in many other streams and rivers, including urban ones, where suspended sediment concentrations typically peak during the rising limb. In the Tame, increased turbidities during the falling limb co-occurred with elevated ammonium concentrations, which suggests that overflow spills from combined sewer overflows or wastewater treatment plants release sediments into the river. Such overflows typically occur late in the hydrograph, during intense storm events when rainwater delivery exceeds the storage capacity of the sewer system or wastewater treatment plant. Biofilms may also play a role in the delayed sediment transport. The presence of biofilms may considerably increase the critical shear stress for erosion of streambed sediments. Therefore, biofilms could suppress bed and sewer sediment entrainment early in the hydrograph, but when the biofilm breaks up due to the larger shear stresses around peak flow, more bed material is released during the later part of the event (Lawler *et al.*, 2006).

18.3.5 Concentration rating curves

Because the hysteresis behaviour of the Q–C relationship is often not ambiguous, a long-term Q–C relationship is frequently used for the prediction of unmeasured substance concentrations from water discharge (Walling, 1977; Ferguson, 1986; Horowitz, 2002). Such a long-term relationship is called a rating curve. Rating curves have especially been adopted to estimate sediment concentrations or loads (e.g. Asselman, 1998; Horowitz, 2002). Usually a concentration rating curve takes the form of a power function:

$$C = a Q^b \tag{18.1}$$

where C = the concentration of the substance under consideration (mg l^{-1}), Q = water discharge (m^3 s^{-1}), and a and b are coefficients. Log transformation of Equation (18.1) yields a linear equation:

$$\log C = \log a + b \log Q \tag{18.2}$$

where the logarithms are to base 10. Fitting a line to the log–log plot of C against Q using least squares linear regression provides the values for a and b. However, predictions of C using these coefficients are statistically biased because linear regression ensures that the mean sample residual of log C equals zero. Using Equation (18.1) to predict C, however, yields the geometric – not arithmetic – mean of the statistical distribution of C given a value for Q. The geometric mean is always less than or equal to the arithmetic mean; to estimate the arithmetic mean of C, Ferguson (1986) derived the following bias correction for Equation (18.1):

$$C = a Q^b e^{2.65 s^2} \tag{18.3}$$

where s^2 = mean square error of the regression Equation (18.2), which is a measure of the degree of scatter around the rating curve.

Figure 18.11 shows the uncorrected (Equation 18.1) and corrected (Equation 18.3) sediment rating curve for sediment concentration measured in the river Rhine near Lobith, the Netherlands. This figure shows that the scatter around the rating curve is considerable,

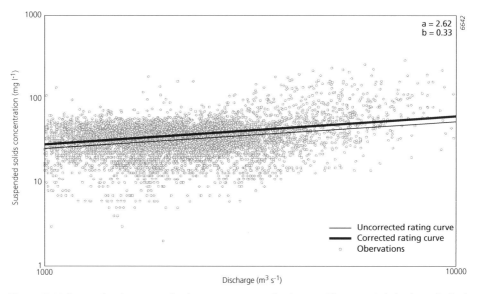

Figure 18.11 Corrected and uncorrected sediment rating curves for the river Rhine near Lobith, the Netherlands (source: Waterbase, 2006).

which can largely be explained by the hysteresis behaviour of the Q–C relationship and the different concentration responses between hydrological events as discussed above.

According to Morgan (1995), the *a* coefficient of the rating curve (Equations 18.1–18.3) represents the erodibility of the soils. Large values of the *a* coefficient occur in areas characterised by intensively weathered materials that can easily be eroded and transported. The *b* coefficient represents the erosive power of the river, with large values being indicative for rivers where an increase in discharge brings about a strong increase in erosive power and sediment transport capacity. Asselman (2000) argued that it is more appropriate to use the steepness of the rating curve, which is a combination of the *a* and *b* values, as a measure of soil erodibility and erosivity of the river. Steep rating curves with a small value for *a* and a large value for *b* are characteristic for river sections with little sediment transport during low discharge. An increase in discharge results in a large increment of suspended sediment concentrations. This suggests that during high discharge periods the river's power to erode material is large or that important sediment sources become available. Flat rating curves usually occur in river sections with intensively weathered materials or loose sedimentary deposits that are transportable during low discharges.

Figure 18.12 shows some rating curves for different locations in the catchment of the river Rhine (Asselman, 2000). The steepest rating curves are found in the river's tributaries, suggesting that there is a limited amount of fine sediment that can be eroded from the bed at low discharge. Once a certain discharge threshold has been exceeded, sediment supply to the river increases, and sediment can be picked up from the river bed, resulting in a rapid increase in suspended sediment concentrations. The presence of weirs in most tributaries also results in steep rating curves. During low discharge, much suspended sediment settles behind the weirs, but it is flushed out during periods of high discharge, causing suspended sediment concentrations to rise sharply. Along the river Rhine, the steepness of the rating curves decreases in downstream direction. The flattest rating curve is found near Rees, indicating that near the German–Dutch border large quantities of fine material are available for transport at low discharge.

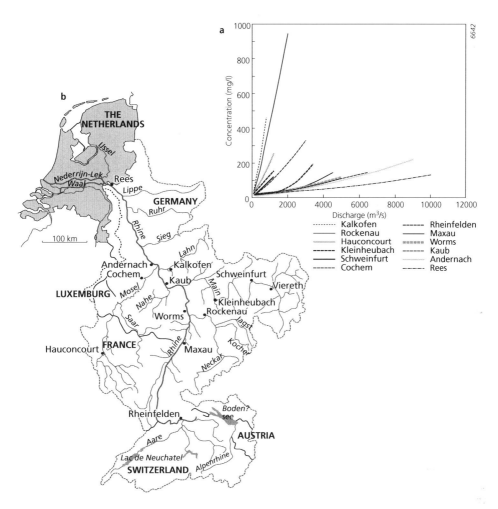

Figure 18.12 a) Sediment rating curves for locations along the Rhine and its main tributaries; b) Measurement locations in the Rhine catchment. Adapted from Asselman (2000).

Similar to the contribution of the different source areas of water, the source areas of sediment also vary in time during hydrological events. Consequently, the quality of suspended matter may also vary with discharge. For example, Figures 18.13a and 18.13b show the N and Zn concentrations in suspended sediment as a function of the discharge of the river Rhine near Lobith, the Netherlands. The N and Zn concentrations in sediment are highly variable when discharge is below about 4000 m^3 s^{-1}. Although few measurements are available for high discharges, the results indicate that when discharge exceeds 4000 m^3 s^{-1}, the concentrations decrease to about one-half to two-thirds of the concentrations during low flow, and show less variation (Middelkoop *et al.*, 2002). The N and Zn concentrations appear to be positively correlated with the organic carbon content of the sediment (Figures 18.13c and 18.13d), but do not correlate with clay content (not shown). Figure 18.13e shows that the relationship of the organic carbon content of the suspended sediment to the discharge is similar to the relationships between N and Zn and discharge; this would account for the response of N and Zn in sediment to increased discharge. It seems that during low flow the suspended sediment contains substantial amounts of organic matter that is highly

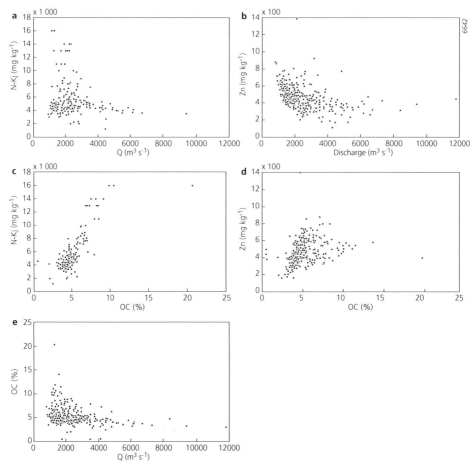

Figure 18.13 Relationships amongst total Kjehdahl N (N-Kj), Zn, organic carbon (OC) concentrations in sediment, and discharge (Q) in the river Rhine near Lobith, the Netherlands: a) N-Kj versus Q (1995–2002); b) Zn versus Q (1988–2002); c) N-Kj versus OC (1995-2002); d) Zn versus OC (1988–2002); e) OC versus Q (1988–2002) (source: Waterbase, 2013).

enriched in contaminants such as heavy metals, nutrients, and pesticides from industrial and domestic point discharges and diffuse urban runoff. During hydrological events, soil erosion on the terrestrial part of the catchment and river bank erosion generate a supply of relatively uncontaminated, predominantly clastic sediment to the river. This input of fresh sediment dilutes the contaminated sediment already present in the river system, which is why the contaminant concentrations in sediment decrease during periods of high discharge.

Despite the decrease in contaminant concentrations in sediment during high discharge, the sediment concentrations in river water increase so much that total transport or load (i.e. the product of concentration and discharge; see Section 11.2.2) of sediment-associated contaminants is almost always greater during high discharge than during low discharge.

18.3.6 Significance of hydrological events for substance transport

The increase in sediment transport during periods of high discharge has significant implications for the importance of hydrological events for the export of sediment-associated contaminants from catchments. However, during periods of high discharge there is a dramatic increase not only in the load of sediment-associated substances, but also in the load of sediment itself and the load of dissolved substances that exhibit a positive Q–C relationship. Hence hydrological events often account for the bulk of the export of these substances from catchments, even though they only occur during a short time of the year. For example, Baker and Richards (2003) quantified the relative contributions to total sediment and nutrient load from the Maumee River at Waterville, Ohio, USA, to Lake Erie due to runoff events. In a 17-year period, there were 226 discrete storm events, which lasted for 37.2 percent of the time but accounted for 76.4 percent of the water discharge, 91.6 percent of the sediment export, 86.4 percent of the total phosphorus export, 78.3 percent of the soluble reactive phosphorus export, and 80.5 percent of the nitrate-nitrogen export. This contribution of episodic hydrological events to the total export of substances from a catchment can be especially large in small catchments, due to their flashy nature.

18.3.7 Seasonal dynamics of nutrient concentrations

As mentioned above, the supply of substances to watercourses varies with the seasons, due to plant uptake and litter fall. Because dissolved inorganic N and P species in soil are taken up during the growing season, plant uptake particularly affects the transfer of nutrients to surface water. In autumn, some of the nutrients are released again through litterfall and subsequent mineralisation. Though most of the P is retained in soil, dissolved N species (i.e. NO_3^- and to a lesser extent NH_4^+) can readily be flushed to the stream by surface and shallow subsurface pathways during winter storms or spring snowmelt. Besides the seasonal variation in supply of nutrients from the terrestrial parts of the catchment, there is seasonal variation resulting from algae and aquatic macrophytes: during the growing season, these take up nutrients dissolved in surface water. Furthermore, the seasonal variation of the water temperature has a pronounced effect on the rates of biochemical decay. The rates of nitrification and denitrification decrease with decreasing temperature. When the water temperature drops below about 10 °C, nitrification and denitrification virtually cease. This means that during winter, NO_3^- and NH_4^+ are less efficiently removed from the water, so that their concentrations remain relatively high. When the water temperature rises, the nitrification and denitrification processes resume, at rates increasing with temperature, thereby lowering the in-stream NO_3^- and NH_4^+ concentrations.

Table 18.2 shows an example of the seasonality of nitrification and denitrification in the South Platte river, Colorado, USA. Sjodin *et al.* (1997) analysed the seasonal variation of nitrification and denitrification rates by means of a mass balance approach. The nitrification rate constant varied from 0.66 d^{-1} in February to 5.20 d^{-1} in early June and the denitrification rate constant from 0.30 in January to 5.15 in late June.

Table 18.2 Seasonal variation of nitrification and denitrification rate constants in the upper reach of the South Platte river, Colorado, USA in 1994–1995 (source: Sjodin *et al.*, 1997).

	Jan	Feb	Jun (early)	Jun (late)	Jul	Aug	Nov
Nitrification							
NH_4^+ concentration; top of the reach (mg l^{-1})	0.79	1.26	0.12	0.43	0.03	0.16	0.55
Actual nitrification rate constant (d^{-1})	1.00	0.66	5.20	3.04	3.22	4.18	1.56
Denitrification							
NO_3^- concentration; top of the reach (mg l^{-1})	6.57	5.94	4.00	2.63	6.45	5.29	9.45
Actual denitrification rate constant (d^{-1})	0.30	0.40	1.06	5.15	1.36	1.12	0.82
Distance to 90% reduction (km)	292	221	84	10.5	52	71	102

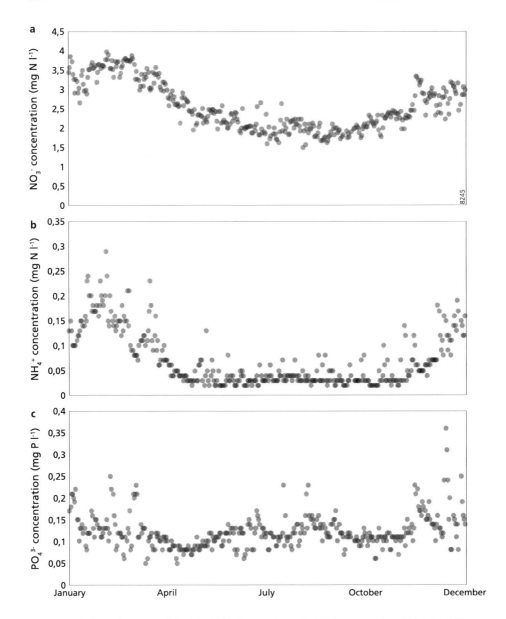

Figure 18.14 Seasonal variation of the a) NO_3^-, b) NH_4^+, and c) ortho-PO_4^{3-} concentrations in the river Rhine near Lobith in 2010 (source: Waterbase, 2013).

All the above processes result in a considerable seasonal variation in stream nutrient levels, as can also be observed from Table 18.2, which shows that NH_4^+ and NO_3^- concentrations increased from low levels during the growing season (late spring and summer) and peaked during winter. Figure 18.14 shows the seasonal trend of NO_3^-, NH_4^+, and ortho-PO_4^{3-} concentrations in the river Rhine near Lobith in 2010. In all cases, the minimum concentrations were observed during late spring or summer and maximum concentrations during late autumn or winter. The ortho-PO_4^{3-} concentration exhibits the largest short-term variation, probably because of its great susceptibility to variations in flow. In fact, the seasonality of in-stream nutrient concentrations is generally most apparent during baseflow conditions, when the short-term effects of increased runoff are absent.

18.3.8 Long-term dynamics

In the long term, stream water quality reflects the trends in point and diffuse pollution sources in the upstream catchment. Among the first and best known long-term monitoring programmes to study linkages between water and nutrient fluxes and cycling in forests and associated aquatic ecosystems, was the Hubbard Brook Ecosystem Study (HBES) (Likens and Bormann, 1995). The Hubbard Brook Experimental Forest (HBEF) is a 3160 ha reserve located in the White Mountain National Forest, near Woodstock, New Hampshire, USA, established in 1955 for hydrological research. The HBEF is operated and maintained by the Northeastern Research Station, US Department of Agriculture, Newtown Square, Pennsylania. In 1963, the actual HBES started to investigate trends in water and element cycling in response to natural and human disturbance, such as climatic factors, landuse changes, forest cutting, and air pollution. The ongoing monitoring programme has resulted in one of the most extensive and longest continuous databases on the hydrology, biology, geology, and chemistry of natural ecosystems. The HBEF contains nine small, gauged experimental subcatchments in which a series of controlled field experiments have been performed, ranging from calcium or herbicide treatment to various extents of forest clear-cutting. One subcatchment ('Watershed 6') was set aside as a biogeochemical reference catchment, in which no treatment was applied. Since 1917, when the last logging operation took place, the forest composed of about 80 to 90 percent hardwoods and 10 to 20 percent conifers has developed without direct human interference. Figure 18.15 shows the observed trends in discharge-weighted monthly pH, and SO_4^{2-} and NO_3^- concentrations in stream water draining this reference catchment in the period 1963–2010. There has been a noticeable decline in concentrations of SO_4^{2-} and NO_3^- and a slight increase in pH in stream water since the mid- to late 1960s. The pH and SO_4^{2-} trends in stream water have been consistent with the decreased emissions and subsequent acid atmospheric deposition of SO_2 (Driscoll *et al.*, 1989). However, the concentration or flux of inorganic N in bulk deposition in the HBEF have not changed significantly over the past several decades, so the explanation of the decline in NO_3^- concentrations in stream water is less clear-cut. In fact, an increase of stream NO_3^- could have been expected in response to prolonged N deposition and forest maturation. When the forest canopy has reached its maximal size, tree biomass accumulation and the N utilisation efficiency decrease and therefore the excess N may be leached from the root zone (Gundersen and Bashkin, 1994). Several hypotheses have been put forward to explain the increase of stream NO_3^- in the early 1970s. Eshleman *et al.* (1998) suggested that the high NO_3^- concentrations were due to heavy insect infestation and defoliation in 1969–1971. Soil frost can also trigger losses of NO_3^- to streams by disturbing soil structure and causing microorganisms and fine roots to die (Groffman *et al.*, 2001). However, enhanced stream NO_3^- concentrations have also been observed prior to freezing events, and the decline in stream NO_3^- concentrations since the early 1970s occurred simultaneously in catchments in the regions that were not affected by insect defoliation (Goodale *et al.*, 2003). This implies that insect defoliation and soil frost are not the only factors contributing to interannual NO_3^- fluctuations. Goodale *et al.* (2003) argued that although NO_3^- losses did not correlate with any particular climate variable, the decrease in NO_3^- concentrations may also be attributed to subtle differences in the rate or timing of plant uptake and N mineralisation in response to temperature and moisture conditions (see also Bernhardt *et al.*, 2005). This example illustrates the complexity of element cycling at the catchment scale and emphasises the importance of long-term monitoring to detect environmental changes and to help in formulating and testing hypotheses that explain these changes.

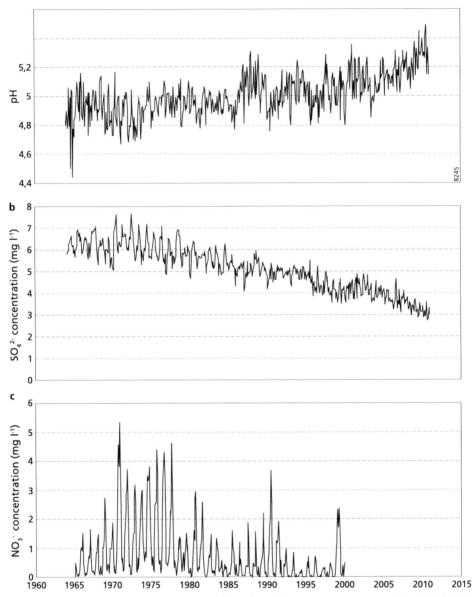

Figure 18.15 Trends in stream water a) pH, and b) SO$_4^{2-}$, and c) NO$_3^-$ concentrations in Watershed 6 of the Hubbard Brook Experimental Forest, New Hampshire, USA, in the period 1963–2010 (source: Gene E. Likens, personal communication)).

In contrast to the Hubbard Brook Ecosystem Study and other research initiatives, most national and regional water quality monitoring programmes have been established to evaluate actual water quality against statutory standards and targets. Data from these monitoring programmes also provide valuable insight into the long-term evolution of water quality in response to changing point source and diffuse emissions. For example, the regular, biweekly monitoring of dissolved nutrients, dissolved oxygen, and chloride at the main monitoring station in the Dutch reach of the river Rhine near Lobith dates back to the 1950s. The monitoring of heavy metals started in the late 1960s and was expanded to include many organic pollutants in the 1970s. The pollution history, however, often goes much further back in time than the available water quality measurements. To reconstruct the pollution history before regular water quality monitoring commenced, several proxies can be used, such as concentration depth profiles in lake or floodplain sediments (e.g. Ayres and Rod, 1986; Winkels, 1997; Middelkoop, 2000). In Section 16.4.3 an example of pollution history of the river Rhine as documented by overbank deposits on a section of washland was discussed.

Figure 18.16 shows the trends in measured dissolved ortho-PO_4^{3-} and Zn concentrations in the river Rhine since regular monitoring began. From the start of the monitoring until the late 1970s the ortho-PO_4^{3-} concentrations rose because intensified agricultural practices

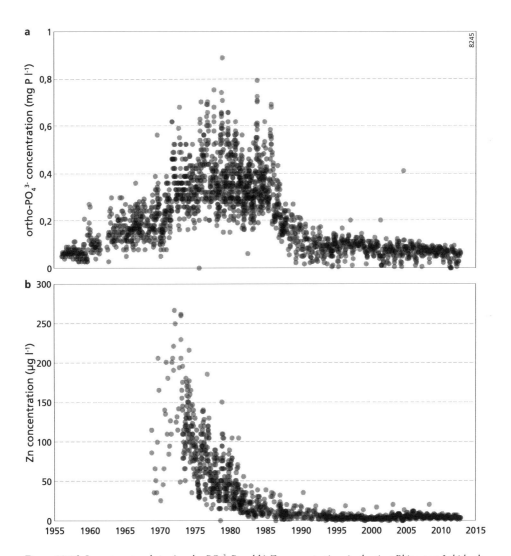

Figure 18.16 Long-term trends in a) ortho-PO_4^{3-}-P and b) Zn concentrations in the river Rhine near Lobith, the Netherlands (source: Waterbase, 2013).

resulted in increased diffuse P sources, and because enhanced connection to wastewater treatment systems caused an increase in direct P emissions to the river. Since the 1980s the improvement of waste water treatment by implementing tertiary treatment (additional P removal) and the phasing out of phosphorus-containing detergents have led to a noticeable decline of ortho-PO_4^{3-} concentrations in the river water (Van Dijk *et al.*, 1996). The maximum dissolved Zn concentrations in river water were measured in the early 1970s. Since then, various measures to reduce heavy metal loads from industrial and municipal waste water discharges have resulted in lower – though still elevated – Zn levels in the river Rhine. Currently, the bulk of Zn originates from diffuse sources (see Table 18.1).

The above example shows a relatively quick response of water quality to improved sanitation of wastewater discharges. However, the response of water quality to changes in diffuse emissions is often much slower, as illustrated by the trends in river water quality in central and eastern Europe following the collapse of communism in the late 1980s and the consequent agricultural crisis and accompanying drastic decrease in the use of fertilisers. For example, in Estonia, there was heavy use of mineral fertilisers in agriculture during the Soviet period, i.e. before 1991. In 1987 and 1988, when this use was highest, 270 000 t y^{-1} was being applied. By 2001, the fertiliser consumption rate in Estonia had fallen to 29 700 t y^{-1}. Livestock numbers also fell, from 800 000 (0.82 livestock units ha^{-1} of agricultural land) in 1988 to less than 390 000 (0.34 livestock units ha^{-1}) in 1994. In addition, the area of abandoned land increased substantially. Mourad and Van der Perk (2004) estimated the decline in mean annual diffuse source emissions in Estonia to be from 5953 kg km^{-2} y^{-1} for 1985–1989 to 2009 kg km^{-2} y^{-1} for 1995–1999. Because the N in most Estonian rivers originates largely from diffuse source emissions (see Mourad and Van der Perk, 2004), the possible effects of the decreased diffuse inputs should ultimately become noticeable in the stream N concentrations. Figure 18.17 shows the decline in total N concentrations in the Emajõgi river near Kavastu located about 30 km downstream from Tartu. At this site, diffuse source emissions contributed 85–90 percent to the total N load in 1995–1999 (Mourad *et al.*, 2006).

Figure 18.17 shows a slow and limited but significant decrease of total N concentrations in response to the decreased diffuse N emissions. Iital *et al.* (2005) examined 22 monitoring stations in 16 Estonian rivers (including the site shown in Figure 18.17). They found that 20 of the 22 sites exhibited statistically significant downward trends in total N between 1986 and 2001. However, in a similar study on Latvian rivers, Stålnacke *et al.* (2003) found that only four of the 12 monitoring stations showed statistically significant downward trends

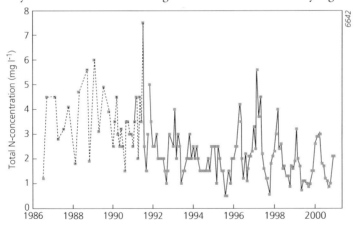

Figure 18.17 Long-term trend in total N concentrations in the Emajõgi river Kavastu, Estonia, between 1986 and 2001 (source: Estonian Hydrometerological Institute).

in total inorganic N. They hypothesised that the weak, delayed, or even absent response might be attributed to long water transit times in the soil water and groundwater, or to mineralisation of organic nitrogen pools that had accumulated during several years before the dramatic decline in agricultural inputs.

The total P concentrations in the Emajõgi river did not show a response to the decreased fertiliser inputs, but this is probably due to the relatively high contribution of 40–60 percent from point source emissions in this reach of the river, mainly from Tartu (Mourad *et al.*, 2006). The trends in other Estonian rivers could also be explained by changes in point source emissions from wastewater treatment plants (Iital *et al.*, 2005).

18.4 VARIATION IN LAKE WATER COMPOSITION

18.4.1 Role of lakes in catchment sediment and nutrient budgets

When a river discharges into a lake, its flow velocity diminishes rapidly and the suspended matter starts to settle out from the water column to the bed sediments. Only suspended particles that are too small to have settled out from the inflows or those that have been introduced near the lake outflow are transferred to the lake outflow. Lakes are therefore very efficient sediment traps and represent an important sink in catchment sediment budgets.

As well as being supplied by inflowing rivers, suspended particles may originate from within the lake itself as a result of resuspension of bed sediments, erosion of shores, or the production of autochthonous organic material. Bed sediments are eroded and resuspended by current and wave action. In deep lakes this only occurs in the shallow parts along the shoreline. When allochthonous sediment sources to lakes (river inputs, eroded or resuspended material, and wind-borne dust) are limited, most of the suspended particulate matter is autochthonous material, i.e. living organisms, organic detritus, calcite, and diatomite. Organic detritus consists of the remains of dead organisms (e.g. algae, aquatic macrophytes, zooplankton, fish) and faecal material. Photosynthesis causes an increase in pH, which may trigger the precipitation of calcite in hard-water lakes rich in Ca^{2+} and HCO_3^-. Diatomite consists of the debris of algal diatoms, which is rich in silica. The processes of settling, resuspension, and river inflow usually bring about an inshore to offshore decline in the concentrations and mean particle size of the suspended matter, whereas the organic matter content and carbonate content of the suspended matter increase. These trends of particle size and organic matter and carbonate content are also often found in the lake bed sediments.

The settling of particulate matter and the sediment characteristics with respect to particle size and organic matter content have important implications for the functioning of lakes in nutrient retention in catchments (Hillbricht-Ilkowska, 1999). The sedimentation of particle-bound P (ortho-PO_4^{3-} and organic forms of phosphorus) causes lakes often to be a sink for total P load, except if the lake is shallow and eutrophic: in the latter case, there is usually net P export during summer, when ortho-PO_4^{3-} is released from sediments (see Section 18.4.4). The anaerobic, organic bed sediments abundant in most lakes provide important sites for effective denitrification, which reduces the total N load.

Marion and Brient (1998) quantified the retention of suspended particulate matter and N and P load in Grand-Lieu, a shallow wetland lake discharging into the Loire estuary, western France, for two hydrologically contrasting seasons. Grand-Lieu is 4000 ha in extent during summer and 6300 ha during winter and is 1.2 to 1.7 m deep in its centre. The retention of suspended particulate matter and nutrients were measured from October 1993 to May 1994, a wet season with a total lake inflow of 292 10^6 m³, and from October 1995 to May 1996, a dry season with a lake inflow of 76 10^6 m³ (Table 18.3). The retention rate (i.e.

Table 18.3 Lake inflow and retention rates in Grand-Lieu, France, in two hydrologically contrasting seasons (source: Marion and Brient, 1998).

	1993–1994	1995–1996
Lake inflow (m^3)	$292 \cdot 10^6$	$76 \cdot 10^6$
Particulate matter (%)	14	20
Total P (%)	40	18
Ortho-PO_4^{3-} (%)	79	72
Total N (%)	32	60
NO_3^- (%)	61	86
NH_4^+ (%)	72	66

lake outflow relative to inflow) of particulate matter was lower with the large flow of 1993–94 than with the small flow in 1995–1996. The lower retention rate might be attributed to residence times being shorter during periods of high inflow, so less time is available for the settling of sediment. However, the sediment exported from the lake consisted primarily of organic, autochthonous particulate matter and resuspended particles from the bed sediment (Marion and Brient, 1998). The resuspension and autochthonous particulate matter production are higher during wet periods, because there is more wind and because the volume and surface area of the lake are larger thanks to the high inflow. Thus the sediment exported from the lake is of different origin and composition than the sediment transported to the lake. During the wet season the retention rates for NO_3^- and total N were also lower. This can largely be attributed to the shorter residence time during periods of high inflow. The retention rates for ortho-PO_4^{3-} and NH_4^+ were similar for both seasons. In contrast, with the high outflow of 1993–94, the retention rate of total P was double. This may be attributed to the increased inflow of particulate P during high flow, which is effectively retained in the lake.

18.4.2 Lateral variation

The water composition of a lake, like that of rivers, is primarily controlled by the geology, climate, and land use in the catchment. Müller *et al.* (1998) studied the influence of climate and land use on the variation in water composition between 68 lakes in carbonaceous catchments in Switzerland, Italy, and France. They found strong relationships between water composition and altitude of the lakes: the lakes below 700 m above sea level contain substantially higher concentrations of major dissolved constituents and nutrients. Obviously, there is a strong relation between altitude and climate on the one hand, and agricultural land use and the degree of urbanisation on the other hand. Although the proportion of agricultural land does not differ much between the lake catchments studied, intensive agriculture is more common at lower altitudes. Urban areas are also mainly located in the lower altitude valleys. The concentrations of Na and K in the lakes investigated correlated positively with the presence of urban areas in the catchment. The effect of intensive agriculture was particularly evident in the increased concentrations of nutrients (N and P) in lake waters at lower altitude. The higher Ca and Mg concentrations and alkalinity in the lower altitude lakes can be attributed to more rapid weathering. At higher altitudes, the low temperatures, thin or absent soils, and short contact times between rock and water result in less chemical weathering. Atmospheric deposition of acidifying compounds contributes to the increased weathering rates at lower altitudes. Air pollution from traffic, industry, and agriculture can become trapped underneath an atmospheric temperature inversion, which frequently occurs at 700–800 m above sea level in the alpine and peri-alpine regions in

Switzerland during winter months. This atmospheric inversion enhances the atmospheric deposition at the lower altitudes. Furthermore, nitrification of manure-derived NH_4^+ also produces acidity, which enhances the weathering of the carbonaceous bedrock, and so increases the runoff of Ca, Mg, and alkalinity to the lakes in areas of intensive agriculture.

Differences in geology and land use in different subcatchments of a lake can also give rise to within-lake variation of water composition. A good example is the distribution of chlorophyll concentrations in Loch Lomond, Scotland, reported by George and Jones (1987) and George (1993). Loch Lomond is the largest freshwater lake in the UK: it is 36.4 km long and at its deepest point is 190 m deep, albeit the average depth is only 37 m. The northern part of the lake is long (22.3 km), narrow, and deep (average just over one km), but the southern basin is much wider (maximum width of 8.8 km) and shallower (Figure 18.18a). There are pronounced differences in geology between the subcatchments of the northern and southern basins of Loch Lomond. Base-rich rock comprises 98 percent of the sub-catchment area of the southern basin, whereas it only occupies approximately 3 percent of the sub-catchment area of the northern basin. Related to the catchment geology there is also a distinct difference in land use in both subcatchments (see Figure 18.18a), with arable fields concentrated on the base-rich rock in the southern part of the catchment. Consequently, the southern part of the lake receives bigger loads of solutes – including nutrients from the inflowing streams – than the northern part. This explains the higher electrical conductivity (EC) of the water in the southern part of the lake (Figure 18.18c and e). The increased supply of nutrients promotes the growth of algae, which becomes expressed in increased chlorophyll concentrations in the southern part of Loch Lomond (Figure 18.18d and f). The difference in depth between the northern and southern parts of the lake emphasises the differences in EC and chlorophyll concentrations.

Figure 18.18 also demonstrates that wind exerts an important influence on the spatial distribution of solutes and particulates in lakes. On 14 July 1978, the wind was blowing strongly from the north-west along the main axis of Loch Lomond. As a consequence, the EC and chlorophyll concentrations increased gradually from north to south (Figure 18.18c and d). In contrast, on 12 October the wind blew lightly from the east (i.e. perpendicular to the main axis of the lake) and there was a sharp increase in EC and chlorophyll concentrations between the two basins (Figure 18.18e and f).

18.4.3 Vertical variation during summer stratification

Deep lakes in the temperate climate regions with distinct cold and warm seasons become thermally stratified during summer, which is manifested in the formation of two water layers: a warm upper epilimnion, and a cool, lower hypolimnion (see Section 3.4.4). Differences in temperature and light conditions, and the resulting differences in biological activity, cause distinct vertical patterns of concentrations of substances, notably of oxygen and nutrients, to emerge in the stratified lake water. Figure 18.19 presents an example of the vertical variation in oxygen concentration, pH, redox potential and N and P compounds during summer stratification in the mesotrophic Lake Dudinghausen, located in Mecklenburg-Vorpommern (northern Germany) (Selig *et al.*, 2004). This lake has a surface area of 18.8 ha and a maximum depth of 15.2 m. The vertical water sampling was performed at the deepest point of the lake. From the temperature profile in Figure 18.19 it can be seen that the epilimnion is approximately 6 m deep. The thermocline, the transition zone between the epilimnion and the hypolimnion, extends from 6 to 9 m depth. The hypolimnion encompasses the zone below 9 m depth.

The well-mixed epilimnion is rich in oxygen due to reaeration from the atmosphere and the production of oxygen through photosynthesis by algae. The pH in the epilimnion is high, due to photosynthetic activity. Dead algae or other organic matter produced in the

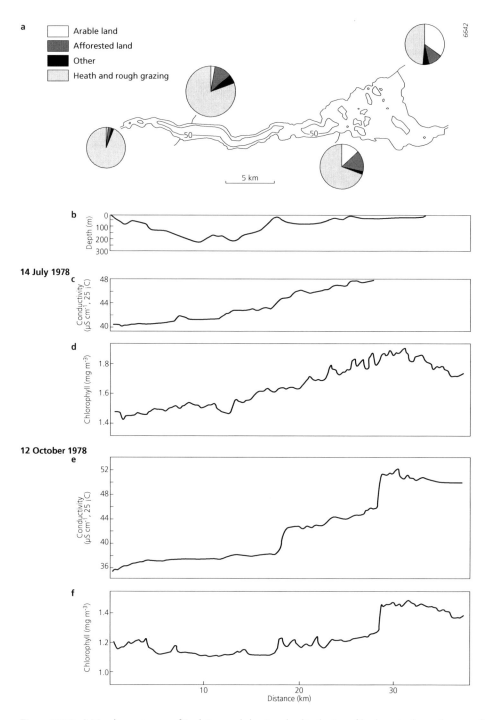

Figure 18.18 a) Morphometric map of Loch Lomond showing the distribution of land use in the catchments of the inflowing rivers; b) Longitudinal profile of the deepest section; c, d) longitudinal distribution of the electrical conductivity and chlorophyll concentration recorded on 14 July 1978; e, f) longitudinal distribution of the electrical conductivity and chlorophyll concentration recorded on 12 October 1978. Adapted from George and Jones (1987).

epilimnion are readily decomposed in the epilimnion, but some sinks to the hypolimnion. Here, the continued decomposition of organic matter depletes the oxygen and lowers the redox potential (Eh) while regenerating dissolved inorganic nutrients and carbon dioxide. Because the presence of the thermocline prevents oxygen from diffusing into the hypolimnion, the oxygen saturation drops gradually from 100 percent to 0 percent. In Lake Dudinghausen this occurs between 5 and 7 m depth (Figure 18.19). In the hypolimnion, oxygen production occurs only if light penetrates below the thermocline. In lakes where light does not penetrate below the thermocline, as apparently is the case in Lake Dudinghausen, there is no internal source of oxygen.

Summer stratification also causes typical vertical profiles of N and P compounds. Nitrate occurs in relatively high concentrations in the epilimnion, but in the thermocline and the deeper hypolimnion it drops to zero due to denitrification by denitrifying bacteria. In contrast, NH_4^+ concentrations are very low in the epilimnion, but high in the hypolimnion. In the epilimnion, NH_4^+ released by the decomposition of organic matter is oxidised to NO_3^-. During stratification, nitrification of NH_4^+ does not occur in the hypolimnion because of the lack of oxygen. The reactive phosphorus (RP (< 0.45); mainly consisting of ortho-PO_4^{3-}) is relatively low in the epilimnion due to uptake by algae. In contrast, the RP (< 0.45) concentrations are relatively high in the hypolimnion as a consequence of P release from the anaerobic sediments. The release of P from anaerobic sediments will be further elaborated upon in Section 18.4.4. Oxygen depletion in the hypolimnion and reducing conditions at the sediment/water interface may, in addition, result in iron, manganese and related trace elements being released from the bed sediments (Balistrieri *et al.*, 1991; Thomas *et al.*, 1996).

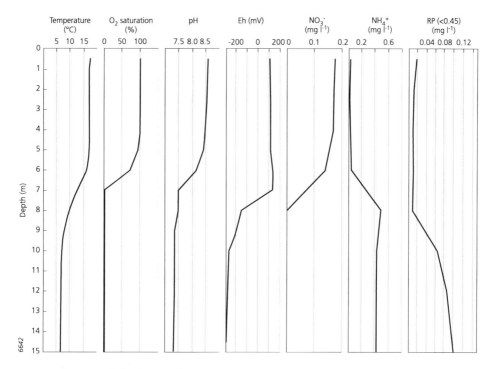

Figure 18.19 Vertical distribution of temperature, oxygen saturation, pH, Eh, dissolved inorganic N and P compounds in Lake Dudinghausen, northern Germany, in September 2000. Adapted from Selig *et al.* (2004).

18.4.4 Temporal variation

As already touched upon in the previous sections, lake water composition varies considerably during the seasons. Figure 18.20 shows the seasonal variations in the concentrations of chlorophyll, soluble reactive P (RP (< 0.45)), total P, and NO_3^- in the near-surface water (epilimnion) of Lake Windermere, Cumbria, UK, in 1973 and 1974. Lake Windermere is 17 km long and 0.4 m to 1.5 km wide and reaches a maximum depth of about 65 m. During spring and summer, the temperature rises and algae start to grow, as can be seen from the chlorophyll-α concentrations. The growth of algae increases the rate of uptake of N and P and reduces the concentration of RP (< 0.45) and NO_3^- in the water. Total P concentrations remain more stable than soluble reactive P fraction (RP (< 0.45)). Total P includes soluble P and the P in living organisms and detritus suspended in the lake water. Primary production by algae occurs from April to late September. The rapid decline of the RP (< 0.45) in April suggests that algal growth is P-limited. In autumn, primary production is followed by death, settling of the detritus to the bed sediment, and decomposition. The decomposition of detritus causes the N and P to be recycled back into the dissolved phase.

In shallow lakes, the dynamics of P are different from those in deep lakes such as Lake Windermere discussed above. Figure 18.21 shows the seasonal variation of total

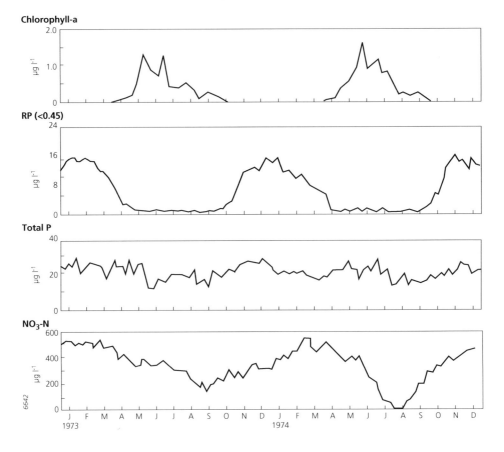

Figure 18.20 Seasonal variation of chlorophyll-α and nutrients in the surface waters of Lake Windermere, UK. Adapted from Thomas *et al.* (996.).

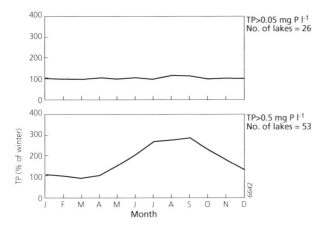

Figure 18.21 Seasonal variation in total P (monthly mean ± standard deviation) as percentage of mean winter concentrations (1 Jan–31 March) in shallow Danish lakes with mean summer total P concentrations below 0.05 mg l⁻¹ and above 0.5 mg l⁻¹. Adapted from Søndergaard *et al.* (1999).

P concentrations in shallow, relatively small Danish lakes for lakes low in P load (mean summer total P concentration < 0.05 mg l⁻¹) and lakes high in P load (mean summer total P concentration > 0.5 mg l⁻¹) (Søndergaard *et al.*, 1999). Figure 18.21 demonstrates that in shallow lakes the seasonal variations in total P depend on nutrient levels. In lakes with mean summer total P concentration below 0.05 mg l⁻¹, seasonal variation is small and summer concentrations do not differ much from winter values. In hypertrophic lakes with total P concentrations of more than 0.5 mg l⁻¹, summer concentrations are typically triple the winter values. This increase in summer P concentrations may be due to increased P concentrations in the lake inflows, because wastewater effluents may constitute a larger proportion of the inflow water during low discharges in summer. However, in most cases the seasonal variations of total P can only be attributed to increased P release from the bed sediments during summer: a process generally referred to as internal P loading.

There are several mechanisms responsible for the seasonal variations in P release from the sediment. The fundamental mechanism of P release from bed sediments is diffusion-controlled by the concentration gradient across the sediment/water interface. In the interstitial water of reduced bed sediments, the soluble P concentrations are generally higher than in the overlying lake water, due to the decomposition and mineralisation of organic matter – even though these processes proceed slowly under the reduced conditions. The ortho-PO_4^{3-} concentration in the interstitial water is usually controlled by the solubility product of vivianite ($Fe_3(PO_4)_2 \cdot (H_2O)_8$). Figure 18.22 shows the seasonal variations in the RP (< 0.45) concentration depth profiles in the interstitial water of bed sediment of Lake Finjasjön, Sweden (Eckerrot and Pettersson, 1993). The RP (< 0.45) concentrations in the interstitial water varied from less than 0.5 mg P l⁻¹ in March to about 5 mg P l⁻¹ in August. The RP (< 0.45) values in the overlying water were much lower: less than 0.01 mg P l⁻¹ in March and about 0.16 mg P l⁻¹ in August. During winter, the mineralisation processes in the sediment slow down because of the low temperatures. Moreover, there is little sedimentation of detritus because there are few algae in the lake water during winter. These conditions allow oxidisers like oxygen and NO_3^- to diffuse deeper into the bed sediments, creating a step-wise gradient in redox potential in the top layer of the sediment (see Section 4.3.4). In the thin oxidised top layer of the sediment, dissolved Fe(II) precipitates as Fe(III) oxyhydroxides, which can adsorb substantial quantities of ortho-PO_4^{3-} (Jensen and Andersen, 1992). The iron oxyhydroxides thus hamper the diffusive transport of ortho-PO_4^{3-} across

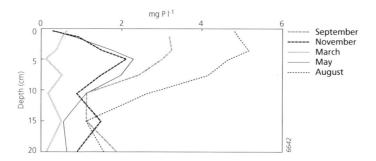

Figure 18.22 Vertical profiles of RP (< 0.45) concentrations (mg l^{-1}) in the interstitial water of bed sediment at 3.5 m depth in Lake Finjasjön, Sweden, from September 1989 to August 1990. Adapted from Eckerrot and Pettersson (1993).

the sediment/water interface. During spring and summer, higher temperatures, biological activity, and increased sedimentation of detritus accelerate the decomposition of sediment organic matter and so the oxidised surface layer dwindles (Søndergaard *et al.*, 2001). The resulting reduced conditions in the sediment top layer lead to the dissolution of Fe oxyhydroxides and, consequently, to the release of the ortho-PO_4^{3-} adsorbed to them and a decrease of the sediment adsorption capacity for P. As a consequence, inorganic ortho-PO_4^{3-} derived from the mineralisation of detritus can diffuse freely across the sediment/ water interface. Mineralisation of detritus is considered to be a quantitatively more important source of P released from lake bed sediments than the release of ortho-PO_4^{3-} adsorbed to Fe oxyhydroxides (Golterman, 2004). Bear in mind that higher temperatures also promote the mineralisation of organic matter in the sediment, thereby releasing ortho-PO_4^{3-} to the interstitial water, and – depending on the sorption capacity of the sediment – to the overlying water (Søndergaard *et al.*, 1999) (see Figure 18.22). Other mechanisms of P release from sediments include resuspension of bed sediments (and the accompanying release of interstitial water rich in soluble P) by wind action and the dissolution of hydroxyapatite ($Ca_5OH(PO_4)_3$) by a decrease in pH (Søndergaard *et al.*, 2001; Golterman, 2004). However, these mechanisms barely affect the seasonal patterns of total P as depicted in Figure 18.21.

As noted in the previous section, the primary production by algae increases the pH of the lake water during summer. In hard-water lakes, the pH increase may cause the solubility

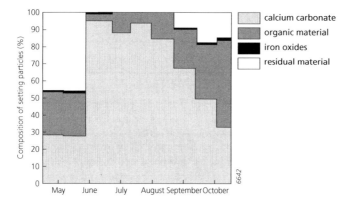

Figure 18.23 Seasonal variations in the composition of settling material collected at 40 m depth in Lake Zurich, Switzerland. Adapted from Wieland *et al.* (2001).

product of calcium carbonate (calcite) to be exceeded. As a result, the calcium carbonate precipitates and starts settling out of the water column to the lake bed sediments. This process is illustrated by Figure 18.23, which shows the seasonal variation in the composition of settling particles collected at 40 m depth in Lake Zurich, Switzerland, from April to October 1989 (Wieland *et al.*, 2001). Lake Zurich is a pre-alpine hard-water lake along the northern rim of the Alps. The lake is approximately 29 km long and 2.5 km wide and its maximum depth is 137 m. Most of the suspended particulate matter at the sampling location is autochthonously produced, since the riverine input of allochthonous sediment is effectively trapped in the upstream sections of the lake. The particles are mainly comprised of calcium carbonate, organic matter, iron oxyhydroxides, and residual material, which contains silica and aluminosilicates. In June, the calcium carbonate content of the suspended particles increases dramatically from less than 30 percent to approximately 95 percent (Figure 18.23). The precipitation of calcium carbonate also increases the total mass flux towards the lake bottom by a factor of two to four. In September and October, the calcium carbonate content gradually declines to the spring levels.

18.5 FURTHER READING ON CONTAMINANTS IN SURFACE WATER

- Miller, J.R. and Orbock Miller, S.M., 2007, *Contaminated rivers: A geomorphological-geochemical approach to site assessment and remediation*, (Dordrecht NL: Springer).
- Westrich, B. and Förstner, U. (Eds.), 2007, *Sediment dynamics and pollutant mobility in rivers: An interdisciplinary approach*, (Berlin: Springer).
- Plate, E.J. and Zehe, E. (Eds.), 2008, *Hydrologie und Stoffdynamik kleiner Einzugsgebiete: Prozesse und Modelle*, (in German), (Stuttgart: Schweizerbart).

EXERCISES

1. Explain briefly:
   a. why the nitrate concentration in river water increases downstream of a wastewater discharge;
   b. why the chloride concentration in streams decreases during a storm event;
   c. why relatively large amounts of groundwater are entrained during runoff events;
   d. why hysteresis occurs in the relationship between river discharge and suspended solids concentrations during single flood events;
   e. why the phosphate concentration in rivers decreases during summer;
   f. why calcium carbonate precipitation occurs in hard water lakes during summer.

2. Explain the role of hydrological events for the export of contaminants from catchments.

3. Describe the timing of the peak or trough concentration relative to peak discharge for:
   a. positive, clockwise Q–C response;
   b. negative, clockwise Q–C response;
   c. positive, anti-clockwise Q–C response;
   d. negative, anti clockwise Q–C response.

4. a. Give a definition of concentration rating curves.
   b. Describe how the coefficient *a* in Equation (18.1) differs between catchments on intensively weathered rocks and catchments on bare unweathered rock.

c. Describe how the coefficient b in Equation (18.1) changes from headwaters to lowland rivers.

5. Describe how anaerobic conditions in the hypolimnion promote the release of phosphorus from lake bed sediments

References

Aarkrog, A., 1995, Source terms and inventories of anthropogenic radionuclides. In *Radioecology; Lectures in Environmental Radioactivity*, edited by Holm, E., (Singapore: World Scientific), pp. 21–38.

Aarkrog, A., 2001, Manmade radioactivity. Section 2.2 in *Radioecology, Radioactivity & Ecosystems*, edited by Van der Stricht, E. and Kirchmann, R., (Liège: International Union of Radioecology), pp. 55–78.

Addiscott, T.M. and Mirza, N.A., 1998, Modelling contaminant transport at catchment or regional scale. *Agriculture, Ecosystems & Environment*, **67**, pp. 211–221.

Admiraal, W. and Botermans, Y.J.H., 1989, Comparison of nitrification rates in three branches of the lower river Rhine. *Biogeochemistry*, **8**, pp. 135–151.

Alexakhin, R.M., Cigna, A. and Kirchmann, R., 2001, General introduction to radioecology; a brief historical perspective. Section 1.1 in *Radioecology, Radioactivity & Ecosystems*, edited by Van der Stricht, E. and Kirchmann, R., (Liège: International Union of Radioecology), pp. 1–15.

Alexander, R.B., Smith, R.A. and Schwarz, G.E., 2000, Effect of stream channel size on the delivery of nitrogen to the Gulf of Mexico. *Nature*, **403**, pp. 758–761.

Alexander, R.B., Johnes P.J., Boyer, E.A. and Smith R.A., 2002, A comparison of methods for estimating the riverine export of nitrogen from large watersheds. *Biogeochemistry*, **57/58**, pp. 295–339.

Alloway, B.J. and Ayres, D.C. 1997, *Chemical Principles of Environmental Pollution*, 2nd edition, (London: Blackie Academic & Professional).

Anderson, W.G. 1986. Wettability literature survey – part 1: rock–oil–brine interactions, and the effects of core handling on wettability. *Journal of Petroleum Technology*, **38**, pp. 1125–1144.

Andrady, A.L., 2011, Microplastics in the marine environment. *Marine Pollution Bulletin*, **62**, pp. 1596–1605.

Appelo, C.A.J. and Postma, D., 1996, *Geochemistry, Groundwater and Pollution*, (Rotterdam: Balkema).

Appelo, C.A.J. and Postma, D., 2005, *Geochemistry, Groundwater and Pollution*, 2nd edition (Leiden: Balkema).

Aqualarm, 2013, *Aqualarm; Measurement Data of the Ministry of Infra Structure and the Environment, the Netherlands*, Available online at http://www.rws.nl/en/watersystems/aqualarm/ [verified 1 July 2013].

Asselman, N.E.M, 1999, Suspended sediment dynamics in a large drainage basin: the River Rhine. *Hydrological Processes*, **13**, pp. 1437–1450.

Asselman, N.E.M., 2000, Fitting and interpretation of sediment rating curves. *Journal of Hydrology*, **234**, pp. 228–248.

ATSDR, 2013, *Toxicological Profile Information Sheet*, (Atlanta: U.S. Agency Toxic Substances and Disease Registry). Available online: http://www.atsdr.cdc.gov/toxprofiles/[verified 1 July 2013].

Ayres, R.U. and Rod, S.R., 1986, Patterns of pollution in the Hudson-Raritan basin. *Environment*, **28**, pp. 14–20, 39–43.

Bach, H.H., Brink, H., Olesen, K.W. and Havnø, K., 1989, *Application of PC-Based Models in River Water Quality Modelling*, (Hørsholm: Water Quality Institute, Denmark).

Bachhuber, H., Bunzl, K. and Schimmack, W., 1987, Spatial variability of fallout – [137]Cs in the soil of a cultivated field. *Environmental Monitoring and Assessment*, **8**, pp. 93–101.

Baker, D.B. and Richards, R.P., 2003, The role of flood timing and intensity in controlling nutrient loading and bioavailability in Lake Erie. In *3rd Biennial Conference of the Lake Erie Millennium Network*, May 6–7, 2003, (Windsor, Canada: The University of Windsor).

Bakker, M.I., 2000, *Atmospheric Deposition of Semivolatile Organic Compounds to Plants,* PhD thesis, (Utrecht: Research Institute of Toxicology, Utrecht University).

Balistrieri, L.S., Murray, J.W. and Paul, B., 1992, The biogeochemical cycling of trace metals in the water column of Lake Sammamish, Washington: Response to seasonally anoxic conditions: *Limnology and Oceanography*, **37**, pp. 529–548.

Banks, R.B., 1975, Some features of wind action on shallow lakes. *Journal of the Environmental Engineering Division, ASCE,* **101**, pp. 813–827

Banks, R.B. and Herrera, F.F., 1977, Effect of wind and rain on surface reaeration. *Journal of the Environmental Engineering Division, ASCE*, **103**, pp. 489–504.

Basher, L.R. and Mathews, K.M., 1993, Relationship between ^{137}Cs in some undisturbed New Zealand soils and rainfall. *Australian Journal of Soil Research*, **31**, pp. 655–663.

Bernard, C., Mabit, L., Laverdière, M.R. and Wicherek, S., 1998, Césium-137 et érosion des sols, (in French). *Cahiers Agricultures*, 7, pp. 179–186.

Bernhardt, E.S., Likens, G.E., Hall Jr., R.O., Buso, D.C., Fisher, S.G., Burton, T.M., Meyer, J.L., Mcdowell, W.H., Mayer, M.S., Bowden, W.B., Findlay, S.E.G., Macneale, K.H., Stelzer, R.S. and Lowe, W.H., 2005, Can't see the forest for the stream? In-stream processing and terrestrial nitrogen exports. *BioScience*, **55**, pp. 219–230.

Beven, K.J. and Kirkby, M.J., 1979, A physically based variable contributing area model of basin hydrology. *Hydrological Sciences-Bulletin-des Sciences Hydrologiques*, **24**, pp. 43– 69.

Beven, K. and Germann, P., 1982, Macropores and water flow in soils. *Water Resources Research*, **18**, pp. 1311–1325.

Beven, K.J., Heathwaite, L., Haygarth, P.M., Walling, D., Brazier, R. and Withers, P., 2005, On the concept of delivery of sediment and nutrients to stream channels. *Hydrological Process*, **19**, pp. 551–556.

Bierkens, M.F.P., Finke, P.A. and De Willigen, P., 2000, *Upscaling and Downscaling Methods for Environmental Research*, (Dordrecht: Kluwer Academic Publishers).

Bini, C., Dall'Aglio, M., Ferretti, O. and Gragnani, R., 1988, Background levels of microelements in soils of Italy. *Environmental Geochemistry and Health*, **10**, pp. 63–69.

Bishop, K., Seibert, J., Köhler, S., and Laudon, H., 2004, Resolving the Double Paradox of rapidly mobilized old water with highly variable responses in runoff chemistry. *Hydrological Processes*, **18**, pp. 185–189.

Bissen, M. and Frimmel, F.H., 2003, Arsenic – a review; part I: occurrence, toxicity, speciation, mobility. *Acta Hydrochimica et Hydrobiologia*, **31**, pp. 9–18.

Bjerg, P.L., Rügge, K., Pedersen, J.K. and Christensen, T.H., 1995, Distribution of redox sensitive groundwater quality parameters downgradient of a landfill (Grindsted, Denmark). *Environmental Science and Technology*, **29**, pp. 1387–1394.

Bleuten, W., 1988, A method for the separation of total discharge into base flow, overland flow and channel precipitation for water quality modelling of a small watershed in the Netherlands. *Soil Technology*, **1**, pp. 373–382.

Bleuten, W., 1990, *De Verwatering van Meststoffen; Analyse en Modellering van de Relaties tussen Landgebruik en Waterkwaliteit in het Stroomgebied van de Langbroeker Wetering*, (in Dutch with English summary), (Amsterdam/Utrecht: Knag/Faculty of Geographical sciences, Utrecht University), Netherlands Geographical Studies 105.

Blust, R., 2001, Radionuclide accumulation in freshwater organisms: concepts and methods. Section 4.2.1B in *Radioecology, Radioactivity & Ecosystems*, edited by Van der Stricht, E. and Kirchmann, R., (Liège: International Union of Radioecology), pp. 275–307.

Bolt, G.H. and Bruggenwert, M.G.M., 1978, *Soil chemistry, A. Basic Elements*, (Amsterdam: Elsevier).

Bourg, A.C.M. and Loch, J.P.G., 1995, Mobilization of heavy metals as affected by pH and redox conditions. Chapter 4 in *Biogeodynamics of Pollutants in Soils and Sediments, Risk assessment of Delayed and Non-linear Responses*, edited by Salomons, W. and Stigliani, W.M., (Berlin: Springer), pp. 87–102.

Bouwman, A.F., Bierkens, M.F.P., Griffioen, J., Hefting, M.M., Middelburg, J.J., Middelkoop, H. and Slomp, C.P., 2013, Nutrient dynamics, transfer and retention along the aquatic continuum from land to ocean: towards integration of ecological and biogeochemical models. *Biogeosciences*, **10**, pp. 1–23.

Bowen, H.J.M., 1979, *Environmental Chemistry of the Elements*, (New York: Academic Press).

Bowman, B.T., Brunke, R.R., Reynolds, W.D. and Wall, G.J., 1994, Rainfall simulator – grid lysimeter system for solute transport studies using, large, intact soil blocks. *Journal of Environmental Quality*, **23**, pp. 815–822.

Brady, P.V., Borns, D.J. and Brady, M.V., 1998, *Natural Attenuation; CERCLA, RBCA's, and the Future of Environmental Remediation*, (Boca Raton: CRC).

Bricker, O.P. and Rice, K.C., 1989, Acidic deposition to streams. *Environmental Science and Technology*, **23**, pp. 379–385.

Broers, H.P., 2002, *Strategies for Regional Groundwater Quality Monitoring*, (Utrecht: Knag/Faculty of Geographical sciences, Utrecht University), Netherlands Geographical Studies 306.

Broers, H.P. and Buijs, E.A., 1997, *Origin of Trace Metals and Arsenic at the Oostrum Well Field*, (in Dutch), (Delft: Netherlands Institute of Applied Geoscience TNO), report NITG 97-189A.

Bronswijk, J.J.B. and Prins, H.F., 2001, Stikstofbemesting en nitraatconcentraties in het diepere grondwater van Nederland, (in Dutch). *H₂O*, **34**, pp. 25–29.

Browne M.A., Crump, P., Niven, S.J., Teuten, E., Tonkin, A., Galloway, T. and Thompson, R., 2011, Accumulation of Microplastic on Shorelines Woldwide: Sources and Sinks. *Environmental Science & Technology*, **45**, pp. 9175–9179.

Brutsaert, W., 2005, Hydrology, An Introduction, (Cambridge: Cambridge University Press).

Burkart M.R. and Stoner, J.D., 2002, Nitrate in aquifers beneath agricultural systems. *Water Science and Technology*, **45(9)**, pp. 19–28.

Burrough, P.A., 1993, Soil variability: a late 20th century view. *Soils and Fertilisers*, **56**, pp. 529–562.

Burrough, P.A., 1996, Opportunities and limitations of GIS-based modeling of solute transport at the regional scale. In *Application of GIS to the Modeling of Non-Point Source Pollutants in the Vadose Zone*, edited by Corwin, D.L. and Loague, K., (Madison, WI: Soil Science Society of America), SSSA Special Publication 48, pp. 19–47.

Burrough, P.A. and McDonnell, R.A., 1998, *Principles of Geographical Information Systems*, (Oxford: Oxford University Press).

Burrough, P.A., van Rijn, R. and Rikken, M., 1996, Spatial data quality and error analysis issues: GIS functions and environmental modelling. In *Environmental Modelling and GIS, Progress and Research Issues*, edited by Goodchild, M. Steyaert, L., Parks, B.O., Johnston, C., Maidment, D., Crane, M. and Glendinning, S., (Santa Barbara, CA: NCGIA), GIS World Publications, pp. 29–34.

Butler, G.C., 1978, *Principles of Ecotoxicology*, (New York: Wiley).

Campbell, C.G., Borglin, S.E., Green, F.B., Grayson, A. Wozei, E. and Stringfellow, W.T., 2006. Biologically directed environmental monitoring, fate, and transport of estrogenic endocrine disrupting compounds in water: A review. *Chemosphere*, **65**, pp. 1265–1280.

Carson, R., 1962, *Silent Spring*, (Boston: Houghton Mifflin Co.).

Caritat P. de and Cooper, M., 2011, *National Geochemical Survey of Australia: The Geochemical Atlas of Australia*. (Canberra: Geoscience Australia) Record 2011/20, Data available online at http://www.ga.gov.au/energy/projects/national-geochemical-survey/atlas.html [verified 1 July 2013].

Choppin, G.R., Liljenzin, J.O. and Rydberg, J., 1995, *Radiochemistry and Nuclear Chemistry*, 2nd edition, (Oxford: Butterworth-Heinemann).

Chow, V.T., Maidment, D.R. and Mays, L.W., 1988, *Applied Hydrology*, (New York: McGraw-Hill).

Christensen, T.N., Bjerg, P.L., Banwart, S.A., Jakobsen, R., Heron, G. and Albrechtsen, H.-J., 2000, Characterization of redox conditions in groundwater contaminant plumes. *Journal of Contaminant Hydrology*, **45**, pp. 165–241.

Christensen, T.N., Kjeldsen, P., Bjerg, P.L., Jensen, D.L., Christensen, J.B., Baun, A., Albrechtsen, H.-J. and Heron, G., 2001, Biogeochemistry of landfill leachate plumes. *Applied Geochemistry*, **16**, pp. 659–718.

Churchill, M.A., Elmore, H.L. and Buckingham, R.A., 1962, The prediction of stream reaeration rates. *International Journal of Air and Water Pollution*, **6**, pp. 467–504.

Copius Peereboom, J.W., 1976, *Chemie, Mens en Milieu*, (in Dutch), (Assen, the Netherlands: Van Gorcum).

Corbett, D.R., Dillon, K., Burnett, W. and Schaefer, G., 2002, The spatial variability of nitrogen and phosphorus concentration in a sand aquifer influenced by onsite sewage treatment and disposal systems: a case study on St. George Island, Florida. *Environmental Pollution*, **117**, 337–345.

Coulibaly, L., Labib, M.E. and Hazen, R., 2004, A GIS-based multimedia watershed model: development and application. *Chemosphere*, **55**, pp. 1067–1080.

Coupe, R.H., Kalkhoff, S.J., Capel, P.D. and Gregoire, C. , 2012, Fate and transport of glyphosate and aminomethylphosphonic acid in surface waters of agricultural basins. *Pest Management Science*, **68**, pp. 16–30.

Cox, M.S., Gerard, P.D., Wardlaw, M.C. and Abshire, M.J., 2003, Variability of selected soil properties and their relationships with soybean yield. *Soil Science Society of America Journal*, **67**, pp. 1296–1302.

Crawford, C.G., 1991, Estimation of suspended-sediment rating curves and mean suspended-sediment loads. *Journal of Hydrology*, **129**, pp. 331–348.

Crutzen, P.J., Steffen, W., 2003, How long have we been in the Anthropocene era? An editorial comment. *Climatic Change*, **61**, pp. 251–257.

Daniels, M.B., Delaune, P. Moore Jr., P.A., Mauromoustakos, A., Chapman, S.L. and Langston, J.M., 2001, Soil phosphorus variability in pastures; Implications for sampling and environmental management strategies. *Journal of Environmental Quality*, **30**, pp. 2157–2165.

De Bakker, H., 1979, *Major Soils and Soil Regions in the Netherlands*, (Wageningen: Pudoc).

De Caritat, P., Reimann, C., Äyräs, M., Niskavaara, H., Chekushin, V.A. and Pavlov, V.A., 1996, Stream water geochemistry from selected catchments on the Kola Peninsula (NW Russia) and in neighbouring areas of Finland and Norway: 1. Elements levels and sources. *Aquatic Geochemistry*, **2**, pp. 149–168.

De Cort, M., Dubois, G., Fridman, Sh.D., Germenchuk, M.G., Izrael, Yu.A., Janssens, A., Jones, A.R., Kelly, G.N., Kvasnikova, E.V., Matveenko, I.I., Nazarov, I.M., Pokumeiko, Y.M., Sitak, V.A., Stukin, E.D., Tabachny, L.Ya., Tsaturov, Y.S. and Avdyushin, S.I., 1998, *Atlas of Caesium Deposition on Europe after the Chernobyl Accident*, (Luxembourg: European Commission), Report EUR 16733.

De Fré, R., Geuzens, P., Rymen, Th., Wevers, M. and Elslander, H., 1992, Bodemverontreiniging rond een huisvuilverbrandingsoven, (in Dutch). *Bodem*, **2**, pp. 64–67.

Delsman, J., 2002, *Reconstructing Floodplain Sedimentation Rates from Heavy Metal and Caesium-137 Profiles: An Inverse Modelling Approach*, Unpublished MSc thesis, (Utrecht: Department of Physical Geography, Utrecht University).

De Roo, A.P.J., Wesseling, C.G. and Ritsema, C.J., 1996, LISEM: a single event physically-based hydrologic and soil erosion model for drainage basins: I. Theory, input and output. *Hydrological Processes*, **10**, pp. 1107–1117.

DeSimone, L.A., Howes, B.L. and Barlow, P.M., 1997, Mass-balance analysis of reactive transport and cation exchange in a plume of wastewater-contaminated groundwater. *Journal of Hydrology*, **203**, pp. 228–249.

Destouni, G., Hannerz, F., Prieto, C., Jarsjö, J. and Shibuo, Y., 2008, Small unmonitored near-coastal catchment areas yielding large mass loading to the sea. *Gobal Geochemical Cycles*, **22**, GB4003.

Destouni, G., Persson, K., Prieto, C. and Jarsjö, J., 2010, General quantification of catchment-scale nutrient and pollutant transport through the subsurface to surface and coastal waters. *Environmental Science & Technology*, **44**, pp. 2048–2055.

De Vos, W., Tarvainen, T., Salminen, R., Reeder, S., De Vivo, B., Demetriades, A., Pirc, S., Batista, M.J., Marsina, K., Ottesen, R.-T., O'Connor, P.J., Bidovec, M., Lima, A., Siewers, U., Smith, B., Taylor, H., Shaw, R., Salpeteur, I., Gregorauskiene, V., Halamic, J., Slaninka, I., Lax, K., Gravesen, P., Birke, M., Breward, N., Ander, E.L., Jordan, G., Duris, M., Klein, P., Locutura, J., Bel-lan, A., Pasieczna, A., Lis, J., Mazreku, A., Gilucis, A., Heitzmann, P., Klaver, G. and Petersell, V., 2006, *Geochemical Atlas of Europe. Part 2 - Interpretation of Geochemical Maps, Additional Tables, Figures, Maps, and Related Publications*, (Espoo, Finland: Forum of European Geological Surveys (FOREGS)), Available online at http://weppi.gtk.fi/publ/foregsatlas/ [verified 1 July 2013].

De Wit, M.J.M., 1999, *Nutrient Fluxes in the Rhine and Elbe Basins*, (Utrecht: Knag/Faculty of Geographical sciences, Utrecht University), Netherlands Geographical Studies 259.

De Wit, M.J.M., 2001, Nutrient fluxes at the river basin scale. I: the PolFlow model. *Hydrological Processes*, **15**, pp. 743–759.

De Zorzi, P., Belli, M., Barbizzi, S., Menegon, S. and Deluisa, A., 2002, A practical approach to assessment of sampling uncertainty. *Accreditation and Quality Assurance*, **7**, pp. 182–188.

Domenico, P.A. and Schwarz, F.W., 1998, *Physical and Chemical Hydrogeology*, 2nd edition, (New York: Wiley).

Doppler, T., Camenzuli, L., Hirzel, G., Krauss, M., Lück, A. and Stamm, C., 2012, Spatial variability of herbicide mobilisation and transport at catchment scale: insights from a field experiment. *Hydrology and Earth System Sciences*, **16**, pp. 1947–1967.

Draaijers, G.P.J., 1993, *The Variability of Atmospheric Deposition to Forests; The Effects of Canopy Structure and Forest Edges*, (Utrecht: Knag/Faculty of Geographical sciences, Utrecht University), Netherlands Geographical Studies 156.

Draaijers, G.P.J., Van Ek, R. and Bleuten, W., 1994, Atmospheric deposition in complex forest landscapes. *Boundary Layer Meteorology*, **69**, pp. 343–366.

Drever, J.I., 1982, *The Geochemistry of Natural Waters*, 1st edition, (Upper Saddle River NJ: Prentice-Hall).

Drever, J.I., 2000, *The Geochemistry of Natural Waters: Surface and Groundwater Environments*, 3rd edition, (Upper Saddle River NJ: Pearson/Prentice-Hall).

Driscoll, C.T., Likens, G.E., Hedin, L.O., Eaton, J.S. and Bormann, F.H., 1989, Changes in the chemistry of surface waters: 25-year results at the Hubbard Brook Experimental Forest, N.H. *Environmental Science and Technology*, **23**, pp. 137–143.

Duffus, J.H., 2002, "Heavy metals" – A meaningless term (IUPAC Technical Report). *Pure and Applied Chemistry*, **74**, pp. 793807.

Dufour, G.C., 2000, *Groundwater in the Netherlands; Facts and Figures*, (Delft/Utrecht: The Netherlands: Netherlands Institute of Applied Geosciences TNO).

EC, 2011, *Report from the Commission to the Council and the European Parliament - On Implementation of Council Directive 91/676/EEC Concerning the Protection of Waters against Pollution Caused by Nitrates from Agricultural Sources Based on Member State Reports for the Period 2004-2007*, (Brussels: European Commission). SEC(2011) 909, Available online at http://ec.europa.eu/environment/water/water-nitrates/pdf/sec_2011_909.pdf [verified 1 July 2013].

Eckerrot, Å. and Pettersson, K., 1993, Pore water phosphorus and iron concentrations in a shallow, eutrophic lake – indications of bacterial regulation. *Hydrobiologia*, **253**, pp. 165–177.

Edmunds, W.M. and Shand, P., 2008a, The Chalk Aquifer of Dorset. Chapter 9 in *Natural Groundwater Quality* edited by Edmunds, W.M., Shand, P., (Oxford: Blackwell Publishing), pp. 195–215.

Edmunds, W.M. and Shand, P. (Eds.), 2008b, *Natural Groundwater Quality*, (Oxford: Blackwell Publishing).

EDS, 1995. *DUFLOW, A Micro Computer Package for the Simulation of One Dimensional Unsteady Flow and Water Quality in Open Channel Systems*, (Leidschendam, The Netherlands: EDS).

Edwards, E.A., Wills, L.E., Reinhard, M. and Grbić-Galić, D., 1992, Anaerobic degradation of toluene and xylene by aquifer microorganisms under sulfate-reducing conditions. *Applied Environmental Microbiology*, **58**, pp. 794–800.

Edwards, R.T., 1998, The hyporheic zone. In *River Ecology and Management: Lessons from the Pacific Coastal Ecoregion*, edited by Naiman, R.J. and Bilby, R.E., (New York: Springer). pp. 399–429.

EEA, 1998. *Europe's Environment: The Second Assessment; A Report on the Changes in the Pan-European Environment as a Follow-up to 'Europe's Environment: The Dobris Assessment (1995)* (Copenhagen: European Environmental Agency), requested by the environment Ministers for the whole of Europe to prepare for the fourth ministerial conference in Aarhus, Denmark, June 1998. Available online: http://reports.eea.eu.int/92-828-3351-8/en [verified 15 April 2006].

EEA, 1999. *Environment in the European Union at the Turn of the Century; Environmental Assessment Report No. 2*, (Copenhagen: European Environment Agency).

EEA, 2005, *Source apportionment of nitrogen and phosphorus inputs into the aquatic environment*, (Copenhagen: European Environmental Agency), EEA Report No 7/2005, Available online at http://www.eea.europa.eu/publications/eea_report_2005_7 [verified 1 July 2013].

EEA, 2006a, *EEA Data Service: Reference WATERBASE*, (Copenhagen: European Environmental Agency). Available online: http://www.eea.europa.eu/data-and-maps [verified 1 July 2013].

EEA, 2006b, *EEA Multilingual Environmental Glossary*, (Copenhagen: European Environmental Agency). Available online: http://glossary.eea.eu.int/EEAGlossary [verified 1 July 2013].

Eghball, B., Schepers, J.S., Negahban, M. and Schlemmer, M.R., 2003, Spatial and temporal variability of soil nitrate and corn yield. *Agronomy Journal*, **9**, pp. 339–346.

Engelen, G.B. and Kloosterman, F.H., 1996, *Hydrological Systems Analysis, Methods and Applications*, (Dordrecht: Kluwer), Water Science and Technology Library 20.

EPA, 1985, *Rates, Constants, and Kinetics Formulations in Surface Water Quality Modeling*, 2nd edition, written by Bowie, G.L. *et. al.* (Athens, GA: U.S. Environmental Protection Agency), Environmental Research Laboratory. Report No. EPA/600/3-85/040.

EPA, 1999a, *Understanding Oil Spills and Oil Spill Response*, (Washington DC: U.S. Environmental Protection Agency, Office of Emergency and Remedial Response), Report No. EPA 540-K-99-007.

EPA, 1999b, *Anaerobic Biodegradation Rates of Organic Chemicals in Groundwater: A Summary of Field and Laboratory Studies*, (Washington DC: U.S. Environmental Protection Agency, Office of Soil Waste Report), Available online at http://www.epa.gov/wastes/hazard/wastetypes/wasteid/hwirwste/pdf/risk/reports/s0535.pdf [verified 1 July 2013].

EPA, 2013, *Chemicals in the Environment: OPPT Chemical Fact Sheets*. Washington DC: U.S. Environmental Protection Agency, Office of Pollution Prevention & Toxics. Available online: http://www.epa.gov/opptintr/chemfact/[verified 1 July 2013].

Erisman, J.W., Vermetten, A.W.M. and Asman, W.A.H., 1988, Vertical distribution of gases and aerosols: The behaviour of ammonia and related components in the lower atmosphere. *Atmospheric Environment*, **22**, pp. 1153–1160.

Eshleman, K.N., Pollard, J.S. and Kuebler, A., 1992, *Interactions between Surface Water and Groundwater in a Virginia Coastal Plain Watershed*. (Blacksburg: Virginia Water Resources Research Center, Virginia Polytechnic Institute and State University), Bulletin 174.

Eshleman K.N., Morgan, R.P., Webb, J.R., Deviney, F.A. and Galloway, J.N., 1998, Temporal patterns of nitrogen leakage from mid-Appalachian forested watersheds: role of insect defoliation. *Water Resources Research*, **34**, pp. 2005–2016.

Evans, C. and Davis, T.D., 1998, Causes of concentration/discharge hysteresis and its potential as a tool for analysis of episode hydrochemistry. *Water Resources Research*, **34**, pp. 129–137.

Fair, G.M., Geijer, J.C. and Okun, D., 1968, *Water and Waste Water Engineering*, Volume 2, (New York: Wiley).

Farmer, A.M., 1999, *Implementation of the 1991 EU Urban Waste Water Treatment Directive and its Role in Reducing Phosphate Discharges*, (London: Institute for European Environmental Policy).

Faure, F., Corbaz, M., Baecher, H. and De Alencastro, L.F., 2012, Pollution due to plastics and microplastics in Lake Geneva and in the Mediterranean Sea. *Archives des Sciences*, **65**, pp. 157–164.

Faure, G., 1998, *Principles and Applications of Geochemistry*, 2nd edition, (Upper Saddle River NJ: Prentice-Hall).

Fawell, J.K. and Hunt, S., 1988, *Environmental Ecotoxicology, Organic Pollutants*, (Chichester: Ellis Horwood).

Ferguson, R.I., 1986, River loads underestimated by rating curves. *Water Resources Research*, **22**, pp. 74–76.

Fischer, H.B., List, E.J., Koh, R.C.Y., Imberger, J. and Brooks, N.H., 1979, *Mixing in Inland and Coastal Waters*, (New York: Academic Press).

Förstner, U. and Wittman, G.T.W., 1983, *Metal Pollution in the Aquatic Environment*, (Berlin: Springer-Verlag).

Franzaring, J., Bierl, R. and Ruthsatz, B., 1992, Active biological monitoring of polycyclic aromatic hydrocarbons using kale (*Brassica oleracea*) as a monitor-species. *Chemosphere*, **25**, pp. 827–834.

Frapporti, G., Vriend, S.P. and Van Gaans, P.F.M., 1994, Qualitative time trend analysis of ground water monitoring networks; example from the Netherlands. *Environmental Monitoring and Assessment*, **30**, pp. 81–102.

Frapporti, G., Hoogendoorn, J.H. and Vriend, S.P., 1995, Detailed hydrochemical studies as a useful extension of national ground-water monitoring networks. *Ground Water*, **33**, pp. 817–828.

Freeze, A.R., 1974, Streamflow generation. *Reviews of Geophysics and Space Physics*, **12**, pp. 627–647.

Froelich, P.N., 1988, Kinetic control of dissolved phosphate in natural rivers and estuaries: A primer on the phosphate buffer mechanism. *Limnology and Oceanography*, **33**, pp. 649–668.

Galán, E., Gonzalez, I. and Fernandez-Caliani, J.C., 2002, Residual pollution load of soils impacted by the Aznalcóllar Spain mining spill after clean-up operations. *The Science of the Total Environment*, **286**, pp. 167–179.

Galloway, J.N., Schlesinger, W.H., Levy II, H., Michaels, A., Schnoor, J.L., 1995, Nitrogen fixation: anthropogenic enhancement – environmental response. *Global Biogeochemical Cycles*, **9**, pp. 235–252.

Galloway, J.N., 1998, The global nitrogen cycle: changes and consequences. *Environmental Pollution*, **102 S1**, pp. 15–24.

Garland, J.H.N., 1978, Nitrification in the river Trent. Chapter 7 in *Mathematical Models in Water Pollution Control*, edited by James, A., (Chichester: Wiley), pp. 167–191.

Gburek, W.J. and Sharpley, A.N., 1998, Hydrologic Controls on Phosphorus Loss from Upland Agricultural Watersheds. *Journal of Environmental Quality*, **27**, pp. 267–277

George, D.G., 1993, Physical and chemical scales of pattern in freshwater lakes and reservoirs. *The Science of the Total Environment*, **135**, pp. 1–15.

George, D.G. and Jones, D.H., 1987, Catchment effects on the horizontal distribution of phytoplankton in five of Scotland's largest freshwater lochs. *Journal of Ecology*, **75**, pp. 43–59.

Gerrard, J., 2000, *Fundamentals of Soils*, (Oxford: Routledge).

Golterman, H.L., 2004, *The Chemistry of Phosphate and Nitrogen Compounds in Sediments*, (Dordrecht: Kluwer).

Goodale, C.L., Aber, J.D. and Vitousek, P.M., 2003, An unexpected nitrate decline in New Hampshire streams. *Ecoystems*, **6**, pp. 75–86.

Govers, G., Quine, T.A. and Walling, D.E., 1993, The effects of water erosion and tillage movement on hillslope profile development: A comparison of field observations and model results. In *Farm Land Erosion in Temperate Plains and Hills*, edited by Wicherek, S., (Amsterdam: Elsevier), pp. 285–300.

Gowda, T.P.H., 1983, Modelling nitrification effects on the dissolved oxygen regime of the Speed river. *Water Research*, **17**, pp. 1917–1927.

Griffioen, J., 1999, Comment on 'Kinetic modelling of microbially-driven redox chemistry of subsurface environments: Coupling transport, microbial metabolism and geochemistry' by K.S. Hunter, Y. Wang and P. Van Cappellen. *Journal of Hydrology*, **226**, pp. 121–124.

Griffioen, J., 2001, Potassium adsorption ratios as an indicator for the fate of agricultural potassium in groundwater. *Journal of Hydrology*, **254**, pp. 244–254.

Grimalt, J.O., Ferrer, M. and Macpherson, E., 1999, The mine tailing accident in Aznalcollar. *The Science of the Total Environment*, **242**, pp. 3–11.

Groffman, P.M., Driscoll, C.T., Fahey, T.J., Hardy, J.P., Fitzhugh, R.D. and Tierney, G.L., 2001, Colder soils in a warmer world: a snow manipulation study in a northern hardwood forest ecosystem. *Biogeochemistry*, **56**, pp. 135–150.

Gundersen, P. and Bashkin, V.N., 1994, Nitrogen cycling. Chapter 11 in *SCOPE 51 – Biogeochemistry of Small Catchments*, edited by Moldan, B. and Cerný, J., (New York: Wiley), pp. 255–283.

Gustavsson, N., Bølviken, B., Smith, D.B. and Severson, R.C., 2001, Geochemical Landscapes of the Conterminous United States — New Map Presentations for 22 Elements, (Denver : U.S. Geological Survey), U.S. Geological Survey Professional Paper 1648, Available online at http://pubs.usgs.gov/pp/p1648/ [verified 1 July 2013].

Håkanson, L., 2004, Modelling the transport of radionuclides from land to water. *Journal of Environmental Radioactivity*, **73**, pp. 267–287.

Hatch, D, Goulding, K. and Murphy, D., 2002, Nitrogen. Chapter 1 in *Agriculture, Hydrology and Water Quality*, edited by Haygarth, P. and Jarvis, S., (Wallingford: CAB International).

Hartman, M.J., Morasch, L.F. and Webber W.D. (eds.), 2002, *Hanford Site Groundwater Monitoring for Fiscal Year 2001*, (Richland, WA: Pacific Northwest National Laboratory) PNNL-13788.

Hartog, N., Bergen, P.F. van, Leeuw, J.W. de, and Griffioen, J.B., 2004, Reactivity of organic matter in aquifer sediments: geological and geochemical controls. *Geochimica et Cosmochimica Acta*, **68**, pp. 1281–1292.

Hayes, W.J. and Laws, E.R., 1991, *Handbook of Pesticide Toxicology*, (San Diego: Academic Press).

Haygarth, P.M. and Sharpley, A.N., 2000, Terminology for phosphorus transfer. *Journal of Environmental Quality*, **29**, pp. 10–15.

Haygarth, P., Turner, B.L., Fraser, A., Jarvis, S., Harrod, T., Nash, D., Halliwell, D., Page, T. and Beven, K., 2004, Temporal variability in phosphorus transfers: classifying concentration–discharge event dynamics. *Hydrology and Earth Systems Sciences*, **8**, pp. 88–97.

Heathwaite, A.L., Quinn, P.F. and Hewett, C.J.M., 2005, Modelling and managing critical source areas of diffuse pollution from agricultural land using flow connectivity simulation. *Journal of Hydrology*, **304**, pp. 446–461.

Heckrath, G., Djurhuus, J., Quine, T.A., Van Oost, K., Govers, G.,and Zhang, Y., 2005, Tillage erosion and its effect on soil properties and crop yield in Denmark. *Journal of Environmental Quality*, **34**, pp. 312–324.

Hellmann, H., 1999, *Qualitative Hydrologie – Wasserbeschaffenheit und Stoff-Flüsse*, (in German), (Berlin: Borntraeger), Lehrbuch der Hydrologie Band 2.

Hem, J.D., 1989, *Study and Interpretation of the Chemical Characteristics of Natural Water*, 3rd edition, (Washington DC: United States Geological Survey) USGS – Supply Paper 2254.

Hemker, K., Van den Berg, E. and Bakker, M., 2004, Groundwater whirls. *Ground Water*, **42**, pp. 234–242

Hemond, H.F. and Fechner-Levy, E.J., 2000, *Chemical Fate and Transport in the Environment*, 2nd edition, (San Diego: Academic Press).

Hendriks, R.F.A., Oostindie, K. and Hamminga, P., 1999, Simulation of bromide tracer and nitrogen transport in a cracked clay soil with the FLOCR/ANIMO model combination. *Journal of Hydrology*, **215**, pp. 94–115.

Herbst, M. and Van Esch, G.J., 1991, *Lindane*, (Geneva: World Health Organization), Environmental Health Criteria 124, Available online at http://www.inchem.org/documents/ehc/ehc/ehc124.htm [verified 1 July 2013]

Hewlett, J.D. and Hibbert, A.R., 1967, Factors affecting the response of small watersheds to precipitation in humid areas. In *Forest Hydrology*, edited by Sopper, W.E., Lull, H.W., (New York: Pergamonn Press), pp. 275–253.

Hill, A.R., 1982, Phosphorus and major cation mass balances for two rivers during low summer flows. *Freshwater Biology*, **12**, pp. 293–304.

Hillbricht-Ilkowska, A., 1999, Shallow lakes in lowland river systems: Role in transport and transformations of nutrients and in biological diversity. *Hydrobiologia*, **408/409**, pp. 349–358.

Hillel, D. and Baker, R.S., 1988, A descriptive theory of fingering during infiltration into layered soils. *Soil Science*, **146**, pp. 51–56.

Hornberger, G.M., Raffensperger, J.P., Wiberg, P.L. and Eshleman, K.N., 1998, *Elements of Physical Hydrology*, (Baltimore: The Johns Hopkins University Press).

Horowitz, A.J., Elrick, K.A. and Smith, J.J., 2001, Estimating suspended sediment and trace element fluxes in large river basins: methodological considerations as applied to the NASQAN program. *Hydrological Processes*, **15**, pp. 1107–1132.

Hossner, L.R., Loeppert, R.H., Newton, R.J., Szaniszlo, P.J. and Attrep Jr., M., 1998, *Literature Review: Phytoaccumulation of Chromium, Uranium, and Plutonium in Plant Systems*, (Amarillo: Amarillo National Resource Center for Plutonium), Report ANRCP-1998-3.

House, W.H., Denison, F.H. and Armitage, P.D., 1995, Comparison of the uptake of inorganic phosphorus to a suspended and stream bed-sediment. *Water Research*, **29**, pp. 767–779.

Hubbert, M.K., 1940, The theory of groundwater motion. Journal of Geology 48, 785–944.

Hunter, K.S., Wang, Y. and Van Cappellen, P., 1998, Kinetic modeling of microbially-driven redox chemistry of subsurface environments: coupling transport, microbial metabolism and geochemistry. *Journal of Hydrology*, **209**, pp. 56–80.

IAEA, 2011, *Report of Japanese Government to IAEA Ministerial Conference on Nuclear Safety - Accident at TEPCO's Fukushima Nuclear Power Stations*, Transmitted by Permanent Mission of Japan to IAEA, 7 June 2011, (Vienna: International Atomic Energy Agency).

IKSR, 1986, *Verunreinigung de Rheins nach dem Brandunfall bij Sandoz AG – Sachstand*, (in German), (Koblenz: Internationale Kommission zum Schutz des Rheins gegen Verunreinigung).

Iital, A., Stalnacke, P., Deelstra, J., Loigu, E. and Pihlak, M., 2005, Effects of large-scale changes in emissions on nutrient concentrations in Estonian rivers in the Lake Peipsi drainage basin. *Journal of Hydrology*, **304**, pp. 261–273.

Irwin, R.J., VanMouwerik, M., Stevens, L., Seese, M.D. and Basham, W., 1998, *Environmental Contaminants Encyclopedia,* (Fort Collins CO: National Park Service, Water Resources Division), Available online at http://nature.nps.gov/hazardssafety/toxic/ [verified 1 July 2013].

ITOPF, 2011, *Fate of Marine Oil Spills*, (London: The International Tanker Owners Pollution Federation Limited (ITOPF)), Technical Information Paper 2, Available online at http://www.itopf.com/ [verified 1 July 2013].

Ivens, W.P.F.M, Draaijers, G.P.J., Bos, M.M. and Bleuten, W., 1988, *Dutch Forests as Air Pollutant Sinks in Agricultural Areas; A Case Study in the Central Part of the Netherlands on the Spatial and Temporal Variability of Atmospheric Deposition to Forests*, (Utrecht: Utrecht University, Department of Physical Geography).

Jarvie, H.P., Oguchi, T. and Neal, C., 2002, Exploring the linkages between river water chemistry and watershed characteristics using GIS-based catchment and locality analyses. *Regional Environmental Change*, **3**, pp. 36–50.

Jarvie, H.P., Jürgens, M.D., Williams, R.J., Neal, C., Davies, J.J.L., Barrett, C. and White, J., 2005, Role of river bed sediments as sources and sinks of phosphorus across two major eutrophic UK river basins: the Hampshire Avon and Herefordshire Wye. *Journal of Hydrology*, **304**, pp. 51–74.

Jenkins, A., Peters, N.E. and Rodhe, A., 1994, Hydrology. Chapter 2 in *SCOPE 51 – Biogeochemistry of Small Catchments*, (New York: Wiley), pp. 31–54.

Jensen, H.S. and Andersen, F.Ø., 1992, Importance of temperature, nitrate and pH for phosphate release from aerobic sediments of four shallow, eutrophic lakes. *Limnology and. Oceanography*, **37**, pp. 577–589.

Jetten, V.G., 2013, *OpenLISEM, A spatial model for runoff, floods and erosion*, (Enschede NL: Universiy of Twente, ITC), Available online at http://blogs.itc.nl/lisem/ [verified 1 July 2013].

Jha, R., Ojha, C.S.P. and Bhatia, K.K.S., 2001, Refinement of predictive reaeration equations for a typical Indian river. *Hydrological Processes*, **15**, pp. 1047–1060.

Johnson, B.D., Kranck, K. and Muschenheim, D.K., 1994, Physicochemical factors in particle aggregation. In *The Biology of Particles in Aquatic Systems*, 2nd edition, edited by Wotton, R.S., (Boca Raton: Lewis), pp. 75–96.

Jørgensen, S.E., 1988, *Fundamentals of Ecological Modelling*, (Amsterdam: Elsevier), Developments in Environmental Modelling 14.

Karssenberg, D., 2002, The value of environmental modelling languages for building distributed hydrological models. *Hydrological Processes*, **16**, pp. 2751–2766.

Kedziora, A., Ryszkowski L. and Kundzewicz, Z. 1995, Phosphate transport and retention in a riparian meadow – a case study. Chapter 13 in *SCOPE 54 – Phosphorus in the Global Environment – Transfers, Cycles and Management*, edited by Tiessen, H., (New York: Wiley), pp. 229–234.

Keesstra, S.D., Geissen, V., Mosse, K, Piiranen, S., Scudiero, E., Leistra, M., van Schaik, L., 2012, Soil as a filter for groundwater quality. *Current Opinion in Environmental Sustainability*, **4**, pp. 507–516.

Kendall, C. and McDonnell, J.J. (eds.), 1998, *Isotope Tracers in Catchment Hydrology*, (Amsterdam: Elsevier).

Kirchner, J.W., Feng, X. and Neal, C., 2000, Fractal stream chemistry and its implications for contaminant transport in catchments. *Nature*, **403**, pp. 524–527.

Kirchner, J.W., 2003, A double paradox in catchment hydrology and geochemistry. *Hydrological Processes*, **17**, pp. 871–874.

Konikow, L.F. and Bredehoeft, J.D., 1978, *Computer Model of Two-Dimensional Solute Transport and Dispersion in Ground Water*, (Washington DC: United States Geological Survey), Techniques of Water-Resources Investigations, book 7, chapter C2.

Lawler, D.M., Petts, G.E., Foster, I.D.L. and Harper, S., 2006, Turbidity dynamics during spring storm events in an urban headwater river system: The Upper Tame, West Midlands, UK. *Science of the Total Environment*, **360**, pp. 109–126.

Laznik, M., Stålnacke, P., Grimvall, A. and Wittgren, H.B., 1999, Riverine input of nutrients to the Gulf of Riga – temporal and spatial variation. *Journal of Marine Systems*, **23**, pp. 11–25.

Leenaers, H, 1989, *The Dispersal of Metal Mining Wastes in the Catchment of the River Geul, Belgium-the Netherlands*, (Amsterdam/Utrecht: Knag/Faculty of Geographical sciences, Utrecht University) Netherlands Geographical Studies 102.

Leenaers, H., Okx, J.P. and Burrough, P.A., 1990, Employing elevation data for efficient mapping of soil pollution on floodplains. *Soil Use and Management*, **6**, pp. 105–114.

Likens, G.E. and Bormann, F.H., 1995, *Biogeochemistry of a Forested Ecosystem*, (New York: Springer).

Lindinger, H. and Scheidleder, A., 2004, *Indicator fact sheet: (WEU1) Nitrate in groundwater*, (Copenhagen: European Environmental Agency), Available online at http://www.eea.europa.eu/data-and-maps/indicators/ [verified 1 July 2013].

Lindqvist, O., Johansson, K., Bringmark, L., Timm, B., Aastrup, M., Andersson, A., Hovsenius, G., Håkanson, L., Iverfeldt, Å. and Meili, M., 1991, Mercury in the Swedish environment – Recent research on causes, consequences and corrective methods. *Water, Air and Soil Pollution*, **55**, pp. 1–261.

Liu, X., Zhang, Y., Han, W., Tang, A., Shen, J., Cui, Z., Vitousek, P., Erisman, J.W., Goulding, K., Christie, P., Fangmeier, A. and Zhang, F., 2013, Enhanced nitrogen deposition over China. *Nature*, **494**, pp. 459–462.

Lobb, D.A., Kachanoski, R.G. and Miller, M.H., 1995, Tillage translocation and tillage erosion on shoulder slope landscape positions measured using 137Cs as a tracer. *Canadian Journal of Soil Science*, **75**, pp. 211–218.

Locher, W.P. and de Bakker, H., 1990, *Bodemkunde van Nederland, Deel 1: Algemene bodemkunde*, (in Dutch), (Den Bosch: Malmberg).

Lock, M.A., 1994, Dynamics of particulate and dissolved organic matter over the substratum of water bodies. In *The Biology of Particles in Aquatic Systems*, 2nd edition, edited by Wotton, R.S., (Boca Raton: Lewis), pp. 137–160.

Lundstedt, S., Haglund, P. and Öberg, L., 2003, Degradation and formation of polycyclic aromatic compounds during bioslurry treatment of an aged gasworks soil. *Environmental Toxicology and Chemistry*, **22**, pp. 1413–1420.

Luoma, S.N. and Rainbow, P.S., 2008. *Metal Contamination in Aquatic Environments; Science and Lateral Management*, (Cambridge: Cambridge University Press).

Luoma, S.N. and Presser, T.S., 2009. Emerging opportunities in management of selenium contamination. *Environmental Science and Technology*, **43**, pp. 8483–8487.

Lijklema, L., 1991, *Water Quality Control; Lecture notes*, (in Dutch), (Wageningen: Wageningen Agricultural University).

Lymann, W.J., Rheel, W.F. and Rosenblatt, D.H., 1990, *Handbook of Chemical Property Estimation Methods*, 2nd printing, (Washington DC: American Chemical Society).

Lyngkilde, J. and Christensen, T.H., 1992, Redox zones of a landfill leachate pollution plume (Vejen, Denmark). *Journal of Contaminant Hydrology*, **10**, pp. 273–289.

Mackay, D., 1991, *Multimedia Environmental Models; the Fugacity Approach*, (Chelsea: Lewis).

Manshoven, S., De Ceuster, T., Dejonghe, W., Gemoets, J., Gutschoven, K., Van Gestel, G., Gregoir, T., 2010, *Duurzaamheid van anaërobe natuurlijke attenuatie op sites verontreinigd met BTEX*, (in Dutch) (Mechelen: Openbare Vlaamse Afvalstoffenmaatschappij), Available online at: http://www.ovam.be/ [verified 1 July 2013].

Marion, L. and Brient, L., 1998, Wetland effects on water quality: input-output studies of suspended particulate matter, nitrogen (N) and phosphorus (P) in Grand-Lieu, a natural plain lake. *Hydrobiologia*, **373/374**, pp. 217–235.

Marsh, W.M. and Grossa, J.M., 2001, *Environmental Geography, Science, Land Use, and Earth Systems*, 2nd edition, (New York: Wiley).

Matamoros-Veloza, A., Newton, R.J. and Benning L.G., 2011. What controls selenium release during shale weathering? *Applied Geochemistry*, **26**, pp. S222–S226.

Matthess, G., 1994, *Die Beschaffenheit des Grundwassers*, 3. überarbeitete Auflage, (in German), (Berlin: Borntraeger), Lehrbuch der Hydrogeologie Band 2.

Mayer, L.M. and Gloss, S.P., 1980, Buffering of silica and phosphate in a turbid river. *Limnology and Oceanography*, **25**, pp. 12–22.

McArthur, J.M., Ravenscroft, P., Safiullah, S., Thirlwall, M.F., 2001, Arsenic in groundwater: testing pollution mechanisms for sedimentary aquifers in Bangladesh. *Water Resources Research*, **37**, pp. 109–117.

McDonald, M.G. and Harbaugh, A.W., 1988, *A Modular Three-Dimensional Finite-difference Ground-Water Flow Model*, (Washington DC: United States Geological Survey), Techniques of Water-Resources Investigations Book 6, Chapter A1.

McDonnell, J.J., 2003, Where does water go when it rains? Moving beyond the variable source area concept of rainfall-runoff response. *Hydrological Processes*, **17**, pp. 1869–1875.

McGuire, K.J. and McDonnell, J.J., 2006, A review and evaluation of catchment transit time modeling. *Journal of Hydrology*, **330**, pp. 543–563.

Mehta, A.J., Hayter, E.J., Parker, W.R., Krone, R.B. and Teeter, A.M., 1989, Cohesive sediment transport. I: Process description. *Journal of Hydraulic Engineering*, **115**, pp. 1076–1093.

Meybeck, M., 2002, Riverine quality at the Anthropocene: Propositions for global space and time analysis, illustrated by the Seine River. *Aquatic Science*, **64**, pp. 376–393.

Meybeck, M., 2003, Global occurrence of major elements in rivers. Chapter 8 in *Treatise on Geochemistry, Volume 5 Surface and ground water, weathering and soils*, edited by Drever, J.I., (New York: Elsevier), pp. 207–224.

Meybeck, M. and Helmer, R., 1996, An introduction to water quality. Chapter 1 in *Water Quality Assessments – A Guide to Use of Biota, Sediments and Water in Environmental Monitoring* – 2nd edition, edited by Chapman, D., (London: E&FN Spon).

Meyer, J.L., 1979, The role of sediments and bryophytes in phosphorus dynamics in a headwater stream ecosystem. *Limnology and Oceanography*, **24**, pp. 365–375.

Michaelides, K. and Chappell, A., 2009, Connectivity as a concept for characterising hydrological behaviour. *Hydrological Processes*, **23**, pp. 517–522.

Middelkoop, H., 1997, *Embanked Floodplains in the Netherlands; Geomorphological Evolution over Various Time Scales*, (Utrecht: Knag/Faculty of Geographical sciences, Utrecht University), Netherlands Geographical Studies 224.

Middelkoop, H., 2000, Heavy-metal pollution of the river Rhine and Meuse floodplains in the Netherlands. *Netherlands Journal of Geosciences*, **79**, pp. 411–428.

Middelkoop, H. and Asselman, N.E.M., 1998, Spatial variability of floodplain sedimentation on the event scale in the Rhine-Meuse delta, the Netherlands. *Earth Surface Processes and Landforms*, **23**, pp. 561–573.

Middelkoop, H. and Van der Perk, M., 1998, Modelling spatial patterns of overbank sedimentation on embanked floodplains. *Geografiska Annaler*, **80A**, pp. 95–109.

Middelkoop, H., Thonon, I. and Van der Perk, M., 2002, Effective discharge for heavy metal deposition on the lower River Rhine flood plains. In *The Structure, Function and Management Implications of Fluvial Sedimentary Systems* (Proceedings of the Alice Springs Symposium), edited by Dyer, F.J., Thoms, M.C. and Olley, J.M., IAHS Publication no. 276, pp. 151–160.

Middelkoop, H., Van der Perk, M. and Thonon, I., 2003, *Herverontreiniging van Uiterwaarden langs de Rijntakken met Sediment-gebonden Zware Metalen*, (in Dutch), (Utrecht: Department of Physical Geography, Utrecht University), ICG Report 03/3.

Millennium Ecosystem Assessment, 2003, *Ecosystems and Human Well-being; A Framework for Assessment*, (Covelo, CA: Island Press).

Millennium Ecosystem Assessment, 2005. *Millennium Ecosystem Assessment Synthesis Report*, (Covelo, CA: Island Press).

Miller, R.W. and Gardiner, D.T., 2008, *Soils in Our Environment*, 11th edition, (Upper Saddle River NJ: Prentice-Hall).

Morel, F.M.M., 1983, *Principles of Aquatic Chemistry*, (New York: Wiley).

Morgan, R.P.C., 1995, *Soil Erosion and Conservation*, 2nd edition, (Harlow: Longman).

Morgan, R.P.C., 2001 A simple approach to soil loss prediction: a revised Morgan–Morgan–Finney model. *Catena*, **44**, pp. 305–322.

Mourad, D. and Van der Perk, M., 2004, Modelling nutrient fluxes from diffuse and point emissions to river loads: the Estonian part of the transboundary Lake Peipsi/Chudskoe drainage basin (Russia/Estonia/Latvia). *Water Science and Technology*, **49** No. 3, pp. 21–28.

Mourad, D.S.J., Van der Perk, M. and Piirimäe, K., 2006, Nutrient emissions, fluxes and retention under changing emissions in a North-Eastern European lowland drainage basin. *Environmental Monitoring and Assessment* (in press).

Müller, B., Lotter, A.F., Sturm, M. and Ammann, A., 1998, Influence of catchment quality and altitude on the water and sediment composition of 68 small lakes in Central Europe. *Aquatic Sciences*, **60**, pp. 316–337.

Murphy, J. and Riley, J.P., 1962, A modified single solution method for the determination of phosphate in natural water. *Analytical Chimica Acta*, **27**, pp. 31–36.

Myers, J. and Thorbjornsen, K, 2004, Identifying Metals Contamination in Soil: A Geochemical Approach. *Soil and Sediment Contamination: An International Journal*, **13**, pp. 1–16.

Nash, J.E. and Sutcliffe, J.V., 1970, River flow forecasting through conceptual models, Part I – a discussion of principles. *Journal of Hydrology*, **10**, pp. 282–290.

Neal, C., Robson, A.J., Shand, P., Edmunds, W.M., Dixon, A.J., Buckle, D.K., Hil, S., Harrow, M., Neal, M., Wilkinson, J. and Reynolds, B., 1997, The occurrence of groundwater in the Lower Palaeozoic rocks of upland Central Wales. *Hydrology and Earth System Sciences*, **1**, pp. 3–18.

Neal, C., Reynolds, B., Rowland, P., Norris, D., Kirchner, J.W., Neal, M., Sleep, D., Lawlor, A., Woods, C., Thacker, S., Guyatt, H., Vincent, C., Hockenhull, K., Wickham, H., Harman, S. and Armstrong, L., 2012, High-frequency water quality time series in precipitation and streamflow: from fragmentary signals to scientific challenge. *Science of the Total Environment*, **434**, pp. 3–12

Neal, C., Reynolds, B., Kirchner, J.W., Rowland, P., Norris, D., Sleep, D., Lawlor, A., Woods, C., Thacker, S., Guyatt, H., Vincent, C., Lehto, K., Grant, S., Williams, J., Neal, M., Wickham, H., Harman, S. and Armstrong, L., 2013, High-frequency precipitation and stream water quality time series from Plynlimon, Wales: an openly accessible data resource spanning the periodic table, *Hydrological Processes*, **27**, 2531–2539.

Nichols, D.S., 1983, Capacity of natural wetlands to remove nutrients from wastewater. *Journal WPCF*, **55**, pp. 495–505.

Nicholson, F.A., Smith, S.R., Alloway, B.J., Carlton-Smith, C., Chambers, B.J., 2003, An inventory of heavy metals inputs to agricultural soils in England and Wales. *Science of the Total Environment*, **311**, pp. 205–219.

Nordstrom, D.K, Plummer, L.N., Langmuir, D., Busenberg, E., May, H.M., Jones, B.F. and Parkhurst, D.L., 1990, Revised chemical equilibrium data for major major water–mineral reactions and their limitation. In *Chemical Modeling of Aqueous Systems II*, edited by Melchior D.C. and Basset, R.L., ACS Symposium Series 416, pp. 398–413.

Novotny, V., 2003, *Water Quality; Diffuse pollution and Watershed Management*, 2nd edition, (New York: Wiley).

Obbard, J.P. and Jones, K.C., 2000, Measurement of symbiotic nitrogen-fixation in leguminous host-plants grown in heavy metal-contaminated soils amended with sewage sludge. *Environmental Pollution*, **111**, pp. 311–320.

Ocampo, C.J., Sivapalan, M. and Oldham, C.E., 2003, Role of upland and near-stream zones on the export of nitrate from catchments: observations at Susannah brook catchment, Western Australia. *Geophysical Research Abstracts*, **5**, p. 7986.

O'Connor, D.J. and Dobbins, W.E., 1958, Mechanism of reaeration in natural streams. *Transactions of the American Society of Civil Engineers*, **123**, pp. 641–684.

Owens, P.N. and Walling, D.E., 1996, Spatial variability of Caesium-137 inventories at reference sites: an example from two contrasting sites in England and Zimbabwe. *Applied Radiation and Isotopes*, **47**, pp. 699–707.

Pattenden, N., 2001, Natural radioactivity. Section 2.1 in *Radioecology; Radioactivity & Ecosystems*, edited by Van der Stricht, E. and Kirchmann, R., (Liège: International Union of Radioecology), pp. 41–54.

Parchure, T.M. and Mehta, A.J., 1985, Erosion of soft cohesive sediment deposits: *Journal of Hydraulic Engineering*, **111**, pp. 1308–1326.

Parkin, T.B., 1993, Spatial variability of microbial processes in soil – A review. *Journal of Environmental Quality*, **22**, pp. 409–417.

Pauwels, H., Kloppmann, W., Walraevens, K. and Wendland, F., 2009, Aquifer Typology, (Bio) geochemical Processes and Pollutants Behaviour. Chapter 2.2 in *Groundwater Monitoring* edited by Quevauviller, P., Fouillac, A.-M., Grath J., Ward, R., (Chichester: Wiley), pp. 49–66.

Pebesma, E.J. and De Kwaadsteniet, J.W., 1994, *Een Landsdekkend Beeld van de Nederlandse Grondwaterkwaliteit op 5 tot 17 meter Diepte in 1991*, (in Dutch), (Bilthoven: Rijksinstituut voor Volsgezondheid en Milieuhygiene), RIVM rapport 714810014.

Pebesma, E.J. and De Kwaadsteniet, J.W., 1997, Mapping groundwater quality in the Netherlands. *Journal of Hydrology*, **200**, pp. 364–386.

Peierls, B.L. Caraco, N.F., Pace, M.L. and Cole, J.J., 1991, Human influence on river nitrogen. *Nature*, **350**, pp. 386–387.

Pionke, H.B., Gburek, W.J. and Sharpley, A.N., 2000, Critical source area controls on water quality in an agricultural watershed located in the Chesapeake Basin. *Ecological Engineering*, **14**, pp. 325–335.

Poeter, E.P. and Hill, M.C., 1998, *Documentation of UCODE: A Computer Code for Universal Inverse Modeling*, (Reston VA: U.S. Geological Survey), USGS Water-Resources Investigations Report 98-4080.

Press, W.H., Flannery, B.P., Teukolsky, S.A. and Vetterling, W.T., 1992, *Numerical Recipes*, 2nd edition, (Cambridge: Cambridge University Press).

Province of Gelderland, 2013, *Meetnet Grondwaterkwaliteit*, (Arnhem, The Netherlands: Province of Gelderland), Available online at http://www.gelderland.nl/ [verified 1 July 2013].

Qu, C.H., Chen, C.Z., Yang, J.R., Wang, L.Z. and Lu, Y.L., 1993, Geochemistry of dissolved and particulate elements in the major rivers of China (The Huanghe, Changjiang, and Zhunjiang Rivers). *Estuaries*, **16**, pp. 475–467.

Quine, T.A., Walling, D.E. and Zhang, X., 1999, Tillage erosion, water erosion and soil quality on cultivated terraces near Xifeng in the Loess Plateau, China. *Land Degradation & Development*, **10**, pp. 251–274.

Rawlins, B.G., Webster, R. and Lister, T.R., 2003, The influence of parent material on topsoil geochemistry in eastern England. *Earth Surface Processes and Landforms*, **28**, pp. 1389–1409.

Rawlins, B.G., McGrath, S.P., Scheib, A.J., Breward, N., Cave, M., Lister, T.R., Ingham, M., Gowing, C. and Carter, S., 2012, *The advanced soil geochemical atlas of England and Wales*, (Keyworth: British Geological Survey), Available online at http://www.bgs.ac.uk/gbase/advsoilatlasEW.html [verified 1 July 2013].

Redfield, A.C., Ketchum, B.H. and Richards, F.A., 1963, The influence of organisms on the composition of sea-water. In *The Sea; Ideas and Observations on Progress in the Study of the Seas. V. 2. The Composition of Sea–Water; Comparative and Descriptive Oceanography*, edited by Hill, M.N., (New York: Interscience), pp. 26–77.

Reimann, C. and Birke, M. (Eds.), 2010, *Geochemistry of European Bottled Water*, (Stuttgart: Borntraeger Science Publishers).

Reimann, C., Flem, B., Fabian, K., Birke, M., Ladenberger, A., Négrel, P., Demetriades, A., Hoogewerff, J. and The GEMAS Project Team, 2012, Lead and lead isotopes in agricultural soils of Europe – The continental perspective. *Applied Geochemistry*, **27**, pp. 532–542.

Repert, D.A., Barber, L.B., Hess, K.M., Keefe, S.H., Kent, D.B., LeBlanc, D.R. and Smith, R.L., 2006, Long-term natural attenuation of carbon and nitrogen within a groundwater plume after removal of the treated wastewater source. *Environmental Science & Technology*, **40**, pp. 1154–1162.

Rickert, D.A., 1982, Use of dissolved oxygen modelling results in the management of river quality: Case history of the Willamette river, Oregon. In *Effects of Waste Disposal on Groundwater and Surface Water* (Proceedings of the Exeter Symposium), edited by Perry, R. IAHS Publication no. 139, pp. 13–22.

Rijkswaterstaat, 2013, *Effluenten RWZI's (gemeten stoffen) –Industriële en communale bronnen– Emissieregistratie*, Version May 2013, (in Dutch), (Den Haag: Rijkswaterstaat).

Rivett, M.O., Feenstra, S. and Cherry, J.A., 2001, A controlled field experiment on groundwater contamination by a multi-component DNAPL: creation of the emplaced-source and overview of dissolved plume development. *Journal of Contaminant Hydrology*, **49**, pp. 111–149.

RIVM, 2002, *Milieubalans 2002; Het Nederlandse Milieu Verklaard*, (in Dutch), (Bilthoven: Rijksinstituut voor Volkgezondheid en Milieu).

RIVM, 2003, *Milieu en NatuurCompendium* (Dutch Environmental Data Compendium), (Bilthoven: Rijksinstituut voor Volkgezondheid en Milieu).

Robson, A. and Neal, C., 1990, Hydrograph separation using chemical techniques: an application to catchments in Mid-Wales. *Journal of Hydrology*, **116**, pp. 345–363.

Roux, J.-C., 1995, The evolution of groundwater quality in France: perspectives for enduring use for human consumption. *The Science of the Total Environment*, **171**, pp. 3–16.

Runde, W., 2002, Geochemical interactions of actinides in the environment. Chapter 2 in *Geochemistry of soil radionuclides*, edited by Zhang, P.-C. and Brady, P.V., (Madison, WI: Soil Science Society of America), SSSA Special Publication No. 59, pp. 21–44.

Ruys, W.J.B., 1981, *Een Zuurstofmodel voor het Wolderwijd*, (in Dutch), (Lelystad: Rijksdijnst voor de IJsselmeerpolders), Werkdocument RIJP nr 1981-294 Abw.

Salminen, R., Gregorauskien, V., 2000. Considerations regarding the definition of a geochemical baseline of elements in the surficial materials in areas differing in basic geology. Applied Geochemistry, 15, pp. 647–653.

Salminen, R., Batista, M.J., Bidovec, M., Demetriades, A., De Vivo, B., De Vos, W., Duris, M.,Gilucis, A., Gregorauskiene, V., Halamic, J., Heitzmann, P., Lima, A., Jordan, G., Klaver, G., Klein, P., Lis, J., Locutura, J., Marsina, K., Mazreku, A., O'Connor, P.J., Olsson, S.Å., Ottesen, R.-T., Petersell, V., Plant, J.A., Reeder, S., Salpeteur, I., Sandström, H., Siewers, U., Steenfelt, A. and Tarvainen, T., 2005, *Geochemical Atlas of Europe. Part 1 - Background Information, Methodology and Maps*, (Espoo, Finland: Forum of European Geological Surveys (FOREGS)), Available online at http://weppi.gtk.fi/publ/foregsatlas/ [verified 1 July 2013].

Salomons, W. and Förstner, U., 1980, Trace metal analysis on polluted sediments, Evaluation of environmental impact. *Environmental Technological Letters*, **1**, pp. 506–517.

Salomons, W. and Förstner, U., 1984. *Metals in the Hydrocycle*, (Berlin: Springer).

Sauer, T.J. and Meek, D.W., 2003, Spatial variation of plant-available phosphorus in pastures with contrasting management. *Soil Science Society of America Journal*, **67**, pp. 826–836.

Samuelsson, C., 1995. Natural radioactivity. In *Radioecology; Lectures in Environmental Radioactivity*, edited by Holm, E., (Singapore: World Scientific), pp. 1–20.

Scheffer, F. and Schachtschabel, P. 1989, *Lehrbuch der Bodenkunde*, 12. neu bearbeitete Auflage von P. Schachtschabel, H.P. Blume, G. Brummer, K.-H. Hartge und Schwertmann, (in German), (Stuttgart: Enke).

Schnoor, J.L., 1996, *Environmental Modeling; Fate and Transport of Pollutants in Water, Air, and Soil*, (New York: Wiley).

Schwarzenbach, R.P., Gschwend, P.M. and Imboden, D.M., 1993, *Environmental Organic Chemistry*, (New York: Wiley).

Scott, H.D., Selim, H.M. and Ward, L.B., 2000, MLRA 131: Southern Mississippi Valley Alluvium. In: *Water and Chemical Transport in Soils of the Southeastern USA*, edited by Scott, H.D., (Fayetteville, AR: Arkansas Agricultural Experiment Station, University of Arkansas), Southern Cooperative Series Bulletin #395. Available online: http://soilphysics.okstate.edu/S257/index.html [verified 15 April 2006].

Seitzinger, P.S., Styles, R.V., Boyer, E., Alexander, R.B., Billen, G., Howarth, R.W., Mayer, B. and Van Breemen, N., 2002, Nitrogen retention in rivers: model development and application to watersheds in the northeastern U.S.A. *Biogeochemistry*, **57/58**, pp. 199–237.

Selig, U., Hübener, T., Heerkloss, R. and Schubert, H., 2004, Vertical gradient of nutrients in two dimictic lakes – influence of phototrophic sulfur bacteria on nutrient balance. *Aquatic Sciences*, **66**, pp. 247–256.

Simmons, B.L. and Cheng, D.M.H., 1985. Rate and pathways of phosphorus assimilation in the Nepean river at Camden, New South Wales. *Water Research*, **19**, pp. 1089–1095.

Sims, J.T., Simard, R.R. and Joern, B.C, 1998, Phosphorus loss in agricultural drainage: Historical perspective and current research. *Journal of Environmental Quality*, **27**, pp. 277–293

Sjodin, A.L., Lewis Jr., W.M. and Saunders III, J.F., 1997, Denitrification as a component of the nitrogen budget for a large plains river. *Biogeochemistry*, **39**, pp. 327–342.

Sklash, M.G. and Farvolden, R.N., 1979, The role of groundwater in storm runoff. *Journal of Hydrology*, **43**, pp. 45–65.

Smedley, P.L. and Kinniburgh, D.G., 2001, Arsenic in groundwaters across the world. Chapter 2 in *Arsenic Contamination of Groundwater in Bangladesh – Volume 2: Final Report*, edited by Kinniburgh, D.G. and Smedley, P.L., (Keyworth: British Geological Survey), BGS Report WC/00/19, pp. 3–16.

Smith, J.T., Comans, R.N.J., Beresford, N.A., Wright, S.M., Howard, B.J. and Camplin, W.C., 2000, Chernobyl's legacy in food and water. *Nature*, **405**, p. 141.

Smith, S.J., Van Aardenne, J., Klimont, Z., Andres, R. J., Volke, A. and Delgado Arias, S., 2011, Anthropogenic sulfur dioxide emissions: 1850–2005. *Atmospheric Chemistry and Physics*, **11**, pp. 1101–1116.

Søndergaard, M., Jensen, J.P. and Jeppesen, E., 1999, Internal phosphorus loading in shallow Danish lakes. *Hydrobiologia*, **408/409**, pp. 145–152.

Søndergaard, M., Jensen, J.P. and Jeppesen, E., 2001, Retention and internal loading of phosphorus in shallow, eutrophic lakes. *The Scientific World*, **1**, pp. 427–442.

Spijker, J., 2005, *Geochemical Patterns in the Soils of Zeeland; Natural Variability versus Anthropogenic Impact*, (Utrecht: Knag/Faculty of Geosciences Utrecht University), Netherlands Geographical Studies 330.

Spreafico, M. and Van Mazijk, A. (Eds.) 1993. *Alarmmodell Rhein. Ein Modell für die operationelle Vorhersage des Transportes von Schadstoffen im Rhein*, (in German), (Lelystad: The International Commission for the Hydrology of the Rhine basin (CHR)), Bericht Nr. I-12.

Stålnacke, P., Grimvall, A., Libiseller, C., Laznik, M. and Kokorite, I., 2003, Trends in nutrient concentrations in Latvian rivers and the response to the dramatic change in agriculture. *Journal of Hydrology*, **283**, pp. 184–205.

Steur, G.G.L., Brus, D.J. and Van den Berg, M., 1987, *Bodem, Deel 14 van de Atlas van Nederland*, (in Dutch), (Den Haag: Staatsuitgeverij).

Stohl, A., Seibert, P., Wotawa, G., Arnold, D., Burkhart, J.F., Eckhardt, S., Tapia, C., Vargas, A. and Yasunari, T.J., 2012, Xenon-133 and caesium-137 releases into the atmosphere from the Fukushima Dai-ichi nuclear power plant: determination of the source term, atmospheric dispersion, and deposition. *Atmospheric Chemistry and Physics*, **12**, pp. 2313–2343.

Stone, M. (Ed.), 2000, *The Role of Erosion and Sediment Transport in Nutrient and Contaminant Transfer* (Proceedings of the Waterloo Symposium), (Wallingford, UK: International Association for Hydrological Sciences), IAHS Publication no. 263.

Strahler, A.N., 1952, Hyposometric (area-altitude) analysis of erosional topography. *Geological Society of America Bulletin*, **63**, pp. 1117–1142.

Stuart, M., Lapworth, D., Crane, E. and Hart, A., 2012, Review of risk from potential emerging contaminants in UK groundwater . *Science of the Total Environment*, **416**. pp. 1–21.

Stumm, W. and Morgan, J.J., 1996, *Aquatic Chemistry*, 3rd edition, (New York: Wiley).

Stuyfzand, P.J., 1999, Patterns in groundwater chemistry resulting from groundwater flow. *Hydrogeology Journal*, **7**, pp. 15–27.

Sudduth, E.B., Perakis, S.S. and Bernhardt, E.S., 2013, Nitrate in watersheds: Straight from soils to streams? Journal of Geophysical Research: *Biogeosciences*, **118**, pp. 291–302.

Tebbens, L.A., Veldkamp, A. and Kroonenberg, S.B., 2000. Natural compositianal variation of the river Meuse (Maas) suspended load: a 13 ka bulk geochemical record from the upper Kreftenheye and Betuwe Formations in northern Limburg. *Netherlands Journal of Geosciences*, **79**, pp. 391–409.

Ternes, T. and Von Gunten, U. (Eds.), 2010, Emerging Contaminants in water: Occurrence, fate, removal and assessment in the water cycle (from wastewater to drinking water). Special Issue *Water Research*, **44**, pp. 351–668.

Thidobeaux, L.J., 1996, *Environmental Chemodynamics; Movement of Chemicals in Air, Water, and Soil*, 2nd edition, (New York: Wiley).

Thomann, R.V., 1972, *Systems Analysis and Water Quality Management*, (New York: McGraw-Hill).

Thomann, R.V., 1973, Effect of longitudinal dispersion on dynamic water quality response of stream and rivers. *Water Resources Research*, **9**, pp. 355–366.

Thomann, R.V. and Mueller, J.A., 1987, *Principles of Surface Water Quality Modeling and Control*, (New York: Harper Collins).

Thomas, R., Meybeck, M. and Beim, A., 1996. Lakes. Chapter 7 in *Water Quality Assessments – A Guide to Use of Biota, Sediments and Water in Environmental Monitoring* – 2nd edition, edited by Chapman, D., (London: E&FN Spon).

Thonon, I., 2006, *Deposition of sediment and associated heavy metals on floodplains*, (Utrecht: Knag Faculty of Geographical sciences, Utrecht University), Netherlands Geographical Studies 337.

Thonon, I., De Jong, K., Van der Perk, M. and Middelkoop, H., 2007, Modelling floodplain sedimentation using particle tracking. *Hydrological Processes*, **21**, pp. 1402-1412.

Tiktak, A., Boesten, J.J.T.I., van der Linden, A.M.A. and Vanclooster, M., 2006, Mapping ground water vulnerability to pesticide leaching with a process-based metamodel of EuroPEARL. *Journal of Environmental Quality*, **35**, pp. 1213–1226.

Tóth, J., 1963, A theoretical analysis of groundwater flow in small drainage basins. *Journal of Geophysical Research*, **68**, pp. 4795–4812.

Uffink, G.J.M., 2003, *Determination of Denitrification Parameters in Deep Groundwater; A Pilot Study for Several Pumping Stations in the Netherlands*, (Bilthoven: Rijksinstituut voor Volksgezondheid en Milieu). RIVM report 703717011/2003.

UKAEA, 2006, *WAGR School's Project*, (Seascale, Cumbria: UK Atomic Agency Authority).

UNEP, 2002, *Depleted Uranium in Serbia and Montenegro. Post-conflict Environmental Assessment in the Federal Republic of Yugoslavia*, (Geneva: United Nations Environment Programme).

UNEP, 2012, *GEO-5 Global Environment Outlook: Environment for the future we want*, (Nairobi: United Nations Environment Programme).

UNSCEAR, 1993, *Exposures from Man-made Sources of Radiation*, United Nations Scientific Committee on the Effects of Atomic radiation, Report to the General Assembly, Annex B, (New York: United Nations).

UNSCEAR, 2000, *Sources and Effects of Ionizing Radiation, Volume I: Sources*, United Nations Scientific Committee on the Effects of Atomic Radiation, Report to the General Assembly, (New York: United Nations).

USDA, 1995, *USDA – Water Erosion Prediction Project (WEPP) – Technical documentation*, (West Lafayette IN: United States Department of Agriculture, Agricultural Research Service, National Soil Erosion Research Laboratory), NSERL Report No. 10, Available online at http:// http://www.ars.usda.gov/Research/docs.htm?docid=18073 [verified 1 July 2013].

USDA, 1999, *Soil Taxonomy; A Basic System of Soil Classification for Making and Interpreting Soil Surveys*, (Washington DC: United States Department of Agriculture, Natural Resources Conservation Service), Agriculture Handbook No. 436.

USDA, 2012, *WEPP – Water Erosion Prediction Project,* (West Lafayette IN: United States Department of Agriculture, Agricultural Research Service), Available online at http:// http://www.ars.usda.gov/Research/docs.htm?docid=10621 [verified 1 July 2013].

Van Breukelen, B.M. and Griffioen, J., 2004, Biogeochemical processes at the fringe of a landfill leachate pollution plume: potential for dissolved organic carbon, Fe(II), Mn(II), NH4, and CH4 oxidation. *Journal of Contaminant Hydrology*, **73**, pp. 181–205.

Van Breukelen, B.M., Griffioen, J., Röling, W.F.M. and Van Verseveld, H.W., 2004, Reactive transport modelling of biogeochemical processes and carbon isotope geochemistry inside a landfill leachate plume. *Journal of Contaminant Hydrology*, **70**, pp. 249–269.

Van Cappellen, P. and Wang, Y., 1995, Metal cycling in sediments: Modeling the interplay of reaction and transport. In *Metal Contaminated Aquatic Sediments*, edited by Allen, H.E., (Chelsea: Ann Arbor Press), pp. 21–64.

Van den Brink, C. Frapporti, G., Griffioen, J. and Zaadnoordijk, W.J., 2007, Statistical analysis of anthropogenic versus geochemical-controlled differences in groundwater composition in The Netherlands. *Journal of Hydrology*, **336**, pp. 470–480.

Van den Boomen, R.M., Salverda, A.P., Uunk, E.J.B. and Roos, C., 1995, Modellering van het Reggesyteem met DUFLOW, (in Dutch with English summary). *H₂O*, **28**, pp. 107–111.

Van der Molen, D.T. and Boers, P.C.N., 1994, Influence of internal loading on phosphorus concentration in shallow lakes before and after reduction of the external loading. *Hydrobiologia*, **275/276**, pp. 379–389.

Van der Perk, M., 1996, *Muddy Waters: Uncertainty Issues in Modelling the Influence of Bed Sediments on Water Composition*, (Utrecht: Knag/Faculty of Geographical sciences, Utrecht University), Netherlands Geographical Studies 200.

Van der Perk, M., 1997, Effect of model structure on the accuracy and uncertainty of results from water quality models. *Hydrological Processes*, **11**, pp. 227–239.

Van der Perk, M., 1998, Calibration and identifiability analysis to assess the contribution of different processes to short-term dynamics of suspended sediment and dissolved nutrients in the surface water of a rural catchment. *Hydrological Processes*, **12**, pp. 683–699.

Van der Perk, M. and Van Gaans, P.F.M., 1997, Variation in composition of stream bed sediments in a small watercourse. *Water, Air and Soil Pollution*, **96**, pp. 107–131.

Van der Perk, M. and de Groot, I., 2013, *Mineral Waters of the World*, Available online: http://fg.geo.uu.nl/perk/mineralwaters/ [verified 1 July 2013].

Van der Perk, M., Ertsen, A.C.D. and Bleuten, W., 1992, Modellering van slibsedimentatie en zware-metalenbelasting in een uiterwaard langs de Waal, (in Dutch with English summary). *H₂O*, **25**, pp. 233–237.

Van der Perk, M., Slávik, O. and Fulajtár, E., 2002, Assessment of spatial variation of ^{137}Cs in small catchments. *Journal of Environmental Quality*, **31**, pp. 1930–1939.

Van der Perk, M., Jetten, V.G., Heskes, E.J., Segers, M. and Wijntjens, I., 2004, Transport and retention of copper fungicides in vineyards. In *Sediment Transfer through the Fluvial System* (Proceedings of the Moscow Symposium), edited by Golosov, V., Belyaev, V. and Walling, D.E., IAHS Publication no. 288, pp. 437–443.

Van der Perk, M., Owens, P.N., Deeks, L.K., Rawlins, B.G., Haygarth, P.M. and Beven, K.J., 2007, Controls on catchment-scale patterns of phosphorus in soil, streambed sediment, and stream water. *Journal of Environmental Quality*, **36**, pp. 694–708.

Van der Stricht, E. anf Kirchmann, R. (Eds.), 2001, *Radioecology; Radioactivity & Ecosystems*, (Liège: International Union of Radioecology).

Van Dijk, G.M., Stålnacke, P., Grimvall, A., Tonderski, A., Sundblad, K. and Schäfer, K., 1996, Long-term trends in nitrogen and phosphorus concentrations in the Lower River Rhine. *Archives of Hydrobiology Suppl.*, **113**, pp. 99–109.

Van Gaans, P.F.M., Griffioen, J., Mol, G. and Klaver, G., 2011, Geochemical reactivity of subsurface sediments as potential buffer to anthropogenic inputs: a strategy for regional characterization in the Netherlands. *Journal of Soils and Sediments*, **11**, pp. 336–351.

Van Leussen, W., 1988, Aggregation of particles, settling velocity of mud flocs, a review. In *Physical Processes in Estuaries*, edited by Dronkers, J. and Van Leussen, W., (Berlin: Springer).

Van Mazijk, A., Leibundgut, C.H. and Neff, H.-P., 1999, *Rhein-Alarm-Modell Version 2.1, Erweiterung um die Kalibrierung von Aare und Mosel. Kalibrierungsergebnisse von Aare und Mosel aufgrund der Markierversuche 05/92, 11/92 und 03/94*, (in German), (Lelystad: The International Commission for the Hydrology of the Rhine basin (CHR)) Report no. II-14.

Van Rijn, L.C., 1989, *Handbook on Sediment Transport by Current and Waves*, (Delft: Delft Hydraulics), Report H461.

Van Rompaey, A.J.J., Verstraeten, G., Van Oost, K., Govers, G. and Poesen, J., 2001, Modelling mean annual sediment yield using a distributed approach. *Earth Surface Processes and Landforms*, **26**, pp. 1221–1236.

Veldkamp, R.G. and Van Mazijk, A., 1989, *Waterkwaliteitsmodellering Oppervlaktewater*, (in Dutch), (Delft: Sanitary Engineering and Water Management Group, Delft University of Technology, the Netherlands), n11.

Vink, R. and Behrendt, H., 2002, Heavy metal transport in large river systems: heavy metal emissions and loads in the Rhine and Elbe river basins. *Hydrological Processes*, **16**, pp. 3227–3244.

Visser, A., Broers, H.P., van der Grift, B. and Bierkens, M.F.P., 2008, Demonstrating trend reversal of groundwater quality in relation to time of recharge determined by 3H/3He. *Environmental Pollution*, **148**, pp. 797–807.

Vissers, H.J.S.M., De Wit, N.H.S.M. and Bleuten, W., 1985, *Ruimtelijke Effecten van Bemesting via Ondiep Grondwater; Bedreiging van de Kwaliteit van Grond- en Oppervlaktewater en Gevolgen voor de Natuur en Waterwinningen op de Nederlandse Zandgronden*, (in Dutch), (Utrecht: Department of Physical Geography, Utrecht University).

Vissers, M.J.M., 2006, *Patterns of Groundwater Quality in Sandy Aquifers under Environmental Pressure*, (Utrecht: Knag/Faculty of Geographical sciences, Utrecht University), Netherlands Geographical Studies 335.

Vissers, M.J.M., Frapporti, G., Vriend, S.P. and Hoogendoorn, J.H., 1999, The dynamics of groundwater chemistry in unconsolidated aquifers: the Salland section. *Physics and Chemistry of the Earth (B)*, **24**, pp. 529–534.

Vitousek, P.M., Mooney, H.A., Lubchenco, J., Melillo, J.M., 1997a, Human domination of earth's ecosystems. *Science*, **277**, pp. 494–499.

Vitousek, P.M., Aber, J.D., Howarth, R.W., Likens, G.E., Matson, P.A., Schindler, D.W., Schlesinger, W.H., Tilman, D.G., 1997b, Human alteration of the global nitrogen cycle: Sources and consequences. *Ecological Applications*, **7**, pp. 737–750.

Von Bertalanffy, L., 1968, *General System Theory: Foundations, Development, Applications*, (New York: George Braziller).

Vörösmarty, C.J., McIntyre, P.B., Gessner, M.O., Dudgeon, D., Prusevich, A., Green, P., Glidden, S., Bunn, S.E., Sullivan, C.A., Reidy Liermann, C. and Davies, P.M., 2010, Global threats to human water security and river biodiversity. *Nature*, **467**, pp. 555–561.

Walling, D.E., 1977, Assessing the accuracy of suspended sediment rating curves for a small basin. *Water Resources Research*, **13**, pp. 531–538.

Walling, D.E., 1980, Water in the catchment ecosystem. In *Water Quality in Catchment Ecosystems*, edited by Gower, A.M., (New York: Wiley).

Walling, D.E. and Webb, B.W., 1980. The spatial dimension in the interpretation of stream solute behaviour. *Journal of Hydrology*, **47**, pp. 129–149.

Walling, D.E and Webb, B.W., 1986, Solutes in river systems. In *Solute Processes*, edited by Trudgill, S.T., (Chichester: Wiley), pp. 251–327.

Walling, D.E. and Webb, B.W., 1996, Erosion and sediment yield: global overview. In *Erosion and Sediment Yield: Global and Regional Perspectives* (Proceedings of the Exeter Symposium), edited by Walling, D.E. and Webb, B.W., IAHS Publication no. 236, pp. 3–19.

Walling, D.E. and Owens, P.N., 2003, The role of overbank floodplain sedimentation in catchment contaminant budgets. *Hydrobiologia*, **494**, pp. 83–91.

Walling, D.E., Owens, P.N., Carton, J., Lewis, S., Leeks, G.J.L., Meharg, A.A. and Wright, J., 2003, Storage of sediment-associated nutrients and contaminants in river channel and floodplain systems. *Applied Geochemistry*, **18**, pp. 195–220.

Ward, R.C. and Robinson, M., 2000, Principles of Hydrology, 4th edition, (London: McGraw-Hill).

Warren, C., Mackay, D., Whelan, M. and Fox, K., 2005, Mass balance modelling of contaminants in river basins: a flexible matrix approach. *Chemosphere*, **61**, pp. 1458–1467.

Waterbase, 2013, *Waterbase; Measurement Data of the Ministry of Infra Structure and the Environment, the Netherlands*, Available online at http:// http://live.waterbase.nl/ [verified 1 July 2013].

Wauchope, R.D. and McDowell, L.L., 1984, Adsorption of phosphate, arsenate, methanearsonate and cacodylate by lake and stream sediments: Comparison with soils. *Journal of Environmental Quality*, **13**, pp. 499–504.

Weathers, K.C., Cadenasso, M.L. and Pickett, S.T.A., 2001, Forest edges as nutrient and pollutant concentrators: Potential synergisms between fragmentation, forest canopies, and the atmosphere. *Conservation Biology*, **15**, p. 1506.

Wedepohl, K.H., 1995, The composition of the continental crust. *Geochimica et Cosmochimica Acta*, **59**, pp. 1217–1232.

Wegener J.W., Van Schaik, M.J. and Aiking, H., 1992, Active biomonitoring of polycyclic aromatic hydrocarbons by means of mosses. *Environmental Pollution*, **76**, pp. 15–19.

Wels, C., Cornett, J.R and Lazerte, B.D., 1991, Hydrograph separation: a comparison of geochemical and isotopic tracers. *Journal of Hydrology*, **122**, pp. 253–274.

Wendland, F., 1992, *Die Nitratbelastung in den Grundwasserlandschaften der „alten" Bundesländer (BRD)*, (in German), (Jülich: Forschungszentrum Jülich).

Wendland, F., Blum, A., Coetsiers, M., Gorova, R., Griffioen, J., Grima, J., Hinsby, K., Kunkel, R., Marandi, A., Melo, T., Panagopoulos, A., Pauwels, H., Ruisi, M., Traversa, P., Vermooten, J. S. A. and Walraevens, K., 2008, European aquifer typology: a practical framework for an overview of major groundwater composition at European scale. *Environmental Geology*, **55**, pp. 77–85.

Whelan, M.J. and Gandolfi, C., 2002, Modelling of spatial controls on denitrification at the landscape scale. *Hydrological Processes*, **16**, pp. 1437–1450.

WHI, 1999, *WinPEST User Manual*, (Waterloo ON: Waterloo Hydrogeologic Inc.).

Whitehead, D.C., 2000, *Nutrient Elements in Grassland: Soil–Plant–Animal Relationships*, (Wallingford: CAB International).

Whitehead, P.G. and Williams, R.J., 1982, A dynamic nitrogen balance model for river systems. In *Effects of Waste Disposal on Groundwater and Surface Water* (Proceedings of the Exeter Symposium), edited by Perry, R., IAHS Publication no. 139, pp. 89–99.

Whitehead, P.G., Beck, M.B. and O'Connel, E., 1981, A systems model of streamflow and water quality in the Bedford Ouse river system; II. Water quality modelling. *Water Research*, **15**, pp. 1157–1171

WHO, 1989, *DDT and its Derivatives – Environmental Aspects*, (Geneva: World Health Organisation). Environmental Health Criteria 83, Available online at: http://www.inchem.org/documents/ehc/ehc/ehc83.htm [verified 1 July 2013].

Wieland, E., Lienemann, P., Bollhalder, S., Lück, A. and Santschi, P.H., 2001, Composition and transport of settling particles in Lake Zurich: relative importance of vertical and lateral pathways. *Aquatic Sciences*, **63**, pp. 123–149.

Willems, W.J., Fraters, B., Meinardi, C.R., Reijnders, H.F.R. and Van Beek, C.G.E.M., 2002, *Nutrients in Soil and Groundwater: Quality Objectives and Quality Development 1984–2000*, (in Dutch with English summary), (Bilthoven: Rijksinstituut voor Volksgezondheid en Milieu), RIVM rapport 718201004.

Winkels, H.J., 1997, *Contaminant Variability in a Sedimentation Area of the River Rhine*, PhD Thesis Wageningen University, (Lelystad: Ministry of Transport and Public Works, Directorate-General Rijkswaterstaat, Directie IJsselmeelgebied), Van Zee tot Land 64.

Winkels, H.J., Kroonenberg, S.B., Lychagin, M.Y., Marin, G., Rusakov, G.V. and Kasimov, N.S., 1998, Geochronology of priority pollutants in sedimentation zones of the Volga and Danube delta in comparison with the Rhine delta. *Applied Geochemistry*, **13**, pp. 581–591.

Winter, M.J., 2006, *WebElements; The Periodic table on the World-Wide Web*, Available online at http://www.webelements.com/ [verified 1 July 2013].

Winter, T.C., Harvey, J.W., Franke, O.L. and Alley, W.M., 1999, *Ground Water and Surface Water; A Single Resource*, (Denver CO: U.S. Geological Survey), USGS Circular 1139.

Wischmeier, W.H. and Smith, D.D., 1978, *Prediction of Rainfall Erosion Losses: A Guide to Conservation Planning*, (Washington DC: U.S. Department of Agriculture), Agricultural Handbook No. 537.

Wollenberg, H.A. and Smith, A.R., 1990, A geochemical assessment of terrestrial γ-ray absorbed dose rates. *Health Physics*, **58**, pp. 183–189.

Wörman, A., Packman, A.I., Marklund, L., Harvey, J.W., and Stone, S.H., 2007. Fractal topography and subsurface water flows from fluvial bedforms to the continental shield, *Geophysical Research Letters*, **34**, L07402.

Wright-Walters, M., 2009, *Exposure concentrations of pharmaceutical estrogens and xenoestrogens in municipal wastewater treatment plant sources, the aquatic environment and an aquatic hazard assessment of bisphenol-a: implications for wildlife and public health*. Doctoral Dissertation, (Pittsburgh: University of Pittsburgh).

Xiaojun, N., Xiaodan, W., Suzhen, L., Shixian, G. and Haijun, L., 2010, ^{137}Cs tracing dynamics of soil erosion, organic carbon and nitrogen in sloping farmland converted from original grassland in Tibetan plateau. *Applied Radiation and Isotopes*, **68**, pp. 1650–1655.

Yasunari, T.J., Stohl, A., Hayano, R.S., Burkhart, J.F., Eckhardt, S. and Yasunarie, T., 2011, Cesium-137 deposition and contamination of Japanese soils due to the Fukushima nuclear accident. *Proceedings of the National Academy of Sciences*, **108**, pp. 19530–19534.

Zanardo, S., Basu, N.B., Botter, N.B., Rinaldo, A. and Rao, P.S.C., 2012, Dominant controls on pesticide transport from tile to catchment scale: Lessons from a minimalist model. *Water Resources Research*, **48**, W04525.

Zhang, P.-C., Krumhansl, J.L. and Brady, P.V., 2002, Introduction to properties, sources and characteristics of soil radionuclides. Chapter 1 in *Geochemistry of Soil Radionuclides*, edited by Zhang, P.-C. and Brady, P.V., SSSA Special Publication No. 59, pp. 1–20.

Zheng, C. and Gorelick, S.M., 2003, Analysis of the effect of decimeter-scale preferential flow paths on solute transport. *Ground Water*, **41**, pp. 142–155.

APPENDIX I

The periodic table of elements

Key:

| Atomic number |
| Symbol |
| Element name |
| Atomic weight |

Group	1	2	3	4	5	6	7	8	9	10	11	12	13	14	15	16	17	18
Period 1	1 **H** Hydrogen 1.00794																	2 **He** Helium 4.002602
2	3 **Li** Lithium 6.941	4 **Be** Beryllium 9.012182											5 **B** Boron 10.811	6 **C** Carbon 12.0107	7 **N** Nitrogen 14.0067	8 **O** Oxygen 15.9994	9 **F** Fluorine 18.9984032	10 **Ne** Neon 20.1797
3	11 **Na** Sodium 22.989770	12 **Mg** Magnesium 24.3050											13 **Al** Aluminium 26.981538	14 **Si** Silicon 28.0855	15 **P** Phosphorus 30.973761	16 **S** Sulfur 32.065	17 **Cl** Chlorine 35.453	18 **Ar** Argon 39.948
4	19 **K** Potassium 39.0983	20 **Ca** Calcium 40.078	21 **Sc** Scandium 44.955910	22 **Ti** Titanium 47.867	23 **V** Vanadium 50.9415	24 **Cr** Chromium 51.9961	25 **Mn** Manganese 54.938049	26 **Fe** Iron 55.845	27 **Co** Cobalt 58.933200	28 **Ni** Nickel 58.6934	29 **Cu** Copper 63.546	30 **Zn** Zinc 65.409	31 **Ga** Gallium 69.723	32 **Ge** Germanium 72.64	33 **As** Arsenic 74.92160	34 **Se** Selenium 78.96	35 **Br** Bromine 79.904	36 **Kr** Krypton 83.798
5	37 **Rb** Rubidium 85.4678	38 **Sr** Strontium 87.62	39 **Y** Yttrium 88.90585	40 **Zr** Zirconium 91.224	41 **Nb** Niobium 92.90638	42 **Mo** Molybdenum 95.94	43 **Tc** Technetium (98)	44 **Ru** Ruthenium 101.07	45 **Rh** Rhodium 102.90550	46 **Pd** Palladium 106.42	47 **Ag** Silver 107.8682	48 **Cd** Cadmium 112.411	49 **In** Indium 114.818	50 **Sn** Tin 118.710	51 **Sb** Antimony 121.760	52 **Te** Tellurium 127.60	53 **I** Iodine 126.90447	54 **Xe** Xenon 131.293
6	55 **Cs** Caesium 132.90545	56 **Ba** Barium 137.327	71 **Lu** Lutetium 174.967 *	72 **Hf** Hafnium 178.49	73 **Ta** Tantalum 180.9479	74 **W** Tungsten 183.84	75 **Re** Rhenium 186.207	76 **Os** Osmium 190.23	77 **Ir** Iridium 192.217	78 **Pt** Platinum 195.078	79 **Au** Gold 196.96655	80 **Hg** Mercury 200.59	81 **Tl** Thallium 204.3833	82 **Pb** Lead 207.2	83 **Bi** Bismuth 208.98038	84 **Po** Polonium (209)	85 **At** Astatine (210)	86 **Rn** Radon (222)
7	87 **Fr** Francium (223)	88 **Ra** Radium (226)	103 **Lr** Lawrencium (262) **	104 **Rf** Unnil-quadium (261)	105 **Db** Unnil-pentium (262)	106 **Sg** Unnil-hexium (266)	107 **Bh** Unnil-septium (264)	108 **Hs** Unniloctium (269)	109 **Mt** Meitnerium (268)	110 **Ds** Darm-stadtium (271)	111 **Uuu** Unununium (272)	112 **Uub** Unununbium (285)		114 **Uuq** Unun-quadium (289)				

*Lanthanoids	57 **La** Lanthanum 138.9055	58 **Ce** Cerium 140.116	59 **Pr** Praseo-dymium 140.90765	60 **Nd** Neo-dymium 144.24	61 **Pm** Promethium (145)	62 **Sm** Samarium 150.36	63 **Eu** Europium 151.964	64 **Gd** Gadolinium 157.25	65 **Tb** Terbium 158.92534	66 **Dy** Dysprosium 162.500	67 **Ho** Holmium 164.93032	68 **Er** Erbium 167.259	69 **Tm** Thulium 168.93421	70 **Yb** Ytterbium 173.04
**Actinoids	89 **Ac** Actinium (227)	90 **Th** Thorium 232.0381	91 **Pa** Protac-tinium 231.03588	92 **U** Uranium 238.02891	93 **Np** Neptunium (237)	94 **Pu** Plutonium (244)	95 **Am** Americium (243)	96 **Cm** Curium (247)	97 **Bk** Berkelium (247)	98 **Cf** Californium (251)	99 **Es** Einsteinium (252)	100 **Fm** Fermium (257)	101 **Md** Mende-levium (258)	102 **No** Nobelium (259)

Source: Winter, M.J. 2006. WebElements; The Periodic table on the World-Wide Web. Available online at http://www.webelements.com/ [verified 15 April 2006].

APPENDIX 2

Answers to exercises

CHAPTER 2

1. a. 100.2 mg l^{-1}
   b. 2.5 mmol l^{-1}
   c. 5.0 meq l^{-1}

2. a.

Cations	Concentration meq l^{-1}	Anions	Concentration meq l^{-1}
Na^+	0.38	Cl^-	0.06
K^+	0.03	HCO_3^-	4.18
Ca^{2+}	2.18	SO_4^{2-}	0.11
Mg^{2+}	1.78	NO_3^-	0.08
H^+	$10^{-4.5}$		

   b. cations: 4.37 meq l^{-1}; anions: 4.43 meq l^{-1}; there is a surplus of anions (0.6 percent of the total charge of cations and anions); H^+ can be ignored.
   c. 440 μS cm^{-1}
   d.

Cations	Activity coefficient γ_i	Anions	Activity coefficient γ_i
Na^+	0.92	Cl^-	0.92
K^+	0.92	HCO_3^-	0.92
Ca^{2+}	0.72	SO_4^{2-}	0.71
Mg^{2+}	0.73	NO_3^-	0.92

5. Cadmium concentration in:
   water: 0.09 mg l^{-1}
   sediment: 6.9 mg kg^{-1}
   fish: 6.4 mg kg^{-1}

6. a.

Experiment no.	Cs concentration after 12 days $\mu g\ g^{-1}$
1	5
2	9
3	19
4.	30
5.	45

b.

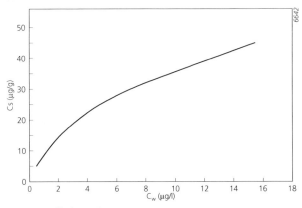

c. Freundlich isotherm.

7. a. endothermic;
    b. $K_s = 10^{-9.97}$
    c. $1.03 \cdot 10^{-2}$ mmol l^{-1}
    d. $9.7 \cdot 10^{-3}$ mmol l^{-1}
    e. The equilibrium shifts to the right.

8. *S.I.* = -0.09, so the sample is slightly undersaturated with gypsum.

9. Total Ca concentration is higher than predicted by the solubility product of gypsum alone.

10. pH = 8: dominant species is HPO_4^{2-}; pH = 6: dominant species is $H_2PO_4^-$; pH = 4: dominant species is $H_2PO_4^-$.

11. pH = 11.2.

12. a. pH = 4.57
    b. pH = 4.38

13. a. $Fe(OH)_3 + 3H^+ + e^- \leftrightarrow Fe^{2+} + 3H_2O$
    b. The solubility of $Fe(OH)_3$ increases;
    c. Eh = $-0.0592 \log[Fe^{2+}] - 0.1776$ pH + 0.9737
    d. Swamps.

14. Sample 1: pe = 10.1; sample 2: pe = 7.9.

CHAPTER 3

3. sandy loam

CHAPTER 4

3.

Material	Surface charge	
	pH = 5.4	pH = 6.8
Al_2O_3	positive	positive
$Al(OH)_3$	negative	negative
Fe_2O_3	positive	negative
$Fe(OH)_3$	positive	positive
Quartz (SiO_2), feldspars	negative	positive
Clay minerals: kaolinite	negative	negative
Montmorillonite	negative	negative

CHAPTER 5

1.

No.	Brand	Location	I.B.	Hardness	Hardness
			%	mg l^{-1}	German degrees
1	Contrex	Contrexeville, France	-1.0	618	34.7
2	Gerolsteiner	Gerolstein, Germany	0.0	2621	147
3	Kaiserbrunnen	Aachen, Germany	-0.1	383	21.6
4	Lete	Pratella, Italy	-0.2	1717	96.5
5	Monchique	Monchique, Portugal	14.8	6.8	0.4
6	Parot	St-Romain-le-Puy, France	-4.5	1084	60.9
7	Ramlösa	Helsingborg, Sweden	-0.1	15.1	0.8
8	Rogaska	Slatina, Slovenia	-0.2	9060	509
9	Sourcy	Bunnik, Netherlands	2.6	221	12.4
10	Spa Barisart	Spa, Belgium	-0.2	29.5	1.7

6. a. 549 mm y^{-1}
   b. No additional Cl inputs are assumed.

8. a. $CaCO_3 + H_2O + CO_2 \leftrightarrow Ca^{2+} + 2\ HCO_3^-$
   b. $CaCO_3 + HNO_3 \leftrightarrow Ca^{2+} + HCO_3^- + NO_3^-$
   c. No;
   d. Average Ca^{2+}: HCO_3^- molar ratio = 0.62;
   e. Approximately 24 percent of the dissolution of $CaCO_3$ is attributable to acidification

9. $SO_4^{2-} + 8\ H^+ + 8\ e^- \rightarrow S^{2-} + 4\ H_2O$
    Sulphide ion (S^{2-}) will precipitate with metal ions, especially Fe^{2+}.

CHAPTER 6

2. N:P mass ratio = 15.4, which implies P limitation.

6   a. Winter;
    b. Dry sandy soil;
    c. Deep water table;
    d. Soils poor in organic matter.

7. a. Summer;
    b. Neutral soils;
    c. Soils rich in organic matter.

CHAPTER 7

8. a. Metal sulphides;
    b. Weathering of metal sulphides;
    c. Dissolved and adsorbed to sediments;
    d. Immediately downstream of the source, the relationship between heavy metal concentrations and clay and organic matter content is weak; it becomes stronger with increasing distance downstream.

CHAPTER 8

5. a. 0.281 g
    b. Pb-210

7. a. ^{90}Sr
    b. ^{137}Cs
    c. ^{131}I

CHAPTER 9

2. a. 1,2-Dichlorocyclohexane:$C_6H_{10}Cl_2$; biphenyl: $C_{12}H_{10}$; anthracene: $C_{14}H_{10}$;
    b.

3-bromo-2-chloro-2-methylbutane   1,2,4-trichlorocyclohexane   cis-1,2-difluoroethene   m-dichlorobenzene

CHAPTER 10

4. a. 10 days
   b. 20 kg
   c. 10 days

8. a. 1 dimension: P transport in soil is mainly in the vertical plane;
   b. 2 dimensions: at least the lateral interaction should be considered;
   c. 0 dimensions may be sufficient; to account for vertical variations in pesticide concentrations or soil organisms in the soil profile, a one-dimensional model may be required.

CHAPTER 11

1. 0.346 tonnes y^{-1}

3. 2500 s

4. 40 mg l^{-1}

5. a. 0.01 $m^2 d^{-1}$
   b. longitudinal dispersion;
   c. 5.15 mg l^{-1}
   d. 5.15 mg l^{-1}
   e. The effects of reduced flow velocity are nullified by the extra time required for dispersion;
   f. The dispersion in porous media is due to variations in flow velocities and flow paths within the medium. Thus, dispersion is proportional to the distance travelled.

6. a.

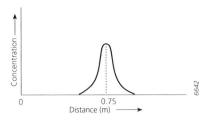

   b. 4.07 mg l^{-1}
   c.

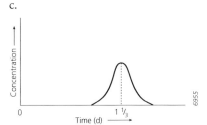

 d. 3.52 mg l^{-1}

 e. Pe = 100; transport is dominated by advection;

 f. The dispersion coefficient is about 2 orders of magnitude greater than molecular diffusion coefficient, which implies that the contribution of molecular diffusion to mixing (and transport) is negligible.

7. 132 m.

9. a. Approximately 0.05

 b. Approximately 0.4 (drainage line).

 c. Approximately 0.11

 d. Because NAPL remains behind in the soil pores (insular saturation)

 e. 1.515 m^3

CHAPTER 12

1. a. 0.98 N m^{-2}

 b. Neither erosion nor deposition.

3. 0.075 N m^{-2}

4. a. 189 kg

 b. The sediment in the water column will become depleted and will settle almost entirely. The concentration of suspended solids in the water column will become almost zero and sedimentation will cease.

7. a.

Section	Slope length (m)	Interrill erosion rate (kg m^{-2} y^{-1})	Rill erosion rate (kg m^{-2} y^{-1})	Transport capacity (kg m^{-1} y^{-1})
1	50	0.135	0.099	11.8
2	100	0.234	0.451	54.1
3	150	0.090	0.107	12.9

b.

Section	Total erosion (kg m^{-2} y^{-1})	Cumulative total erosion (kg m^{-1} y^{-1})	Transport capacity (kg m^{-1} y^{-1})
1	$0.0052\, l^{0.75} + 0.135$	$0.0030\, l^{1.75} + 0.135\, l$	$0.63\, l^{0.75}$
2	$0.0143\, l^{0.75} + 0.234$	$0.0082\, (l - 50)^{1.75} + 0.234\, (l - 50) + 9.6$	$1.71\, l^{0.75}$
3	$0.0025\, l^{0.75} + 0.090$	$0.0014\, (l - 100)^{1.75} + 0.090\, (l - 100) + 29.0$	$0.30\, l^{0.75}$

c. Calculate the cumulative total erosion and the transport capacity at the beginning and end of each section.

Section	Slope length (m)	Cumulative total erosion (kg m⁻¹ y⁻¹)	Transport capacity (kg m⁻¹ y⁻¹)
1	0	0	0
	50	9.6	11.8
2	50	9.6	32.2
	100	29.0	54.1
3	100	29.0	9.5
	150	34.8	12.9

The cumulative total erosion does not exceed the transport capacity until the end of section 2 (100 m from the divide). At the beginning of section 3 the transport capacity is exceeded. It is here that the excess sediment is deposited.

CHAPTER 13

1. a. $C(t) = C_0\, e^{-kt}$

   b. $C(t) = C_{eq} + (C_0 - C_{eq})\, e^{-kt}$

3. a. 0.05 mg l⁻¹
   b. 0.086 mg l⁻¹
   c. uptake of NH4 and/or NO3 by algae or aquatic macrophytes (sink); atmospheric deposition (source); ammonia volatilisation (sink); adsorption of NH4 by suspended solids (sink); decomposition of DOM or POM in the water column (source).

4. a. The concentration of o-xylene in this particular equation; the substrate concentration in general.
   b. The analytical solution of the differential equation is

   $$1000\ln\left(\frac{S_0}{S(t)}\right) + S_0 - S(t) = 45t$$

   The graph can be drawn to solve t for different values of $S(t)$:

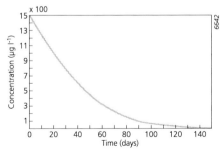

   c. approximately 100 days

5. a.

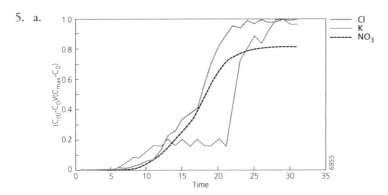

    b. 0.057 m d^{-1}
    c. 0.052 l kg^{-1}
    d. 0.331 g
    e. see a.

6. a. 21.1 m
    b. No, benzene mainly follows the flow paths and does not mix over the entire aquifer depth.

8. a. 1.43 d^{-1}
    b. 3850 m
    c. 2.82 d^{-1}

CHAPTER 14

2. 4716 m at a flow velocity of 10 cm s^{-1}; 4815 m at a flow velocity of 10 cm s^{-1}.

3. a. $k_{L,W}$ = 1.31 m d^{-1} and $k_{L,P}$ = 2.71 m d^{-1}. Thus, precipitation contributes more to reaeration than wind.
    b. 0.435 d

4. 3.16 m d^{-1} (for β = 0.7).

5. a. 0.00345 mg l^{-1} at pH = 7; 0.0328 mg l^{-1} at pH = 8;
    b. Water-side controlled gas exchange:
        pH = 7: 3.37 10^{-3} g m^{-2} d^{-1}
        pH = 8: 3.20 10^{-2} g m^{-2} d^{-1}
        Air-side controlled gas exchange:
        pH = 7: 1.6 10^{-5} g m^{-2} d^{-1}
        pH = 8: 1.5 10^{-4} g m^{-2} d^{-1}
        The air-side controlled gas exchange is two orders of magnitude smaller than the water-side controlled gas exchange. This means that the gas exchange is largely air-side controlled and the fluxes for air-side controlled gas exchange apply.

CHAPTER 15

4. 4^6 = 1296 model runs for four model parameters; 6^6 = 46 656 model runs for six parameters.

5. a. $R^2 = 0.918$
   b. Nash efficiency coefficient = 0.868
   c. $t = 0.988$; $t_0 = 2.069$ for 23 degrees of freedom. Hence, we cannot reject the null hypothesis that the mean difference between the observed and predicted values differs significantly from zero.

CHAPTER 16

2. 37.8 g m^{-2}

3. 382 mg kg^{-1}; the sample is more contaminated than the area average ($Y_m > Y_r$).

4. b. 0.30 mg kg^{-1}
   c. 1) samples from the topsoil and the deeper soil are of the same origin (same parent material) and 2) the deeper soil is not contaminated.

CHAPTER 17

3. 1) Enhanced atmospheric deposition of acids and 2) oxidation of pyrite by agriculturally derived nitrate.

4. Forest: approximately 400 mm y^{-1}; heather: approximately 585 mm y^{-1}.

5.

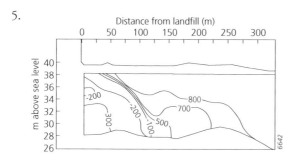

CHAPTER 18

3. a. peak concentration before peak discharge;
   b. trough concentration after peak discharge;
   c. peak concentration after peak discharge;
   d. trough concentration before peak discharge.

4. b. Coefficient a is greater in catchments on intensively weathered rocks than in catchments on bare unweathered rock.
   c. Coefficient b decreases from headwaters to lowland rivers.

Index

Printed and bound by PG in the USA